Synthetic cannabinoids in drug discovery:

Design, synthesis and evaluation of modified coumarins as CB receptor ligands

Zur Erlangung des akademischen Grades eines

DOKTORS DER NATURWISSENSCHAFTEN

(Dr. rer. nat.)

von der KIT-Fakultät für Chemie und Biowissenschaften

des Karlsruher Instituts für Technologie (KIT)

genehmigte

DISSERTATION

von

M. Sc. Florian Mohr
aus Karlsruhe

KIT-Dekan: Prof. Dr. Manfred Wilhelm

1. Referent: Prof. Dr. Stefan Bräse

2. Referent: Prof. Dr. Frank Breitling

Tag der mündlichen Prüfung: 11.12.2019

Band 88
Beiträge zur organischen Synthese
Hrsg.: Stefan Bräse

Prof. Dr. Stefan Bräse
Institut für Organische Chemie
Karlsruher Institut für Technologie (KIT)
Fritz-Haber-Weg 6
D-76131 Karlsruhe

Bild-Illustratorin: Frau Lisa Beuchle

Bibliografische Information der Deutschen Bibliothek

Die Deutsche Nationalbibliothek verzeichnet diese Publikation in der
Deutschen Nationalbibliografie; detaillierte bibliografische Daten sind
im Internet über http://dnb.d-nb.de abrufbar.

ISBN 978-3-8325-5102-5
ISSN 1862-5681

Logos Verlag Berlin GmbH
Comeniushof, Gubener Str. 47,
10243 Berlin
Tel.: +49 030 42 85 10 90
Fax: +49 030 42 85 10 92
INTERNET: http://www.logos-verlag.de

"Zu Risiken und Nebenwirkungen lesen Sie die Packungsbeilage und fragen Sie Ihren Arzt oder Apotheker"

-§4 Abs. 3 HWG

Die vorliegende Arbeit wurde im Zeitraum vom 01. November 2016 bis 6. November 2019 am Institut für Organische Chemie, der Fakultät für Chemie und Biowissenschaften am Karlsruher Institut für Technologie (KIT) unter der Leitung von Prof. Dr. Stefan Bräse angefertigt. Der praktische Teil mit dem Titel "Pharmacological evaluation of synthetic cannabinoids" wurde zwischen dem 18. Februar 2019 und dem 31. Juni 2019 am Leiden Academic Centre for Drug Research unter der Leitung von Prof. Laura Heitman durchgeführt. Der praktische Teil mit dem Titel "Synthesis and evaluation of MAGL inhibitors" wurde zwischen dem 01. März 2018 und dem 17. August 2019 am Leiden Institute of Chemistry unter der Leitung von Prof. Mario van der Stelt durchgeführt.

The present thesis was realized at the Karlsruhe Institute of Technology, Faculty of Chemistry and Biosciences, Institute of Organic chemistry, between the November 1ˢᵗ 2016 and the November 6ᵗʰ 2019 under the supervision of Prof. Dr. Stefan Bräse. A part of the practical work with the titel "Pharmacological evaluation of synthetic cannabinoids" was carried out between February 18ᵗʰ 2019 and June 31ˢᵗ 2019 at Leiden Academic Centre for Drug Research under supervision of Prof. Laura Heitman. Another part of the practical work with the titel "Synthesis and evaluation of MAGL inhibitors" was carried out between March 1ˢᵗ 2018 and August 17ᵗʰ 2018 at Leiden Institute of Chemistry under supervision of Prof. Mario van der Stelt

Hiermit versichere ich, die vorliegende Arbeit selbstständig verfasst und keine anderen als die angegebenen Quellen und Hilfsmittel verwendet sowie Zitate kenntlich gemacht zu haben. Die Dissertation wurde bisher an keiner anderen Hochschule oder Universität eingereicht.

Table of contents

Abstract

Extracts and preparations of *Cannabis sativa* have been used for centuries in human history as part of traditional medicine or in recreational rituals. Cannabinoids evolve their bioactivity throughout strong interactions in the endocannabinoid system, mainly with the two cannabinoid receptors CB_1 and CB_2. Since the discovery of this transmitter system, present in nearly all animals, extensive research on the deeper understanding and utilisation in modern medicine were carried out. Modulating the endocannabinoid system is proposed as promising target in drug discovery. Regarding the huge influences of the system involved in many physiological and pathological processes, deeper understanding of this system can help to improve treatment in multiple diseases, including a variety of neurologic disorders, pain or inflammations. Facing the abusive opioid crisis and consequences for the public health systems in the USA, new chronical pain treatment approaches must be discovered. Selective synthetic cannabinoids with precise target specificity in the endocannabinoid system and simultaneously low abductive potential makes them to the perfect candidate therefore and in drug discovery.

In the presented thesis, the design, synthesis and pharmacologic evaluation of substituted coumarins as potential new drug candidates against the cannabinoid receptors is reported. Based on initial hits, new lead structure for potent cannabinoid ligands were proposed. In a combinatorial approach, several libraries of diversely modified coumarins were synthesised. Pharmacological evaluation by radioligand binding assays to determine receptor affinities and efficacies were used to reveal the corresponding structure-activity relationships, which are necessary for high cannabinoid activity. Additionally, ligand-based docking studies were used to further strengthen the proposed relationships and to get a deeper inside in ligand-receptor interactions.

Furthermore, another project aimed to develop novel highly selective and reversible MAGL inhibitors. MAGL is the key serine hydrolase in the endocannabinoid system for degradation of the endocannabinoid 2-arachidonoyl glycerol. By inhibiting MAGL, cannabinoid signalling can be extended, which was proven as promising therapeutic approach for the treatment of multiple neuropathological conditions. Thereby, several new highly potent inhibitors were successfully synthesised and tested in enzymatic and proteomic assays for potency and selectivity.

Zusammenfassung/ Abstract in German

Extrakte aus den Blüten oder Blättern von *Cannabis sativa* wurden bereits vor Jahrhunderten in traditioneller Medizin oder Freizeit eingesetzt. Dabei entfalten Cannabinoide ihre bioaktive Wirkung im Endocannabinoid-System durch starke Wechselwirkungen mit den beiden Cannabinoidrezeptoren CB_1 und CB_2. Seit der Entdeckung dieses komplexen Signalsystems, welches in nahezu allen Tieren vorhanden ist, wurden intensive Untersuchungen zum besseren Verständnis dieses Systems und deren Verwendung in der modernen Medizin durchgeführt. Durch die Beteiligung des Endocannabinoid-Systems in vielen physiologischen und pathologischen Prozessen, stellt die gezielte Funktionsveränderungen des Systems ein vielversprechendes Ziel im Prozess der Wirkstoffentdeckung dar. Verschiedene neurologische Störungen, Schmerz- oder Entzündungsreaktionen können durch ein tieferes Verständnis der Wechselwirkungen im Endocannabinoid-System besser behandelt werden. Die aktuelle Opiumkrise und dessen Folgen für das öffentliche Krankensystem zeigt, dass neue Ansätze zur Langzeit-Schmerztherapie entwickelt werden müssen. Synthetische Cannabinoide die genaustens auf Wirksamkeit und geringes Abhängigkeitspotential maßgeschneiderten werden, stellen dafür den idealen Wirkstoff in der Medikamentenentwicklung dar.

In der vorliegenden Arbeit ist die Entwicklung, Synthese und biologische Auswertung modifizierter Coumarine und deren Potential als neuer Wirkstoff gegen die Cannabinoidrezeptoren beschrieben. Dazu wurden mehrere Bibliotheken verschieden modifizierter Coumarinderivate synthetisiert. Die Bestimmung der pharmakologischen Aktivitäten erfolgte mit Radioligandenbindungstests und daraus wurden die Struktur-Aktivitätsbeziehungen abgeleitet, welche für eine hohe Cannabinoidwirksamkeit erforderlich sind. Des Weiteren wurde für ein tieferes Verständnis der Liganden-Rezeptor Bindung Liganden-basierte Docking Studien durchgeführt.

In einem weiteren Projekt sollten neue hoch potente und reversible MAGL Inhibitoren entwickelt werden. MAGL stellt das Schlüsselenzym zum Abbau des endogenen Cannabinoids 2-Arachidonylglycerol im Endocannabinoid-System dar. Für verschiedenen neurophatologische Symptome zeigte sich ein vielversprechender therapeutischer Ansatz bei Inhibierung von MAGL. Es konnten dabei erfolgreich mehrere neue Inhibitoren hergestellt und deren Wirksamkeit und Selektivität durch verschiedene biologische Untersuchungen bestätigt werden.

1 Introduction

Cannabis sativa, commonly known as marihuana, is probably the most controversially used plant in human history. For centuries of human history, marihuana has been named greatest benefactor and worst scourge at the same time. However, extracts and preparations of this herb are reported to already have been used for recreational or medical purposes for over thousands of years. Beyond its broad applications as food and as a basic material for clothes, nets or ropes, the well-known intoxicating effects were repurposed in war, love or for personal pleasure.[1] First profound reports of the medicinal use of marihuana date back to ancient China and the world's oldest pharmacopoeia, SHÊN-NUNG's *pen-ts'ao ching* (around 2000 B.C.). Reports indicated its use as an anaesthetic in surgeries or as a treatment of diarrhoea and rheumatism, but if prescribed with its flowers strong hallucinations were observed.[2, 3] Over centuries, the use of cannabis was furthermore reported in other regions such as ancient Assyria, India, Persia or Arabia. Cannabis was first introduced into Western medicine by O'Shaughnessy in 1839, who confirmed the analgesic and sedative properties after administration and successfully used it to relieve the pain of rheumatism or to terminate convulsions or muscle spasms of tetanus.[3, 4] Furthermore, during an epidemic outbreak of cholera, O'Shaughnessy studied a new therapeutic approach of cannabis application to alleviate cholera related nausea and increase appetite. Due to diarrhoea and vomiting, cholera was usually a fatal disease back in the 19th century and providing a successful therapeutic drug was equated with a wonder.[3, 5] In the following decades wide uses for various symptoms or indications, e.g. related to convulsions or pain, were reported. However, after the availability of synthetic analgesics or hypnotics, the interest in cannabis as drug decreased and ultimately came to an end after a big disinformation campaign and the passing of the Marijuana Tax Act to declare the use of cannabis illegal in the US.[6, 7] The renaissance of using cannabis in our modern medicine most likely came from empiric observations by patients themselves. For several symptoms like chemotherapy induced nausea, wasting syndrome associated with HIV infections or convulsive disorders and migraine, patients claimed a higher effectiveness without side effects compared to their conventional medicines, if treated with marihuana preparations.[6] Discovery of cannabinoid receptors (CBRs), the related highly complex endocannabinoid system and its involvement in many pathological conditions pushed the research interests to develop active modulators within this system to a new level (for comprehensive reviews see [8, 9]). However, the remaining *status quo* makes cannabis to the most common illicit drug in the world. According to the annual drug report of the United Nations Office on Drugs and Crime (UNODC) in 2016, around 192 million people, corresponding to 3.9% of the global population, were estimated to consume cannabis at least once in the past year, showing that the public opinion on the use of medicinal cannabis has been recognisably changed. [10] Several US states and other countries such as Portugal have launched more liberal politics with decriminalisation policies of the users or higher possession limits of cannabis products, but simultaneously with drastically more severe punishments for drug dealing. Furthermore, the general legalisation of cannabis products for adults in Uruguay and Canada with strict regulations is worth mentioning.[11] A strong regulatory framework, like recently introduced in guidelines for public health and safety metrics by the Canadian government, can help to evaluate the actual impact of these general legalisations. Issues like public health and safety or an uncontrolled increase of users are often used as counterarguments against legalisation. On the other hand, strong arguments such as exterminating drug markets and providing standardised qualities also have to be considered in this discussion.[12]

1.1 Drug discovery and development

The process of drug discovery and drug development until market launch is a multidimensional endeavour, which can only be achieved with interdisciplinary research. From initial target discovery to final drug approval more than one decade can pass. Per approved drug it takes around 13.5 years on average from target hit to launch and the estimated costs differ in ranges from $150 million to more than $2.5 billion.[13-15] The complex process can be divided in two main parts, drug discovery and drug development (Figure 1). At the beginning of drug discovery, a potential (new) target is identified and validated. Anything that is related to or involved in a pathological condition can serve as target.[13] In total, around 850 proteins (human or pathogen) have been identified by literature and database analyses to act as drug targets and are combined in four major gene families: G-protein coupled receptors (GPCRs), enzymes (mostly kinases), ion channels and nuclear receptors.[16]

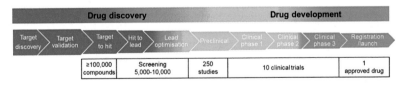

Figure 1: Process of drug discovery and development with various stages new drug candidates have to pass until market launch (modified from literature [13, 17]).

In the process of high-throughput screenings of chemical libraries, a hit (e.g. a small molecule) is identified to interact and modulate the target of interest. Based on the best initial hit a so-called lead structure is developed. This lead structure is then further optimised by repetitive rounds of structural modification and retesting. Several requirements must be taken into account at this stage of process. Structure-activity relationship (SAR) studies can help to examine the chemical functionalities responsible for the target activity. Nevertheless, many other variables like drug likeness, pharmacokinetic, target efficacy, toxicology and off-target activities must be considered as well. The optimised drug candidates will usually be validated for their proof-of-principle in a suitable animal model in a preclinical phase, before selected for drug development.[13, 18, 19]

The process of drug development in general consists of clinical studies (or trials) to assess the potential of the novel drug in humans (Figure 1). Clinical phase 1 is used to set safety margins for the progression of the other two phases by determining the impact and maximum tolerated dose of the drug. It usually includes only a small number of healthy volunteers. If successful, phase 2 can be initiated to prove if the drug candidate provides the desired pharmacological efficacy in a small cohort of patients. Nowadays, due to side effect related safety issues or lack of efficacy most of the clinical drug candidates fail at this stage.[13] Between 2000 and 2015 only 38% of phase 2 clinical candidates have successfully reached phase 3.[20] In this last phase, the effectiveness of the new drug candidate is monitored in a large number of patients and in comparison to the currently available standard care. Often randomized studies are included at this stage. Depending on the clinical endpoint, durations can vary dramatically and the costs usually exceed both previous trials combined. Only one out of ten candidates passes all clinical trials and will be submitted for approval to the respective regulatory administrations.[13]

Due to these rigorous and complex processes from drug discovery to final market introduction, approval rates of new drugs are low. The time-consuming and slow progresses combined with high failure rates in late stage clinical trials makes drug development uncertain and risky for companies. On the other hand, the reward for successfully approved drugs can be substantial and even higher than the initial costs. Collaborations of industry with academic research in drug discovery can strongly contribute to the identification of new drug targets and the development of optimised lead structures. Joint forces can therefore shorten drug discovery processes and help to reduce the expenses necessary to develop new drugs, ultimately resulting in the reduction of therapeutic costs for all patients.

1.2   The endocannabinoid system

The endocannabinoid system (ECS, Figure 2) refers to a complex lipid based (neuro-) transmitter system.[21] It is present in mammals and other animals like vertebrates and invertebrates, indicating the enormous physiological importance of the system.[22] Extensive investigations initially started after the discovery of the cannabinoid receptor 1 and 2 (CB$_1$ or CB$_1$R[23] and CB$_2$ or CB$_2$R[24]) subtypes, which were identified as the target structures of the psychoactive compound Δ^9-tetrahydrocannabinol (THC) derived from plant preparations of *Cannabis sativa*.[25, 26] The discovery of endogenous "cannabinoids" (eCB) or rather endocannabinoids (EC), which interact with both receptors, and of a related machinery of enzymes catalysing the biosynthesis and metabolism of these compounds coined the term of "endocannabinoid system" and summarises nowadays all related or involved processes.[27] The ECS is involved in numerous essential physiological and pathological processes such as food intake, mood, energy balance, pain, anxiety, (neuro-) inflammation, immune function, metabolic regulations, neuronal plasticity or reproduction.[9, 28-37]

The ECS was found to be present throughout the entire body regardless of the cell or tissue differentiation. However, the expression levels of the two receptor subtypes were found to be tissue dependent and differ significantly between anatomical localisations. Generalisation coined the segregated expression of CB$_1$ as "central" and CB$_2$ as "peripheral" receptor. Indeed, the CB$_1$ receptor density is in general very high in tissues and cells of the central nervous system (CNS), but still found in functional quantities in different peripheral tissues as well. In the CNS CB$_1$R is predominantly found in the brain regions related to cognition or movement like basal ganglia, cerebellum, hippocampus and brain stem.[9, 29, 38] The peripheral CB$_2$R was initially identified in immune system related cells such as B lymphocytes or macrophages and tissues like the marginal zone of the spleen or the lymph node cortex.[39-41] The impact of CB$_2$R in the CNS is still controversially debated in the community.[42] Initial studies were presented contradicting the existence of CB$_2$R in the CNS under healthy conditions.[43-45] However, significantly higher expression levels were found in neurological tissues during acute inflammation processes.[46-48]

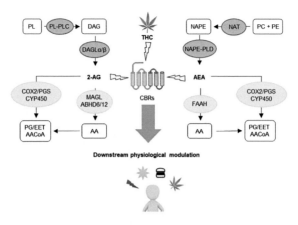

Downstream physiological modulation

Figure 2: Schematic overview of the endocannabinoid system including anabolic and catabolic pathways for the two main EC 2-arachidonoyl glycerol and anandamide. Precursors and metabolites in boxes, ligands in bold, anabolic enzymes in blue, catabolic enzymes in yellow. AA = arachidonic acid; AACoA = AA coenzyme A; ABHD6/12 = α/β-hydrolase domain containing 6/12; AEA = anandamide; 2-AG = 2-arachidonoyl glycerol; CBRs = cannabinoid receptors; COX2 = cyclooxygenase 2; CYP450 = cytochrome P450; DAG = diacylglycerol; DAGL = diacylglycerol lipase α/β; EET = epoxyeicosatrienoic acid; FAAH = fatty acid amide hydrolase; MAGL = monoacylglycerol lipase; NAPE = N-acyl-phosphatidylethanolamine; NAPE-PLD = NAPE phospholipase D; NAT = N-acyl transferase; PC = phosphatidylcholine; PE = phosphatidylethanolamine; PG = prostaglandin; PGS = prostaglandin synthase; PL = phospholipids; PL-PLC = PL phospholipase C (modified from literature[21, 49]).

The ECS consists of two CBRs, two main EC N-arachidonoyl ethanolamide (AEA) and 2-arachidonoyl glycerol (2-AG) and their intermediates, as well as several enzymes, mainly serine hydrolases, performing the biosynthesis or degradation.[50-53] In the early 1990s a N-acylethanolamine (NAE) was identified by Mechoulam and co-workers as first EC to bind to the CBRs and named as "anandamide" (AEA),[50] followed by the identification of a second main endogenous ligand 2-AG, the respective ester of the fatty acid amide a few years later.[51, 54] Several enzymes within the ECS were identified to control the biosynthesis as well as the degradation of the ECs to maintain homoeostasis. 2-AG is mainly produced from diacylglycerol lipase α or β (DAGLα or β) by hydrolysing the precursor diacylglycerol (DAG) into 2-AG.[53] Primary degradation of 2-AG to arachidonic acid (AA) and glycerol (Gly) is mediated by monoacylglycerol lipase (MAGL, MGLL or MGL), whereby the enzymes α/β-hydrolase domain containing 6 or 12 (ABHD6 or 12) are also able to catalyse degradation.[55-57] Moreover, AA can act as substrate for cascades further downstream such as prostaglandin synthesis in inflammation or it is cleared after activation with coenzyme A (CoA). The biosynthesis of AEA on the other hand is not as straightforward as that of 2-AG. Three different pathways for the synthesis of NAEs are known. However, direct conversion of *N*-acyl-phosphatidylethanolamines (NAPEs) mediated by NAPE phospholipase D (NAPE-PLD) is considered as main canonical biosynthetic pathway towards AEA formation.[21, 58] The main catabolic enzymes for AEA degradation into AA and ethanolamine is the fatty acid amide hydrolase (FAAH).[36, 59] Additionally, AEA and 2-AG can be metabolised by oxidative enzymes such as prostaglandin synthase (PGS), cyclooxygenase 2 (COX2) or cytochrome P450 (CYP450) to generate other second messengers like prostaglandins (PG) or epoxy-eicosatrienoic acids (EET), which are involved in other physiological modulation cascades further downstream.[21]

In contrast to traditional neurotransmitters ECs are generally synthesised coupled to the cell membrane due to poor solubility and only on-demand. Neurophysiologically most important to mention is the retrograde mode of signalling of ECs (Figure 3). Hence, this negative-feedback mechanism of interactions must be controlled precisely by regulating the ECs level tightly and terminating their signalling *via* hydrolysis. [9, 60, 61]

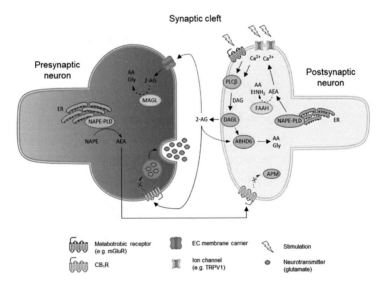

Figure 3: Schematic overview of retrograde and anterograde synaptic modulation by the ECs 2-AG and AEA in the ECS. Red arrows indicate retrograde and anterograde signalling response. Plain lines indicate stimulating pathways and dashed lines degrading pathways (modified from literature[9, 60]). APM = anterograde physiological modulations; ER = endoplasmic reticulum; mGluR = metabotropic glutamate receptor; TRPV1 = transient receptor potential cation channel subfamily V member 1.

After external stimulation (e.g. by glutamate signalling) the ECs, in particular 2-AG, are syn-thesised postsynaptically and released into the synaptic cleft. A still not entirely understood mechanism carried out by transport proteins ensures that the ECs are efficiently distributed to either bind to the presynaptic CB_1R or for re-uptake from the synaptic cleft into the presynaptic neuron. Binding to the CB_1R terminates the presynaptic signalling by inhibition of neurotrans-mitter releases like glutamate. To control the ECs signalling, 2-AG is rapidly catabolized by MAGL after re-uptake into the neuronal cell. Aside from retrograde termination of neurotrans-mitter release, anterograde agonistic EC signalling, e.g. mediated by AEA, can activate numer-ous other signal transduction pathways (Figure 4). Whether a signal transduction pathway is activated or deactivated depends on the surrounding tissue and the nature of the signalling lig-ands. In general, agonistic CBR activation can lead to the inhibition of adenylyl cyclase (AC) and the deactivation of protein kinase A (PKA) or the modulation of extracellular regulated kinases (ERK), resulting in a change of the cellular transcriptome. Other pathways more related to cell physiology control are the activation of mitogen-activated protein kinases (MAPK) *via* ceramide release, the modulation of voltage gated calcium channels (e.g. TRPV1) or the acti-vation of inward-rectifying potassium (IRK) channels and the regulation of several other kinase pathways including focal adhesion kinase (FAK) or phosphatidylinositol-3-kinase (PI3K).[9, 22, 62]

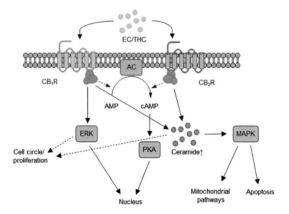

Figure 4: Generic schematic overview of agonistic signal transduction pathways mediated by the ECS and possi-ble downstream cellular responses. Plain lines indicate stimulating pathways and dashed lines inhibiting pathways. AC = adenylyl cyclase; (c)AMP = (cyclic) adenosine monophosphate; ERK = extracellular regulated kinase; PKA = protein kinase A; MAPK = mitogen-activated protein kinase (modified from literature[9]).

1.2.1   Cannabinoid receptors and G protein-coupled receptors

The CBRs belong to the superfamily of G protein-coupled receptors (GPCRs), which is one of the most important drug target class in drug development. More than 700 different GPCRs are known, which are divided in six subclasses and act as targets for more than 30% of the available drugs on the market.[63, 64] The CBRs belong to the rhodopsin related subfamily, described as class A GPCRs. Both CBR subtypes share an overall genetically homology of 48% and a slightly higher homology of 68% in the ligand-binding site.[24] In general, GPCRs are integral cell membrane receptors with an orthosteric ligand binding site located either extracellular or within the membrane and seven α-helical transmembrane segments spanning through the cell membrane. Intracellular the C-terminus is bound to heterotrimeric G proteins, which transduce the external stimuli further.[64] G proteins comprise the $G\alpha$ subunit and the dimeric $G\beta$ and $G\gamma$ subunits. The $G\alpha$ subunit is further divided in four major subtypes depending on their effector interactions: $G\alpha_s$ are activating AC, $G\alpha_{i/o}$ are two subunits which usually act in an inhibitory mode reducing for example the intracellular cAMP level and are characterised as sensitive to the pertussis toxin (PTX). $G\alpha_{q/11}$ are characterised by the activation of phospholipase C-β and lastly, $G\alpha_{12/13}$ regulate other small GTP binding proteins.[65] Within these major subtypes twenty other subunit types were identified for $G\alpha$, as well as six subtypes for $G\beta$ and twelve $G\gamma$ subtypes.[65-67] Concerning the enormous combination possibilities of G subunits, a complex physiological signalling system can be addressed by GPCR modulation.

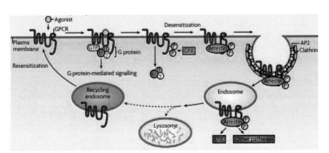

Figure 5: Signalling *via* GPCRs and receptor desensitisation by internalisation mechanisms. Adapted with permission from literature.[68]

To signal *via* GPCRs several other factors must be considered (Figure 5). GPCRs are known to be present in three different states (resting, active or inactive), depending on their structural conformation. In the inactive state the heterotrimeric G protein is bound to the receptor as a complex with GDP. An activating (agonistic) ligand binding event results in a conformational change of the receptor, leading to the dissociation of GDP, allowing GTP to bind. Subsequently the dissociation of the G proteins from the receptor follows as well as the separation of the $G\alpha$ subunit from the $G\beta\gamma$ dimer. Depending on their characteristics, both subunits can act as second messenger molecules to transduce the signal. For terminating G protein signalling, the heterotrimeric conformation and receptor binding must be reformed. Hydrolysis of GTP to GDP by GTPase activity of the $G\alpha$ subunit enables the $G\beta\gamma$ dimer to recombine and bind as heterotrimer to the GPCR again.[66] Alternatively, GPCRs can be modulated by the interaction with G protein-coupled receptor kinases (GRKs) and β-arrestins. Besides their signalling pathways, phosphorylation and β-arrestin recruitment represents an important control mechanism for desensitisation of GPCRs. β-Arrestin-binding to the receptor can initiate clathrin associated internalisation to deactivate the GPCR. Furthermore, trafficking into different compartments allows the receptors to recycle back to the cell surface in re-sensitising processes or to finally degrade the receptors in lysosomes.[68-72] As receptor activity is a result of a specific conformational orientation, it is also possible that GPCRs adopt the active conformation without ligand binding. This type of activity is usually referred to as "constitutive" or "basal" activity.[73] Ligand binding on the other hand can also cause a conformational change towards the inactive state of the receptor or a reduced activity can be observed. Ligands of this type are usually called (neutral) antagonists or inverse agonists, respectively.[74]

Beside the classic CBRs, other GPCRs have also been identified to interact in or with the ECS. Most noteworthy, is the receptor GPR55, often referred as new "subtype 3" CBR (CB_3R), which still is under intensive discussions often.[75] Initially it was discovered as an orphan GPCR within the purine (δ) cluster of the rhodopsin family, sharing only low genetic homology to the classical CBRs (CB_1 13% and CB_2 14%).[76, 77] First evidence of interactions of the cannabinoid ligands with GPR55, were made after the publication of two patents.[78, 79] The respective GPR55 expression levels were found to be tissue dependent. Highest observed mRNA levels were found in the adrenals, parts of the gastrointestinal (GI) tract and the CNS such as frontal cortex or hypothalamus. Still significant levels were identified in peripheral tissues like spleen, testis and endothelial cells.[77, 80-82] Nevertheless, the pharmacology profile of GPR55 is still most controversially debated. Several cannabinoid ligands, either of classical or synthetic nature, were identified to interact with GPR55 in the same or opposite manner compared to the other two CBRs. Other non-classical cannabinoids such as abnormal cannabidiol (aCBD) were reported with high activities at GPR55, but without any recognition at CB_1 or CB_2. However, considering the inconsistent activities of the two main ECs AEA and 2-AG at GPR55 the classification as a new CBR was considered as unjustified for a long time.[80, 82-84] GPR55 was deorphanised after the identification of lysophosphatidylinositol (LPI) and arachidonoyl-LPI as the most likely endogenous ligands.[85, 86] Signal transduction by GPR55 seems to be even more complex and not only dependent on a selective ligand binding, but rather additionally influenced by the addressing system (tissue, intra- and extracellular composition). Several different pathways have been identified, differing either in the coupled $G\alpha$ subunit identity or interacting enzyme cascades. In general, GPR55 initiates a change of intracellular Ca^{2+}-ions, resulting in a rapid phosphorylation of ERKs or activation of the nuclear factor of activated T-cells (NFAT).[85, 87] G protein activation was found to occur by C-terminal interactions with the $G\alpha$ subunits $G\alpha_{12/13}$, $G\alpha_{q/11}$ or $G\alpha_q/G\alpha_{12}$, which subsequently activate small GTPase proteins (RhoA, rac, cdc42) or other kinases (PI3K, ROCK).[80, 83, 85, 87] A noteworthy alternative pathway which includes AEA was observed in human umbilical vein endothelial cells (HUVEC). An AEA mediated GPR55 activation results in the absence of extracellular Ca^{2+}-ions and the formation of clustered integrins. Signal transduction by activation of $G\alpha_{q/11}$ follows the activation of an enzyme cascade comprised of several kinases, the release of intracellular Ca^{2+} from the endoplasmic reticulum and the same downstream effects as mentioned before. However, the presence of AEA can simultaneously terminate the integrin clustering by activating CB_1 and therefore results in a negative feedback mechanism between the classical CB_1R and GPR55.[82] Pharmacologically GPR55 was found to possibly be involved in cardiovascular regulations such as endothelium-dependent vasodilation, hyperalgesia and nociception in combination with inflammation and neuropathic pain.[75, 83, 88] Additionally, key roles in regulation of osteoclast and osteoblast activities, as well as cancer cell proliferation and migration correlating with cancer aggressiveness were observed.[89-91] Overall, further studies are necessary to identify the pharmacological role and involvement within the physiological context of GPR55 mediated pathways. Nevertheless, GPR55 seemed to be a potential target for the development of novel neuroprotective and anti-inflammatory drugs.

1.3   Coumarins as synthetic cannabinoid ligands

Over the last decades intensive research was done on identification and development of cannabinoid ligands as drug candidates for drug development. In general CB ligands can be categorized by their natural or synthetically origin and further divided into herbal derived or endogenous ligands. Synthetic cannabinoids are categorised as classical CBs if based on natural (herbal) CBs or as non-classical if entirely synthetically originated (Figure 6).

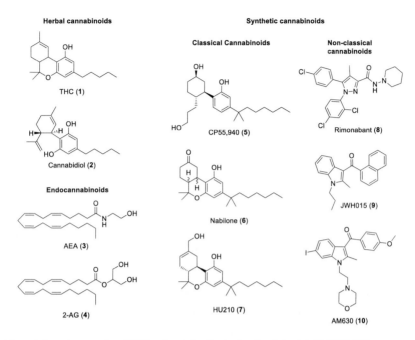

Figure 6: Exemplary overview of CB ligands and classification based on their origin.[25, 50, 51, 54, 92-95]

Besides the described ECs and herbal-derived CBs like THC (**1**) and Cannabidiol (CBD, **2**), the initially developed synthetic cannabinoids (or classical CB[1]) were mostly based on their herbal counterparts.[92] Further research led to the development of non-classical CBs like the pyrazole-based Rimonabant (**8**) or the indol-derivatives JWH015 (**9**) and AM630 (**10**).[95] Rimonabant (**8**) was initially developed for the treatment of chronic obesity through antagonistic blockage of CB signalling at CB_1R, resulting in reduced appetite. However, already two years after market release in 2006 it was withdrawn due to its severe psychiatric side effects like increased anxiety, depression and suicidal thoughts.[96, 97]

Until then, substituted coumarins had not been considered as a potential lead structure for the development of CB ligands. In 2009 our group were able to show a cannabinergic activity for several substituted 3-benzylcoumarins with highest affinities in sub-micromolar ranges.[98] The high potential of 3-benzylcoumarins as lead structure for the development of CB ligands can

---

[1] CP55,940 is usually referred as non-classical CB and is only included to the classical CB to show the similarity towards THC and CBD.

be highlighted by structural comparison with established classical and non-classical CB ligands (Figure 7).

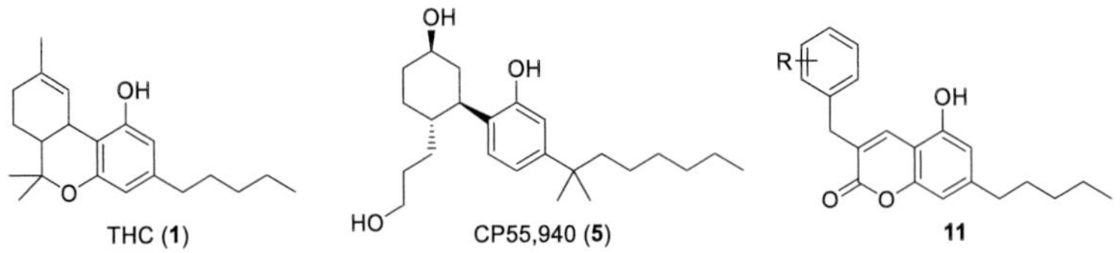

Figure 7: Structural comparison of classical and non-classical CB ligands with 3-benzylcoumarin **11** as lead structure.

Based on this lead structure our group began to develop a comprehensive number of coumarins with different substitution patterns. The research of these first-generation ligands was focused on improving receptor affinities as well as selectivities towards both main CBRs. In subsequent conducted structure-activity studies several ligands were identified with affinities in low nanomolar ranges or high selectivities over one receptor subtype. At the same time, more evidence arose that GPR55 is a CB-related GPCR (see Chapter 1.2.1). Only small structural changes in the lead structure were necessary, more precisely attaching small core substituents instead of long alkyl chains to also address GPR55 and to include related SAR studies into our research. In Figure 8 four coumarin derivatives (**12–15**) of the first generation ligands are shown, which represent the investigated lead structures of these SAR studies.[98-100] Only minor structural changes were needed to partially or completely change the pharmacological profile of the ligand. At CB$_1$R the full range of efficacies from full agonists to antagonists were observed. For example, high CB$_1$R affinity with moderate selectivity was observed for coumarin **12**. Inverse efficacies were observed after increasing side chain bulkiness and benzylic polarity. In contrast to that, most of the active ligands at CB$_2$R showed an agonistic efficacy. As mentioned before, by attaching small substituents to the coumarin core GPR55, active ligands with excellent selectivities, like for example **14**, could be generated. Interestingly, an allosteric antagonistic binding mode was observed by further increasing the side chain length as shown by **15**.

12, CB_1 antagonist
CB_1: K_i = 22 nM
CB_2: K_i = 405 nM
CB_2/CB_1 = >18

13, CB_2 partial agonist
CB_1: K_i = 4890 nM
CB_2: K_i = 49 nM
CB_1/CB_2 = ~100

14, GPR55 antagonist
GPR55: pA_2 0.547 μM
CB_1: K_i = >10 μM
CB_2: K_i = >10 μM

15, CB_2 agonist
GPR55 allosteric antagonist
GPR55: pA_2 0.483 μM
CB_1: K_i = 1580 nM
CB_2: K_i = 68 nM
CB_1/CB_2 = ~23

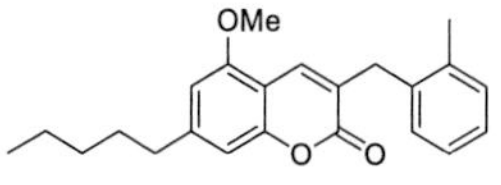

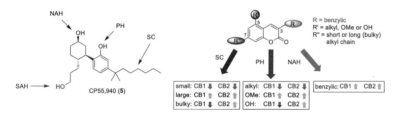

Figure 8: Selected first generation ligands with their affinities and efficacies.[98-100]

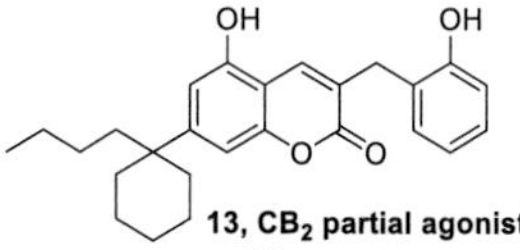

For classical and non-classical CBs extensive SAR studies have already been conducted to identify the pharmacologically relevant structural requirements for a high CBR affinity. Several important pharmacophoric regions could be identified and the evaluation of the first generation SAR studies showed a comparable profile of pharmacophores (Figure 9).[98, 99, 101]

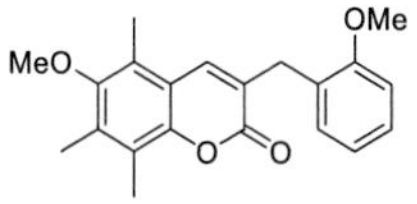

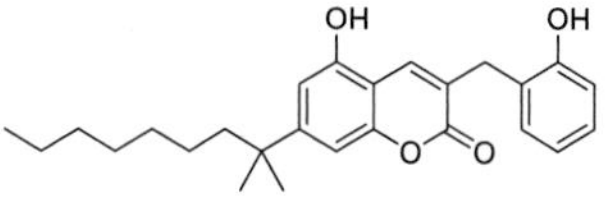

Figure 9: Important pharmacophores identified in high affinity CB ligands compared to the 3-benzylcoumarins. SC = side chain; PH = phenolic hydroxyl; NAH = northern aliphatic hydroxyl group; SAH = southern aliphatic hydroxyl group.[101]

The side chain (SC) and the phenolic hydroxyl group (PH) have been identified as key pharmacophores in traditional CBs. Lipophilic groups at the SC position have been proven to be essential for high interactions of the CB ligand with the CBR binding pockets.[101] Elongation or branching compared to the pentyl side chain in THC (**1**) further increased the affinities towards the CBRs.[102, 103] Similar observations have been made at the 7-position of the coumarins, where increased carbon side chains resulted in increased affinities towards the CBRs. Additionally, branching of the side chain similar to CP55,940 (**5**) resulted in increased selectivity towards CB_2R.[99] The PH have been thought to be essential for a successful cannabimimetic activity as well, based on the observed SARs of classical and non-classical CBs towards CB_1.[104] However, readily after identification of the CB_2R this assumption had to be revised. Several SAR studies showed a significant improvement regarding selectivity towards CB_2.[102, 105, 106] In accordance to that, similar results were obtained after ether cleavage of 5-methoxycoumarins.[99] The northern aliphatic hydroxyl (NAH) group showed a significant increase in potency if introduced into classical CB ligands. Same results can be observed in THC (**1**) and (−)-11-hydroxy-Δ9-THC, a key intermediate in THC metabolism with a much higher potency as THC.[101] However, inconsistent results were observed for the 3-benzylcoumarins after introduction of polar groups at the 3-benzyl moiety.[98, 99] This suggested a high structural dependence and influence of every substituent inside the active pocket to the binding mode, which results in a different pocket occupancy compared to the traditional CB ligands. Lastly, the southern aliphatic hydroxyl (SAH) group was introduced first in non-classical cannabinoids like CP55,940 (**5**), showing another structural modification to improve potencies.[101] In the coumarin this group most likely is represented by the carbonyl functionality at the coumarin core. Based on these results we started to develop a second generation of coumarin CB ligands. To expand the scope and diversity of substitution patterns THOMAS HURRLE, PhD, extensively investigated new methods or postsynthetical core modifications with focus on the 3-position. This led to the synthesis of several libraries according to the new proposed lead structures (Figure 10). The pharmacological evaluation of these ligands was included into the presented SAR studies of this work.

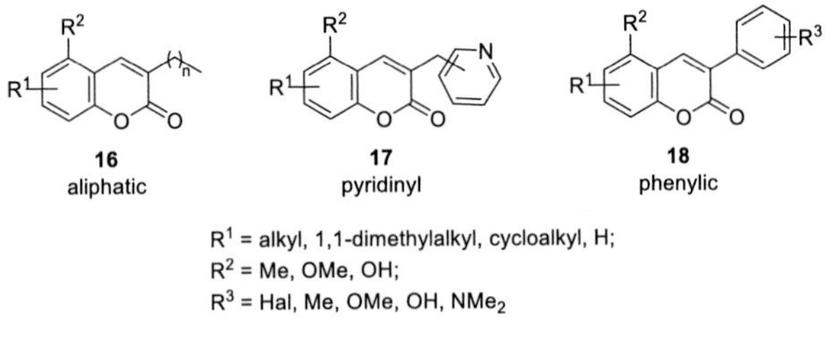

16
aliphatic

17
pyridinyl

18
phenylic

R^1 = alkyl, 1,1-dimethylalkyl, cycloalkyl, H;
R^2 = Me, OMe, OH;
R^3 = Hal, Me, OMe, OH, NMe_2

Figure 10: Selected lead structures **16–18** of second generation coumarin CB ligands developed by THOMAS HURRLE, PhD.[107]

1.3.1   Coumarin synthesis

The 3-benzylcoumarins **21** are synthesised according to an in-house developed method (Scheme 1).[108] This method is based on the reaction of modified salicylic aldehydes **19** and cinnamaldehydes **20**, in the presence of an ionic liquid (IL, **22**) *via* N-heterocyclic carbene (NHC) catalysis. Desired substitutions on the coumarin core structure can be introduced by the choice of the starting materials. Main advantages of this method are the acceptance of a broad substrate scope and the fast reaction time by microwave supported irradiation.

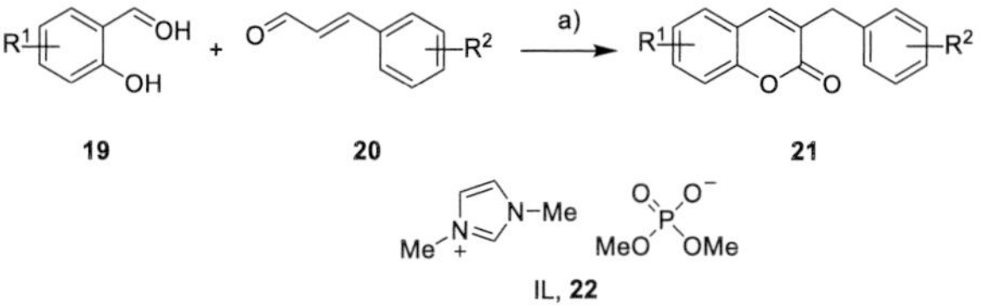

Scheme 1: NHC catalysed, microwave supported coumarin synthesis by BRÄSE *et al.*[108] *a) IL 22, K₂CO₃, toluene, 110°C, MW (230 W max., 7 bar max.), 50 min.*

The proposed mechanism for the catalytic cycle is outlined in Scheme 2. Deprotonation of the NHC precursor (**22**) forms the active carbene catalyst **23**, which performs a nucleophilic attack on the cinnamaldehyde leading to the primary carbene-aldehyde product **24**. A proton shift towards the homoenolate **25** and deprotonation/protonation steps result in the formation of the tautomeric forms **26a** and **26b**. Nucleophilic attack of salicylic aldehyde **19** releases the intermediate **27** under reformation of the catalytic NHC species **23**. Finally, an intramolecular condensation reaction of intermediate **27** yields the coumarin product **21**. Besides, with progressing reaction time a side reaction was observed by hydrolysis towards the acid derivative **28**, resulting in a decrease of the active NHC catalyst. To prevent a premature stop of the reaction, stoichiometric amounts of the IL **22** are necessary.

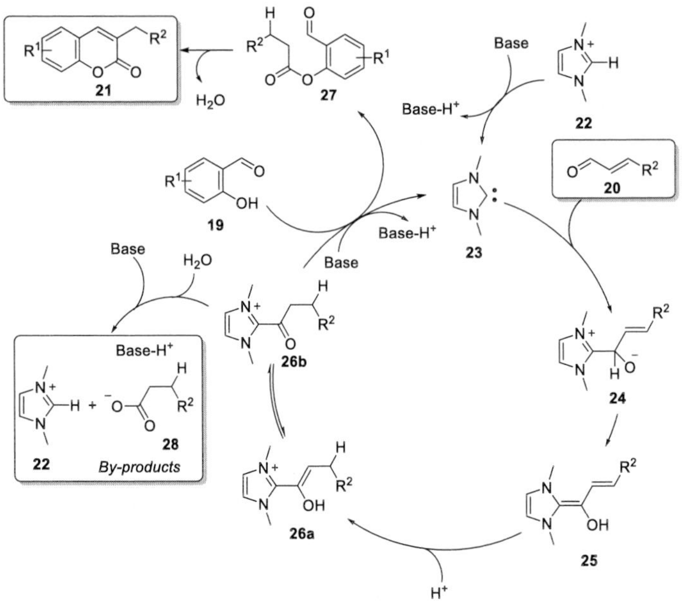

Scheme 2: Proposed catalytic cycle for the synthesis of 3-benzylcoumarins according to Bräse *et al.*[108-110]

1.4 Monoacylglycerol lipase

The responsible enzymes of the ECS are mostly represented by serine hydrolases (SH). This class of enzymes represents ~1% of all proteins in mammals and is characterised as catalytic species by a central serine residue, embedded in a catalytic dyad or triad.[111-113] Within the triad hydrogen bonds increase the nucleophilicity of the serine oxygen and allow the formation of an acyl-enzyme intermediate with the substrate, followed by a water-induced saponification in the catalytic circle.[112] Besides the catalytic triad, metabolic SHs show a GXSXG consensus sequence bordering the active serine. The tertiary structure often consists of a canonical α/β-hydrolase fold, which is characterised by a central β-sheet conformation, surrounded by α-helices of the hydrolytic parts.[114]

Monoacylglycerol lipase (MAGL) is one these SHs and represents a key enzyme in the ECS and lipid metabolism. It is expressed as a ~33 kDa enzyme and several isoforms ,either species- or tissue-dependent, could be isolated.[115] Highest expression levels were found in the brain, white adipose tissue and liver.[116, 117] In the brain, MAGL was found in the hippocampus, cerebellum, cortex and thalamus, mainly expressed at the presynaptic axon terminal, to rapidly terminate 2-AG signalling at co-localised CB_1R.[115, 118] Additionally, MAGL was identified as a soluble serine hydrolase, peripherally associated with cell membranes and present in membrane and cytosolic fractions.[115] MAGL is the central regulating enzyme in controlling the cellular levels of endocannabinoid 2-AG and pro-inflammatory precursor AA.[117] Several studies indicated, that MAGL is responsible for >85% of the 2-AG metabolism in the brain. The remaining 15% are catalysed by ABHD6 and ABHD12.[55, 56, 119-121] Within MAGL the amino acid residues Ser^{122}, Asp^{239} and His^{269} were identified as the catalytic triad.[116] The catalytic triad is furthermore surrounded by three cysteines (Cys^{201}, Cys^{208} and Cys^{242}), suggesting their possible key role in stabilising the conformation of MAGL and improving substrate interaction.[120, 122, 123] The active binding pocket comprises three regions: the large hydrophobic tunnel forming the acyl-chain binding pocket (ABP) with the catalytic triad at the top, the alcohol-binding channel (ABC) and the glycerol exit channel.[124, 125]

Biological data suggested a huge therapeutic potential by MAGL inhibition for multiple diseases. Promising results were found for the treatment of acute neuropathic pain and neuro-inflammation by reduced antinociception, antiallodynic effects or decreased synthesis of pro-inflammatory metabolites.[8, 126, 127] Chronical neuro-inflammation ultimately can result or accelerate pathogenesis of neurodegenerative diseases like Alzheimer, Parkinson, Huntington's chorea or demyelinating diseases including multiple sclerosis.[126, 128-132] Besides neurodegenerative diseases, targeting MAGL showed a promising potential for the treatment of mental disorders like anxiety or depression with novel anxiolytic drugs.[8] In October 2018 Abide Therapeutics announced the initiation of clinical phase 2 of the novel MAGL inhibitor ABX-1413 for the treatment of Gilles de la Tourette (Tourette) syndrome.[133] Classical cannabinomimetics like Nabilone (6) are already approved and in use for the treatment of chemotherapy induced nausea and vomiting. However, undesired side effects like paranoia or hallucinations limit their widespread application. Enhancement of endocannabinoid signalling by inhibiting MAGL was proven to be a rational alternative to more effectively treat chemotherapy induced nausea or vomiting in humans.[134] Additionally, by elongation of EC signalling a decreased appearance of addiction withdrawal effects were observed or elevated MAGL activities in autism could be targeted.[8, 126, 135] Finally, the distinct role of MAGL in aggressive human cancer cells or primary tumours by regulating fatty-acid signalling and other fatty-acid derived oncogenic signalling metabolites is noteworthy. Therefore, inhibition of the highly expressed MAGL can help to reduce cancer pathogenesis and improve early stage cancer therapies.[117, 126]

2 Aim and outline

Since the discovery of the highly complex endocannabinoid signalling system, scientist all over the world try to selectively modulate this system using small molecules.[136, 137] Studying this system over decades led to a deep understanding in interactions and dependencies, pharmacological relevance in normal circumstances and most important the consequences of disrupting the balance of this system by diseases.[9, 130] However, lack of selectivity and off-target activities are still main reasons to disqualify potential drug candidates in the process of drug development. Therefore, new approaches and lead structures for potential drug candidates with more precise pharmacological properties must be developed.

Aim of the research described in this thesis was the design, synthesis and pharmacological evaluation of potential drug candidates.

Based on the results of the first-generation coumarin based CB ligands, new lead structures were included to increase the diversity of the second generation (Figure 11). By introducing heteroatoms at position 7 (library 1 and 2), fluorescent coumarin derivatives can be generated.[138] Cannabinergic activity would make them to a powerful tool as intrinsic fluorescent probe, to study receptor localisation and distribution in different tissues or *in vivo*. Alternatively, they could serve as reference compounds for fluorescence-based receptor binding studies to avoid the handling with radioactive material.

The 1,1-dimethyl motif in library 3 was already proven to increase receptor potency in traditional CB ligands and similar observations were already made in the previous studies (see Figure 9). However, given the small number of tested derivatives containing this motif and the promising results, further derivatives should be developed.

Lastly, the synthesis of more derivatives with small substituents attached to the coumarin core came into research focus (library 4). New results of collaboration partner BERND FIEBICH PhD demonstrated a neuroprotective effect of the GPR55 antagonist by reducing the inflammatory response in lipopolysaccharide (LPS) treated primary microglia cells.[34]

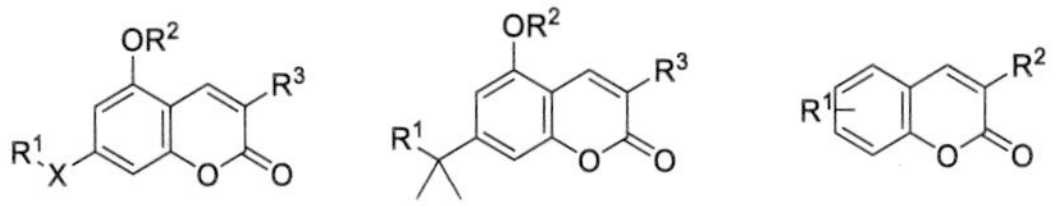

X = O **29**
X = N **30**
Library 1/2

31
Library 3

32
Library 4

Figure 11: Generic lead structures for the development of second generation coumarin based synthetic cannabinoids.

In the second part of this thesis, the pharmacologic activities of the developed and synthesised second-generation coumarins had to be evaluated. During a research stay in the laboratories of Prof. LAURA HEITMAN at Leiden University (the Netherlands), a comprehensive SAR study should be performed to determine the receptor dependent potencies and pharmacologic efficacies of ligand receptor binding.

During another stay abroad in the laboratories of Prof. MARIO VAN DER STELT at Leiden University (the Netherlands) in a joint project based on the work of MING JIANG, further optimisations of reversible MAGL inhibitors (Figure 12) and their pharmacological evaluation should be performed.

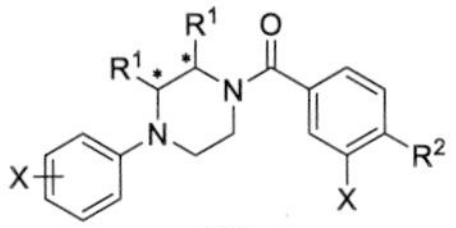

±33

Figure 12: Lead structure core of promising drug candidates as new reversible MAGL inhibitors.

As previously described (see Chapter 1.4) inhibition of MAGL bears a huge therapeutic potential for multiple diseases. However, to get access to this potential the severe side effects resulting from a covalent MAGL inhibition must be overcome first. Therefore, new approaches have to be focused on, with inhibitors having a reversible binding mode and comparable high potencies as their reported covalent analogues (see also Chapter 1.4).[129, 139, 140]

3 Results and discussion

The major focus of this work was the synthesis and characterisation of novel synthetic cannabinoids. The present thesis is divided into three parts. In the first part, the library synthesis of
coumarin-based synthetic cannabinoids is described. Secondly, the conducted pharmacological
evaluation is presented in form of a structure-activity relationship (SAR) study, enhanced with
further investigations to functionality, computational docking experiments and pharmacology.
Finally, the third part covers all efforts towards the design, synthesis and pharmacological evaluation of selective MAGL inhibitors.

3.1 Synthetic work on coumarin-based cannabinoids

In this subchapter, the synthetic work on the coumarin-based cannabinoids is summarised. First,
the synthetic routes towards the necessary starting materials for the syntheses of lead structures
29-32 (Figure 11) had to be designed and accomplished. Essential requirements for the synthetic
routes were simple, commercially available starting materials, modular late-stage derivatisation
to increase diversity much easier, and few reaction steps. Subsequently the building blocks were
applied in a library synthesis approach to generate the novel substituted coumarin ligands.

3.1.1 Synthesis of salicylic aldehydes

The modularity of the coumarin synthesis is mainly based on the introduction of functionalities
within the salicylic aldehyde **19** and cinnamaldehyde **20** building blocks (see Scheme 1). The
following paragraph summarises the design and synthetic work for all later used salicylic aldehydes. For some synthetic routes, previous established work was applied as starting point for
the development and two salicylic aldehydes were fully reproduced according to the established
procedures (see literature[141]).

For the first lead structure **29**, alkoxy substituted salicylic aldehydes **19** had to be prepared.
Based on the previous SAR studies, primarily it was decided to use linear alky chains first, in
order to obtain comparable pharmacological results. After verification of the activity, further
optimisations for example by introduction of bulkier groups for a higher space demand inside
the active side of the receptor were planned. Differences in the alkyl chain lengths were applied
to determine the shortest and longest length necessary for any activity as well as the ideal
lengths, for either the highest activity or highest selectivity.

The proposed synthetic route towards the alkoxy salicylic aldehydes **19a-g** is shown in Scheme 3. Starting from 3,5-dimethoxyphenol (**34**) with a derivatisation of the phenol group by WIL-LIAMSON ether synthesis[142] to introduce an aliphatic functionality in the final coumarin ligand at position 7. Followed by *para* (*p*)-position formylation of the phenol and lastly a selective mono-deprotection to obtain the salicylic aldehydes **19a-g**. Thereby, the formylation as well as the selective-deprotection should be performed according to previous protocols. [143]

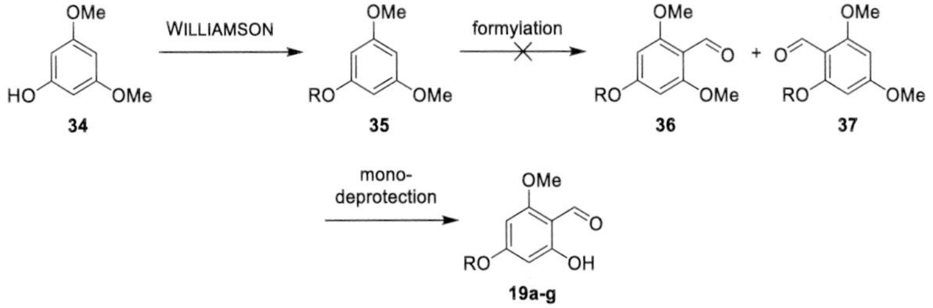

Scheme 3: Proposed synthetic strategy towards alkoxy salicylic aldehydes **19a-g**.

Unexpectedly, a successful isolation of the salicylic aldehydes **19a-g** with this approach was not achieved, due to non-selectivity of the formylation reaction. The product was only obtained as an inseparable mixture of the product **36** and its corresponding regioisomere **37**. To overcome the selectivity issue in another approach a formylation procedure of a VILSMEIER reaction was applied to avoid the usage of *n*-butyl lithium (*n*-BuLi) as strong base. Additionally, the order of the reactions was changed, starting with the *p*-formylation, followed by derivatisation by WILLIAMSON ether synthesis, to use the non-selectivity as an advantage. The results of the respective reaction steps for the syntheses of salicylic aldehydes **19a-g** are summarised in Table 1. By modifying the purification protocols published by ALBERICIO *et al.* [144] and JIN *et al.* [145] the products **38** and **19a** were obtained in sufficient yield and excellent purity. Undesired di-formylation of **34** was only observed in a low quantity (<5%) and therefore negligible.

Table 1: Summary of syntheses towards alkoxy salicylic aldehydes **19a-g**: *a) POCl₃, DMF, 0°C to rt. o.n., then H₂O, 42%; b) n-alkylbromide, K₂CO₃, 18-crown-6, acetone, reflux, o.n.; c) AlCl₃, NaI, ACN, DCM, 1 h, 0°C to rt.*

Entry	R	Yield[b] [%]	Yield[c] [%]
1	methyl	-	**19a**, 34[a]
2	ethyl	**36b**, 82	**19b**, 34
3	propyl	**36c**, 79	**19c**, 94
4	butyl	**36d**, 87	**19d**, 63
5	pentyl	**36e**, 84	**19e**, quant.
6	hexyl	**36f**, 86	**19f**, 90
7	heptyl	**36g**, 76	**19g**, 58

[a]directly obtained in reaction step a) as regioisomere.

In the WILLIAMSON ether synthesis *n*-alkyl bromides from ethyl to heptyl were decided to use for the introduction of alkyl side chains with different lengths and all products (**19a-g**) were obtained in good yields. The yields for the selective mono-deprotection ranged between low (**19b**, 34%) and quantitative (**19e**, quant.), whereby no trend for lower yields was observed.

Furthermore, to increase the spectra of directly attached heteroatoms at the 7-position of the coumarins, in the second lead structure **30** (Figure 11), the phenolic ether functionality was substituted by an anilinic amine functionality. Again, it was primarily decided to use linear alky chain substituents first. Additionally, in contrast to the ether function, the amine group bears another position for substitutions. To reduce the theoretically possible variations between different length combinations, it was decided to first limit this second position with a methyl group to a fixed length. From the previous studies it was known already, that methyl groups at this position are well tolerated or even enhance the CB receptor affinities in some cases.[99, 100]

In Scheme 4 the modified proposed synthetic route towards the unsymmetrical dialkylamino functionalised salicylic aldehydes **19h-j** is shown, as a straightforward two-step unsymmetrical alkylation of aniline **39** was not feasible (see Appendix Scheme 21). Starting with the protection of 3,5-dimethoxyaniline (**39**) with an allyloxycarbonyl (alloc) protecting group (PG), followed by methylation with methyl iodine to the corresponding N-methyl aniline **41**. The alloc protecting group was chosen over the benzyloxycarbonyl group (Cbz) to allow a deprotection of the formylated aniline **42** without any risk of aldehyde reduction. This reaction was followed the individual introduction of the second derivatisation, using alkyl halogens as alkylation reagents to gain the unsymmetrically *N,N*-dialkylated aniline **44**. The synthesis strategy is finalised by the selective mono-deprotection to the desired salicylic aldehydes **19h-j**. The second alkylation was placed to the end of the route, to achieve modularity, but retain a straightforward synthesis to safe reaction steps and time.

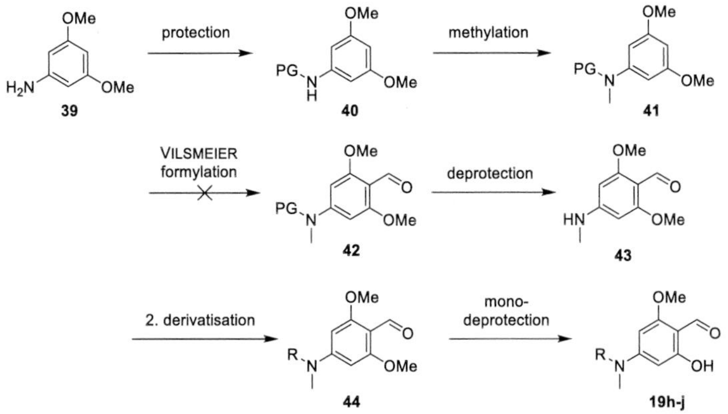

Scheme 4: Modified proposed synthesis route towards unsymmetrically substituted *N*-dialkylamino salicylic aldehydes **19h-j**, PG = alloc.

The execution of the proposed synthesis strategy was done as part of a bachelor thesis by AARON BIRK.[146] The first proposed strategy was not immediately successful, again because of non-selectivity and low yields in the formylation reaction. Apparently, the electron-withdrawing effect of the carbamate group was high enough to sufficiently deactivate the aromatic system and lose *para*-position selectivity. Minor adjustments of the reaction order resulted in the final synthetic route, which is shown in Scheme 5. As variable alkyl chains an ethyl, butyl and hexyl group were used to investigate the different structure-activity relationships between short and long chain substituents, as well as to ensure comparability with the previous studies.

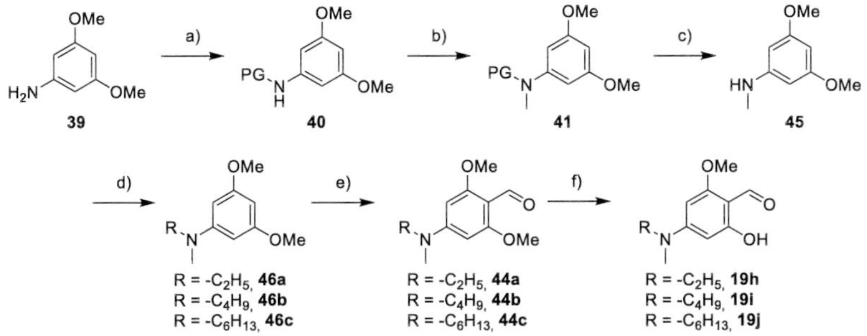

Scheme 5: Summary of syntheses towards *N,N*-dialkylamino salicylic aldehydes **19h-j**, PG = alloc: *a) allyl chloroformate, K₂CO₃, THF, 2 h , 0°C, 97%; b) i) NaH, THF, 30 min, 0°C, ii) MeI, 1.5 h, rt., 95%; c) Pd(Ph₃P)₄, 1,3-dimethylbarbituric acid, THF, o.n., 50°C, 90%; d) n-alkylbromide, Et₃N, 1,4-dioxane, o.n., reflux, 46a 73%, 46b 85%, 46c 99%; e) i) POCl₃, DMF, DCM, 1 h, 0°C, then 4 h, rt., ii) H₂O, 30 min, 0°C, 44a 29%, 44b 73%, 44c 52%; f) AlCl₃, NaI, ACN, DCM, 1 h, 0°C to rt, 19h 71%, 19i 80%, 19j 57%.*

The intermediate products of protection, methylation and deprotection (**40**, **41** and **45**) were obtained with excellent yields of 97%, 95%, and 90%, respectively. Yields for the second alkylation reaction range from good (**46a**, 73%) to excellent (**46c**, 99%), while an increase of alkyl length strongly correlates with an increase of yield. This can be explained by the higher available concentrations of the longer alkyl bromide, due to their higher boiling points. Alternatively, reactions with low boiling alky bromides could be transferred into pressure resistant crimp vials to overcome this problem. In contrast, the formylation reactions resulted in a higher variety of yields. While the ethyl substituted (**44a**) or hexyl substituted aniline (**44b**), resulted in low (29%) and moderate (52%) yields, respectively, a good yield of 73% was achieved for the butyl aniline **44b**. In the final selective mono-deprotection, salicylic aldehyde **19h** and **19i** were obtained in good yields of 71% and 80%, salicylic aldehyde **19j** in a moderate yield of 57%.

The next lead structure **31** (Figure 11) bears a 1,1 geminal dimethylated alkyl chain as main core element variation. In previous studies the dimethyl alkyl moiety exhibited good affinities towards both CBRs, one example as dual ligands and two examples as slightly more CB₂ selective ligand, all showing an agonistic efficacy.[99] Even more interesting, however, were the affinities found towards the orphan GPCR GPR55. Three examples containing the deprotected 5-hydroxy group showed an affinity in the low μM range and an antagonistic efficacy was observed. Additionally, due to a rightward shift and a depression of the maximal (max.) agonistic effect response curve in the β-arrestin assays, an allosteric mode of action was proposed for these ligands.

The previously applied protocol for the synthesis of the dimethyl alkyl substituted salicylic aldehydes was lacking a modular way to introduce different substituents next to the dimethyl moiety. Therefore a new synthesis strategy was proposed, that involves as a key step a simultaneous direct geminal dimethylation by literature know protocols of REETZ et al.[147-149] to reduce reaction steps (for synthesis strategy and proposed mechanism see Appendix Scheme 22 and Scheme 23). However, in every applied reaction condition a β-elimination occurred as competing reaction, resulting in an inseparable mixture of desired product and elimination product. Mechanistically it appeared, that after release of the GRIGNARD-type adduct towards the intermediate carbocation species, competing kinetic rates of the β-elimination resulted in the side product formation. As in literature similar issues were published [150, 151], a modification to a two-step geminal dimethylation was done by converting the ketone to a tertiary alcohol by GRIGNARD reaction. The second geminal methyl group was introduced following the protocol of HARTSEL et al.[151] The applied synthetic strategy towards the salicylic aldehydes 19k–19m is shown in Scheme 6 and was partially included into a bachelor thesis done by MAXIMILIAN KNAB.[152]

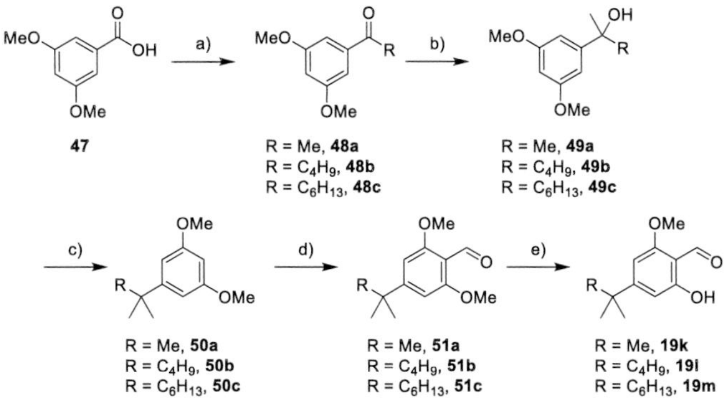

Scheme 6: Syntheses route towards 1,1-dimethylalkyl salicylic aldehydes 19k-m: a) alkyllithium, THF, 2 h, –78°C to rt., 48a 94%, 48b 85%, 48c 98%; b) MeMgBr, 2 h, rt., 49a 74%, 49b 72%, 49c 82%; c) i) SOCl₂, DCM, 2 h, 0°C, ii) AlMe₃ (2 M in toluene), DCM, 49a 2 h, 49b and 49c 3 h, –78°C, then o.n., rt., 50a 55%, 50b 70%, 50c, 54%; d) i) N,N,N,N-tetramethylethylendiamine (TMEDA), 50a Et₂O, 50b and 50c THF, 0°C, n-BuLi, 2 h, rt., ii) 0°C, DMF, 4 h, rt., 51a 60%, 51b 82%,51Xc 88%; e) AlCl₃, NaI, ACN, DCM, 1 h, 0°C to rt, 19k 90%, 19i 96%, 19m 96%.

A modification of protocols from COMPTON et al.[153], by treating 3,5-dimethoxy benzoic acid (47) with two equivalents of alkyl lithium reagents (48a = methyl, 48b = butyl and 48c = hexyl) at –78°C and stirring for 2 h, resulted in the pure products with excellent yields. To avoid lower yields due to side reactions like a disubstitution, thawing the reaction mixture to rt was observed as crucial. In the next step the ketone groups of **48a–c** were converted to a tertiary hydroxyl group by a GRIGNARD reaction with methyl magnesium bromide in good yields. Following the earlier mentioned protocol of HARTSEL *et al.* all three dimethylated products were obtained in moderate (**50a** 55% and **50c** 54%) to good (**50b** 70%) yields. The salicylic aldehydes **19k-m** were obtained in moderate to good yields (54% – 88% over two steps) by applying the estab-lished formylation and selective mono-deprotection protocols as previously described.[108]

Besides, within the second methylation reaction a side product formation was observed for the tertiary alcohols **49b** and **49c** with 8% and 21% yield, respectively. With structural elucidation by NMR, the side products were identified as the benzothiophens **52a** and **52b** and the assigned structures were confirmed by X-Ray analysis.

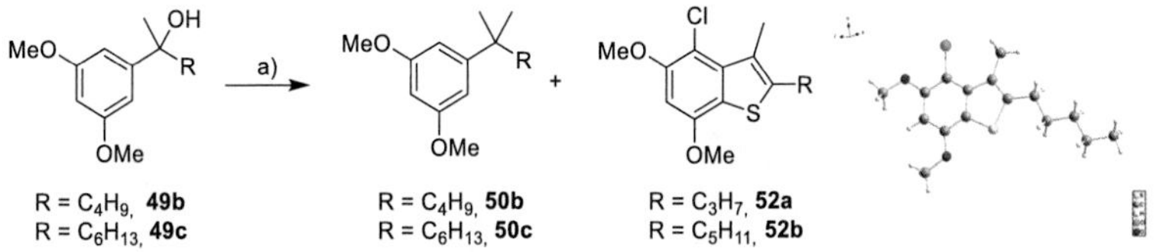

| R = C$_4$H$_9$, **49b** | R = C$_4$H$_9$, **50b** | R = C$_3$H$_7$, **52a** |
| R = C$_6$H$_{13}$, **49c** | R = C$_6$H$_{13}$, **50c** | R = C$_5$H$_{11}$, **52b** |

Scheme 7: Identified side products from the geminal alkylation on **49b** and **49c**: *a) i) SOCl$_2$, DCM, 2 h, 0°C, ii) AlMe$_3$ (2 M in toluene), DCM, 3 h, –78°C, then o.n., rt.*

Comparable preparations of benzothiophenes by introducing the thiol functionality using thio-nylchloride were reported in literature so far only once by BLATT *et al.* The authors described the formation of 3-arylbenzothiophens by treatment of diarylcarbinols and diarylolefins with thionylchloride under heating to reflux. Furthermore, it was concluded, for the formation of benzothiophenes the starting molecule has to be a 1,1-diarylethylene with at least one hydrogen at C$_2$ and no aryl group at this position.[154] Therefore, further synthetic and mechanistic inves-tigations could lead to a new reaction method for the synthesis of this highly functionalised benzothiophenes. Several reasons indicated that the reaction occurs in more than one mecha-nistic pathway. A plausible mechanistically sequence would most likely start after substituting the hydroxyl group to the chlorinated species, as this is described by HARTSEL and co-workers to be completed within the first 2 h reaction time. As thionylchloride was added in an excess, an insufficient removal of it after the first reaction step had to be the reason for the side product formation. Regarding the investigations of BLATT and co-workers, the addition of the trimethyl-aluminium, must have led to the formation of a catalytic species that promotes the benzothio-phenes reaction to occur at lower temperatures. Furthermore, these specie or species probably lead to a LEWIS acid catalysed electrophilic aromatic chlorination, resulting in the formation of the products **52a** and **52b**. Due to time reasons, no further investigations were carried out within this thesis.

In the last two sequences, salicylic aldehydes with small substituents were in the focus of investigations. In contrast to the long and bulky substituents as shown in previous studies, small or no alkyl substitutions at position 7 appeared to be favourable for an antagonistic GPR55 activity and at the same time high selectivity.[100] The synthesis strategies were performed according to previously established protocols[107, 155] and executed in a bachelor thesis by LUKAS LANGER.[156]

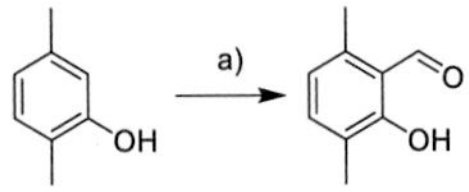

Scheme 8: Synthesis towards salicylic aldehyde **19n**: *a) paraformaldehyde, MgCl₂, Et₃N, ACN, 4 h, reflux, then H₃O⁺, rt., 30 min, 44%.*

Salicylic aldehyde **19n** was obtained directly by an *ortho* (*o*)-formylation of phenol **53** with paraformaldehyde in an MgCl$_2$-Et$_3$N system with moderate a yield of 44%.

For the synthesis of trimethylated salicylic aldehyde **19o**, a modified protocol from THOMAS HURRLE was used.[107] Straightforward formylation, dimethylation and selective mono-deprotection resulted in the trimethylated salicylic aldehyde **19o**, with an improved overall yield of 32% (before 28%[107]). A faster attempt of formylating **54** by the previous used paraformaldehyde approach resulted in a lower yield of 10%.

Scheme 9: Synthetic route towards salicylic aldehyde **19o**: *a) i) 0°C, TiCl₄, DCM, 1 h, rt., ii) dichloromethyl methyl ether, 2 d, rt., 46%; b) dimethyl sulphate, K₂CO₃, acetone, 4 h, reflux, 76%; c) AlCl₃, NaI, ACN, DCM, 1 h, 0°C to rt, 92%.*

With completion of the last to reaction sequences, all for the synthesis of the coumarins needed salicylic aldehydes were prepared successfully. In the next section the synthetic effort towards the cinnamaldehydes will be discussed.

3.1.2 Synthesis of cinnamaldehydes

As second precursors for the synthesis of the substituted coumarins several cinnamaldehydes (**20**) had to be prepared, besides the commercially available. As substituents, either electron-donating groups like methyl or methoxy groups or electron-withdrawing groups like halogens, as well as the neutral unsubstituted cinnamaldehyde (**20a**) were planned to be part of the SAR investigations. Furthermore, sterically effects of substitution positions (*o* or *p*) or hydrogen to fluorine exchange came into consideration.

The synthesis of cinnamaldehydes were performed *via* cross coupling reactions previously described by HECK reaction.[141, 157] In total nine different cinnamaldehydes were used, whereby six were commercially available and three were synthesised in good yields of 64% – 73% (Scheme 10).

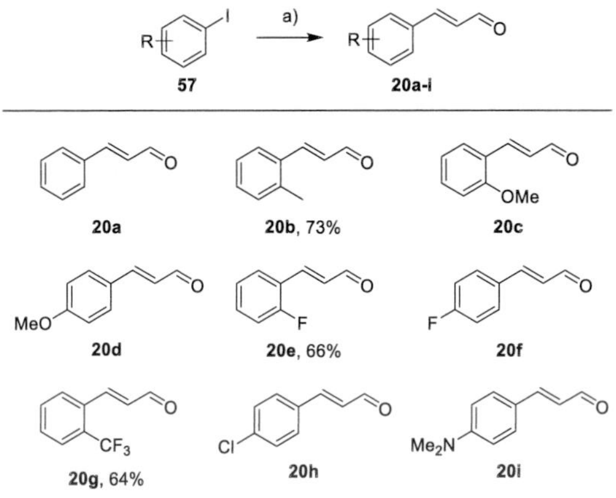

Scheme 10: Overview of used cinnamaldehydes **20a-j**: *a) acrolein, TBAI, NaHCO₃, 10mol% Pd(OAC)₂, DMF, 60°C, 24 – 72 h.*

With all starting materials in hand the library synthesis of the substituted coumarins were performed and will be described in the following sections.

3.1.3 Synthesis of 7-alkoxy-3-benzylcoumarins

The syntheses of substituted 3-benzylcoumarines were performed as described in Chapter 1.3.1, following the method of BRÄSE *et al.*[108] With the prepared salicylic aldehydes **19a-g** and the cinnamaldehydes **20a-h**, a considerable number of combinations towards different 7-

alkoxycoumarines according to lead structure **29** were possible and the synthetic results are summarised in Table 2.

Table 2: Summary of yields [%] for the library synthesis of 7-alkoxy-3-benzylcoumarins **21aa-gi**: *a) IL 22, K$_2$CO$_3$, toluene, 110°C, MW (230 W max., 7 bar max.), 50 min.*

Entry	20 19	20a	20b	20c	20d	20f	20h	20i
		R^2 = H	*o*-Me	*o*-OMe	*p*-OMe	p-F	*p*-Cl	*p*-NMe$_2$
1	19a	17	11	28	30	19[a]	30[a]	18
2	19b	18	23	26	23	25	23	25
3	19c	27[a]	15	22	15	15[a]	15	27[a]
4	19d	15	21	30	22	15	10	21
5	19e	16	10	25	7[b]	9	15	13
6	19f	10	12	23	15	15	18	10
7	19g	13	16	22	15	7	15	12

[a]purity 90% – 95%; [b]loss of yield during purification

In general, the obtained yields of all synthesised 7-alkoxycoumarins were in similar ranges to previous results by this method.[107, 141, 158] Lower isolated quantities not only resulted from lower conversions within the reactions itself, moreover different purification issues had to be overcome. The in general lower yield of the reactions can be explained by a significant change of electron density in the salicylic aldehydes. Due to the strong mesomeric effect of the alkoxy groups, a mesomeric more stable phenolic anion could be generated, resulting in a lower reaction kinetic of the nucleophilic attack to the NHC complex **26b** and increased side product formation (Scheme 2). Asides from that, a low resolubility during purification by flash column chromatography was observed. Even with several combinations of different solvents, for example cyclohexane/ethyl acetate, cyclohexane/acetone or cyclohexane/diethyl ether, no improvement was achieved. In addition, the use of a solvent gradient or application as dry-load did not improve the purification. Because of that, a change of purification method was done towards a high-performance liquid chromatography (HPLC) approach, resulting in excellent product purities.

Furthermore, several minor trends within the table were observable. Firstly, an elongation of the alkoxy side chain at position 7 resulted in a decreased yield of the products. Cinnamaldehydes bearing a more electron donating substituent like *o*-methyl or *o*-/*p*-methoxy, showed a higher yield. In accordance to that, coumarins with an electron-withdrawing benzyl substituent, for example *p*-fluorine or *p*-chlorine, were only obtained in lower yields and with the neutral unsubstituted cinnamaldehyde (**20a**) yields between both of them were achieved. In addition, it was observed that the *p*-dimethylamino substituent had a much higher propensity to undergo side reactions compared to the rest.

Several synthesised coumarins were selected for a post-synthetic modification by demethylation of the 5-methoxy group, as in the previous studies more polar groups at this position appeared to be favourable for the activity. The deprotection, however, had to be performed with more caution, as the 7-alkoxy side chain potentially could be cleaved as well (Table 3).

Table 3: Summary of yields [%] from demethylation reactions of 7-alkoxy-3-benzylcoumarins **58a-i**: *a) BBr₃,* *DCM, –78°C, 30 min, then rt., o.n.*

Entry	Coumarin	Yield[a] [%]
1	**21cb**	27
2	**21cc**	13
3	**21ci**	_[a]
4	**21eb**	20
5	**21ec**	26
6	**21ei**	_[a]
7	**21gb**	45
8	**21gc**	19
9	**21gi**	10

[a]no product obtained.

As expected, lower yields correlated with the appearance of double dialkylated and unreacted starting material were observed. Either no conversion (entry 3 and 6) or lowest yield (entry 9) were obtained with the dimethylamino substituted coumarins, indicating again a higher reactivity towards side reactions of the amine functionality. Other approaches of selective deprotection with sodium thiomethoxide or 2-(diethylamino)ethanethiol were unsuccessful.

3.1.4   Synthesis of 7-dialkylamino-3-benzylcoumarins

In this section, the library synthesis for the second lead structure **30** is summarised. In the first attempts to synthesise 7-dialkyamino-3-benzylcomarins **21hb-jg**, using the standard method conditions, a drastically decrease of yields (<5%) was observed. In accordance with previous observations described above, an increase of electron-density in the salicylic aldehyde results in a decreased reaction kinetics. This theory was supported, as most of the starting material **19h** was recovered. On the other hand, as an additional acceptor, the amino group could lead to an imin isomerisation at the elevated reaction temperatures, removing the availability of the salicylic aldehyde for the reaction.

To prove a possible imin formation, several NMRs at 60°C or 100°C of aldehyde **19h** were measured to simulate the elevated reaction temperatures, as well as neutral or basic conditions (see Chapter 6.3). None of the measured NMRs, at either 60°C or 100°C or under neutral or basic conditions showed a rearrangement towards the imin. Consequently, before any further syntheses an optimisation of the reaction conditions adjusted towards the salicylic aldehydes had to be done. In accordance to similar reaction systems in literature[109, 159-164], alternative NHC catalysts, prolonged reaction times and addition of supplements like crown ether or acetate buffer were tested (see Appendix Table 21). Thereby, the most promising conditions were with the same NHC catalyst, addition of acetate buffer instead of potassium acetate and supplemented with molecular sieve (3Å powder), at slightly elongated reaction time (75 min). The results of the realised syntheses for 7-aminocoumarins **21hb-jg** are shown in Table 4.

Table 4: Summary of yields [%] for the library synthesis of 7-dialkylamino-3-benzylcoumarins **21hb-jg**: *a) IL 22, KOAc, AcOH, toluene, 110°C, MW (230 W max., 7 bar max.), 75 min.*

Entry	20 19	20a	20b	20c	20d	20e	20f	20g
		R^2 = H	*o*-Me	*o*-OMe	*p*-OMe	*o*-F	*p*-F	*o*-CF$_3$
1	19h	–	5	18	–	6	4[a]	3
2	19i	8	8	13[b]	17	5[b]	8	6[b]
3	19j	5	10	10	7	6	6	3

[a]purity <90%; [b]product started to decompose by exposure to CDCl$_3$ for longer than 24 h.

Even with optimised reaction conditions, the obtained yields were low (3% – 17%). In most of the reactions, a conversion of only 10% – 20% of the salicylic aldehyde was observed. Considering this, much higher yields would be possible, if more of the salicylic aldehyde were pushed to contribute to the reaction progress. Again, electron-rich cinnamaldehydes (**XXhc–ib**) end up in higher yields compared to electron-deficient cinnamaldehydes. On the other hand, a drastically reduced stability of this class of coumarins was detected, as degradation processes within the NMR samples never were observed before. Due to the higher propensity to undergo side reactions by coumarins containing an amino functionality, further modification by 5-position demethylation was not considered.

3.1.5 Synthesis of 7-(1,1-dimethylalkyl)-3-benzylcoumarins

In the next part, the library synthesis of the 1,1-dimethylalky coumarins was realised and the results are shown in Table 5. As mentioned before, parts of the syntheses were included into a bachelor thesis carried out by MAXIMILIAN KNAB. [152]

Table 5: Summary of yields [%] for the library synthesis of 7-(1,1-dimethylalkyl)-3-benzylcoumarins **21ka-mg**: *a) IL 22, K₂CO₃, toluene, 110°C, MW (230 W max., 7 bar max.), 50 min.*

Entry	20 / 19	20a	20b	20c	20d	20e	20f	20g
		R² = H	*o*-Me	*o*-OMe	*p*-OMe	*o*-F	*p*-F	*o*-CF₃
1	**19k**	32	68	63	13[a]	30[b]	45	31
2	**19l**	19	48	37	26	23	13	16
3	**19m**	46	32[c]	46	35	47[c]	43	41

[a]loss of yield during purification; [b]yield calculated by ¹H NMR; [c]purity <90%.

All coumarins were obtained in moderate to good yields. Electron-rich cinnamaldehydes resulted in higher yields than electron-deficient cinnamaldehydes. Interestingly, in general, lower yields for the butyl substituted salicylic aldehyde **19l** were observed. No products were obtained with the *p*-dimethylaminocinnamaldehyde **20i** (not shown). Moreover, the synthesised coumarins were modified further by deprotection of the 5-methoxy group (Table 6).

Table 6: Summary of yields [%] from demethylation reactions of 7-(1,1-dimethylalkyl)-3-benzylcoumarins **59ka-mg**: *a) BBr₃, DCM, −78°C, 30 min, then rt., o.n.*

21ka-mg   **59ka-mg**

Entry	Coumarin	21ya	21yb	21yc	21yd	21ye	21yf	21yg
		R^2 = H	*o*-Me	*o*-OH	*p*-OH	*o*-F	*p*-F	*o*-CF$_3$
1	21kx	86	58	49	–	71	98	86[a]
2	21lx	58	89	34	97	92	72[b]	98[a]
3	21mx	71	66	69	quant.	74	95	58

[a]purity <90%; [b]yield calculated by ^{1}H NMR.

The obtained yields ranged between good (**59la**) and excellent (**59md**). For the trifluoromethyl (CF$_3$) substituted coumarins (**59kg-mg**), a formation of a not further assigned side product with a more downfield chemical shift in the fluorine NMR was observed. Nevertheless, only for the coumarin **59mg** it was possible to isolate the pure product by HPLC.

3.1.6 Synthesis of 3-benzylcoumarins

Preliminary studies in collaboration with the group of BERND FIEBICH[2] revealed an anti-neuroinflammatory effect by attenuation of prostaglandin E$_2$ (PGE$_2$) production and microsomal prostaglandin E synthase-1 (mPGES-1)/cyclooxygenase-2 (COX-2 levels), mediated by GPR55 antagonists in primary microglial cells.[34] Therefore, a small library of the coumarins with small substituents had been synthesised by FRANZISKA GLÄSER.[141] Besides the limited number, main disadvantage, however, was the lack of GPR55 selectivity over CB$_2$R affinity. To investigate further the potential neuroprotective effects of the small substituted coumarins encouraged by the promising results and to improve the GPR55 selectivity, completion of the libraries **21na** to **61c** (Table 7 and Table 8) were done within the bachelor thesis of LUKAS LANGER.[156]

[2] Research group Dr. Bernd Fiebich, Neurochemistry and Neuroimmunology, University Clinic Freiburg

Table 7: Summary of yields [%] for the library synthesis of 5,8-dimethyl-3-benzylcoumarins **21na-ni** and **60a-b**: *a) IL 22, K_2CO_3, toluene, 110°C, MW (230 W max., 7 bar max.), 50 min; b) BBr_3, DCM, –78°C, 30 min, then rt., o.n.*

Entry	R^2 =	Yield[a] [%]	Yield[b] [%]
1	H	60	–
2	*o*-OMe	42	74
3	*p*-OMe	52	64
4	*p*-F	81	–
5	*p*-Cl	33	–
6	*p*-NMe$_2$	33	–

Compared to the previous work, a similar range of yields between low (33%) and good (81%) were achieved for both library syntheses, as well as the 5-position deprotections. The influences of electron-donating or electron-withdrawing substituents at the cinnamaldehyde were, in general, lower as for the previous described results and no clear trend was observed.

Table 8: Syntheses of 5,7,8-trimethyl-3-benzylcoumarins **21oc-oi** and **61a-c**: *a) IL 22, K_2CO_3, toluene, 110°C, MW (230 W max., 7 bar max.), 50 min; b) BBr_3, DCM, –78°C, 30 min, then rt., o.n.*

Entry	R^2 =	Yield[a] [%]	Yield[b] [%]
1	*p*-OMe	61	20[a]
2	*p*-Cl	65	70[b]
3	*p*-NMe$_2$	41	69

[a]purity <90%; [b]yield calculated by ^{1}H NMR;

Completion of these two tables 7 and 8 finalised the synthetic effort done for the syntheses of substituted 3-benzylcoumarines as new cannabinoid ligands.

3.1.7 Synthesis of 3-alkylcoumarins

As the last synthetic part of this project, the syntheses of 3-alkylcoumarins as another compound class is reported. In collaboration with the group of Prof. JONATHAN SLEEMAN[3] 3-alkycoumarins were identified to act in a CBD (**2**) like behaviour against inhibitor of DNA binding proteins 1 and 3 (ID1 and ID3).[165] Suppression of ID related proteins play a crucial role in carcinogenesis and metastasis.[166]

In previous work, THOMAS HURRLE established in our group an adapted reaction protocol for a microwave supported PERKIN reaction based on an optimised protocol of FLOREKOVÁ et al.[167] A summary of realised experiments is given in Table 9, whereby entry 7 and 8 were performed by Lukas Langer during his bachelor thesis.[156]

Table 9: Summary of yields [%] for the library synthesis of 3-alkylcoumarins **16aa-oh**: a) K_2CO_3, 180°C, MW (230 W max., 7 bar max.), 65 min.

Entry	19 / 62	62a	62b	62c	62d	62e	62f	62g	62h
	$R^2 =$	H	CH_3	C_2H_5	C_3H_7	C_4H_9	C_5H_{11}	C_6H_{13}	iPr
1	19a	88	64	85	74	78[a]	82	58	
2	19b	80	78	76	69	86	72	33[b]	
3	19c	77	–	–	–	15[b]	–	–	
4	19d	–	–	–	32	–	–	–	
5	19e	–	–	86	–	–	–	–	
6	19f	–	46	–	–	–	–	–	

---

[3] Research group Prof. Jonathan Sleeman, Microvascular Biology and Pathobiology, Medical Faculty Mannheim, University Mannheim.

| 7 | 19n | 64 | 79 | 93 | quant. | 62 | 98 | 80 | 97[c] |
| 8 | 19o | – | – | 82 | 94 | 69 | 74 | 46 | 27[c] |

[a]purity 90% – 95%; [b]loss of yield during purification; [c]reaction time 3 h.

In total 33 new 3-alkylcoumarins were synthesised, mostly in good to excellent yields (15% to quant.). Lower yields always resulted from difficulties during purification. No trend in yields between short or long linear carboxylic acid anhydrides was observed, only for the branched isovaleric acid (62h) longer reaction times were necessary to obtain comparable yields. The combinations of side chain lengths at the 3- and 7-position (entry 1 to 6) were chosen to inves-tigate either a small or a big substituent at these positions as well as the influences of their combinations. Filling of the missing blanks was considered after activity validation and selec-tion of the most active substitutions.

Further modification by 5-methoxy deprotection was performed, if applicable, according to the previous described protocol (Table 10). The synthesised 3-alkylcoumarins 63a-f were obtained in good yields of 65% to 83%, whereby no influence on the yield depending on the side chain length was observed.

Table 10: Summary of yields [%] for the deprotection of 5,7,8-trimethyl-3-alkylcoumarins 63a-f: *a) BBr₃, DCM, –78°C, 30 min, then rt., o.n.*

Entry	$R^2 =$	Yield[b] [%]
1	C_2H_5	83
2	C_3H_7	79
3	C_4H_9	69
4	C_5H_{11}	78
5	C_6H_{13}	81
6	*i*Pr	65

3.2   Pharmacological evaluation of synthetic cannabinoids[4]

For the bioactivity determination the novel coumarin derivatives were investigated in a SAR study using displacement radioligand-binding assays to determine the receptor affinities and [^{35}S]GTPγS assays as functional assays for the receptor efficacies. In the next paragraph the basic principles of these kind of assays will be discussed first.

3.2.1   Radioligand binding assays

Common methods in binding studies on pharmacological receptors are radioligand binding assays at an equilibrium state.[168] Depending on the target and outcome several approaches can be used. To determine ligand affinities towards a target receptor usually a competitive displacement assay between the unlabelled ligand of interests and a radiotracer with known affinities is used. In another approach activation or deactivation of a receptor based to its basal activity can be used to first determine the ligand of interest affinity towards the receptor and simultaneously a functional outcome of ligand efficacy can be determined.[66, 168]

In receptor-binding studies, usually cell membrane homogenates containing high quantities of receptor of interest per protein concentration are used. They can be easily produced by receptor transfection and overexpression in cell culture, isolation of the cell membrane fraction after lysis and several centrifugation steps at a high gravitational force equivalent (g). Aliquoted in suitable buffer this membrane preparations can be stored at −80°C until further use.

In a radioligand displacement assay, receptor membrane preparation is incubated together with different concentrations of the ligand of interest and a fixed concentration of radiotracer is used for quantification. Radiotracers are usually tailored ligands against the receptor of interest with an equilibrium dissociation constant $K_d \leq 10^{-9}$ M (1 nM) to achieve sufficient receptor occupancy and usage of low specific radioactivity at the same time.[168] After a defined incubation time the formation of radioligand-receptor complex will reach an equilibrium. Trapping the cell membrane by rapid filtration through a filter plate allows to remove free radioligand from the incubation medium and the bound concentration of radioligand can be measured by a scintillation counter. Depending on applied ligand concentrations or potency more or less radioligand will be displaced resulting in lower (or higher) radioligand-receptor complex concentrations. For

---

[4] The presented SAR studies were performed during a research internship in the laboratories of Prof. Laura Heitman at Leiden Academic Centre for Drug Research, Leiden University, The Netherlands.

normalisation and quantification two additional determinations are necessary. First, the maximum possible concentration of radioligand-receptor complex must be determined by incubation of membrane preparation with used concentration of radioligand only and is expressed as total binding (TB). To assure that only signals coming from occupying the binding pocket are used in the calculations, non-specific binding (NSB) of the radioligand to other components in the assay system must be determined. This can be achieved by full blockage of the binding pocket using a high concentration (usually 10 μM) of a selective receptor ligand and measurement of the remaining non-specific bound radioligand. In the radioligand-binding assays presented in this work [3H]-(−)-*cis*-3-[2-Hydroxy-4-(1,1-dimethylheptyl)phenyl]-*trans*-4-(3-hydroxypropyl)cyclohexanol (CP55.940) was used as radioligand and Rimonabant (CB$_1$) and AM630 (CB$_2$) were used as NSB ligands. The characteristic sigmoidal dose-response curves are obtained by fitting a non-linear regression model to the logarithmic (log) ligand concentrations vs. the normalised data for specific binding (Figure 13A). The half max. inhibitory constant (IC$_{50}$) can be determined directly from the fit using the turning point. Transformation to the experimental independent inhibition constant (K_i) can be achieved by applying the CHENG-PRUSOFF equation (1).[169]

$$K_i = \frac{IC_{50}}{1 + \frac{[L]}{K_d}} \tag{1}$$

Whereby, [L] is defined as radioligand concentration (determined for each individual experiment) and K_d describes the receptor and temperature dependent dissociation constant of the used radioligand. In the presented work K_d values for CP55,940 of 0.414 nM (CB$_1$) and 1.24 nM (CB$_2$)[170] were used. The K_d for CB$_1$ was determined by standard association and dissociation kinetic assays (three individual experiments in duplicates, at 25 °C) and following observed rate constants (K_{obs}) were determined: K$_{on}$ = 4.5 ± 0.2 × 10^7 (M^{-1}s^{-1}) and K$_{off}$ = 1.9 ± 0.4 × 10^{-2} (s^{-1}). Regarding the previous SAR study results, initial experiments were performed as single–point experiments at a final ligand concentration of 1 μM. Full dose-response experiments were conducted after radioligand-binding inhibition >50% in the single-point experiment.

For determination of the intrinsic efficacy several of the ligands were selected and tested in a [35S]GTPγS assay. In this radioligand-binding assay G protein activation as functional consequence of receptor occupancy is measured. One of the earliest events in GPCRs after an agonistic receptor occupancy is the dissociation of the heterotrimeric G protein unit into Gα followed by the formation of a GαGTP complex and modulation of downstream effectors. Hydrolysis of GTP by GTPase activity of the α-subunit results in a reformation of the heterotri-

meric G protein and deactivation of the receptor signalling. Substitution of GTP to the non-

hydrolysable and ^{35}S labelled analogue GTPγS in the assay medium allows to measure the G

protein activation after incubation with different concentrations of ligand of interest and deter-

mination of pharmacological parameters like potency or efficacy. In advantage to other func-

tional assays, a determination of an antagonistic or inverse agonistic activity compared to the

basal receptor activity is also possible (Figure 13B).[66]

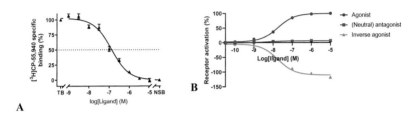

Figure 13: (A) Exemplary dose-response curve resulting from radioligand-binding experiment. **(B)** Illustration of observable efficacies in the [^{35}S]GTPγS assay compared to the basal receptor activity. Blue: typical sigmoidal dose-response curve resulting from agonist activation. Red: antagonistic activity of the ligand results in no disso-ciation of the G protein subunits and no enhanced radioactivity is detected. Green: decreased receptor activity resulting from an inverse agonistic ligand binding (modified from Literature [171]).

3.2.2   Results of competitive radioligand binding studies

The results of only a small selection for the tested coumarin-derivatives are summarised in **Table** 11, due to limited space. For further information concerning the results of the tested coumarin-derivatives and SAR interpretations see Appendix (Table 22). For coumarins tested either protected or deprotected at 5-position the results for both compounds are shown, whereby the first line refers to the protected and the second line to the deprotected compound. Finally, in the last column the respective CB$_2$ selectivity is given as well. Besides the already presented coumarin-derivatives described in Chapter 3.1, additional so far untested coumarin-derivatives synthesised by THOMAS HURRLE[107] (group **4–7** and partially **9**) were included into the studies. Nevertheless, as the total number of available untested compounds exceeded the scope of this study, selections based on structural considerations were made before execution.

Table 11: Results of radioligand-binding assay for selected coumarins of several classes.

Entry	Compound	hCB$_1$	hCB$_2$	CB$_2$/CB$_1$ [c]
		pK$_i$ ± SEM (K$_i$ in nM[a] or displacement at 1 μM)[b]		
Group 1: 3-Benzyl-7-alkylether coumarins				
1	21ab	<6.00 (19%)	<6.00 (21%)	–
2	21ac	≪6.00 (4%)	≪6.00 (7%)	–
3	R = CH$_3$, 21cb; R = H, 58a	<6.00 (28%) / <6.00 (27%)	<6.00 (38%) / ≪6.00 (–9%)	–
4	R = CH$_3$, 21cc; R = H, 58b	<6.00 (23%) / ≪6.00 (7%)	6.63±0.02 (236) / ≪6.00 (7%)	≥4.27 / –
5	21cd	<6.00 (19%)	<6.00 (31%)	–
6	21eb	<6.00 (24%)	≪6.00 (7%)	–
7	R = CH$_3$, 21ec; R = H, 58e	<6.00 (24%) / ≪6.00 (–4%)	~6.00 (52%) / ≪6.00 (0%)	–
8	R = CH$_3$, 21gb; R = H, 58g	≪6.00 (8%) / ≪6.00 (–4%)	≪6.00 (7%) / ≪6.00 (1%)	–

9	R = CH₃, **21gc** R = H, **58h**	<6.00 (27%) <6.00 (12%)	≪6.00 (–4%) ≪6.00 (–6%)	–
10	**64c**	<6.00 (33%)	6.83±0.09 (147)	≥6.76
11	**64d**	≪6.00 (–19%)	≪6.00 (–15%)	–
12	**64g**	<6.00 (19%)	~6.00 (50%)	–

Group 2: 3-Benzyl-7-(dialkylamino)coumarins

13	**21hg**	<6.00 (39%)	~6.00 (46%)	–
14	**21ia**	~6.00 (48%)	<6.00 (28%)	–
15	**21ja**	<6.00 (28%)	<6.00 (23%)	–
16	**21je**	~6.00 (47%)	<6.00 (21%)	–

Group 3: 3-Benzyl-7-(1,1′-dimethylalkyl)coumarins

17	R = CH₃, **21ka** R = H, **59ka**	<6.00 (19%) <6.00 (17%)	≪6.00 (2%) ≪6.00 (0%)	–

Note: The coumarin structures referenced by compound codes (21gc/58h, 64c, 64d, 64g, 21hg, 21ia, 21ja, 21je, 21ka/59ka) are drawn as chemical structure diagrams.

#	Structure			
18	R = CH$_3$, **21kc** R = H, **59kc**	≪6.00 (5%) ≪6.00 (2%)	<6.00 (12%) ≪6.00 (7%)	–
19	R = CH$_3$, **21la** R = H, **59la**	6.31±0.22 (486) <6.00 (12%)	<6.00 (39%) <6.00 (24%)	<0.49 –
20	R = CH$_3$, **21lb** R = H, **59lb**	6.66±0.15 (217) <6.00 (30%)	<6.00 (32%) <6.00 (41%)	<0.22 –
21	R = CH$_3$, **21lc** R = H, **59lc**	6.71±0.11 (196) <6.00 (24%)	6.64±0.003 (231) <6.00 (37%)	<0.85 –
22	R = CH$_3$, **21mb** R = H, **59mb**	<6.00 (27%) ~6.00 (47%)	≪6.00 (9%) 6.65±0.08 (222)	– ~4.47
23	R = CH$_3$, **21mg** R = H, **59mg**	<6.00 (26%) ~6.00 (49%)	≪6.00 (3%) <6.00 (42%)	– –

Group 4: 3-Fluorobenzyl-7-pentylcoumarins

#	Structure			
24	R = CF$_3$, **65a** R = F, **65b**	~6.00 (47%) 6.69±0.12 (202)	6.82±0.07 (152) 6.49±0.02 (326)	~6.61 0.63

Group 5: 3-Phenylcoumarins

#	Structure			
25	R = CH$_3$, **18c** R = H, **18d**	≪6.00 (–10%) ≪6.00 (–43%)	≪6.00 (–37%) ≪6.00 (2%)	– –

26	R = CH$_3$, **18e** R = H, **18f**	≪6.00 (–18%) ≪6.00 (–17%)	≪6.00 (–9%) ≪6.00 (–14%)	– –
27	R = CH$_3$, **18g** R = H, **18h**	≪6.00 (–26%) <6.00 (42%)	≪6.00 (–13%) ~6.00 (53%)	– –

Group 6: 3-Pyridylcoumarins

28	R = CH$_3$, **17a** R = H, **17b**	7.15±0.06 (70.3) ≪6.00 (–15%)	7.08±0.14 (82.4) ≪6.00 (–17%)	0.85 –
29	R = CH$_3$, **17c** R = H, **17d**	6.77±0.12 (171) ≪6.00 (–21%)	7.25±0.04 (56.5) ≪6.00 (–21%)	3.02 –
30	R = CH$_3$, **17e** R = H, **17f**	≪6.00 (0%) ≪6.00 (–48%)	<6.00 (11%) ≪6.00 (–5%)	– –
31	R = CH$_3$, **17g** R = H, **17h**	<6.00 (29%) <6.00 (34%)	<6.00 (20%) <6.00 (44%)	– –
32	R = CH$_3$, **17i** R = H, **17j**	<6.00 (29%) <6.00 (19%)	<6.00 (40%) 6.51±0.07 (310)	– >3.24
33	R = CH$_3$, **17m** R = H, **17n**	~6.00 (47%) ≪6.00 (–7%)	<6.00 (12%) 7.14±0.13 (71.9)	– >13.8

Group 7: 3-Alkyl-7-(butylcycloalkyl)coumarins

#	Compound			
34	R = CH₃, 66a	<6.00 (16%)	≪6.00 (1%)	–
	R = H, 66b	≪6.00 (−10%)	~6.00 (49%)	–
35	R = CH₃, 66c	<6.00 (33%)	≪6.00 (0%)	–
	R = H, 66d	<6.00 (34%)	7.22±0.08 (60.6)	>16.6
36	R = CH₃, 66e	<6.00 (32%)	<6.00 (38%)	–
	R = H, 66f	~6.00 (47%)	7.73±0.01 (18.6)	>53.7
37	R = CH₃, 66g	<6.00 (15%)	≪6.00 (−1%)	–
	R = H, 66h	~6.00 (50%)	7.86±0.11 (13.7)	~72.4
38	R = CH₃, 66i	≪6.00 (9%)	≪6.00 (−34%)	–
	R = H, 66j	<6.00 (19%)	6.98±0.03 (106)	>9.55
39	R = CH₃, 66k	≪6.00 (6%)	≪6.00 (−1%)	–
	R = H, 66l	<6.00 (39%)	7.41±0.04 (39.1)	>25.7
40	R = CH₃, 66m	<6.00 (18%)	≪6.00 (−4%)	–
	R = H, 66n	6.80±0.22 (159)	8.19±0.12 (6.5)	24.5
41	R = CH₃, 66o	≪6.00 (−11%)	≪6.00 (2%)	–
	R = H, 66p	~6.00 (48%)	7.90±0.03 (12.5)	~79.4

[a] Data expressed as means ± SEM of at least three individual experiments in duplicates; [b] data expressed as means of at least two individual experiments in duplicates; [c] CB₂R selectivity calculated as follows: 10^(pKi CB₂R − pKi CB₁R);

Hereinafter the results of the radioligand-binding studies will be described more detailed. For better clarity the results will be discussed regarding their structural classification. A comprehensive discussion between the structural groups and more detailed explanation for the reasons of the observed binding affinities will follow in subchapter 3.2.4 which includes the observations made within the docking studies. Variations of the coumarin core were made at position 3, 5, 6, 7 and 8. In general, different substituents were introduced on most of the positions, ranging from 3-benzyl, phenyl, pyridyl or alkyl groups to 5-methoxy or hydroxyl groups and 7-heteroalkyl, linear alkyl or branched alkyl groups (group **1–8**). Coumarins containing small groups as substituents were combined into group **9**. Within the respective groups the coumarins were ordered according to their substituent at position 7.

Initially, 3-benzylcoumarins with alkyl ether groups at the 7-position (lead structure 1, **29**, Table 22 1–64) were tested on their CBR affinities. In general, within this group only low or no affinities towards the CB receptors were observed. Only for two compounds low affinities against CB_2 on a sub-micromolar level were observed (entry 3 K_i = 236 nM and 10 K_i = 147 nM, respectively). Structural similarity was an o-methoxy group at position 3, a methoxy group at position 5 and a propyl side chain as ether group. Against CB_1 no tested compound exceeded a radioligand displacement higher than 44% (entry 32, appendix) at 1 μM.

Next, the 3-benzylcoumarins containing a dialkylamino substitution at position 7 were investigated (lead structure 2, 30, Table 22 65–83). However, no significant affinity (>1 μM) towards either one the receptors were observed.

Therefore, the screening of the 3-benzyl-7-(1,1′-dimethylalkyl)coumarins was performed subsequently (lead structure 3, **31**, Table 22 84–125). A few compounds bearing either no substituent at the 3-benzyl ring (entry 98) or an o-methyl group (entry 100 and 114) showed a selective sub-micromolar activity against CB_1 (entry 98, 100 and 114, appendix). Additionally, a sub-micromolar potent slightly more CB_2 selective (entry 123, appendix) and one active compound with similar potency against both receptors (entry 102, appendix) were identified. All active compound had only small substituents at the 3-position.

Concerning the previous studies [99] two new fluorinated coumarins (lead structure 4, **65**, entry 126 and 127, Table 22) based on similar high potent structures at that time were included to explore the influences of lipophilic fluorine substitutions further. Both coumarins exhibited a moderate potency at sub-micromolar level, whereby higher selectivities were achieved with the smaller fluorine substituent in *ortho* position.

Exchanging the substituent at the 3-position to a phenyl group (lead structure 5, **18**, Table 22 128–141) abolished the activity against the CBRs nearly completely and only one compound with an affinity about 1 μM against CB_2 was identified (**18c**).

Further variations of the 3-position by introducing pyridylmethyl groups (lead structure 6, **17**, Table 22 143–159) resulted in three active compounds (entry 28, 29 and 33) with the highest potency (**17c**, $K_i^{CB2} = 56.5$ nM) and best selectivity (32, $CB_2/CB_1 > 13.8$) so far. Within this group a nice correlation between pyridyl configuration and potency at the CB_1R was observed (entry 27 – 29). Highest potency was shown if the pyridyl nitrogen was introduced at o-position (**17a**), leading to a drastic decrease if changed to the m-position (**17c**) and finally resulting in a complete loss of potency at p-position (**17e**). In contrast to that, however, on CB_2 best selectivity (within this three) and highest potency were observed with the m-pyridylmethyl substituent (**17c**).

Substitutions at the 3-position with alkyl groups and simultaneously butylcycloalkyl chains at the 7-position were investigated in the next group (lead structure 7, **66**, Table 22 160–175). Interestingly, all protected compounds (at 5-position) within this group showed neither activity at the CB_1R nor CB_2R (entry 34–41). But in contrast to that, all deprotected compounds (entry 33–40) showed potencies between low ($K_i^{CB2} \sim 1.00$ μM, entry 34, **66b**) and very high (K_i^{CB2} <60.6 nM, entry 35–37 and 39–41) towards CB_2 as well as excellent selectivity over CB_1 ranging from 10-fold (**66j**) up to 79-fold (**66p**) difference. However, decreasing the steric demand of the bulky cycloalkyl side chain at position 7 (lead structure 8, **67**, Table 22 176–179) abolished the activity against both CBRs completely.

Finally, the screening of coumarins containing small substituents at the core moiety were tested (lead structure 9, **68**, Table 22 180–250). As this group of derivatives initially were not planned to act as CBR ligands and moreover as ligands for the orphan cannabinoid related GPR55, no activity against both CBRs was desired. In the results significant differences in affinities and selectivities towards both receptors were observed. None of the tested compounds exceeded a displacement of the radioligand at CB_2 higher than 18%, whereby a few compounds at CB_1 showed a higher displacement up to low binding affinities (max 53% at 1 μM).

As mentioned previously, full competition curves with ten different concentrations were performed after showing a radioligand displacement >50% in the single-point experiment to determine the exact K_i values. In Figure 14, the representative curves for selected coumarins with the highest shown activity within their group are shown (see Figure 46 for rest).

A   B

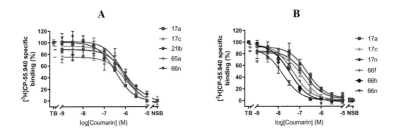

Figure 14: Competitive concentration-dependent curves of [^{3}H]-CP55,940 displacement vs. coumarin ligands at hCB$_1$ (**A**) and hCB$_2$ (**B**), respectively. Data expressed as mean ± SEM of at least three individual experiments in duplicates.

3.2.3   Functional properties

As already described (see Chapter 3.2.1), receptor ligands can show different downstream signalling effects after receptor occupancy dependent on their efficacy. In order to investigate whether the novel coumarin derivatives show an agonistic, antagonistic or inverse agonistic effect functional [^{35}S]GTPγS binding assays were performed. Initially, among the active coumarins tested in the radioligand binding assays, several ligands were selected and tested in a single-point experiment at 1 μM final ligand concentration to determine their efficacy (E$_{max}$) compared to the maximal response of full agonist CP55.940. The results of the performed assays are summarised in Table 12 (for graph see Appendix Figure 47).

Table 12: E$_{max}$ results from [^{35}S]GTPγS binding assay and respective EC$_{50}$ values for selected coumarins.

En-try	Cpd.	E$_{max}$ effect on [^{35}S]GTPγS binding to hCB$_1$[a] (EC$_{50}$ ± SEM [μM])[b]	E$_{max}$ effect on [^{35}S]GTPγS binding to hCB$_2$[a] (EC$_{50}$ ± SEM [μM])[b]
1	**CP55.940**	100 ± 0 (0.00151 ± 0.00013)	100 ± 0 (0.000540 ± 0.000012)
2	**64c**	n.d.	21 ± 14[****]
3	**17a**	46 ± 4 (1.01 ± 0.20)[***]	34 ± 1 (0.188 ± 0.090)[****]
4	**17c**	40 ± 3[***]	40 ± 5[****]
5	**17j**	n.d.	82 ± 2[ns]
6	**17n**	n.d.	68 ± 4 (0.042 ± 0.007)[*]
7	**66d**	n.d.	87 ± 14[ns]

8	66f	n.d.	91 ± 3^{ns}
9	66h	n.d.	$85 \pm 1 \ (0.018 \pm 0.008)^{ns}$
10	66j	n.d.	$66 \pm 6^{*}$
11	66l	n.d.	$65 \pm 3^{**}$
12	66n	$23 \pm 6 \ (1.12 \pm 0.49)^{****}$	$62 \pm 3 \ (0.00451 \pm 0.00279)^{**}$
13	66p	n.d.	$65 \pm 1^{**}$

[a] E_{max} expressed as means $\pm$ SEM relative to the max effect of full agonist CP55,940 at 1 µM (= 100%) of two individual experiments in duplicates; [b] EC_{50} expressed as means $\pm$ SEM relative to the max effect of full agonist CP55,940 of three individual experiments in duplicates; n.d. = not determined; Statistics were performed using a one-way ANOVA with Dunnett's post-test for multicomparison analysis, ns = not significant, * $p<0.05$, ** $p<0.01$, *** $p<0.001$, **** $p<0.0001$.

All tested coumarins whether CB_2 selective or dual ligands showed only agonistic intrinsic activities. Whereby, four coumarins were identified to behave like a full agonist (entry 5 and 7–9) and the rest as partial agonists (entry 2–4, 6 and 10–13). Additionally, full dose-responsive curves (Figure 15) were performed to determine the EC_{50} values for the most active compounds from group 5 and 6 (**Table** 11).

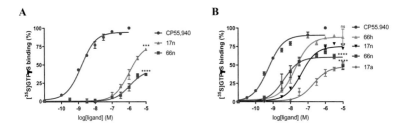

Figure 15: Agonist induced [^{35}S]GTPγS binding to stably expressed recombinant hCB_1 (**A**) and hCB_2 (**B**) in CHO cell membrane preparation. Data were normalised to the maximal response of reference full agonist CP55.940. Statistical significance was determined using a one-way ANOVA with Dunnett's post-test for multicomparison analysis: ns = not significant, ** $p<0.01$, **** $p<0.0001$.

The experimentally obtained pK_i for the selected set of CB_2 agonists were correlated with the experimental pEC_{50} values (Figure 16). A strong correlation ($R^2 = 0.86$) between both experimental values was observed, indicating a strong relation between potency and efficacy.

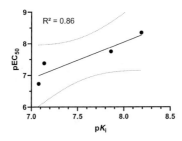

Figure 16: Correlation analysis of determined pK_i values in radioligand binding studies and measured pEC$_{50}$ values for selected potent CB$_2$ agonists. Regression line with 95% confidence interval (dotted lines) is shown.

3.2.4 Docking studies and Structure-activity-relationships[5]

To obtain a deeper understanding and improved visualisation of dependencies between side chain modifications and structure-activity relationships of the tested coumarins, *in silico* ligand-based docking studies were performed. For this purpose, crystal structures of the receptor subtypes (PDB CB$_1$: 5XRA[172] and CB$_2$: 5ZTY[173]) co-crystallised with ligands were used to dock the new ligands into. Initially, all the tested coumarins were docked into both receptor subtypes without concerning binding preferences and affinity. For both receptor subtypes general key regions were determined, which were presumed to be crucial for high receptor binding (Figure 17). The used crystal structure of the CB$_1$R refers to an active state of the receptor induced by the ligand and therefore, comprehensive SARs studies were performed and will be discussed in detail. In the case of CB$_2$R, no crystal structure in an active state is available yet. However, as the tested CB$_2$ ligands showed only agonistic efficacy, clear and rational docking poses could not be obtained for CB$_2$R as the ligands might adopt different poses in an active state co-crystallised receptor. Nevertheless, within rigid docking and visually discussion a general understanding in the CB$_2$R binding of the new ligands was achieved, allowing to conclude some minimal structural requirements for more favourable receptor binding and higher affinities. Superscript denotation at amino acid residues indicates the respective Ballesteros-Weinstein numbering[174].

---

[5] Docking studies were performed by LINDSEY BURGGRAAFF and MARTIJN BEMELMANS in the group of Prof. Gerard van Westen at Leiden Academic Centre for Drug Research, Leiden University, The Netherlands.

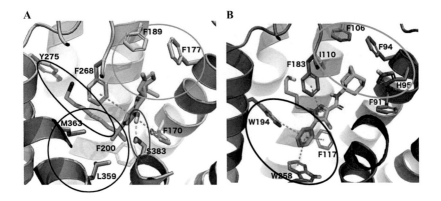

Figure 17: General overview of used crystal structure of both CBR (**A** CB$_1$, **B** CB$_2$) subtypes with their co-crystallised ligands. Important regions for interactions are highlighted.

As no SARs study for the CB$_2$R was performed the general structural observations for a favourable binding will be discussed first, followed by the comprehensive CB$_1$R SARs study. In the receptor binding site of CB$_2$R two important regions were identified to have most significant impact for ligand binding (Figure 17B, black and blue cycles). The upper region (blue circle) is located at the top of the receptor binding site showing a hydrophobic pocket mainly defined by the amino acid (AA) residues of F91$^{2.61}$, F94$^{2.64}$, H95$^{2.65}$, F106$^{3.25}$, and I110$^{3.29}$. The second region (black circle) is located at the bottom of the receptor binding site, showing an ambivalent hydrophobic and amphiphilic characteristic, restricted by the AA residues of F117$^{3.36}$, W194$^{5.43}$, W258$^{6.48}$, and V261$^{6.51}$. To achieve a sufficient binding between ligand and receptor the measured binding affinities suggested that both pockets have to be occupied. Coumarins with large lipophilic groups pointing bidirectional away from the coumarin core showed high affinities, whereby derivatives containing smaller or more polar substituents showed only low or no activity. Another considerable observation was the increased affinity for coumarins with a free hydroxyl group at position 5 compared to no activity for methoxy substituted coumarins. The presence of polar groups at this central position appears to be beneficial for a higher affinity. Structurally this might be explained by polar interactions of the hydroxyl group *via* hydrogen bonds (H-bond) by pointing towards either one of centrally located AA residues of S285$^{7.38}$ or T114$^{3.33}$. However, due to multiple possible spatial orientations always obtained during the docking process and inactive receptor population of the crystal structure no conclusive substantiations can be made at this point.

Similar observations were also made for the CB$_1$R ligand binding and three important regions were identified. Shown by a similar hydrophobic pocket at the upper end (Figure 17A, blue circle) of the binding site mainly encompassed by the AA residues of F177$^{2.64}$ and F189$^{3.25}$. The second hydrophobic pocket at the lower end (Figure 17A, black circles) of the binding site, however, was mainly defined by F200$^{3.36}$, L359$^{6.51}$ and M363$^{6.55}$ forming an extended hydrophobic tunnel towards the residue of Y275$^{5.39}$. Coumarins containing longer side chains which were able to fill this hydrophobic tunnel showed in general higher affinities up to a maximal tolerated length. Lastly, a third region (Figure 17A, red circle) around S383$^{7.39}$ was found and a trend of increased affinities by interactions via hydrogen bonds towards this residue was observed and may even be required for a best possible receptor ligand binding. For better clarity the SAR conclusions made from the docking studies will be described again in relation to the structural classification of the ligands. Additionally, during the docking experiments a different binding behaviour between 5 position protected and unprotected coumarins were observed. Therefore, the results of both substitution patterns within the respective group must be discussed separately.

Coumarins according to lead structure 1 (**29**, Table 22) in general showed only low affinities towards the CB1R, whereby, substitutions at position 7 were identified as main contributor. An increasing alkyl chain length resulted in a higher measured affinity in the radioligand binding assays. This may be explained by an improved interaction in the hydrophobic tunnel of the side chain and the Y275$^{5.39}$ residue. However, limitations inside chain lengths were observed comparing the results to the heptyl ether showing a decreased affinity (e.g. radioligand displacement of **21ea** 44% vs. **21gb** 8%). Additional groups attached to the 3-benzyl position resulted in a decreased affinity towards the CB$_1$R suggesting that no substitution to the benzyl ring is more preferred. This becomes apparent by comparing for example **21ea** (no substitution, 44%), **21eb** (*o*-methyl, 25%) and **21ef** (*o*-F, 11%). Lastly in this set, the free hydroxyl group at position 5 was considered, however, only showing a drastically decrease of binding affinity compared to the ether substituted ligands. Nevertheless, it is worth to mention, that the tested set also contained derivatives with substitutions at the 3-benzyl position and these as already discussed also contribute to a decreased affinity. The unsubstituted 3-benzylcoumarin with a hydroxyl group at position 5 was unfortunately not available in this set.

Similar observations were made while comparing the results for the 7-(dialkylamino)coumarins (lead structure 2, **30**, Table 22) showing an even further decreased affinity in the presence of the amino group compared to the ether moiety (**21ia** 28% vs **21ea** 44%). In general, for both coumarins containing heteroalkyl substituents the decreased affinity might be explained by a closer proximity of the heteroatomic lone pairs at the lower hydrophilic pocket around F200[3.36], L359[6.51] and M363[6.55] resulting in repulsive forces pushing the ligands out of the binding pocket.

In conclusion to that, the next group according to lead structure 3 (**31**, Table 22) should be more suitable to fit in this hydrophobic environment. Indeed, radioligand displacement of this set was in general higher and as mentioned previously a few coumarins showed affinities in the sub-micromolar range. Coumarins containing an ether group at position 5 (facing to S383[7.39]) showed similar binding behaviour concerning the alkyl chain length at position 7. For an optimal fit the alkyl chain should contain between three and five carbon atoms to interact best with Y275[5.39]. A longer chain length than a pentyl residue decreased the binding affinity also in this case (**21ka** 19% vs. **21la** K_i = 486 nM and Ki = 1.43 µM for the hexyl residue[99]). The same observations were made for substitutions at the 3-benzyl ring. Therefore, it can be concluded that only small groups like methyl groups are tolerated and no substituents are preferred. Interestingly, replacing a hydrogen by a less polar fluorine was not tolerated as well (**21la** K_i = 486 nM vs. **21le** 30% or **21lf** 18%). In case of the substituents at the benzyl ring, no clear trend was observed concerning which position might be more favourable as only low tolerance for substitutions in general was observed. With focus on the free 5-position hydroxyl group a different trend was observed. Again, the length of the side chain positively correlated with the affinity, but in contrast to the previous observations in this set an optimal length was achieved by the hexyl group (**59kf** –3% vs. **59lf** 15% and **59mf** 47%). This directly leads to the next observation of a beneficial influence of substituents at the 3-benzyl ring in contrast to the previous observations. For instance, comparing the o-F substitution of **59me** and **59ke** to **59ma** and **59ka**, higher displacement for the substituted ring was observed. For better visualisation a rational pose of each of four ligands (**59kf** –3%, **59md** 39%, **59me** 35%, and **59mf** 47%) was overlapped while remaining the coumarin aligned in the same plane (Figure 18). Firstly, the hydrophobic tunnel towards Y275[5.39] and the ideal occupancy by longer alkyl chains as well as the H-bond interaction in closer proximity to S383[7.39] was observed. Focusing only on the residues at the 3-benzyl ring only slight differences in the poses were observed, which correlate with significant changes in binding affinity. Possible interactions of the free p-hydroxyl group with the hydrophobic F177[2.64] and F189[3.25] seems to weaken the attractive interactions in the binding site, pulling more upwards and so reducing the binding affinity. Substitution with a more hydrophobic p-fluorine was therefore better tolerated (47%). In contrast to that, the o-fluorine was less tolerated, proving again that stronger hydrophobic interactions within this region are beneficial for higher affinities.

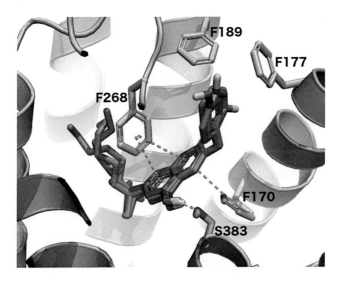

Figure 18: Docking results of exemplary rational poses and overlapping of coumarin 59kf (magenta), 59md (purple), 59me (orange) and 59mf (rose).

Further investigations by an induced-fit screening revealed a most likely explanation for these observations. Due to the closer proximity of the free hydroxyl group to S383$^{7.39}$ stronger interactions in the hydrophobic pocket encompassed by F177$^{2.64}$ and F189$^{3.25}$ and a more distal position towards the Y275$^{5.39}$ was observed which results in a higher stabilisation of the ligand inside the binding pocket.

As mentioned before, introduction of fluorinated groups at the benzyl residue as shown in lead structure 4 (**65**, Table 22) resulted in sub-micromolar active compounds. In comparison to the non-fluorinated structures (see [99]) higher affinities at CB$_1$ were observed for the o-F group, whereby significantly lower affinities were observed substituting a methyl by a CF$_3$ group. Both ligands showed docking poses (not shown) facing their benzyl group closer to F177$^{2.64}$ and F189$^{3.25}$ and therefore, indicating a beneficial effect of these interactions. Nevertheless, higher lipophilicity through fluorine groups was only tolerated for rather small changes.

In contrast to that, introduction of 3-phenyl groups (lead structure 5, **18**, Table 22) abolished the activity completely, suggesting a crucial role of the linking methylene group in the active binding. The reduced structural flexibility by the out of plane configuration of the phenyl group lead to an incapability to face the F177$^{2.64}$ and F189$^{3.25}$ region in an interactive conformation. Furthermore, the docking poses (not shown) suggested a disrupting conformation between both phenylalanine residues. Thus, the linking methylene group seemed to be essential for the compounds to be able to adept a structural conformation which results in a beneficial binding towards this region without disrupting their conformation at the same time.

Comparing these observations to rational poses of two 3-pyridyl coumarins (**17a** and **17c**) the different binding behaviour between 5-protected and unprotected coumarins can be visualised (Figure 19). Both coumarins showed a binding pocket occupancy more distant from S383$^{7.39}$. An explanation for the different affinities between the pyridyl configurations can be seen by a slightly shifted fit of **17c** resulting in weakened interactions with the F268 residue from the extracellular loop. Furthermore, a more distal pyridyl conformation seemed to disrupt the hydrophobic binding towards the upper located F177$^{2.64}$ and F189$^{3.25}$ and decrease the affinity. Due to uncertainties in the ether docking poses unfortunately this could not be confirmed finally.

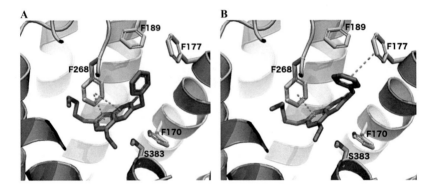

Figure 19: Possible CB$_1$R docking poses of 3-pyridyl coumarins **17a** (**A**) and **17c** (**B**).

Next, coumarins according to lead structure 7 (**66**, Table 22), which are showing the highest activities in this study were docked. Focusing on the size of the cycloalkyl ring significant differences in binding affinities were observed, showing higher activities for cyclohexyl substitutions (**66f** vs. **66n**). By comparing possible rational docking poses of these two ligands a more profound occupancy of the lower hydrophobic pocket towards the surrounding AA residues F200$^{3.36}$, L359$^{6.51}$ and M363$^{6.55}$ by the cyclohexyl ring was observed (Figure 20). However, resulting at the same time in a more distal positioning of the 3-alkylchain at the upper hydrophobic pocket and less interactions with $F177^{2.64}$ and $F189^{3.25}$. In contrast to that, the cyclopentyl substituent showed a higher motility and weaker interactions at the lower part. Despite stronger interactions with the upper binding pocket this motility might be the reason for the lower general binding affinities of derivatives with cyclopentyl substituents against CB_1. Interestingly, comparing the measured efficacies at CB_2 to the cycloalkyl size all cyclopentyl substituted coumarins showed full agonistic efficacies, whereby cyclohexyl substituted only reached partial agonistic responses. Thus, the data suggests that a minimum of flexibility inside the binding pocket might be more important than higher binding affinity to reach a full agonistic response. Due to the missing CB_2R crystal structure in an active state, unfortunately this theory could not finally be confirmed.

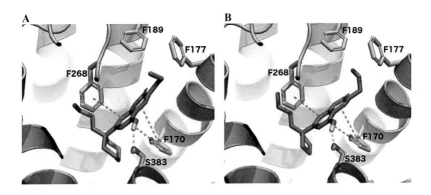

Figure 20: Comparison of binding mode between 7-cyclopentyl and -cyclohexyl substituted coumarins **66n (A)** and **66f (B)**.

Uncertainties in the docking of the 5-position protected coumarins did not allow to rationalise any possible binding pose over the other and therefore was avoided.

The importance of bulky hydrophobic substituents for high activity towards CB_1 can be comprehended by comparing the 3-alkyl-7-pentyl substituted coumarins of lead structure 8 (**67**, Table 22). Decreasing the steric demand towards an unsubstituted 7-pentyl chain abolished the affinity completely, suggesting that the hydrophobic interactions with $F200^{3.36}$, $L359^{6.51}$ and $M363^{6.55}$ are very important for a CB_1R binding.

Concerning the described requirements for high CB_1R binding like lipophilic and/or bulky groups at 3- or 7-position coumarins containing small substituents according to lead structure

9 (**68**, Table 22) were expected to occupy the binding pocket insufficiently and therefore were excluded of the docking studies.

Summary Structure-Activity-Relationships

In the presented SAR study different structural correlations and requirements were observed allowing to propose following general conclusions (Figure 21). An alkyl chain at position 7 was identified to be an essential requirement for a general binding affinity against both receptors. Higher binding affinities were achieved by filling the hydrophobic tunnel towards $Y275^{5.39}$ with longer alkyl chains, whereby the length shouldn't exceed six carbon atoms. Bulky substituents connected to the C1' of the side chain additionally enhanced the binding affinity by stronger hydrophobic interactions with the surrounded binding pocket. At the same time, introduction of more polar heteroatoms at this position wasn't tolerated. Different binding behaviours inside the pocket were observed by comparing coumarins with an ether or free hydroxyl group at position 5, whereby less certain docking poses were obtained for the ether substituted coumarins. Higher affinities were found for the free hydroxyl groups by stronger interactions with $S383^{7.39}$ and higher motility for the ether substitutions. A functional response depending on more profound interactions with $F200^{3.36}$, $L359^{6.51}$ and $M363^{6.55}$ were observed for the tested CB_2 ligands, whereby less motility resulted in a partial agonistic response and higher motility in a full agonistic. Furthermore, structural flexibility at the 3-position was found to be essential for a successful receptor binding, shown by complete loss of activity if the linking methylene group was removed. Derivatives containing alkyl chains at this side of the molecule only showed high affinities if another bulky group was present. For high affinities at least one bulky group either at 3- or 7-position is necessary. Benzyl groups are tolerated best if left unsubstituted or only substituted with small hydrophobic groups preferred in descending order from o > m > p. Introduction of fluorine can be beneficial. Heterocycles are tolerated if the polar substituents aren't orientated in p direction. In general filling the binding pocket with hydrophobic and sterically demanding substituents at the lower and upper side of the binding pocket correlate with low nanomolar activities. Higher selectivities can be achieved by introducing a free hydroxyl group at the core structure, as a hydrophilic group is better tolerated at CB_2.

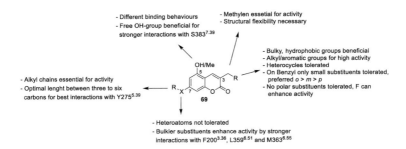

Figure 21: Summary of structure-activity relationships at the CB₁R resulting from binding assays and docking studies.

3.2.5   ADME and physicochemical properties

Aside from target affinity, efficacy and selectivity, further parameters like physicochemical properties and metabolic stability are crucial for the optimisation in drug development processes. Absorption, distribution, metabolism and elimination (ADME) studies combine these parameters to provide crucial data to drug developers to select only candidates with desired properties for late preclinical and early clinical phase 0 *in vivo* trials and help to shorten the time until new effective drugs are available.[175, 176] One of the early stage ADME determination to focus on is the *in vitro* metabolic stability of a drug candidate. Metabolic stability means the percentage loss of drug candidate concentration in a metabolic active screening system over a period of time.[177] For example, early *in vitro* metabolic clearance determination of main elimination route *via* hepatic metabolism can help to improve pharmacokinetic data concerning both metabolic phases (I and II). Phase I metabolism usually refers to oxidation of parent drug by all isoforms of the cytochrome P450 (CYPs) family, including coumarin 7-hydroxylase CYP2A6[178]. Whereby, metabolism phase II includes enzymes involved in the conjugation of phase I metabolites with highly polar molecules like glutathione, glucose, sulphate and others. These transformations are mainly mediated by UDP-dependent glucuronosyl transferase (UGT), phenol sulfotransferase (PST), oestrogen sulfotransferase (EST), and glutathione-S-transferase (GST).[179] Metabolic stability is expressed as half-life time ($t_{1/2}$ in min) and allows to calculate an estimation of the *in vitro* intrinsic clearance (CL_{int} in µl/min/mg protein) for the selected screening system.

In the presented work it was decided to screen the metabolic stability of four of the ligands (Figure 22), based on their functional groups, against enzymes in human liver S9 fraction (the

9000 *g* supernatant of liver homogenate[180]). Liver S9 fractions as metabolic system was chosen, since it includes enzymes of both metabolic phases I and II and therefore provide extended
metabolic data within the same assays. Standard procedure for a metabolic stability screening
typically includes incubation of the test compound together with liver homogenate and respective cofactors like nicotinamide adenine dinucleotide phosphate (NADPH) regeneration systems for phase I enzymes or uridine diphosphate glucuronic acid (UDPGA) and 3'-phospho-
adenosine-5'-phosphosulfate (PAPS) for phase II enzymes. Within the incubation period end
point analysis or kinetic investigations are possible by analysing the ligand of interest at several
time points or after completed incubation period only. The remaining concentration of test compound is determined after reaction quenching (usually by addition of organic solvents) by LC-
MS or tandem LC/MS (higher resolution at lower concentrations) and compared to a standard
curve of the test compound with known concentrations.

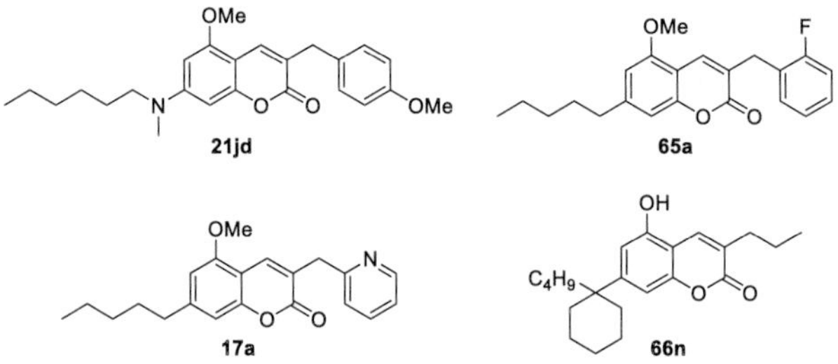

Figure 22: Chosen representative ligand structures from structural group 2, 4, 6 and 7 (**Table 22**) for metabolic
stability determination.

In the case of ligand **66n** only insufficient resolution by single LC-MS analysis or fully breakdown within the tandem LC-MS analysis was observed during the tandem LC-MS method optimisation. Hence, no determination of the hepatic metabolic stability was possible. The normalised results of the tandem LC-MS data for ligand **21jd**, **65a** and **17a** are shown in Figure 23
(for standard curve data see Chapter 6.6).

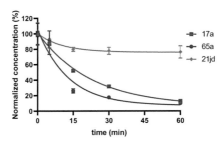

Figure 23: Time-dependent hepatic metabolic stability assay of coumarin-derivatives **17a**, **21jd** and **65a** after incubation for several time points in liver S9 fraction. Data expressed as means ± SD from four tandem LC-MS analysis normalised to the t = 0 area as 100% and analysed by a "one-phase exponential decay" fit.

For all three tested ligands $t_{1/2}$ and CL_{int} (Table 13) were calculated according to (2) and **(3)**, whereby the elimination rate constant (k) is defined as the negative gradient of the log peak area ratio (compound peak area/internal standard peak area) and the volume (V) as total incubation volume per incubated protein (µl/mg).[180, 181]

$$t_{1/2}[\text{min}] = \frac{0.693}{k} \tag{2}$$

$$CL_{int}\ [\text{µl/min/mg}] = \frac{V \times 0.693}{t_{1/2}} \tag{3}$$

Two investigated ligands showed a low (**65a**) or moderate (**17a**) metabolic stability with a $t_{1/2}$ of 15.5 and 8.87 min, respectively. Whereby, only small differences were observed comparing the intrinsic clearance. Most interesting was the nearly linear shaped decay of **21jd** with still more than 75% remaining ligand concentration after 60 min. For high metabolic stable compounds, a steady decrease would be expected, which suggests an almost complete inhibition of the degrading CYP enzymes by **21jd** as no change of remaining concentration was observed roughly after 20 min anymore.

Table 13: Metabolic stability data for representative coumarin-derivatives against liver S9 enzymes.

Entry	17a	65a	21jd
$t_{1/2}$ [min]	15.5	8.87	≫60
CL_{int} [µl/min/mg protein]	34.3	38.6	4.1

Additional to the metabolic stability, other physicochemical properties were determined for a selected set of coumarins and are summarised in Table 14, including the calculated partition coefficient (CLogP) and lipophilic efficiency (LipE) among others.

Table 14: Summary of physicochemical properties for tested and most active coumarin-derivatives.

Entry	Cpd.	M.W. [g/mol][a]	CLogP[a]	LipE (CB$_2$)	tPSA (Å^2) [b]	BBB permeant[b]
1	66d	314.43	6.82	0.40	50	Yes
2	66f	328.45	7.35	0.38	50	Yes
3	66h	342.48	7.88	−0.03	50	No
4	66l	328.45	7.38	0.03	50	Yes
5	66n	342.48	7.91	0.28	50	No
6	66p	356.51	8.44	−0.54	50	No
7	17a	337.42	4.93	2.15	52	Yes
8	17c	337.42	4.93	2.32	52	Yes
9	17n	391.51	6.92	0.22	63	No

[a] Calculated in ChemDraw® Professional 16.0 [b] Determined using SwissADME[182].

In medicinal chemistry the listed descriptors are considered to keep in mind as early as possible and an optimisation can contribute to a successful drug development.[183] Limitations for example in molecular weight (M.W.) and lipophilicity (CLogP) are mainly considered to define the later possible administration of the developed drug. Mostly for common drugs an oral application is desired due to the easiest possible application by self-administration. Nowadays these considerations are better known by "*Lipinski's Rule of Five*", describing the necessary physicochemical properties for a potential oral administration.[184] Nevertheless, regarding the necessary high lipophilicity to show high affinities towards the CBRs reducing lipophilic properties for an oral administration are not an option.

Moreover, for the development of CNS and peripheral active drugs as presented in this work blood-brain barrier (BBB) permeability is a more important physicochemical property to focus on. Especially, as CBR subtype expressions are tissue dependent and concerning a possible anti-inflammatory application of the tested CB$_2$ agonists in the peripheral immune system, no BBB permeability was desired to reduce potential CB$_1$ mediated side effects. As a rough rule of thumb molecules with a total polar surface area (tPSA) lower 60–70 Å^2 are considered to be able to cross the BBB[185]. In addition, for a more detailed description, the BBB permeabilities were calculated using the SwissADME web tool reported by DAINA *et al.*[182] The obtained results suggest an independence of the respective tPSA for a successful BBB crossing (entry 1–2, 4 and 7–8) and a high dependence on the molecule's size, as molecules with larger side chains are more likely to cross the BBB (entry 3, 5–6 and 9). Nevertheless, these results should be considered as possible *in vitro* predictions and can differ significantly from the results obtained by *in vivo* observations.

At last, to screen for potential off-target effects within the ECS related enzymes the coumarins were tested in a chemical proteomics experiment (for detailed description see Figure 28). For this purpose, a gel based competitive ABPP analysis using the membrane proteome of mouse brain was performed. To guarantee the visibility of a potential off-target interaction coumarins were tested at a high concentration of 10 µM. For the fluorescence analysis samples were labelled using either TAMRA-FP (Figure 24 left) or MB064 (Figure 24 right) as ABP.

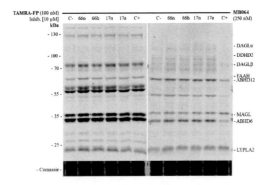

Figure 24: Competitive ABPP in mouse brain membrane proteome for off-target screening selected coumarin-derivatives. Mouse proteome was incubated (30 min, rt) with inhibitor, DMSO as vehicle (C–) or THL as positive control (C+). Samples were labelled (10 min, rt) with ABP TAMRA-FP (left) or MB064 (right). Coomassie staining was used as protein loading control. Important EC related enzymes are highlighted on the right site.

Despite the high ligand concentration, no reduced band intensity was observed in the fluorescence analysis, suggesting the tested coumarins either do not interact or even do not bind within the active sites of enzymes in this proteome. Therefore, the tested coumarins showed a high selectivity profile in the ECS as no off-target activity again any EC related enzyme was identified.

3.3   Synthesis and evaluation of MAGL inhibitors[6]

Over the last decade, an increased interest in the development of selective inhibitors targeting relevant metabolic enzymes of the ECS has been seen (for reviews see [137, 186]). For the treatment of neurological disorders like inflammation, pain, Tourette syndrome, anxiety, depression, drug addiction, as well as other related disease like multiple sclerosis, or Parkinson, selective inhibition of MAGL was considered as a promising approach. [187-191] Since CRAVATT et al.[192, 193] reported the development of JZL184 as a potent and highly selective MAGL inhibitor, other more or less selective inhibitors with different chemical core structures and nanomolar to sub-nanomolar potencies were reported.[123, 125, 194-197]

Besides several off-target activities, mainly against FAAH, all the described inhibitors were reported as covalent inhibitors, reacting via an irreversible or partially irreversible inhibition mode. In general, due to drug safety concerns, a covalent mode of inhibition is less favoured than a reversible inhibition.[198] A potential drug acting via a covalent MAGL inhibition had to face several pharmacological and therapeutic challenges. *In vivo* studies by CRAVATT et al.[199] reviled a functional antagonism in mice after prolonged inactivation of MAGL by administration of JZL184. Furthermore, the extended blockage of MAGL resulted in a loss of analgesic activity of JZL184, an increased cross-tolerance to CB_1R agonist WIN55,212-2, as well as desensitised CB_1R, dysfunctional endocannabinoid-dependent synaptic plasticity and physical dependence. To overcome these potential side effects and enhance the therapeutic window a reversible inhibition approach could lead to a promising drug candidate.

In recent literature, this approach attracted more attention and several new reversible MAGL inhibitors were reported.[129, 140, 200, 201] However, high potencies in a low nanomolar dose compared to the covalent inhibitors weren´t achieved yet. Therefore, the development of a reversible, highly selective MAGL inhibitor with sub-nanomolar potency was an encouraging aim of the last project of this work and the results will be discussed in the following part.

Design and retrosynthesis

The joint project of the development of new reversible MAGL inhibitors was based on preliminary work from MING JIANG.[202] Initially started from a high-throughput screening hit, a new lead structure was proposed. In several consecutive SAR studies, a structure optimisation had

---

[6] The presented work was performed within a research internship in the laboratories of Prof. van der Stelt at Leiden Institute of Chemistry, Leiden University, The Netherlands.

been performed towards the structure moiety **33a** with good potency and excellent selective (Figure 25). Retrosynthetically, the compound was divided into a left part and a right part. Former bearing a *trans*-(2,3)-dimethylpiperazin moiety substituted with an *m*-chlorophenyl and the right part containing a fluorinated benzoic acid derivative, substituted further with a β-sulfoxide ethyl ester side chain. Both parts were connected by an amide functionality, which simultaneously serves as "warhead-group" to reversibly bind to the catalytic triad serine SER122 in the active site.

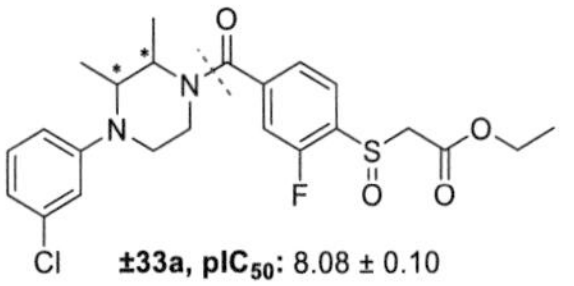

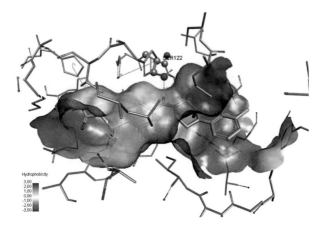

±33a, pIC$_{50}$: 8.08 ± 0.10

Figure 25: Lead structure of new reversible MAGL inhibitors. Indicated stereocenters were racemic mixture of *trans* diastereomers only. Retrosynthetic cleavage indicated by the red line.

Visual docking experiments had been performed to show a possible binding mode of **33a** within the active site of MAGL (Figure 26). The active site of MAGL is split into a more hydrophobic cave (right) and a hydrophilic tube (left) resulting in a perfect fit of **33a** with the amide functionality in near proximity (~3Å) towards the catalytic SER122 as described above.

Figure 26: Docking studies and binding mode within the active site of **33a** (visualisation was performed with DISCOVERY STUDIO, 3DS, France, version 3.5 and crystal structure PDB: 3PE6[203]). The active site of MAGL is split into a more hydrophobic cave (right) and a hydrophilic tube (left) resulting in a perfect fit of **33a** with the amide functionality in near proximity (~3Å) of the catalytic SER122.

Furthermore, in combination with the preliminary SAR studies several crucial and more favourable functionalities for higher affinities were revealed. Substituting the left part with the
free piperazine, resulted in only low affinities, whereby moderate affinities were observed with
3-methylpiperazine or *cis*-(2,3)-dimethylpiperazine. Therefore, a higher three-dimensional
(3D) demand in the hydrophobic cave (Figure 26 right) seemed to be favourable to stabilise the
ligand within the active site. Expanding the steric demand was achieved by introduction of
halogens at both aromatic rings. Initially introduced fluorine substituents on both phenyl rings
resulted in elevated affinities, which were further increased by substitution of one fluorine to
chlorine (Figure 25). If multiple halogens were introduced at the left phenyl ring, affinities
decreased again. On the other side of the molecule, saponification of the ester to the free acid,
as well as reduction of the sulfoxide to the thioether resulted in a complete loss of affinity.
Hence, existence of hydrogen-bond acceptors at this part of the compound were crucial for a
strong binding, but together with free acidic protons or anionic species were not tolerated.

Based on these results we decided several structural changes as interesting and worth to investigate further. First, to exchange the right fluorine with a higher steric demanding halogen
group. Second, substitution of the sulfoxide to a ketone functionality to reduce structural com-
plexity. Third, to reduce the amide functionality to an amine for a higher flexibility and to prove
possible interactions of SER122 with the carboxylic ester group. Furthermore, to stabilise the
ester group against hydrolysis, the substitution with iso-alkyl esters was planned.

The syntheses of building blocks and new inhibitors was following an established protocol developed by MING JIANG[202] (Scheme 11). The strategy started with the syntheses of the benzoic
acid part by protecting the free benzoic acid precursor **70** with a *tert*-butyl group (**71**). Followed
by nucleophilic aromatic substitution (S$_N$Ar) in *p*-position to obtain the thioether **72**. Subsequent mono-oxidation by Oxone® to the sulfoxide **73** and hydrolysis of the *tert*-butyl ester fi-
nalises the syntheses of the first building block **74**. The second building block was realised by
a straightforward approach according to a transition metal-free amination reported by BOL-
LINGER *et al.* [204] The final compound **33a** was obtained in the last step by simple peptide coupling of both building blocks.

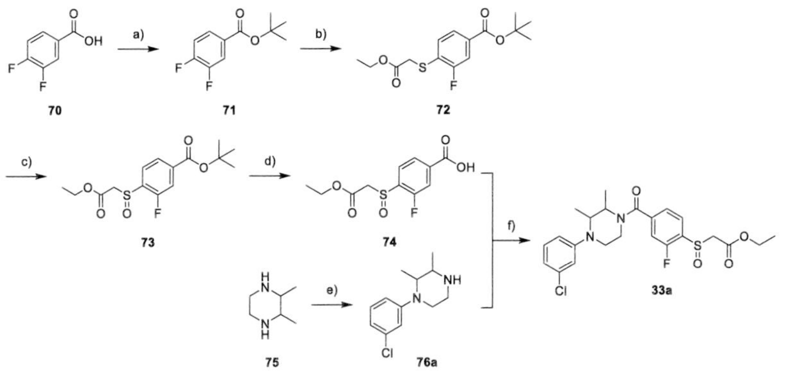

Scheme 11: Established protocol for the synthesis of lead structure **33a**: *a) di-tert-butyldicarbonat (Boc₂O), 4-(dimethylamino)pyridine (DMAP), tert-butanol, 60 °C, o.n., 80%; b) ethyl thioglycolate, K₂CO₃, DMF, rt, o.n.,71%; c) Oxone®, MeOH/H₂O, 0 °C then rt, 1 h, 90%; d) TFA, DCM, rt, 6 h, 72%; e) potassium bis(trime-thylsilyl)amide (KHMDS), 1-bromo-3-chlorobenzene, toluene, 100 °C, o.n., 13%; f) i) HATU, N,N-diisopro-pylethylamine (DIPEA), DCM, rt, 1 h, ii) 76a, rt, o.n., 22%.*

3.3.1 Syntheses of benzoic acid building blocks

The synthetic work on the project began with the syntheses of the benzoic acid building blocks **77a** and **77b**, whereby the *m*-fluorine was substituted to either a CF₃ group or a chlorine (Scheme 12).

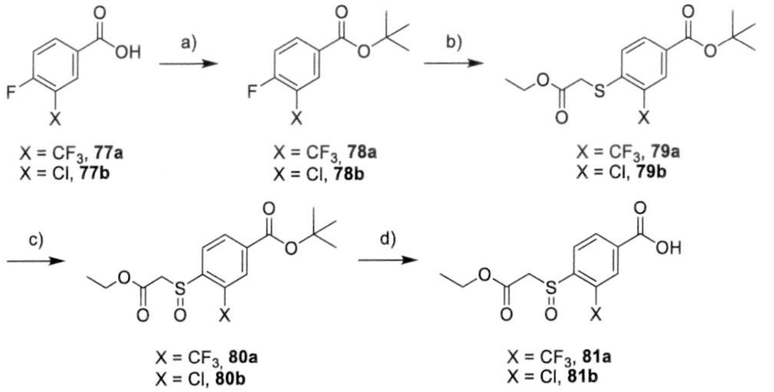

Scheme 12: Syntheses of benzoic acid building blocks **81a** and **81b**: *a) di-tert-butyldicarbonat, DMAP, tert-buta-nol, 60°C, 4 h, 78a 91%, 78b 88%; b) ethyl thioglycolate, K₂CO₃, DMF, rt, o.n., 79a 90%, 79b 91%; c) Oxone®, MeOH/H₂O, 0°C then rt, 1 h, 80a and 80b quant.; d) TFA, DCM, rt, o.n., 81a 59%, 81b 90%.*

As both benzoic acid precursors (**77a** and **77b**) were commercially available, a straightforward synthesis according to the previous protocol was performed. All intermediates were obtained in good to excellent yields (59% – 91%). The reduced yields of **81a** can be explained by stronger interferences of the CF$_3$-group.

For the sulfoxide exchange and the amide reduction, the established synthesis protocol for the benzoic acid building blocks was not applicable and new synthetic strategies had to be developed. The amide reduction was achieved by slight modification of the reaction order, using a reductive amination approach as coupling step (see Chapter 3.3.3). The required aldehyde **83** for the reductive amination was synthesised in excellent yields (92%) right away from the commercially available aldehyde **82** by the established S$_N$Ar reaction.

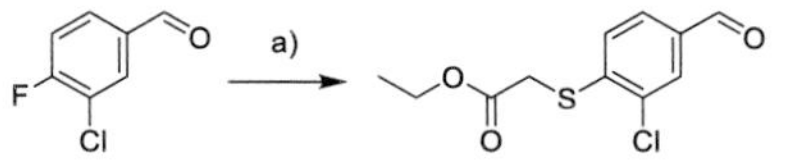

Scheme 13: Synthesis of aldehyde precursor **83** for reductive amination: *a) ethyl thioglycolate, K$_2$CO$_3$, DMF, rt, o.n., 92%.*

The building block for the sulfoxide exchange was synthesised by treating 4-bromo-3-chlorobenzoic acid (**84**) with GRIGNARD reagent (isopropylmagnesium chloride lithium chloride complex, iPrMgCl·LiCl) and subsequently quenching with DMF to obtain the formylbenzoic acid **85** (70% yield).

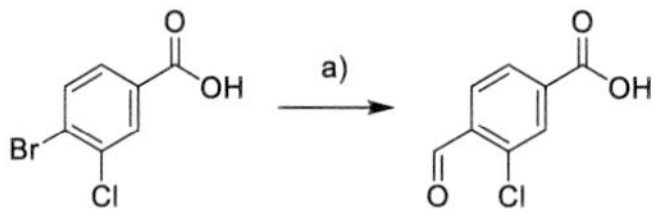

Scheme 14: Turbo-GRIGNARD reaction towards formylbenzoic acid XX: *a) i) iPrMgCl·LiCl, abs. THF, −78 °C, 10 min, then 0 °C, 1 h; ii) DMF, rt, 1.5 h, 70%.*

3.3.2   Synthetic work related to the *trans*-(2,3)-dimethylpiperazine building blocks

Concurrent to the syntheses of the benzoic acid building blocks, syntheses on the *trans*-(2,3)-dimethylpiperazine building block were performed. Concerning the low yield of the amine coupling reaction and considering that only small quantities of *trans*-(2,3)-dimethylpiperazine were commercially available, optimising the synthesis strategy beforehand was highly preferable.

For the *trans*-(2,3)-dimethylpiperazine syntheses, the optimisation was performed, at the same time, within the research internship of ENRICO VERPALEN[205] and was continued in accordance. The strategy started with a condensation reaction of 2,3-butanedione (**86**) with ethylenediamine to form the 5,6-dimethyl-2,3-dihydropyrazine (**87**). Subsequently followed by sodium reduction in ethanol to yield mostly the *trans*-isomers ±**75** in excellent yield (compared to literature [205, 206]).

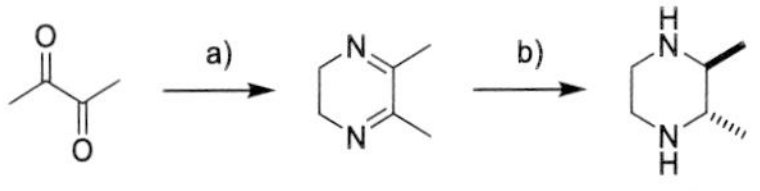

Scheme 15: Optimised syntheses strategy towards *trans*-(2,3)-dimethylpiperazine (±**75**): *a) ethylenediamine, Et₂O, 0 °C then rt, o.n., 85%; b) Na (solid), EtOH, reflux, 16 h, 42%.*

Improving the purification conditions for flash column chromatography resulted in an easy and fast feasible procedure to separate the minor formed *cis* isomer from the desired *trans*-enantiomers. Formation of the *trans*-isomers as major product can be explained by the alternating electrochemical-chemical-mechanism of sodium reductions. In general, consecutively single-electron-transfer (SET) and protonation of the resulting radical anion, twice, results in the formation of the 2,3-dimthylpiperazines. Hence, before receiving the second SET, the pyrazine forms the most stable conformation. Adopting the most stable conformations of piperazine, the chair[207], both methyl groups of **87** will be arranged in an equatorial position, resulting in the formation of the *trans*-isomers as main product. Due to the increased temperature, however, a small quantity of *cis* isomer will still be formed. Furthermore, main advantages of the established synthetic strategy were cheap starting materials and reaction scales at molar level.

With enough of ±**75** in hand, the initial planned optimisation of the amine coupling reaction was executed. Different approaches of couplings according to BUCHWALD-HARTWIG cross coupling reactions with either unprotected or protected dimethylpiperazine, as well as transition metal (TM) catalysed or TM free were prepared for comprehensive investigations.

At first, a TM catalysed protocol according to literature, reported by REILLY *et al.*[208], was investigated. Providing a fast and simple methodology with excellent yields. The transformed reaction conditions to the new system and the results are summarised in Table 15.

Table 15: Screened reaction conditions for the BUCHWALD-HARTWIG reaction of ±**75** or **88** and **89a-b**:

Entry	R (equiv.)	89	Yield	Product
1	CH₃ (2.00)	1-bromo-3-chloro	–	76a
2	H (4.00)	1-bromo-3-chloro	–	76a
3	CH₃ (4.00)	1,2-dichloro	–	76b
4	CH₃ (1.00)	1,2-dichloro	–	76b
5	H (4.00)	1,2-dichloro	75%	76b

Conditions: a) indicated piperazine (±**75** or **88**), 1.00 eq. of indicated aryl halides (**89a-b**), 1 mol% tris(dibenzyli-denacetone)dipalladium (Pd₂(dba)₃), 2 mol% RuPhos, sodium *tert*-butoxide (NaOtBu), 1,4-dioxane, 100 °C, 10 min.

In entry 1, the literature protocol was directly transferred to the desired system with dimethylpi-perazine (±**75**) and 1-bromo-3-chlorobenzene (**89a**), but no product was isolated. To investigate the influence of an additional bromine substituent (in literature only chlorine substituents), the literature conditions and piperazine (**88**) substrate were used (entry 2). Again, no product was obtained, indicating that bromines are not tolerated within this catalysis. Next (entry 3 and 4), a dichloro-substituted benzene (**89b**) was used with different equivalents of dimethylpiperazine (±**75**), to prove the general applicability of the methodology. However, neither product nor any side product were formed during the reaction. At last (entry 5), as proof of concept the original reaction conditions were adopted and the product **76b** was isolated with the same yield as reported. Thus, these results concluded that the dimethylpiperazine (±**75**) instead of unsubstituted piperazine (**88**) was not applicable in this methodology.

In consequence of that, a new approach was started, whereby single *tert*-butyloxycarbonyl (Boc) protected dimethylpiperazine should be used. For the protection reaction, several conditions were screened and are summarised in Table 16. The synthesis of catalyst 1-methy-1H-imidazol-3-ium tetrafluoroborate ([HMIm]⁺BF₄⁻) was prepared according to literature[209] and used in a protocol published by SUNITHA et al.[210], reporting a facile method for selective single Boc protection of unsubstituted piperazine (**88**).

Table 16: Screened reaction conditions for the mono-protection of ±**75**.

Entry	Additive (equiv.)	Conc. [mM]	Temp. [°C]	Time	Yield [%]
1	[(HMIm)BF4] (0.10)	_[a]	35	30 min	–
2	[(HMIm)BF4] (0.10)	10.0	35	30 min	15[b]
3	–	16.0	rt	20 h[c]	29[d]
4[e]	Et₃N (3.00)	20.0	rt	3 d[c]	26[d]
5[e]	Et₃N (3.00)	10.5	rt	20 h[c]	28
6	Et₃N (3.75)	18.0[f]	rt	1 d[c]	45

Conditions: a) 1.00 equiv. ±**75**, 1.00 equiv. Boc₂O, DCM. [a] IL as solvent; [b] contaminated with catalyst; [c] addition of Boc₂O over the entire time; [d] double protected **91b** major product; [e] ±**75** was used as TFA salt; [f] MeOH was used as solvent.

In contrast to the results reported by SUNITHA *et al.*, no desired product was obtained with the dimethylpiperazine (entry 1). Instead, only double protected (**91b**) and starting material (±**75**) was isolated. To avoid a double protection again, the reaction was highly diluted (entry 2), resulting in isolation a low yield of the desired product ±**91a**. Unfortunately, the product was contaminated with residual catalyst and any attempt of purification failed. Therefore, in the next entry 3, no catalyst was added, the reaction time was prolonged and Boc₂O was slowly added over the entire period, leading to a nearly doubled isolated yield of ±**91a** (29%). However, the major formed product was still **91b** (34%). Deprotecting of **91b** in an excess of TFA was performed to regain the starting material ±**75** as TFA salt. Further elongation of the reaction time at higher concentrations decreased the yield of ±**91a** only slightly. Though, proving that the usage of the TFA salt resulted in similar yields as before, if supplemented with Et₃N (entry 4, 35% ±**91a**). Thus, shorter reaction times with, at the same time, lower concentrations (entry 5), resulted in the same yield of ±**91a** (28%) and slightly lower yields of **91b** (27%). Lastly, supplementation with Et₃N while not using the TFA salt and elongation of the Boc₂O addition, resulted in the highest achieved yield of 45% ±**91a**. Due to issues later in the synthesis sequence, no further attempts of optimising the coupling conditions were conducted. Nevertheless, it can be summarised that high dilution and very slow addition of Boc$_2$O was crucial for better yields. Moreover, the addition of Et$_3$N was beneficial to achieve higher yields.

With the protected dimethylpiperazine ±91a in hand, further optimisation attempts using a TM catalysed BUCHWALD-HARTWIG coupling were executed (Table 17).

Table 17: Screened conditions for the BUCHWALD-HARTWIG coupling with protected dimethylpiperazine ±91a.

Entry	Cat. (mol%)	Ligand (mol%)	Base	Temp. [°C]	Time [h]	Yield [X]
1	PdOAc$_2$ (4)	BINAP (6)	NaOtBu	85	20	13
2	PdOAc$_2$ (4)	RuPhos (8)	NaOtBu	85	24	16[a]
3	PdOAc$_2$ (4)	BINAP (6)	Cs$_2$CO$_3$	85	24	11[a]
4	Pd$_2$(dba)$_3$ (4)	BINAP (6)	NaOtBu	85	24	5[b]
5	Pd$_2$(dba)$_3$ (4)	BINAP (6)	Cs$_2$CO$_3$	85	24	43[b]
6	Pd$_2$(dba)$_3$ (4)	BINAP (8)	Cs$_2$CO$_3$	100	24	31[c]
7	Pd$_2$(dba)$_3$ (4)	Xantphos (6)	Cs$_2$CO$_3$	85	72	–
8	Pd$_2$(dba)$_3$ (4)	dppf (6)	Cs$_2$CO$_3$	85	72	–
9	Pd$_2$(dba)$_3$ (4)	BINAP (6)	Cs$_2$CO$_3$	85	72	53[c]

Conditions: a) 1.00 equiv. ±91a, 1.20 equiv. 89a, 1.50 equiv. base, 1,4-dioxane. [a] purity <90%; [b] purity 90% – 95%; [c] purity >95%;

As the previous used catalyst-ligand system was not suitable, other ligands and catalyst were considered to investigate. First conducted experiments were performed according to a reported protocol[211] (entry 1 and 4), differing only in the used catalyst (palladium acetate (PdOAc$_2$) vs. Pd$_2$(dba)$_3$) and slightly longer reaction times (entry 4). However, both reactions resulted in a successful conversion into the desired product ±92, but with unsatisfactory yields. Even if the isolated yield was low, this was a good starting point for the following experiments. Ligand substitution towards the previous applied RuPhos, resulted in a slightly higher yield, but puri-fication of RuPhos contaminations in the product was unsuccessful. In entry 3 and 5, changing the base towards caesium carbonate (Cs$_2$CO$_3$) increased the isolated yield drastically, if

Pd$_2$(dba)$_3$ was used as catalyst (entry 5, 43%). Increasing the ligand concentration and reaction temperature (entry 6) gave only lower isolated yields. No product was obtained, if other bidentate diphopshine ligands such as Xantphos or dppf were applied (entry 7 and 8). Another improvement of isolated yield (53%) was achieved with extended reaction time (entry 9 vs. entry 5).

In the last step of the synthesis sequence, ±**92** was deprotected in the presence of an excess of TFA (Scheme 16).

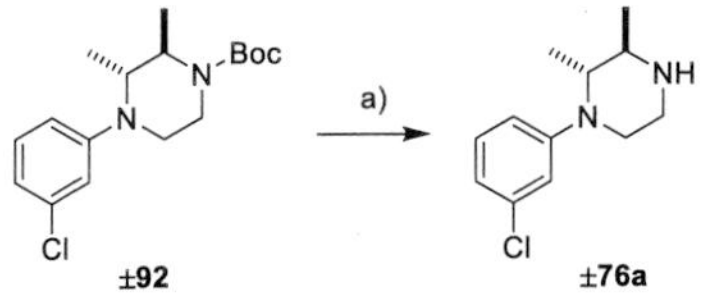

Scheme 16: Deprotection of ±**92**: *a) TFA, DCM, rt, o.n., 51%.*

The reaction resulted in moderate yields (51%). Due to low overall yields (12%) even after optimisation, no improvement compared to the yields obtained from the initial used TM free BUCHWALD-HARTWIG coupling (13%) was achieved. Therefore, no further attempts to optimise this synthetic sequence was made.

Instead, further screening of conditions using the TM free BUCHWALD-HARTWIG coupling were performed (Table 18).

Table 18: Screened conditions for TM free BUCHWALD-HARTWIG coupling.

Entry	Equiv. base	Temp. [°C]	Time [min]	Yield [%]
1[a]	3.00	100	90	32[b]
2[a]	2.00	90	90	23[b]
3[c]	2.00	90	25	22[d]
4[c]	2.00	90	15	23[e]
5[c]	2.00	10	o.n.	17[e]
6[c]	2.00	rt	o.n.	25[e]

Conditions: a) 1.00 equiv. ±**75**, 1.00 equiv. **89a**, base KHMDS (0.5 M in toluene), 1,4-dioxane/toluene. [a] 1,4-dioxane/toluene 3:1; [b] purity 77%; [c] 1,4-dioxane/toluene 4:1 [d] purity 84%; [e] purity >90%.

Already in the first attempt (entry 1), a drastically increase of isolated yield to 32% was achieved by change of solvent to the widely used and more polar 1,4-dioxane as main solvent in a mixture with toluene. Reaction control by LC-MS reviled the formation of the dimer product as impurity. Unfortunately, a sufficient separation of the dimer during purification was not achieved and the product was obtained in a low purity (77%). Reducing the amount of base, at slightly lower temperatures, resulted in a decreased yield without any purity improvement (entry 2). Further reaction control by LC-MS reviled, that longer reaction times were not necessary. Instead, leading to the formation of more dimer impurity. To investigate this in more detail (entry 3) the reaction process was monitored after several time points (5, 12 and 25 min). Already after 5 min, no difference in the reaction progress was detected anymore. Reducing the reaction time resulted in similar yields, but an improved purity was observed directly (entry 3 and 4). As the applied high temperature seemed to enhance the formation of dimers, drastically lower temperatures were tested (entry 5 and 6). Necessarily at lower temperatures, the reaction time had to be adjusted again and stirring overnight at room temperature resulted in a doubled isolated yield (25% vs. 13%) with excellent purities. With enough quantities of the trans-2,3-dimethylpiperatine building block in hand, the syntheses of final compounds were carried out.

3.3.3 Syntheses of final MAGL inhibitors

Last step in the synthesis procedure was the final coupling of previous synthesised building blocks towards the finally substituted inhibitors. As described before, a peptide coupling approach of the building blocks was chosen. Depending on the changed substitutions, adjustments in the syntheses order had to be done and will be described here in more details. For the halogen exchange, standard procedure was used and the syntheses of final compounds bearing a trifluoromethyl group or a chlorine instead of the initial fluorine is shown in Scheme 17 (±**33b** 29% and ±**33c** 75%).

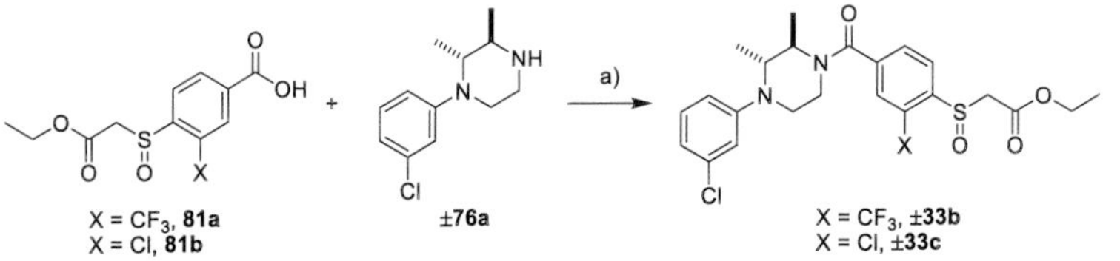

Scheme 17: Syntheses of final compounds **33b** and **33c**: *a) HATU, DIPEA, DCM, rt, 1 h, ii) ±76a, rt, o.n., 33b 29%, 33c 75%.*

For the reduction of the amide towards an amine group as central coupling element, the synthesis strategy was alternated in the coupling and oxidation step (Scheme 18).

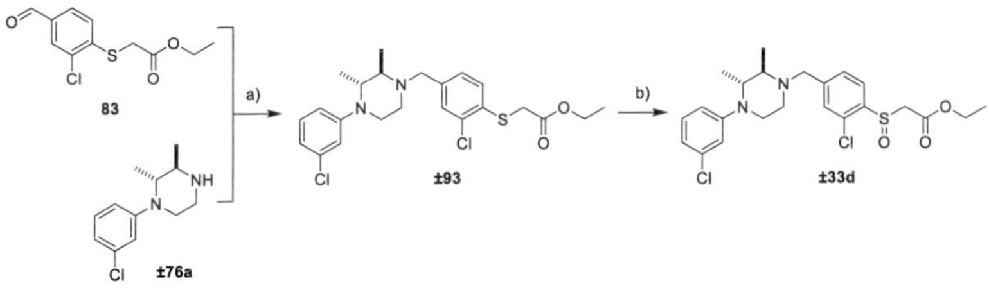

Scheme 18: Alternated synthesis strategy for amine modified final compound ±**XX**: *a) i) 1,2-dichloroethane, rt, 30 min; ii) sodium triacetoxyborohydride, rt, o.n., 40%; b) Oxone®, MeOH/H₂O, 0°C then rt, 2.5 h, 32%.*

The coupling of both building blocks **83** and ±**76a** was done by reductive amination resulting in the thioether ±**93** (yield 40%). Subsequently followed by oxidising towards the final inhibitor ±**33d** (yield 32%).

For the next substitution, the carboxylic acid of ±**33c** had to be synthesised first. Thereafter, two different, bulkier, esters were implemented by conversion of the carboxylic acid ±**94** into the acid chloride and substituting with the respective alcohols.

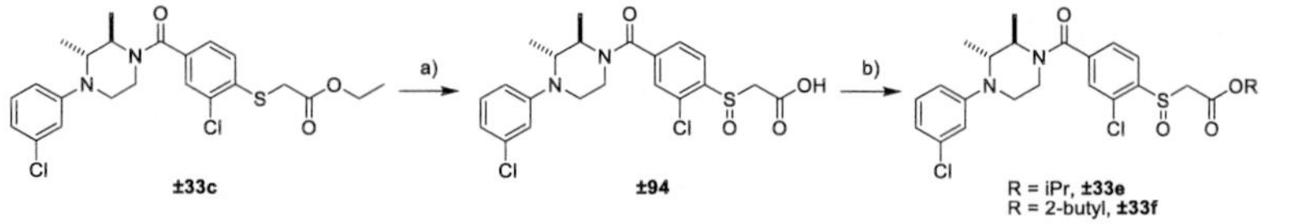

Scheme 19: Synthesis strategy for ester substituted final compounds ±**33e** and ±**33f**: *a) TFA, DCM, 62%; b) i) oxalyl chloride, DMF, DCM, reflux (±33e), rt (±33f), 2 h, ii) ROH, DIPEA, DCM, rt, o.n., ±33e 43%, ±33f 25%.*

However, transformation of the carboxylic acid ±**94** was not as easy as initially planned and either resulted in no conversion at all or complete breakdown of the starting material. After careful addition of only a small excess of oxalyl chloride, solved in DCM (2 M), and transfer into an excess of the respective pre-dried alcohol a successful synthesis of both inhibitors (±**33e** 43% and ±**33f** 25%) was achieved.

For the synthesis of the β-ketone ester ±**33g**, the synthetic route had to be adjusted as well. Coupling of aldehyde precursor **85** and the second building block ±**76a** resulted in aldehyde ±**95**, which then was mildly oxidised with Oxone® to the carboxylic acid ±**96**. Final transformation was done by activation of the carboxylic acid with carbonyldiimidazole (CDI) and subsequently coupling with potassium malonate and rearrangement towards the final compound ±**33g**.

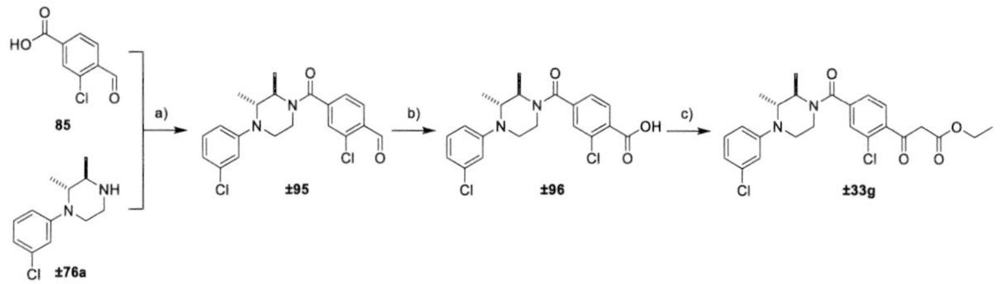

Figure 27: Applied synthesis strategy for the sulfoxide-ketone exchanged final compound ±**33g**: *a) HATU, DIPEA, DCM, rt, 1 h, ii) ±76a, rt, o.n., 50%; b) Oxone®, DMF, rt, o.n., 50%; c) i) CDI, THF, rt, 2 h, ii) potassium malonate, MgCl₂, TEA, THF, rt, 24 h, 6%.*

The final compound of ±**33g** was obtained in an isolated yield of 6%. Additionally, prolonged reaction times did not improve the isolated yield, suggesting a limited activation of the carboxylic acid by CDI. Nevertheless, sufficient quantities were obtained to proceed with the pharmacological evaluation of the synthesised MAGL inhibitors.

3.3.4 Pharmacological evaluation

To investigate the SAR of the prepared inhibitors, their activities against MAGL were tested either in an enzyme-based natural substrate assay or proteome-based competitive activity-based protein profiling (ABPP). Furthermore, the selectivity and possible off-target activities against other relevant enzymes of the ECS and the ability to bind only reversibly in the active site of MAGL were investigated.

The natural substrate assay (NSA) is a biochemical screening method to measure the inhibitory potency of the compound of interest by an enzyme coupled approach and was developed in house by VAN DER WEL et al.[212] Initially the assay was designed to investigate the activity of selective DAGL-α inhibitors, but only small modifications towards a natural-substrate-based fluorescence assay for MAGL inhibitors were necessary. As the enzyme cascade originally already involves MAGL as second step, hydrolysing its natural substrate 2-AG, it is only necessary to use cell lysate preparations of MAGL overexpressing cells and supplementing the reaction mixture with synthetic 2-AG instead. The adjusted enzymatic cascade would look like as described in Scheme 20. Initially 2-AG is hydrolysed by MAGL into AA and glycerol. Phosphorylation by glycerol kinase (GK) and oxidation by glycerol-3-phosphate oxidase (GPO) results in the formation of hydrogen peroxide (H_2O_2). In the final enzymatic step, horseradish peroxidase (HRP) uses the H_2O_2 as substrate to oxidise commercially available, non-fluorescent Ampliflu™ Red into fluorescent resorufin. As one fluorescent molecule will be formed for each degraded 2-AG molecule, a quantitative read out in real time is easy feasible by using kinetic fluorescence analysis.

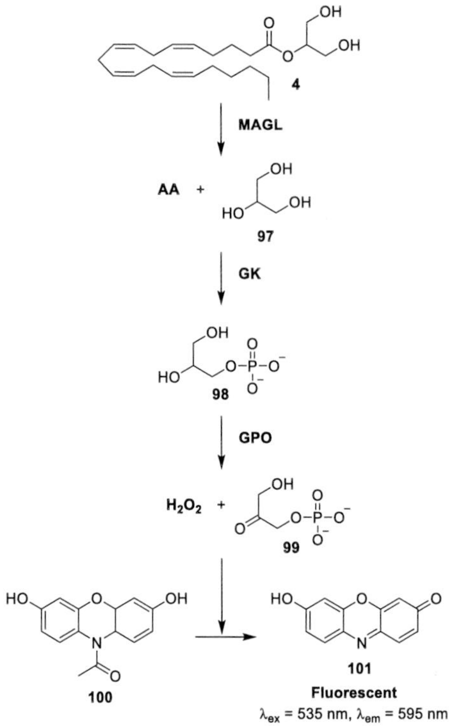

Scheme 20: Principle of MAGL natural substrate assay. Substrate 2-AG is hydrolysed by MAGL into arachi-donic acid and glycerol. Followed by phosphorylation of glycerol by glycerol kinase (GK) and oxidation into dihydroxyacetone phosphate and hydrogen peroxide (H_2O_2) by glycerol-3-phosphate oxidase (GPO). In the final step, the produced H_2O_2 will be used by horseradish peroxidase (HRP) to oxidise commercially available Ampli-flu™ Red into fluorescent resorufin. Quantitative kinetic fluorescent measurement is monitored in real-time us-ing a plate reader (λ_{ex} = 535 nm, λ_{em} = 595 nm).[212]

To confirm the tested activities of the inhibitors measured by NSA, activity-based protein pro-filing (ABPP) was used as second chemical proteomic method. This technique can be used to screen a complex biological proteome with a mixture of several different enzymes by precisely labelling only the enzyme of interest. Therefore, specific activity-based probes (ABPs) are de-signed, which selectively interact in the active site of their target enzyme to form a covalent bond. In general, ABPs consist of three parts (Figure 28A): a reactive head group, often referred as "warhead", to form the covalent bond in the enzyme´s active site, a linker group or chemical spacer to ensure an unaffected interaction of the "warhead", and a reporter tag to visualise or detect the enzyme target. [213, 214] The reporter tag can be either a fluorophore for direct visual-isation by fluorescence analysis after resolving by sodium dodecyl sulphate polyacrylamide gel electrophoresis (SDS-PAGE) or a biotin tag for a pull-down assay approach and identification of the enriched proteins by tandem LC-MS. [215] CRAVATT and co-workers introduced the application of this technique for SHs in the late ΄90s. [216] Since then, the ECS related enzymes have been extensively studied and many ABPs either with a broad or selective labelling spectra of enzymes were developed. [217-223] As this method is usually dependent on the activity of the targeted enzyme, it is the perfect tool to study inhibitory functionalities. In this modified so-called competitive ABPP (Figure 28B), in a first step, the potential inhibitor will be incubated with the biological sample to compete in a second step with the ABP. The higher the inhibitory effect the lower labelling of the target enzyme will occur. Additionally, off-target activities against other undesired enzymes within the biological sample can be studied.

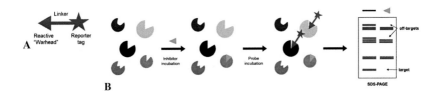

Figure 28: General workflow of competitive activity-based protein profiling (ABPP). (**A**) General structure of an activity-based probe (ABP), containing a reactive group as "warhead" to bind covalently in the active side of the enzyme, coupled to a reporter tag (e.g. a fluorophore for gel-based SDS-PAGE analysis) by a chemical linker. (**B**) General scheme of a competitive ABPP experiment. The investigated proteome is incubated with inhibitor (or vehicle as control) and remaining active enzymes are labelled by ABP incubation. Reporter visualizing by using SDS-PAGE and fluorescence analysis.

Like most of the enzymes in the ECS, MAGL belongs to the superfamily of SHs. For this enzyme class several types of ABPs were developed based on different "warheads" linked to the reporter or tailored covalent inhibitors as recognition element to achieve higher selectivity. In this work, three different ABPs were used (Figure 29). For a broad visualisation of most SHs in the mouse brain proteome, TAMRA-FP (**102**) and MB064 (**103**) were used, whereby the latter labels slightly more selectively. Precise visualisation of MAGL was achieved by using the in-house developed ABP LEI463 (**104**).[224]

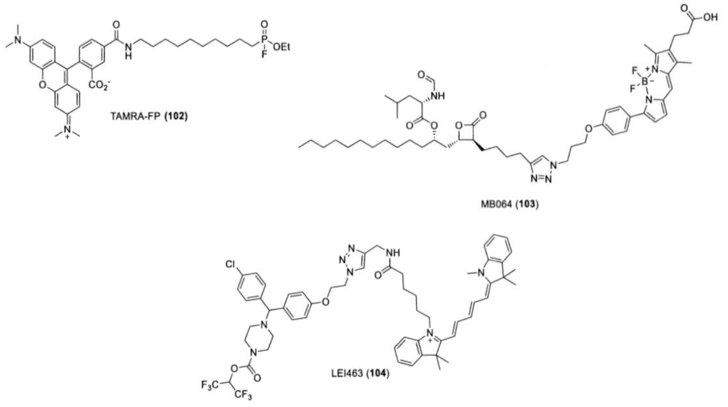

Figure 29: In this work applied ABPs to target serine hydrolases in gel based ABPP. Fluorophosphonate (FP)-based probe TAMRA-FP [216] were used for broad unspecific labelling of serine hydrolases. The tailored, THL-derived, ABP MB064[218] reveals, among other unselective labelling, a more selective labelling against DAGLs. For the selective labelling against MAGL ABP LEI463[224] was used.

As previously described, to investigate the SARs the synthesised MAGL inhibitors were first analysed by the natural substrate assays (Table 19). As a reference control, inhibitor ±**33a** (entry 1) was included as the most active inhibitor at that time. Maintaining most of the core structure elements, while varying the centred fluorine substitution gave already a significant increase of potency, if changed to the bigger chlorine (entry 3). Whereby, higher steric demand at this position slightly decreased the affinity (entry 2). Removing the central amide bond did not result in an expected decrease of affinity (entry 4), but rather a higher affinity in between of previous best inhibitor (entry 1) and improved inhibitor (entry 3) was observed. This suggested, that the central amide element is not crucial for the activity and structural flexibility or rigidity plays a less significant role. Tailoring the ester moiety with more steric residues resulted in a slightly higher (entry 5) to similar potency (entry 6) compared to entry 3. Substitution of the sulfoxide (entry 7) abolished the low nanomolar potency against MAGL, suggesting a significant importance of the sulfoxide.

Table 19: Potencies of final compounds ±33a-g in the natural substrate assay.

Entry	Cpd.	M.W.	CLogP	tPSA (Å^2)	pIC$_{50}$[a]	LipE
1	±33a	480.98	4.36	67	8.08±0.10	3.72
2	±33b	530.99	5.14	67	7.83±0.10	2.69
3	±33c	497.43	4.93	67	8.50±0.10	3.57
4	±33d	483.45	–[b]	50	8.21±0.11	–
5	±33e	511.46	5.23	67	8.57±0.12	3.34
6	±33f	525.49	5.76	67	8.41±0.08	2.65
7	±33g	477.38	5.19	67	5.94±0.08	0.75

[a] Data expressed as mean ± SD of at least two individual experiments in duplicates; [b] CLogP not applicable.

Next, the obtained potencies should be confirmed by dose-response competitive ABPP experiments. However, beforehand it must be mentioned, that due to the fluorescent analysis and normalisation procedure in this method, in general lower potencies and lower accuracies are to be expected. As inhibitor ±33b already showed slightly lower potencies in the NSA, no ABPP experiments were conducted. Proceeding instead with inhibitor ±33c together with ±33a as reference (Figure 30).

A

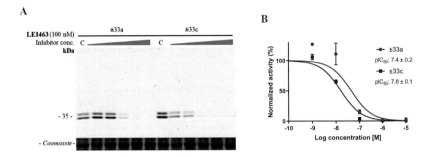

Figure 30: Dose response inhibition profile of inhibitor ±**33c** by competitive ABPP of mouse brain membrane proteome. (**A**) Mouse proteome was incubated with different concentrations of inhibitors (1 nM – 10 μM) or DMSO as vehicle (30 min, rt). Samples were labelled with ABP LEI463 (10 min, rt). Coomassie staining was used as protein loading control. Inhibitor ±**33a**, showing the highest activity until then, was used as reference. (**B**) Dose-response curves of ±**33a** and ±**33c** from measured gel-based ABPP (n=3 and pIC_{50} ± SD).

The higher potency of ±**33c** was already noticeable by optical comparison in the fluorescence measurement (Figure 30A). Quantification after normalisation against the protein loading control and plotting into a graph resulted in the dose-response curve shown in Figure 30B, where a significant potency improvement (pIC_{50}: 7.4 ± 0.2 vs. 7.8 ± 0.1) can be observed. To further prove the selectivity, another competitive ABPP experiment was conducted applying the broad-spectra ABPs MB064 and TAMRA-FP (Figure 31).

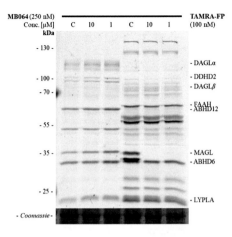

Figure 31: Selectivity profile of ±**33c** by competitive ABPP of mouse brain membrane proteome. Mouse proteome was incubated with inhibitor or DMSO as vehicle (30 min, rt). Samples were labelled (10 min, rt) with ABP MB064 (left) or TAMRA-FP (right). Coomassie staining was used as protein loading control. Important EC related enzymes are highlighted on the right site.

Neither with MB064 nor TAMRA-FP, at both inhibitor concentrations (10 µM and 1 µM), any off-target activity was detected, showing the high selectivity of the developed inhibitor ±**33c**.

Next in sequence to test was the inhibitor ±**33d** containing the reduced amine bond as central connection (Figure 32). In contrast to the halogen exchanged inhibitors, reducing the central amide connection had a drastically effect on the selectivity of the inhibitor and resulting in an off-target activity at higher concentrations (10 µM and 1 µM). The off-target enzyme was identified to be FAAH, as shown in the red boxes (Figure 32A). Not only resulting in a dual inhibitor, no improvement of potency compared to ±**33a** was determined.

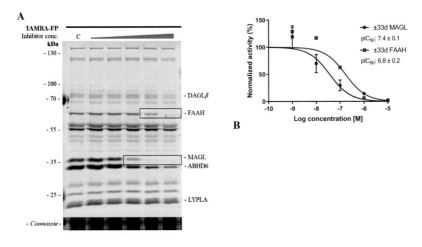

Figure 32: Dose-response and selectivity profile of ±**33d** by competitive ABPP of mouse brain membrane prote-ome. (**A**) Mouse proteome was incubated with inhibitor (1 nM – 10 µM) or DMSO as vehicle (30 min, rt). Samples were labelled with ABP TAMRA-FP (100 nM) (10 min, rt). Coomassie staining was used as protein loading con-trol. Red boxed denoted analysed bands. Important EC related enzymes are highlighted on the right site. (**B**) Dose-response curves of inhibitor ±**33d** from measured gel-based ABPP (n=3 and pIC_{50} ± SD).

Subsequent, inhibitor ±**33e** and ±**33f** were tested for their affinities (Figure 33) and selectivities (Figure 34).

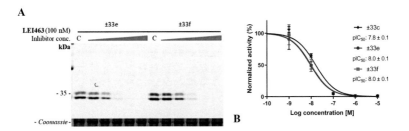

Figure 33: Dose response inhibition profile of inhibitor ±**33e** and ±**33f** by competitive ABPP of mouse brain membrane proteome. (**A**) Membrane proteome was incubated with different concentrations of respective inhibitor (1 nM – 10 µM) or DMSO as vehicle (30 min, rt). Samples were labelled with ABP LEI463 (10 min, rt). Coomassie staining was used as protein loading control. (**B**) Dose-response curves of ±**33e** and ±**33f** from measured gel-based ABPP (n=3 and pIC_{50} ± SD). Results from inhibitor ±**33c** were included as reference.

Modifications of the ester moiety resulted in a slightly higher affinity compared to ±**33c**. In contrast to the NSA results, in the competitive ABPP experiment, both inhibitors showed the same affinities, if compared to each other. No influence on the selectivity of the bulkier esters was observed, as shown by the selectivity profile in Figure 34.

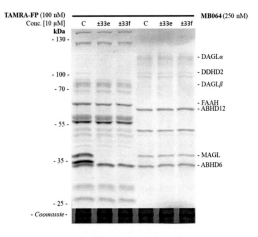

Figure 34: Selectivity profile of ±**33e** and ±**33f** by competitive ABPP of mouse brain membrane proteome. Mouse proteome was incubated with respective inhibitor (10 µM) or DMSO as vehicle (30 min, rt). Samples were labelled (10 min, rt) with ABP TAMRA-FP (left) or MB064 (right). Coomassie staining was used as protein loading control. Important EC related enzymes are highlighted on the right site.

Next, the ability of the inhibitors binding to MAGL in a reversible mode was the focus of interests. To prove the reversibility, a time-dependent competitive ABPP experiment with prolonged incubation times for the ABP was planned. Thereby, to guarantee a sufficient competition higher concentration of inhibitor (used conc. >IC_{80}) and ABP (1 µM = 10 × standard conc.) were applied, resulting in a full displacement of the inhibitor after 120 min. As inhibitors, ±**33c** and ±**33d** were chosen and the covalent inhibitor ABX was used as negative control (Figure 35).

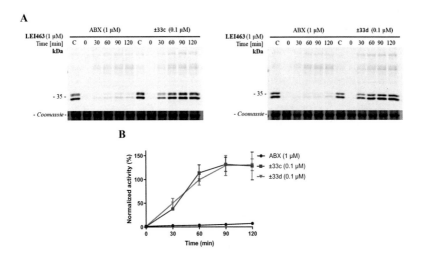

Figure 35: Reversibility profile of selected inhibitor ±**33c** and ±**33d**. (A) Gel-based kinetic reversibility experiment by competitive ABPP on mouse brain membrane proteome. Mouse brain proteome was incubated (30 min, rt) with inhibitors (100 nM) or DMSO (control) and subsequently labelled with ABP LEI463 (1 µM) at different time points (0, 30, 60, 90, 120 min, rt). Coomassie staining was used as protein loading control. The covalent inhibitor ABX (1 µM) was used as negative control. (B) Corresponding curves of ±**33c** and ±**33d** from measured gel based kinetic ABPP (n=3 and mean ± SD). Results from covalent inhibitor ABX were included as negative control.

Compared to the covalent inhibitor ABX a displacement of the inhibitors by the ABP was observed after ~30 min (Figure 35A). Data analysis determined a $t_{1/2}$ of 36 min (±**33c**) and 30 min (±**33d**), showing an excellent reversibility profile in the binding mode of the MAGL inhibitors (Figure 35B).

Lastly, screening of a broader selectivity profile, to prove no off-target effects on other ECS unrelated enzymes came into the research focus. As most EC enzymes are connected to or incorporated into the cell membrane, the first additional proteome to consider was the cytosolic fraction of the same tissue cells. Therefore, another competitive ABPP experiment was conducted using the cytosolic proteome fraction of the same mouse brain tissue (Figure 36). To assure the appearance of possible off-targets, high concentrations of inhibitor (10 µM) were applied and the broad spectra ABPs MB064 and TAMRA-FP were used for labelling. Despite the high inhibitor concentration, no off-target activity in the cytosolic mouse brain proteome was identified.

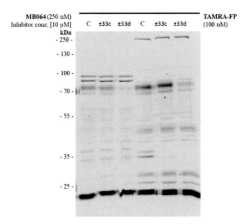

Figure 36: Off-target screening for ±**33c** and ±**33d** in cytosolic mouse brain proteome by competitive ABPP. Mouse cytosol proteome was incubated with respective inhibitor (10 μM) or DMSO as vehicle (30 min, rt). Samples were labelled (10 min, rt) with ABP MB064 (left) or TAMRA-FP (right).

Discussion

Despite the limited number of tested inhibitors, important SARs were identified, which allow to conclude several beneficial effects or less favoured substitutions. Including more 3D space demanding groups at either the centre or side chain of the molecule had a positive effect on potency and no influence on selectivity. Also confirming the initial hypothesis of remaining space inside the active site as seen in the visual docking experiment. However, limitations in the tolerated expanding space were shown by further structural increase at the molecule centre. The connecting amide moiety was identified to have a significant influence on selectivity, suggested by main interaction to the catalytic SER122. Removing this group drastically reduced selectivity and an off-target activity at higher concentrations against FAAH was observed. Higher structural flexibility, resulting in a simplified induced binding in the active site of FAAH could be the reason therefore. Moreover, strong binding interactions with the catalytic SER122 was not possible anymore and led to lower potencies. Further structural simplification by removing the sulfoxide moiety resulted in a drastically decrease of potency, suggesting an important influence of this moiety towards binding stabilisation in the active site. This might be mediated by the tetragonal expanding of the sulfoxide compared to a planar structure of a carbonyl group and, furthermore, interactions through the lone pair could additionally enhance the binding affinities.

4 Conclusion and future prospects

In the presented work new approaches and evaluations of modified coumarins as promising structure motif for the drug development of novel synthetic cannabinoids were investigated. Successful targeting and modulation of the ECS bears a great chance in treatment of multiple diseases associated with pain, neurological disorders or cancer. However, achieving balance between high potency, selectivity and pharmacokinetics are still the main challenges to overcome in this field.

The synthetic afford of this work started with the development of feasible synthetic routes towards substituted salicylic aldehydes, which serve as starting building blocks in the coumarin synthesis based on the methods from BRÄSE *et al.*[108] and THOMAS HURRLE[107]. Thereafter, in total 170 substituted coumarins, based on four different lead structures were synthesised (Figure 11).

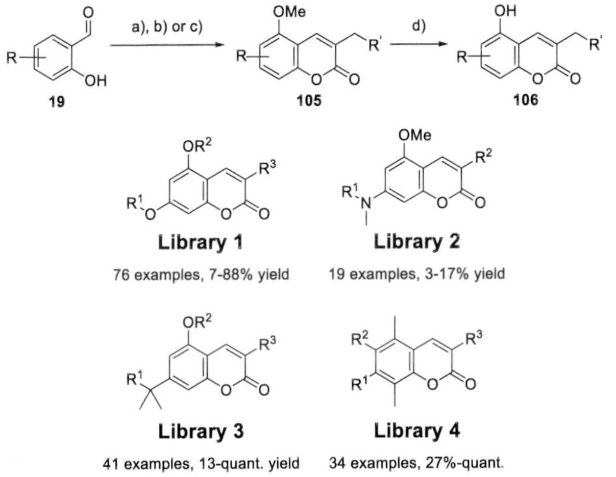

Figure 37: Lead structures **29-32** of the developed second-generation coumarin-based synthetic cannabinoids. Conditions: *a) cinnamaldehyde, IL **22**, K₂CO₃, toluene, 110 °C, 50 min, MW; b) cinnamaldehyde, IL **22**, KOAc, HOAc, 3Å MS, toluene, 110 °C, 75 min, MW; c) acid anhydride, K₂CO₃, 180 °C, 65 min, MW; d) BBr₃, DCM, –78 °C to rt, 1 h to o.n.*

During a stay abroad, in the laboratories of Prof. LAURA HEITMAN[7], the second generation of coumarin ligands were tested in comprehensive *structure-activity-relationship* (SAR) studies

---

[7] Prof. Laura Heitman, Leiden Academic Centre for Drug Discovery, Leiden University, the Netherlands.

for their cannabimimetic activity. Therefore, several representatives of different modification classes were selected for the investigations in radioligand binding assays (Figure 38).

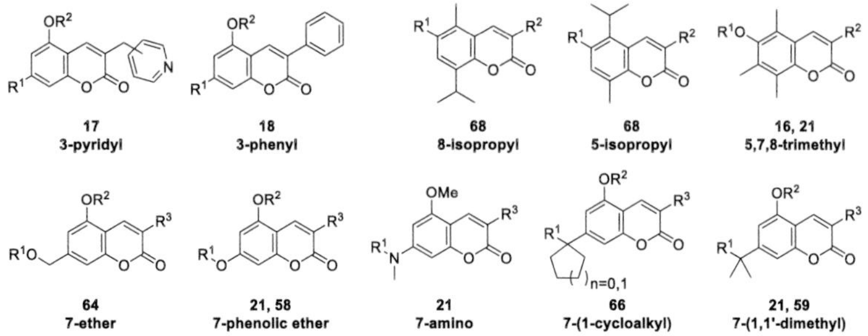

Figure 38: Lead structures of coumarin ligands selected for the new *SAR* studies.

In total, around 200 different compounds were tested within this SAR study and several structure-activity relationships were concluded (Figure 39). Additionally, to strengthen the proposed relationships, ligand-based docking experiments were performed as well. For strong binding affinities a high LipE value was determined to be necessary. This was achieved by big aliphatic or aromatic groups attached to the coumarin core structure. In previous studies a long alkyl chain at position 7 was determined to be crucial for the activity.[100] In the presented study these requirements were also observed, but furthermore new requirements to improve the binding affinities were identified (Figure 39). In general, an alkyl chain at position 7 was essential for any affinity and by introduction of bulkier lipophilic groups higher potencies were observed. Heteroatoms attached to the coumarin core as proposed for lead structure 1 and 2 (Figure 11) were not tolerated. On the other side of the coumarin (3-position) (hetero-)aromatic groups as well as aliphatic chains showed high affinities. Structural flexibility at this position by a methylene spacer group between the core and aromatic groups was essential for any activity. In contrast to previous observations, only small substituents preferably of apolar nature were well tolerated by both receptor subtypes. Similar observations were made for different polarities at position 5, whereby a free hydroxyl group was more tolerated by CB_2R, providing by deprotection a strong possibility to control selectivity towards CB_2.

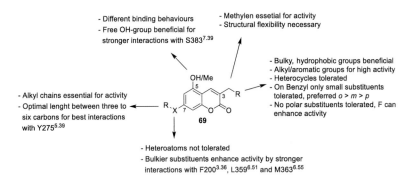

Figure 39: Derived structure-activity relationships based on the results of the conducted SAR studies.

Furthermore, the most potent ligands were tested in a functional $[^{35}S]GTP\gamma S$ assay for their pharmacological efficacy after receptor binding. All of the tested coumarin ligands showed an agonistic effect in relation to the basal activity. In comparison to the full agonist CP55,940, four of the ligands showed a full agonistic efficacy. A partial agonistic efficacy was observed by the remaining eight ligands. In summary, 20 of the tested coumarin ligands showed potencies in a medium to low nanomolar range. The majority (13 compounds) were only active against the CB_2 receptor with varying selectivities, five coumarins showed an activity against both receptors and only two were identified to be CB_1 selective.

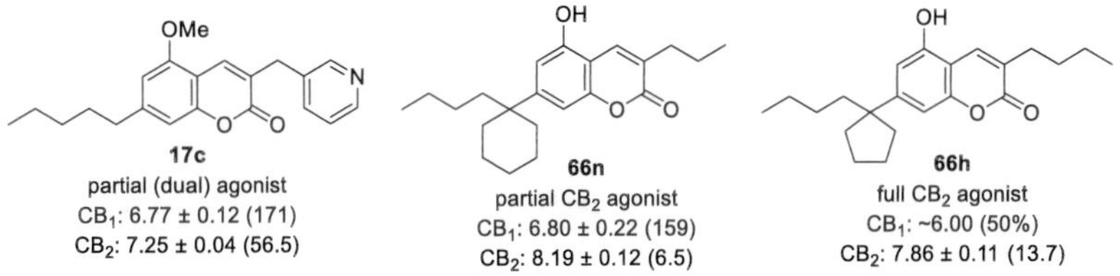

17c
partial (dual) agonist
CB_1: 6.77 ± 0.12 (171)
CB_2: 7.25 ± 0.04 (56.5)

66n
partial CB_2 agonist
CB_1: 6.80 ± 0.22 (159)
CB_2: 8.19 ± 0.12 (6.5)

66h
full CB_2 agonist
CB_1: ~6.00 (50%)
CB_2: 7.86 ± 0.11 (13.7)

Figure 40: Depicted coumarins **17c**, **66h** and **66n** with highest observed potencies in low nanomolar ranges.

In future work based on these results several structural as well as pharmacological suggestions might be considered. The beneficial effects of bulky lipophilic groups, as shown by the SAR and docking studies can be utilised to further increase the affinity. A bicyclopentane moiety for example as bioisosteric group for the benzyl group should be beneficial for the affinity by higher steric filling of the upper binding pocket and stronger binding interactions. Alternatively, introduction of chirality by branched alkyl chains at the 3-position might also be an interesting modification to further investigate.

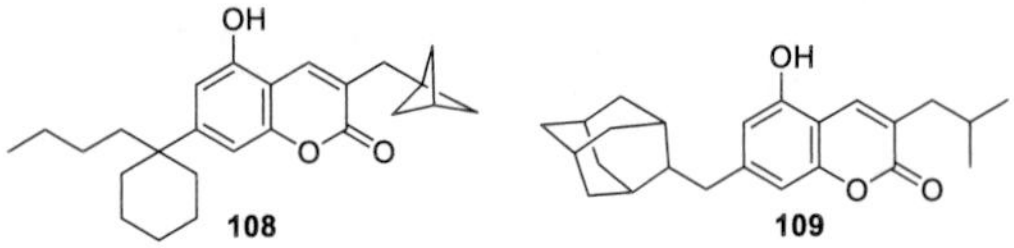

Figure 41: Proposed structures of new coumarin ligands with bulky or chiral substituents.

Furthermore, with the most potent ligands additional pharmacologic assays like cAMP accumulation or β-arrestin recruitment assays should be performed to determine the triggered signal transduction cascade after receptor binding. Additional investigations to off-target activities, for example against p-ERK 1/2 or hERG, should also be considered as next steps to proof. Lastly, the synthesised coumarins with small substituents must be tested towards their pharmacological affinity and efficacy against GPR55 to show their potential as new neuroprotective drug candidates.

During the stay abroad in the laboratories of Prof. Mario van der Stelt[8] in a joint project with MING JIANG, several optimisations and new reversible MAGL inhibitors were developed, synthesised and pharmacologically evaluated. In total six different inhibitors with four different changes of substitution moieties were synthesised.

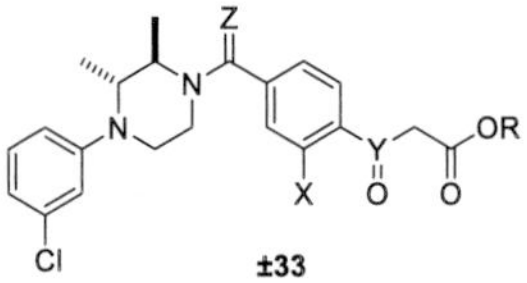

Figure 42: Generic structure of synthesised reversible MAGL inhibitors. Changed substitution moieties are indicated by different letters. X = halogen exchange, Y = sulfoxide-ketone exchange, Z = amine reduction, R = ester substituent exchange.

Pharmacologic evaluation by natural substrate assays and ABPP revealed different correlations between structure and activity. First, exchange of the halogen resulted in a significant higher potency, if changed to a chlorine. Further steric expanding by a trifluoromethyl group strongly decreased the affinity towards MAGL. Second, structural rigidity seems to be very important for the selectivity, as a dual inhibitor with off-target activity against FAAH was obtained after reduction of the central amide group to an amine. Furthermore, the sulfoxide functionality was identified to be essential for higher activities as well. On the other hand, only slight differences

---

[8] Prof. Mario van der Stelt, Leiden Institute of Chemistry, Leiden University, the Netherlands.

of potency were observed after introduction of other ester substituents. Overall by introduction of the chlorine substituent the highly selective and reversible MAGL inhibitor ±**33c** was synthesised with the highest observed potency at that time.

In ongoing research on the project by MING JIANG, further optimisations on pharmacokinetic and –dynamic properties were performed and ultimately an inhibitor candidate for *in vivo* investigations was developed (Figure 43).

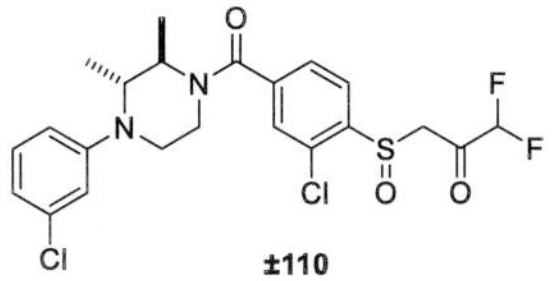

Figure 43: Representing reversible MAGL inhibitor for further *in vivo* screenings.

The optimised inhibitor showed a sufficient hepatic metabolic stability and even further improved potency towards a subnanomolar range. Comprehensive *in vivo* studies to evaluate the drug candidate potential of the developed unique reversible MAGL inhibitors are already in preparation.

5   Experimental part

5.1   Synthetic work on coumarin-based cannabinoids

5.1.1   General remarks chemistry

Air or moisture sensitive reactions were carried out in oven dried glass devices, sealed with rubber seals, under an Argon atmosphere according to SCHLENK-techniques. Liquids were added *via* plastic syringes and V2A-needles, volumes at small µl scale were added with full glass syringes from HAMILTON or if not easy applicable exactly weighed. Solids were added in pulverized form. Reactions at 0 °C were cooled with a mixture of ice/water. Reactions at deeper temperatures were tempered with brine/ice mixture (–10 °C) or isopropanol/dry ice mixture (–78 °C). All reactions were monitored either by thin layer chromatography (TLC), LC-MS or GC-MS. Solvents were removed at 40 °C with a rotary evaporator. Used solvent mixtures were measured volumetrically. If not stated otherwise, solutions of inorganic salts are saturated aqueous solutions. If not otherwise specified, the crude products were purified by flash column chromatography following the concepts of STILL *et al.*[225] using silica gel (SIGMA-ALDRICH or SCREENING DEVICES B.V., pore size 60 Å, mesh 40 – 63 µm) and sand (calcined and purified with hydrochloric acid) as stationary phase. Solvents were distilled prior use or p.a. grade solvents were used. Solvent mixtures were prepared individually, volume ratios are given as volumetric. The use of a gradient is indicated in the experimental procedures. If not stated otherwise, purities of the synthesized compounds were determined as ≥95% and if lower calculated from ^{1}H NMR.

5.1.2   Analytics and equipment

Nuclear Magnetic Resonance (NMR)

NMR spectra have been recorded on following spectrometer:

	^{1}H MHz	^{13}C MHz	^{19}F MHz
BRUKER Avance 300	300	75	
BRUKER Avance 400	400	101	376
BRUKER Avance 500 DRX 500	500	126	471

If not stated otherwise, all spectra were obtained at room temperature. As solvents, products obtained from EURISOTOP were used: chloroform-d_1, acetone-d_6 and dimethylsulfoxide-d_6. The chemical shift δ is expressed in parts per million (ppm) where the residual signal of the solvent has been used as secondary reference: chloroform (δ = 7.26 ppm) acetone (δ = 2.05 ppm) and dimethyl sulfoxide (δ = 2.50 ppm) in ^{1}H spectra and chloroform (δ = 77.0 ppm), acetone (δ = 30.8 ppm) and dimethyl sulfoxide (δ = 39.4 ppm) in ^{13}C spectra. The spectra were analysed according to first order. For central symmetrical signals the midpoint is given, for multiplets the range of the signal region is given. The multiplicities of the signals were abbreviated as follows: s = singlet, d = doublet, t = triplet, q = quartet, quin. = quintet, sex. = sextet, hept. = heptet, bs = broad singlet, m = multiplet and combinations thereof. All coupling constants J are stated as modulus in Hertz [Hz] and the number of bonds is given as superscripted index. The assignment ensued for the ^{1}H NMR directly *via* the chemical shifts and for the ^{13}C NMR through the DEPT-technique (DEPT = Distortionless Enhancement by Polarization Transfer) and is stated as follows: DEPT: "+" = primary or tertiary carbon atoms (positive DEPT-signal), "–" = secondary (negative DEPT-signal), C_q = quaternary carbon atoms (no DEPT-signal). Carbon assignments based on a CF coupling are always given in relation to the fluorine coupled carbon and literature[226, 227]. If necessary, for the assignment different 2D NMR techniques (HSQC = Heteronuclear Single Quantum Coherence or HMBC = Heteronuclear Multiple Bond Correlation) were used. For ^{19}F NMR only internal reference by the spectrometer was used. Analysis of the NMR raw data was done with MESTRENOVA (Version 9.0).

Mass Spectrometry (EI-MS, FAB-MS, ESI-MS)

EI- and FAB mass spectra were recorded on a FINNIGAN MAT 95, ESI-mass spectra were recorded on a THERMO SCIENTIFIC Q EXACTIVE PLUS with a THERMO SCIENTIFIC LT ORBITRAP XL. Ionization was achieved through either EI (Electron Ionization), FAB (Fast Atom Bombardment) or ESI (Electron Spray Ionization). Notation of molecular fragments is given as mass to charge ratio (*m/z*); the intensities of the signals are noted in percent relative to the base signal (100%). As abbreviation for the ionized molecule [M]$^+$ (for all EI and FAB spectra) respectively [M+H]$^+$ (for all ESI spectra) was used. Characteristic fragmentation peaks are given as [M–fragment]$^+$ and [fragment]$^+$. For HRMS (High Resolution Mass Spectrometry) following abbreviations were used: calc. = expected value (calculated); found = value found in analysis.

Gas Chromatography Mass Spectrometry (GC-MS)

GC-MS measurements have been recorded with an AGILENT TECHNOLOGIES model 6890N (electron impact ionization), equipped with an AGILENT 19091S-433 column (5% phenyl methyl siloxane, 30 m, 0.25 μm) and a 5975B VL MSD detector with turbo pump. As a carrier gas helium was used.

Preparative High-Performance Liquid Chromatography (prepHPLC/MS)

For purification by preparative high-performance liquid chromatography a JASCO LC-NetII / ADC HPLC system, equipped with two PU-2087 Plus pumps, a CO-2060 Plus thermostat, a MD-2010 Plus diode array detector (195 – 650 nm) and a CHF-122SC fraction collector from ADVANTEC was used. The stationary phase was a preparative C18 column (VDSpher C18-M-SE, C18, 10 μm, 250 × 20 mm from VDS OPTILAB) at a temperature of 25 °C and a flow rate of 15 mL/min, with ACN, bidistilled water (ddH$_2$O) and 0.1% TFA as mobile phase. The fractions were detected at λ = 205 nm, 330 nm, or 380 nm. Following linear gradients were applied: a) 5% – 95% ACN in ddH$_2$O + 0.1% TFA, 40 min, detection at λ = 330 nm (5–95_STD); b) 5% – 95% ACN in ddH$_2$O + 0.1% TFA, 40 min, λ = 330 nm (5–95_MOD); c) 5% – 95% ACN in ddH$_2$O + 0.1% TFA, 40 min, λ = 218 nm (5–95_amino); and d) 50% – 95%% ACN in ddH$_2$O + 0.1% TFA, 32 min, λ = 330 nm (50–95_alkyl).

Infrared Spectroscopy (IR)

IR-spectra were recorded on a BRUKER *Alpha P*. Measurement of the samples was conducted *via* Attenuated Total Reflection (ATR). Position of the absorption bands is given as wavenumber $\tilde{v}$ with the unit cm^{-1}.

Microwave (MW)

All microwave assisted reactions were carried out in a single mode CME Discover LabMate microwave, operated with CEM´s Synergy^{TM} software with pressurized closed microwave *vi-a*ls. The microwave irradiation was produced by a power source with an energy per time between 0 – 300 W and an optimized focus through reaction chamber. The irradiation power was adjusted by temperature, which was monitored by an infrared sensor or a maximum of 230 W.

Thin Layer Chromatography (TLC)

Analytical and preparative thin layer chromatography (TLC, prepTLC) were done on TLC plates purchased from MERK (silica gel 60 on aluminium plate, fluorescence indicator F254, layer thickness analytical: 0.25 mm, preparative: 1.00 mm). Detection was carried out under UV-light at $\lambda = 254$ nm and $\lambda = 366$ nm. Alternatively, the TLC plates were stained by dipping solutions of SEEBACH (2.5% phosphor molybdic acid, 1.0% cer(IV)sulphate-tetrahydrate, 6.0% conc. H_2SO_4, 90.5% H_2O), potassium permanganate (1.50 g $KMnO_4$, 10.0 g K_2CO_3 and 1.25 mL 10% NaOH in 200 mL H_2O), dinitrophenylhydrazine (DNP, 1.20 g 2,4-dinitrophenylhydrazine, 6.0 mL conc. H_2SO_4, 8.0 mL H_2O in 20 mL EtOH) and dried in a hot air stream. PrepTLC fractions were scratched of with a spatula, extracted and filtrated via vacuum filtration.

Analytical Scale

As analytical scales a SARTORIUS BASIC BA310 ($\Delta = 1$ mg) and a RADWAG AS220.X2 ($\Delta = 0.1$ mg) were used.

Solvents and Reagents

Solvents of technical quality were purified by distillation prior to use. Solvents of the grade p.a. (*pro analysi*) were purchased (ACROS, FISHER SCIENTIFIC, SIGMA-ALDRICH, ROTH, RIEDEL–DE HAËN, HONEYWELL) and used without further purification. Absolute solvents were dried either by using the methods listed in Table 20, by the solvent purification system MBRAUN SPS 800 or were purchased from a commercial supplier (abs. ACN (ACROS, < 0.005% water), abs. chloroform (FISCHER, over molecular sieves), abs methanol (FISCHER, < 0.005% water). All absolute solvents were stored under argon afterwards.

Table 20: Methods for absolutizing of solvents. All distillations were carried out under argon atmosphere.

Solvent	Method
Dichloromethane:	heating to reflux over calcium hydride, distilled over a packed column
Ethanol:	drying several times over activated 3 Å molecular sieve
Tetrahydrofuran:	heating to reflux over sodium metal (benzophenone as an indicator), distilled over a packed column
Toluene:	heating to reflux over sodium metal (benzophenone as an indicator) distilled over a packed column

Reagents have been purchased from ABCR, ACROS/FISCHER, ALFA AESAR, FISHER SCIENTIFIC, MERCK, SIGMA-ALDRICH, TCI, and VWR and, if not stated otherwise, were used without further purification.

5.1.3   General procedures

General Procedure: WILLIAMSON ether synthesis (**GP1**)

A *vial* was charged with the respective phenol (1.00 equiv.), K_2CO_3 (1.50 equiv.), 18-crown-6 (2 mol%) and acetone (2.00 mL/mmol phenol) and stirred for 15 min at rt. Then the respective alkyl bromide (1.50 equiv.) was added and the mixture was stirred under reflux overnight. The reaction was quenched by addition of H_2O and extracted with EtOAc or Et_2O (min 3 ×). The combined organic layers were dried over Na_2SO_4, vacuum filtrated over a frit, and the volatiles were removed under reduced pressure. The crude product was purified by flash chromatography.

General Procedure: Formylation of 1,3-dimethoxybenzenes (**GP2**)

To a solution of the respective 1,3-dimethoxybenzene (1.00 equiv.) in abs. Et_2O or abs. THF (10.0 mL/g phenol), at rt TMEDA (1.50 equiv.) was added dropwise. The mixture was cooled to 0 °C and followed by addition of *n*-BuLi (2.5 M in *n*-hexane, 1.50 equiv.). After stirring for 2 h at rt the solution was cooled to 0 °C, DMF (3.00 equiv.) was added and the mixture was stirred for another 4 h at rt. Then the reaction was quenched by the addition of brine and extracted with Et_2O (min 3 ×). The combined organic layers were dried over Na_2SO_4, vacuum filtrated over a frit, and the volatiles were removed under reduced pressure. The crude product was purified by flash column chromatography.

General Procedure: Selective *mono*-deprotection of *o*-methoxy benzaldehydes (**GP3**)

To a solution of the respective *o*-methoxy aldehyde (1.00 equiv.), dissolved in a mixture of abs. ACN and abs. DCM (2:1, 7.60 mL/mmol aldehyde) at 0 °C, $AlCl_3$ (2.00 – 2.50 equiv.) and NaI (2.00 – 2.50 equiv.) were added in portions. The reaction was stirred for 1 h or until TLC showed complete conversion and quenched with water. Followed by extraction with DCM (min 3 ×), the combined organic layers were dried over Na_2SO_4, vacuum filtrated over a frit, and the volatiles were removed under reduced pressure. The crude product was then purified by flash chromatography.

General Procedure: Alkylation of benzoic acids (**GP4**)

Under inert atmosphere, a round-bottom flask was charged with 3,5-dimethoxybenzoic acid (1.00 equiv.) in abs. THF (4.00 mL/mmol benzoic acid) and cooled to –78 °C. At this temperature the respective alkyl lithium (2.10 equiv.) was added dropwise. After the addition the reaction was allowed to warm to 0 °C and stirred for 2 h. Subsequently the reaction was quenched with 1 M aq. HCl, extracted with Et_2O (min 3 ×), the combined organic layers were dried over Na_2SO_4 and vacuum filtrated over a frit. After removing of the volatiles under reduced pressure, the crude product was purified by flash column chromatography.

General Procedure: GRIGNARD reaction towards tertiary aromatic alcohols (**GP5**)

A flame dried round-bottom flask was charged under inert atmosphere with the respective ketone (1.00 equiv.) in abs. THF (0.25 M), cooled to 0 °C and followed by the addition of MeMgBr (3 M in Et_2O, 1.20 – 2.00 equiv.). Then the reaction mixture was slowly warmed up to rt and stirred for 2 h. By addition of 2 M aq. HCl the reaction was quenched, extracted with Et_2O (min 3 ×) and the combined organic layers were dried over Na_2SO_4 and vacuum filtrated over a frit. After removing the volatiles under reduced pressure, the crude product was purified by flash column chromatography.

General Procedure: Geminal alkylation of tertiary aromatic alcohols (**GP6**)

Under inert atmosphere, $SOCl_2$ (2.00 equiv.) was added to a solution of the respective tertiary alcohol (1.00 equiv.) in dry DCM (5 × volume of $SOCl_2$) at 0 °C and stirred for 2 h. Then the volatiles were removed under reduced pressure and the residue was taken up in abs. DCM (3.33 ml/mmol alcohol) again. The reaction mixture was cooled to –78 °C, $AlMe_3$ (2 M in toluene, 2.00 equiv.) was added slowly, and stirred for 3 h at this temperature. While further stirring overnight the reaction mixture was allowed to warm up to rt and then quenched at 0 °C by careful addition of 1 M aq. HCl. After extraction with DCM (min 3 ×), the combined organic layers were dried over Na_2SO_4, vacuum filtrated over a frit, and the volatiles were removed under reduced pressure. The crude product was then purified by flash column chromatography.

General Procedure: Synthesis of cinnamaldehydes (**GP7**)

Under an inert atmosphere, a flask or *vial* was charged with the respective aryl iodine (1.00 equiv.), acrolein (1.50 equiv.), $NaHCO_3$ (2.50 equiv.), TBAI (1.00 equiv.) and $Pd(OAc)_2$ (10 mol%) in abs. DMF (4.00 mL/mmol aryl iodine) and stirred for 24 h at 60 °C. After cooling to rt, H_2O was added, followed by extraction with EtOAc (min 3 ×). The combined organic layers were dried over Na_2SO_4, vacuum filtrated over a frit, and the volatiles were removed under reduced pressure. The crude product was purified by flash chromatography.

General Procedure: Synthesis 3-benzylcoumarins (**GP8**)

Under argon atmosphere, a microwave *vial* was charged with the respective salicylic aldehyde (1.00 equiv.), the respective cinnamaldehyde (2.50 equiv.), K_2CO_3 (1.20 equiv.) and 1,3-dimethylimidazolium dimethyl phosphate (1.20 – 1.50 equiv.) and suspended in abs. toluene (3.30 mL/mmol salicylic aldehyde). The reaction mixture was then heated at 110 °C for 50 min at max. 230 Watt microwave irradiation and a maximum pressure of 7 bars. After cooling to rt, H_2O was added, followed by extraction with EtOAc (min 3 ×). The combined organic layers were dried over Na_2SO_4, vacuum filtrated over a frit, and the volatiles were removed under reduced pressure. The crude product was purified by flash chromatography or HPLC.

General Procedure: Demethylation of coumarins (**GP9**)

Under argon atmosphere, the respective methoxy coumarin was dissolved in abs. DCM (20.0 mL/mmol coumarin). The mixture was cooled to –78 °C and BBr_3 (1 M in DCM, 5.00 equiv./methoxy group) was added dropwise. The mixture was stirred for 30 min at this temperature, then allowed to warm to rt and further stirred overnight. The reaction was quenched by addition of $NaHCO_3$ solution, extracted with DCM (min 3 ×) and the combined organic layers were dried over Na_2SO_4, and vacuum filtrated over a frit. After removal of the volatiles under reduced pressure, the crude product was purified by flash column chromatography.

General Procedure: Synthesis 7-amino-3-benzylcoumarins (**GP10**)

Under argon atmosphere, a microwave *vial* was charged with the respective salicylic aldehyde (1.00 equiv.), the respective cinnamaldehyde (2.50 equiv.) KOAc (2.50 – 3.50 equiv.), catalytic amount of glacial acetic acid (0.25 equiv.), 1,3-dimethylimidazolium dimethyl phosphate (1.20 equiv.) and molecular sieve (3 Å powder), and suspended in abs. toluene (3.30 mL/mmol salicylic aldehyde). The reaction mixture was then heated at 110 °C for 75 min at max. 230 Watt microwave irradiation and a maximum pressure of 7 bars. After cooling to rt, H_2O was added, followed by extraction with EtOAc (min 3 ×). The combined organic layers were dried over Na_2SO_4, vacuum filtrated over a frit, and the volatiles were removed under reduced pressure. The obtained crude product was filtered over a small pad of silica (*c*Hex/EtOAc 1:1) and concentrated again under reduced pressure. The filtrated product was then purified by prepHPLC (5–95_amino).

General Procedure: PERKIN reactions (**GP11**)

A microwave *vial* was charged with the respective salicylic aldehyde (1.00 equiv.) and K_2CO_3 (0.05 equiv.) and suspended in the respective carbonic acid anhydride (3.50 equiv.). The reaction mixture was then heated at 180 °C for 65 min at max. 230 Watt microwave irradiation and a maximum pressure of 7 bars. Then after cooling to rt, by addition of aq. NaHCO$_3$, the reaction was quenched and the pH value was adjusted to 7, followed by extraction with EtOAc (min 3 ×). The combined organic layers were dried over Na$_2$SO$_4$, vacuum filtrated over a frit, and the volatiles were removed under reduced pressure. The crude product was purified by flash chromatography.

5.1.4   Synthesis and characterisation of salicylic aldehydes

1-Butoxy-3,5-dimethoxybenzene (35)[142]

According to **GP1**, to a solution of 3,5-dimethoxyphenol (733 mg, 4.76 mmol, 1.00 equiv.), K$_2$CO$_3$ (986 mg, 7.13 mmol, 1.50 equiv.) and 18-crown-6 (25.1 mg, 0.10 mmol, 0.02 equiv.) in acetone (2.38 ml), 1-bromobutane (0.61 ml, 782 mg, 5.71 mmol, 1.20 equiv.) was added and the reaction was stirred under reflux overnight. Filtration over a small silica plug (*n*Hex/EtOAc 30:1) gave the product as a pale-yellow oil (2.10 g, 9.99 mmol, 70%).

^{1}H NMR (400 MHz, CDCl$_3$) δ = 6.08 (s, 3H, 3 × C*H*$_{ar}$), 3.92 (t, ^3J = 6.5 Hz, 2H, OC*H*$_2$), 3.77 (s, 6H, 2 × OC*H*$_3$), 1.83 – 1.68 (m, 2H, OCH$_2$C*H*$_2$), 1.55 – 1.42 (m, 2H, C*H*$_2$CH$_3$), 0.97 (t, ^3J = 7.4 Hz, 3H, C*H*$_3$) ppm. – ^{13}**C NMR** (101 MHz, CDCl$_3$) δ = 161.6 (C$_q$, 2 × C$_{ar}$OCH$_3$), 161.2 (C$_q$, C$_{ar}$OCH$_2$), 93.5 (+, 2 × C$_{ar}$H), 93.0 (+, C$_{ar}$H), 67.9 (–, OCH$_2$), 55.5 (+, 2 × OCH$_3$), 31.42 (–, OCH$_2$CH$_2$), 19.4 (–, CH$_2$CH$_3$), 14.0 (+, CH$_3$) ppm. – **IR** (ATR, ṽ) = 2955, 2934, 2870, 2839, 1593, 1458, 1427, 1389, 1202, 1192, 1148, 1063, 944, 927, 816, 681, 538, 408 cm^{-1}. – **MS** (70 eV, EI, 20 °C) *m/z* (%) = 210 (46) [M]$^+$, 155 (10), 154 (100), 126 (33), 125 (44), 111 (10). – **HRMS** (EI, C$_{12}$H$_{18}$O$_3$) = calc.: 210.1250; found: 210.1251.

2,6-Dimethoxy-4-butoxybenzaldehyde (36d), 2-butoxy-4,6-dimethoxybenzaldehyde (37)

A) According to **GP2**, to a solution of 1-butoxy-3,5-dimethoxybenzene (411 mg, 2.10 mmol, 1.00 equiv.) and TMEDA (0.48 ml, 366 mg, 3.15 mmol, 1.50 equiv.) in abs. Et$_2$O (2.38 ml), at 0 °C *n*-butyl lithium (2.5 M in hexane, 1.26 ml, 202 mg, 3.15 mmol, 1.50 equiv.) were added and the reaction was stirred at rt for 2 h. Subsequently followed by addition of DMF (0.49 ml, 460 mg, 6.30 mmol, 3.00 equiv.) at 0 °C and stirred for another 4 h at rt. Purification by flash column chromatography (*c*Hex/EtOAc 5:1 to 1:2) resulted in an inseparable mixture of the two isomers **36d** and **37** as a pale yellow oil (219 mg, 0.92 mmol, 44%).

B) According to **GP1**, to a solution of 4-formyl-3,5-dimethoxyphenol (182 mg, 1.00 mmol, 1.00 equiv.), K_2CO_3 (207 mg, 1.50 mmol, 1.50 equiv.) and 18-crown-6 (5.28 mg, 0.02 mmol, 0.02 equiv.) in acetone (1.00 ml), 1-bromobutane (0.19 ml, 237 mg, 1.50 mmol, 1.50 equiv.) was added and the reaction was stirred under reflux for 22 h. Purification by flash column chromatography (cHex/EtOAc 2:1) resulted in the title compound **36d** as a yellow oil (208 mg, 0.87 mmol, 87%).

36d) R_f(cHex/EtOAc 1:1) = 0.38. – **XXa)** **^{1}H NMR** (400 MHz, CDCl$_3$) δ = 10.35 (s, 1H, C*H*O), 6.07 (s, 2H, 2 × C*H*$_{ar}$), 4.02 (t, ^3J = 6.5 Hz, 2H, OC*H*$_2$), 3.87 (s, 6H, 2 × OC*H*$_3$), 1.79 (quin., ^3J = 8.0, 6.5 Hz, 2H, OCH$_2$C*H*$_2$), 1.56 – 1.45 (m, 2H, C*H*$_2$CH$_3$), 0.99 (t, ^3J = 7.4 Hz, 3H, C*H*$_3$) ppm. – **^{13}C NMR** (101 MHz, CDCl$_3$) δ = 187.9 (+, *C*HO), 166.0 (C$_q$, *C*$_{ar}$OCH$_2$), 164.2 (C$_q$, 2 × *C*$_{ar}$OCH$_3$), 108.8 (C$_q$, *C*$_{ar}$CHO), 90.8 (+, 2 × *C*$_{ar}$H), 68.2 (–, O*C*H$_2$), 56.1 (+, 2 × O*C*H$_3$), 31.2 (–, OCH$_2$*C*H$_2$), 19.3 (–, *C*H$_2$CH$_3$), 13.9 (+, *C*H$_3$) ppm. – **IR** (ATR, ṽ) = 2957, 2937, 2872, 1712, 1675, 1651, 1597, 1573, 1461, 1450, 1409, 1334, 1228, 1197, 1163, 1124, 1068, 1049, 1027, 969, 923, 809, 770, 626, 581, 562, 503, 493 cm^{-1}. – **MS** (EI, 70 eV, 60 °C) m/z (%) = 239 (16) [M+H]$^+$, 238 (100) [M]$^+$, 226 (16), 221 (11), 192 (12), 182 (22), 181 (45), 170 (36), 167 (12), 166 (11), 165 (45), 164 (28), 155 (14), 154 (13), 153 (32), 151 (12), 137 (14), 136 (17), 125 (23), 123 (11), 69 (20). 57 (14). – **HRMS** (EI, C$_{13}$H$_{18}$O$_4$) = calc.: 238.1200; found: 238.1199. – Analytical data are consistent with literature.[141]

37)[9] **^{1}H NMR** (400 MHz, CDCl$_3$) δ = 10.38 (s, 1H, C*H*O), 6.12 – 6.00 (m, 2H, 2 × C*H*$_{ar}$), 4.04 – 3.97 (m, 2H, OC*H*$_2$), 3.87 (s, 3H, OC*H*$_3$), 3.86 (s, 3H, OC*H*$_3$), 1.88 – 1.72 (m, 2H, OCH$_2$C*H*$_2$), 1.60 – 1.44 (m, 2H, C*H*$_2$CH$_3$), 0.91 – 1.05 (m, 3H, C*H*$_3$) ppm. – **^{13}C NMR** (101 MHz, CDCl$_3$) δ = 187.9 (+, *C*HO), 166.2 (C$_q$, *C*$_{ar}$OCH$_2$), 164.5 (C$_q$, *C*$_{ar}$OCH$_3$), 164.2 (C$_q$, *C*$_{ar}$OCH$_3$), 109.1 (C$_q$, *C*$_{ar}$CHO), 90.9 (+, *C*H$_{ar}$), 90.4 (+, *C*H$_{ar}$), 68.8 (–, O*C*H$_2$), 56.1 (+, O*C*H$_3$), 55.6 (+, O*C*H$_3$), 31.2 (–, OCH$_2$*C*H$_2$), 19.4 (–, *C*H$_2$CH$_3$), 13.9 (+, *C*H$_3$) ppm.

2,6-Dimethoxy-4-hydroxybenzaldehyde (38) 4,6-Dimethoxy-2-hydroxybenzaldehyde (19a) 4,6-Dimethoxy-2-hydroxyisophthalaldehyde (111)

A solution of 3,5-dimethoxyphenol (**34**, 3.00 g, 19.5 mmol, 1.00 equiv.) in POCl$_3$ (16.6 ml, 5.98 g, 39.0 mmol, 2.00 equiv.) was cooled to 0 °C and DMF (2.27 ml, 2.14 g, 29.3 mmol, 1.50 equiv.) was slowly added to the mixture, which was then stirred overnight at rt. After

---

[9] NMR analytical data were acquired only due to inseparable isomeric mixture.

cooling to 0 °C again, the reaction was quenched by careful addition of ice and stirred for further 30 min. Subsequently followed by careful neutralization (~pH 6) of the very acidic aqueous solution with NaOH (solid). The formed precipitate was then filtrated off and washed with H_2O (2 × 30 ml). After drying in vacuo, the obtained residue was resuspended in $CHCl_3$ (~50 ml), filtrated again and washed with $CHCl_3$ (2 × 25 ml) to remove a brown solid, which, after re-crystallization in hot EtOH, was identified by NMR as the desired product **38** (1.51 g, 8.51 mmol, 42%). The filtrate was concentrated in vacuo, the remaining residue resuspended in as few as possible of hot EtOH and the formed precipitate was filtered of again. The ethanolic phase was concentrated and the obtained beige brown solid was identified by NMR as pure product **19a** (1.22 g, 6.63 mmol, 34%). Recrystallization of the remaining precipitate in hot MeOH provided after cooling NMR pure product **111** (golden flakes, 131 mg, 0.78 mmol, 4.4%).

38) **^{1}H NMR** (400 MHz, DMSO-d_6) δ = 10.15 (s, 1H, CHO), 6.11 (s, 2H, 2 × CH_{ar}), 3.75 (s, 6H, 2 × OCH_3) ppm. – **^{13}C NMR** (101 MHz, DMSO-d_6) δ = 185.2 (+, CHO), 165.2 (C$_q$, C_{ar}OH), 163.7 (C$_q$, 2 × C_{ar}OCH$_3$), 106.8 (C$_q$, C_{ar}CHO), 91.8 (+, 2 × C_{ar}H), 55.7 (+, 2 × OCH$_3$) ppm. – **MS** (70 eV, EI, 130 °C) m/z (%) = 182 (100) [M]$^+$, 181 (58) [M–H]$^+$, 166 (18), 165 (41), 164 (19), 153 (15), 151 (10), 138 (13), 137 (15), 136 (13), 131 (26), 125 (11), 123 (11), 95 (11), 93 (23), 69 (77). – **HRMS** (EI, C$_9$H$_{10}$O$_4$) = calc.: 182.0574; found: 182.0572. – Analytical data are consistent with literature.[144, 145]

19a) **^{1}H NMR** (400 MHz, DMSO-d_6) δ = 12.37 (s, 1H, OH), 10.01 (s, 1H, CHO), 6.14 (d, ^4J = 2.2 Hz, 1H, CH_{ar}), 6.09 (d, ^4J = 2.2, 1H, CH_{ar}), 3.86 (s, 3H, OCH_3), 3.83 (s, 3H, OCH_3) ppm. – **^{13}C NMR** (101 MHz, DMSO) δ = 191.5 (+, CHO), 168.1 (C$_q$, C_{ar}OH), 165.2 (C$_q$, C_{ar}OCH$_3$), 163.5 (C$_q$, C_{ar}OCH$_3$), 105.4 (C$_q$, C_{ar}CHO), 93.2 (+, C_{ar}H), 90.8 (+, C_{ar}H), 56.2 (+, OCH$_3$), 56.0 (+, OCH$_3$) ppm. – **MS** (70 eV, EI) m/z (%) = 182 (100) [M]$^+$, 181 (51) [M–H]$^+$, 165 (14), 164 (29), 151 (17), 136 (18), 122 (11), 95 (13), 69 (22). – **HRMS** (EI, C$_9$H$_{10}$O$_4$) = calc.: 182.0574; found: 182.0575. – Analytical data are consistent with literature.[144]

111) **^{1}H NMR** (400 MHz, DMSO-d_6) δ = 13.36 (s, 1H, OH), 10.13 (s, 2H, 2 × CHO), 6.35 (s, 1H, CH_{ar}), 4.02 (s, 6H, 2 × OCH_3) ppm – **^{13}C NMR** (101 MHz, DMSO) δ = 188.7 (+, 2 × CHO), 168.6 (C$_q$, 2 × C_{ar}OCH$_3$), 167.0 (C$_q$, C_{ar}OH), 105.3 (C$_q$, 2 × C_{ar}CHO), 87.8 (+, C_{ar}H), 57.0 (+, 2 × OCH$_3$) ppm. – **MS** (70 eV, EI, 120 °C) m/z (%) = 210 (74) [M]$^+$, 182 (26), 181 (100), 153 (49), 152 (56), 131 (56), 123 (15), 122 (11), 119 (11), 100 (11), 92 (11), 69 (76). – **HRMS** (EI, C$_{10}$H$_{10}$O$_5$) = calc.: 210.0523; found: 210.0521. – Analytical data are consistent with literature.[144]

2,6-Dimethoxy-4-ethoxybenzaldehyde (36b)

According to **GP1**, to a solution of 4-formyl-3,5-dimethoxyphenol (**38**, 182 mg, 1.00 mmol, 1.00 equiv.), K_2CO_3 (207 mg, 1.50 mmol, 1.50 equiv.) and 18-crown-6 (5.28 mg, 0.02 mmol, 0.02 equiv.) in acetone (1.00 ml), 1-bromoethane (0.11 ml, 163 mg, 1.50 mmol, 1.50 equiv.) was added and the reaction was stirred under reflux for 22 h. Purification by flash column chromatography (cHex/EtOAc 1:1) resulted in the title compound as a white solid (173 mg, 0.82 mmol, 82%).

R_f (cHex/EtOAc 1:1) = 0.25. – **^{1}H NMR** (400 MHz, CDCl$_3$) δ = 10.35 (s, 1H, CHO), 6.07 (s, 2H, 2 × C_{ar}H), 4.10 (q, ^3J = 7.0 Hz, 2H, OCH$_2$), 3.87 (s, 6H, 2 × OCH$_3$), 1.45 (t, ^3J = 7.0 Hz, 3H, CH$_3$) ppm. – **^{13}C NMR** (101 MHz, CDCl$_3$) δ = 187.9 (+, CHO), 165.8 (C$_q$, C_{ar}OCH$_2$), 164.3 (C$_q$, 2 × C_{ar}OCH$_3$), 108.9 (C$_q$, C_{ar}CHO), 90.8 (+, 2 × C_{ar}H), 64.0 (–, OCH$_2$), 56.1 (+, 2 × OCH$_3$), 14.8 (+, CH$_3$) ppm. – **IR** (ATR, ṽ) = 2988, 2937, 2782, 1666, 1592, 1460, 1407, 1326, 1229, 1196, 1161, 1120, 1052, 997, 930, 882, 806, 783, 722, 635, 575, 501 cm^{-1}. – **MS** (70 eV, EI, 40 °C) m/z (%) = 211 (13) [M+H]$^+$, 210 (100) [M]$^+$, 209 (23) [M–H]$^+$, 193 (39), 181 (58), 165 (18), 164 (15), 69 (30), 58 (42). – **HRMS** (EI, C$_{11}$H$_{14}$O$_4$) = calc.: 210.0892; found: 210.0891.

2,6-Dimethoxy-4-propoxybenzaldehyde (36c)

According to **GP1**, to a solution of 4-formyl-3,5-dimethoxyphenol (**38**, 260 mg, 1.43 mmol, 1.00 equiv.), K_2CO_3 (296 mg, 2.14 mmol, 1.50 equiv.) and 18-crown-6 (7.54 mg, 0.03 mmol, 0.02 equiv.) in acetone (2.85 ml), 1-bromopropane (0.20 ml, 263 mg, 2.14 mmol, 1.50 equiv.) was added and the reaction was stirred under reflux overnight. Purification by flash column chromatography (cHex/EtOAc 1:1) resulted in the title compound as yellow crystals (253 mg, 1.13 mmol, 79%).

R_f (cHex/EtOAc 1:1) = 0.43. – **^{1}H NMR** (400 MHz, CDCl$_3$) δ = 10.33 (s, 1H, CHO), 6.06 (s, 2H, 2 × CH$_{ar}$), 3.97 (t, ^3J = 6.5 Hz, 2H, OCH$_2$), 3.86 (s, 6H, OCH$_3$), 1.91 – 1.77 (m, 2H, CH$_2$CH$_3$), 1.05 (t, ^3J = 7.4 Hz, 3H, CH$_3$) ppm. – **^{13}C NMR** (101 MHz, CDCl$_3$) δ = 187.8 (+, CHO), 166.0 (C$_q$, C_{ar}OCH$_2$), 164.2 (C$_q$, 2 × C_{ar}OCH$_3$), 108.8 (C$_q$, C_{ar}CHO), 90.8 (+, 2 × C_{ar}H), 69.9 (–, OCH$_2$), 56.1 (+, 2 × OCH$_3$), 22.6 (–, CH$_2$CH$_3$), 10.6 (+, CH$_3$) ppm. – **IR** (ATR, ṽ) = 2967, 2941, 2875, 2784, 1667, 1638, 1597, 1575, 1466, 1453, 1446, 1425, 1408, 1332, 1226, 1198, 1167, 1123, 1051, 1020, 967, 925, 906, 813, 793, 769, 726, 681, 644, 636, 576, 498 cm^{-1}. – **MS** (70 eV, EI, 76 °C) m/z (%) = 225 (16) [M+H]$^+$, 224 (100) [M]$^+$, 223 (16) [M–H]$^+$, 207 (17), 182 (17), 181 (45), 165 (32), 164 (19), 153 (14), 69 (12). – **HRMS** (EI, C$_{12}$H$_{16}$O$_4$) = calc.: 224.1043; found: 224.1045.

2,6-Dimethoxy-4-(pentyloxy)benzaldehyde (36e)

According to **GP1**, to a solution of 4-formyl-3,5-dimethoxyphenol (**38**, 260 mg, 1.43 mmol, 1.00 equiv.), K_2CO_3 (296 mg, 2.14 mmol, 1.50 equiv.) and 18-crown-6 (7.54 mg, 0.03 mmol, 0.02 equiv.) in acetone (2.85 ml), 1-bromopentane (0.27 ml, 323 mg, 2.14 mmol, 1.50 equiv.) was added and the reaction was stirred under reflux overnight. Purification by flash column chromatography (*c*Hex/EtOAc 2:1) resulted in the title compound as a tan solid (302 mg, 1.20 mmol, 84%).

R_f (*c*Hex/EtOAc 2:1) = 0.33. – **^{1}H NMR** (400 MHz, CDCl$_3$) δ = 10.33 (s, 1H, C*H*O), 6.04 (s, 2H, 2 × C*H*$_{ar}$), 4.04 – 3.96 (m, 2H, OC*H*$_2$), 3.89 – 3.84 (m, 6H, 2 × OC*H*$_3$), 1.87 – 1.73 (m, 2H, OCH$_2$C*H*$_2$), 1.53 – 1.30 (m, 4H, C*H*$_2$C*H*$_2$CH$_3$), 1.01 – 0.83 (m, 3H, C*H*$_3$) ppm. – **^{13}C NMR** (101 MHz, CDCl$_3$) δ = 188.5 (+, *C*HO), 166.7 (C$_q$, *C*$_{ar}$OCH$_2$), 165.0 (C$_q$, 2 × *C*$_{ar}$OCH$_3$), 109.7 (C$_q$, *C*$_{ar}$CHO), 91.6 (+, 2 × *C*$_{ar}$H), 69.2 (–, O*C*H$_2$), 56.9 (+, 2 × O*C*H$_3$), 29.7 (–, *C*H$_2$), 29.0 (–, *C*H$_2$), 23.3 (–, *C*H$_2$CH$_3$), 14.8 (+, *C*H$_3$) ppm. – **IR** (ATR, ṽ) = 3118, 3016, 2997, 2953, 2919, 2870, 2857, 2786, 1669, 1639, 1594, 1570, 1493, 1460, 1448, 1424, 1409, 1397, 1378, 1340, 1303, 1248, 1231, 1204, 1196, 1187, 1160, 1115, 1061, 1045, 1023, 979, 965, 928, 895, 866, 817, 800, 735, 708, 647, 562, 511, 486, 452, 399 cm^{-1}. – **MS** (70 eV, EI, 50 °C) *m/z* (%) = 253 (13) [M+H]$^+$, 252 (76) [M]$^+$, 235 (13), 231 (13), 224 (16), 182 (37), 181 (92), 170 (14), 169 (10), 168 (12), 165 (33), 164 (24), 154 (41), 153 (17), 136 (13), 131 (39), 125 (18), 119 (13), 69 (100), 58 (40), 57 (13). – **HRMS** (EI, C$_{14}$H$_{20}$O$_4$) = calc.: 252.1362; found: 252.1360.

2,6-Dimethoxy-4-(hexyloxy)benzaldehyde (36f)

According to **GP1**, to a solution of 4-formyl-3,5-dimethoxyphenol (**38**, 260 mg, 1.43 mmol, 1.00 equiv.), K_2CO_3 (296 mg, 2.14 mmol, 1.50 equiv.) and 18-crown-6 (7.54 mg, 0.03 mmol, 0.02 equiv.) in acetone (2.85 ml), 1-bromohexane (0.30 ml, 353 mg, 2.14 mmol, 1.50 equiv.) was added and the reaction was stirred under reflux overnight. Purification by flash column chromatography (*n*Hex/EtOAc 1:1) resulted in the title compound as yellow crystals (328 mg, 1.23 mmol, 86%).

R_f (*n*Hex/EtOAc 2:1) = 0.30. – **^{1}H NMR** (400 MHz, CDCl$_3$) δ = 10.33 (s, 1H, C*H*O), 6.06 (s, 2H, 2 × C*H*$_{ar}$), 4.00 (t, ^3J = 6.5 Hz, 2H, OC*H*$_2$), 3.86 (s, 6H, 2 × OC*H*$_3$), 1.86 – 1.71 (m, 2H, OCH$_2$C*H*$_2$), 1.54 – 1.42 (m, 2H, C*H*$_2$), 1.38 – 1.29 (m, 4H, 2 × C*H*$_2$), 0.90 (t, ^3J = 7.0 Hz, 3H, C*H*$_3$) ppm. – **^{13}C NMR** (101 MHz, CDCl$_3$) δ = 187.8 (+, *C*HO), 166.0 (C$_q$, *C*$_{ar}$OCH$_2$), 164.2 (C$_q$, 2 × *C*$_{ar}$OCH$_3$), 108.8 (C$_q$, *C*$_{ar}$CHO), 90.8 (+, 2 × *C*$_{ar}$H), 68.5 (–, O*C*H$_2$), 56.1 (+, 2 × O*C*H$_3$), 31.6 (–, *C*H$_2$), 29.1 (–, *C*H$_2$), 25.8 (–, *C*H$_2$), 22.7 (–, *C*H$_2$CH$_3$), 14.1 (+, *C*H$_3$) ppm. – **IR** (ATR, ṽ) = 3109, 2996, 2948, 2921, 2870, 2853, 2783, 1673, 1596, 1571, 1466, 1458, 1445, 1424, 1412, 1401, 1340, 1305, 1235, 1198, 1162, 1120, 1057, 1028, 1009, 959, 928, 897, 888, 833, 803, 752, 730, 710, 652, 584, 564, 511, 499, 445, 425 cm^{-1}. – **MS** (70 eV, EI, 60 °C) *m/z* (%) = 266 (100) [M]$^+$, 249 (13), 182 (48), 181 (47), 167 (11), 165 (48), 164 (33), 154 (40), 153 (22), 151 (13), 136 (20), 125 (18), 69 (17), 55 (10). – **HRMS** (EI, C$_{15}$H$_{22}$O$_4$) = calc.: 266.1518; found: 266.1517.

2,6-Dimethoxy-4-(heptyloxy)benzaldehyde (36g)

According to **GP1**, to a solution of 4-formyl-3,5-dimethoxyphenol (**38**, 260 mg, 1.43 mmol, 1.00 equiv.), K$_2$CO$_3$ (296 mg, 2.14 mmol, 1.50 equiv.) and 18-crown-6 (7.54 mg, 0.03 mmol, 0.02 equiv.) in acetone (2.85 ml), 1-bromoheptane (0.34 ml, 383 mg, 2.14 mmol, 1.50 equiv.) was added and the reaction was stirred under reflux overnight. Purification by flash column chromatography (*n*Hex/EtOAc 2:1) resulted in the title compound as an off-white solid (302 mg, 1.08 mmol, 76%).

R$_f$ (*n*Hex/EtOAc 2:1) = 0.33. – **^{1}H NMR** (400 MHz, CDCl$_3$) δ = 10.33 (s, 1H, C*H*O), 6.18 – 5.95 (m, 2H, 2 × C*H*$_{ar}$), 4.05 – 3.96 (m, 2H, OC*H*$_2$), 3.90 – 3.84 (m, 6H, 2 × OC*H*$_3$), 1.86 – 1.74 (m, 2H, OCH$_2$C*H*$_2$), 1.45 (quin., ^3J = 7.9, 7.3 Hz, 2H, C*H*$_2$), 1.40 – 1.25 (m, 6H, 3 × C*H*$_2$), 0.94 – 0.85 (m, 3H, C*H*$_3$) ppm. – **^{13}C NMR** (101 MHz, CDCl$_3$) δ = 187.7 (+, *C*HO), 166.0 (C$_q$, *C*$_{ar}$OCH$_2$), 164.2 (C$_q$, 2 × *C*$_{ar}$OCH$_3$), 108.9 (C$_q$, *C*$_{ar}$CHO), 90.9 (+, 2 × *C*$_{ar}$H), 68.5 (−, O*C*H$_2$), 56.1 (+, 2 × O*C*H$_3$), 31.8 (−, *C*H$_2$), 29.2 (−, *C*H$_2$), 29.1 (−, *C*H$_2$), 26.1 (−, *C*H$_2$), 22.7 (−, *C*H$_2$CH$_3$), 14.1 (+, *C*H$_3$) ppm. – **IR** (ATR, ṽ) = 3105, 2945, 2920, 2850, 2785, 1669, 1597, 1570, 1463, 1454, 1412, 1341, 1304, 1233, 1202, 1163, 1121, 1051, 1034, 1024, 959, 930, 844, 826, 803, 775, 726, 708, 651, 564, 515, 498 cm^{-1}. – **MS** (70 eV, EI, 90 °C) *m/z* (%) = 281 (20) [M+H]$^+$, 280 (100) [M]$^+$, 182 (34), 181 (30), 165 (36), 164 (21), 153 (13), 136 (11). – **HRMS** (EI, C$_{16}$H$_{24}$O$_4$) = calc.: 280.1669; found: 280.1670.

4-Ethoxy-2-hydroxy-6-methoxybenzaldehyde (19b)

According to **GP3**, to a solution of 2,6-dimethoxy-4-ethoxybenzaldehyde (**36b**, 70.0 mg, 0.33 mmol, 1.00 equiv.), dissolved in a mixture of abs. ACN and abs. DCM (2:1, 3.00 ml), AlCl$_3$ (89.0 mg, 0.67 mmol, 2.00 equiv.) and NaI (100 mg, 0.67 mmol, 2.00 equiv.) were added in portions at 0 °C and the reaction was stirred for 1 h at rt. Purification by flash column chromatography (*c*Hex/EtOAc 2:1) resulted in the title compound as colourless to slightly green crystals (22.0 mg, 0.11 mmol, 34%).

R_f (cHex/EtOAc 2:1) = 0.59. – ^{1}H NMR (400 MHz, CDCl$_3$) δ = 12.50 (s, 1H, O*H*), 10.07 (s, 1H, C*H*O), 5.98 (d, ^4J = 2.2 Hz, 1H, C*H*$_{ar}$), 5.89 (d, ^4J = 2.2 Hz, 1H, C*H*$_{ar}$), 4.05 (q, ^3J = 7.0 Hz, 2H, OC*H*$_2$), 3.83 (s, 3H, OC*H*$_3$), 1.41 (t, ^3J = 7.0 Hz, 3H, C*H*$_3$) ppm. – ^{13}C NMR (101 MHz, CDCl$_3$) δ = 191.9 (+, *C*HO), 167.7 (C$_q$, *C*$_{ar}$OH), 166.5 (C$_q$, *C*$_{ar}$OCH$_2$), 163.7 (C$_q$, *C*$_{ar}$OCH$_3$), 106.0 (C$_q$, *C*$_{ar}$CHO), 93.4 (+, *C*$_{ar}$H), 91.0 (+, *C*$_{ar}$H), 64.2 (–, O*C*H$_2$), 55.8 (+, O*C*H$_3$), 14.6 (+, *C*H$_3$) ppm. – IR (ATR, ṽ) = 2978, 2948, 2927, 2897, 2854, 1638, 1612, 1581, 1466, 1452, 1426, 1398, 1375, 1330, 1298, 1217, 1198, 1167, 1118, 1044, 1017, 969, 881, 846, 810, 800, 732, 637, 540, 517, 472 cm^{-1}. – MS (70 eV, EI, 30 °C) *m/z* (%) = 197 (11) [M+H]$^+$, 196 (100) [M]$^+$, 195 (16) [M–H]$^+$, 181 (24) [M–CH$_3$]$^+$, 178 (11), 168 (17), 167 (47) [M–C$_2$H$_5$]$^+$, 152 (11), 151 (11), 150 (36), 140 (11), 139 (13), 137 (15), 131 (20), 122 (12), 108 (12), 69 (58). – HRMS (EI, C$_{10}$H$_{12}$O$_4$) = calc.: 196.0736; found: 196.0735.

2-Hydroxy-6-methoxy-4-propoxybenzaldehyde (19c)

According to GP3, to a solution of 2,6-dimethoxy-4-propoxybenzaldehyde (36c, 200 mg, 0.89 mmol, 1.00 equiv.), dissolved in a mixture of abs. ACN and abs. DCM (2:1, 6.00 ml), AlCl$_3$ (238 mg, 1.78 mmol, 2.00 equiv.) and NaI (267 mg, 1.78 mmol, 2.00 equiv.) were added in portions at 0 °C and the reaction was stirred for 1 h at rt. Purification by flash column chromatography (cHex/EtOAc 2:1) resulted in the title compound as a white powder (176 mg, 0.84 mmol, 94%).

R_f (cHex/EtOAc 5:1) = 0.37. – ^{1}H NMR (400 MHz, CDCl$_3$) δ = 12.51 (s, 1H, O*H*), 10.08 (s, 1H, C*H*O), 5.99 (d, ^4J = 2.1 Hz, 1H, C*H*$_{ar}$), 5.91 (d, ^4J = 2.1 Hz, 1H, C*H*$_{ar}$), 3.95 (t, ^3J = 6.6 Hz, 2H, OC*H*$_2$), 3.85 (s, 3H, OC*H*$_3$), 1.81 (sex., ^3J = 7.4 Hz, 2H, C*H*$_2$CH$_3$), 1.03 (t, ^3J = 7.4 Hz, 3H, C*H*$_3$) ppm. – ^{13}C NMR (101 MHz, CDCl$_3$) δ = 191.9 (+, *C*HO), 167.9 (C$_q$, *C*$_{ar}$OH), 166.5 (C$_q$, *C*$_{ar}$OCH$_2$), 163.7 (C$_q$, *C*$_{ar}$OCH$_3$), 106.1 (C$_q$, *C*$_{ar}$CHO), 93.5 (+, *C*$_{ar}$H), 91.1 (+, *C*$_{ar}$H), 70.2 (–, O*C*H$_2$), 55.9 (+, O*C*H$_3$), 22.5 (–, *C*H$_2$CH$_3$), 10.6 (+, *C*H$_3$) ppm. – IR (ATR, ṽ) = 3077, 3021, 2969, 2952, 2929, 2881, 2859, 1632, 1604, 1494, 1462, 1445, 1424, 1398, 1380, 1367, 1332, 1312, 1295, 1215, 1197, 1174, 1129, 1112, 1040, 993, 945, 912, 902, 839, 815, 809, 762, 731, 694, 659, 640, 545, 517, 463, 449 cm^{-1}. – MS (70 eV, EI, 50 °C) *m/z* (%) = 211 (13) [M+H]$^+$, 210 (100) [M]$^+$, 181 (27) [M–C$_2$H$_5$]$^+$, 168 (74), 167 (53), 151 (17), 150 (52), 140 (34), 139 (16), 137 (13), 131 (28), 125 (14), 112 (13), 108 (11), 69 (50). – HRMS (EI, C$_{11}$H$_{14}$O$_4$) = calc.: 210.0887; found: 210.0888.

4-Butoxy-2-hydroxy-6-methoxybenzaldehyde (19d)

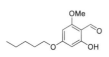

According to **GP3**, to a solution of 4-butoxy-2,6-dimethoxy-benzaldehyde (**36d**, 200 mg, 0.89 mmol, 1.00 equiv.), dissolved in a mixture of abs. ACN and abs. DCM (2:1, 6.00 ml), AlCl₃ (238 mg, 1.78 mmol, 2.00 equiv.) and NaI (267 mg, 1.78 mmol, 2.00 equiv.) were added in portions at 0 °C and the reaction was stirred for 1 h at rt. Purification by flash column chromatography (*c*Hex/EtOAc 2:1) resulted in the title compound as an off-white solid (119 mg, 0.53 mmol, 63%).

R_f (*c*Hex/EtOAc 5:1) = 0.37. – **¹H NMR** (400 MHz, CDCl₃) δ = 12.51 (s, 1H, O*H*), 10.08 (s, 1H, C*H*O), 5.99 (d, ^4J = 2.2 Hz, 1H, C*H*ar), 5.90 (d, ^4J = 2.2 Hz, 1H, C*H*ar), 3.99 (t, ^3J = 6.5 Hz, 2H, OC*H₂*), 3.84 (s, 3H, OC*H₃*), 1.82 – 1.72 (m, 2H, OCH₂C*H₂*), 1.52 – 1.43 (m, 2H, C*H₂*CH₃), 0.98 (t, ^3J = 7.4 Hz, 3H, C*H₃*) ppm. – **¹³C NMR** (101 MHz, CDCl₃) δ = 191.9 (+, *C*HO), 167.9 (*C*q, *C*arOH), 166.5 (*C*q, *C*arOCH₂), 163.7 (*C*q, *C*arOCH₃), 106.1 (*C*q, *C*arCHO), 93.5 (+, *C*arH), 91.1 (+, *C*arH), 68.4 (–, O*C*H₂), 55.9 (+, O*C*H₃), 31.1 (–, O*C*H₂*C*H₂), 19.3 (–, *C*H₂*C*H₃), 13.9 (+, *C*H₃) ppm. – **IR** (ATR, ṽ) = 2958, 2947, 2930, 2865, 1616, 1577, 1468, 1447, 1417, 1403, 1372, 1331, 1297, 1216, 1199, 1161, 1113, 1063, 1041, 995, 946, 899, 841, 824, 804, 794, 741, 728, 694, 654, 637, 545, 517, 484, 438, 402 cm⁻¹. – **MS** (70 eV, EI, 50 °C) *m/z* (%) = 225 (12) [M+H]⁺, 224 (81) [M]⁺, 169 (13), 168 (100), 167 (41) [M-C₄H₉]⁺, 151 (15) [M-C₄H₉O]⁺, 150 (46), 140 (30), 139 (10), 125 (11), 122 (13), 112 (12), 69 (20). – **HRMS** (EI, C₁₂H₁₆O₄) = calc.: 224.1043; found: 224.1042.

2-hydroxy-6-methoxy-4-(pentyloxy)benzaldehyde (19e)

According to **GP3**, to a solution of 2,6-dimethoxy-4-(pentyloxy)benzaldehyde (**36e**, 50.0 mg, 0.20 mmol, 1.00 equiv.), dissolved in a mixture of abs. ACN and DCM (2:1, 3.00 ml), AlCl₃ (53.0 mg, 0.40 mmol, 2.00 equiv.) and NaI (59.0 mg, 0.40 mmol, 2.00 equiv.) were added in portions at 0 °C and the reaction was stirred for 1 h at rt. Purification by flash column chromatography (*c*Hex/EtOAc 5:1) resulted in the title compound as colourless to off-white crystals (47.0 mg, 0.20 mmol, quant.).

R_f (*c*Hex/EtOAc 5:1) = 0.45. – **¹H NMR** (400 MHz, CDCl₃) δ = 12.52 (s, 1H, O*H*), 10.08 (s, 1H, C*H*O), 5.99 (d, ^4J = 2.1 Hz, 1H, C*H*ar), 5.90 (d, ^4J = 2.1 Hz, 1H, C*H*ar), 3.98 (t, ^3J = 6.5 Hz, 2H, OC*H₂*), 3.85 (s, 3H, OC*H₃*), 1.84 – 1.74 (m, 2H, OCH₂C*H₂*), 1.47 – 1.34 (m, 4H, 2 × C*H₂*), 0.93 (t, ^3J = 7.1 Hz, 3H, C*H₃*) ppm. – **¹³C NMR** (101 MHz, CDCl₃) δ = 191.9 (+, *C*HO), 167.9 (*C*q, *C*arOH), 166.5 (*C*q, *C*arOCH₂), 163.7 (*C*q, *C*arOCH₃), 106.1 (*C*q, *C*arCHO), 93.5 (+, *C*arH),

91.1 (+, $C_{ar}H$), 68.7 (–, OCH_2), 55.9 (+, OCH_3), 28.8 (–, CH_2), 28.2 (–, CH_2), 22.5 (–, CH_2CH_3), 14.1 (+, CH_3) ppm. – **IR** (ATR, ṽ) = 2958, 2925, 2879, 2867, 2853, 1614, 1572, 1494, 1461, 1450, 1422, 1396, 1365, 1325, 1323, 1274, 1246, 1216, 1202, 1191, 1168, 1124, 1107, 1059, 1034, 991, 940, 886, 807, 797, 728, 691, 651, 641, 547, 513, 481, 453, 408, 388 cm^{-1}. – **MS** (70 eV, EI, 70 °C) m/z (%) = 238 (58) [M]$^+$, 169 (14), 168 (100) [M-C$_5$H$_{10}$]$^+$, 167 (23) [M-C$_5$H$_{11}$]$^+$, 151 (10) [M-C$_5$H$_{11}$O]$^+$, 150 (30), 140 (18), 69 (11). **HRMS** (EI, C$_{13}$H$_{18}$O$_4$) = calc.: 238.1200; found: 238.1199.

4-(Hexyloxy)-2-hydroxy-6-methoxybenzaldehyde (19f)

According to **GP3**, to a solution of 2,6-dimethoxy-4-(hexyloxy)ben-zaldehyde (**36f**, 55.0 mg, 0.21 mmol, 1.00 equiv.), dissolved in a mixture of abs. ACN and abs. DCM (2:1, 3.00 ml), AlCl$_3$ (55.0 mg, 0.41 mmol, 2.00 equiv.) and NaI (62.0 mg, 0.41 mmol, 2.00 equiv.) were added in portions at 0 °C and the reaction was stirred for 1 h at rt. Purification by flash column chromatography (cHex/EtOAc 5:1) resulted in the title compound as colourless to off-white crystals (47.0 mg, 0.19 mmol, 90%).

R_f (cHex/EtOAc 5:1) = 0.50. – **^{1}H NMR** (400 MHz, CDCl$_3$) δ = 12.51 (s, 1H, O*H*), 10.08 (s, 1H, C*H*O), 5.99 (d, ^4J = 2.1 Hz, 1H, C*H*$_{ar}$), 5.90 (d, ^4J = 2.1 Hz, 1H, C*H*$_{ar}$), 3.98 (t, ^3J = 6.6 Hz, 2H, OC*H*$_2$), 3.85 (s, 3H, OC*H*$_3$), 1.83 – 1.73 (m, 2H, OCH$_2$C*H*$_2$), 1.50 – 1.40 (m, 2H, C*H*$_2$), 1.38 – 1.30 (m, 4H, 2 × C*H*$_2$), 0.91 (t, ^3J = 6.8 Hz, 3H, C*H*$_3$) ppm. – **^{13}C NMR** (101 MHz, CDCl$_3$) δ = 191.9 (+, *C*HO), 167.9 (C$_q$, *C*$_{ar}$OH), 166.5 (C$_q$, *C*$_{ar}$OCH$_2$), 163.7 (C$_q$, *C*$_{ar}$OCH$_3$), 106.1 (C$_q$, *C*$_{ar}$CHO), 93.5 (+, *C*$_{ar}$H), 91.1 (+, *C*$_{ar}$H), 68.7 (–, *O*CH$_2$), 55.9 (+, OCH$_3$), 31.6 (–, OCH$_2$*C*H$_2$), 29.1 (–, *C*H$_2$), 25.8 (–, *C*H$_2$), 22.7 (–, *C*H$_2$CH$_3$), 14.2 (+, *C*H$_3$) ppm. – **IR** (ATR, ṽ) = 2950, 2924, 2866, 2856, 1634, 1617, 1580, 1470, 1449, 1416, 1403, 1369, 1330, 1297, 1259, 1216, 1199, 1163, 1114, 1071, 1059, 1032, 1012, 988, 942, 894, 838, 826, 803, 722, 692, 656, 641, 555, 517, 487, 474, 398 cm^{-1}. – **MS** (70 eV, EI, 70 °C) m/z (%) = 252 (44) [M]$^+$, 169 (15), 168 (100) [M-C$_6$H$_{12}$]$^+$, 167 (17) [M-C$_6$H$_{13}$]$^+$, 150 (23), 140 (15). – **HRMS** (EI, C$_{14}$H$_{20}$O$_4$) = calc.: 252.1356; found: 252.1354.

4-(Heptyloxy)-2-hydroxy-6-methoxybenzaldehyde (19g)

According to **GP3**, to a solution of 2,6-dimethoxy-4-(hepty-loxy)benzaldehyde (**36g**, 200 mg, 0.71 mmol, 1.00 equiv.), dis-solved in a mixture of abs. ACN and abs. DCM (2:1, 6.00 ml), AlCl$_3$ (190 mg, 1.43 mmol, 2.00 equiv.) and NaI (214 mg, 1.43 mmol, 2.00 equiv.) were added in portions at 0 °C and the reaction was stirred for 1 h at rt. Purification by flash column chromatography (*c*Hex/EtOAc 5:1) resulted in the title compound as colourless to pale-green crystals (111 mg, 0.41 mmol, 58%).

R$_f$ (*c*Hex/EtOAc 5:1) = 0.50. – **^{1}H NMR** (400 MHz, CDCl$_3$) δ = 12.51 (s, 1H, O*H*), 10.08 (s, 1H, C*H*O), 5.99 (d, ^4J = 2.1 Hz, 1H, C*H*$_{ar}$), 5.90 (d, ^4J = 2.1 Hz, 1H, C*H*$_{ar}$), 3.98 (t, ^3J = 6.6 Hz, 2H, OC*H*$_2$), 3.84 (s, 3H, OC*H*$_3$), 1.83 – 1.73 (m, 2H, OCH$_2$C*H*$_2$), 1.48 – 1.39 (m, 2H, C*H*$_2$), 1.38 – 1.28 (m, 6H, 3 × C*H*$_2$), 0.89 (t, ^3J = 6.8 Hz, 3H, C*H*$_3$) ppm. – **^{13}C NMR** (101 MHz, CDCl$_3$) δ = 191.9 (+, *C*HO), 167.9 (C$_q$, *C*$_{ar}$OH), 166.5 (C$_q$, *C*$_{ar}$OCH$_2$), 163.7 (C$_q$, *C*$_{ar}$OCH$_3$), 106.1 (C$_q$, *C*$_{ar}$CHO), 93.5 (+, *C*$_{ar}$H), 91.1 (+, *C*$_{ar}$H), 68.7 (–, O*C*H$_2$), 55.9 (+, O*C*H$_3$), 31.9 (–, O*C*H$_2$*C*H$_2$), 29.1 (–, 2 × *C*H$_2$), 26.0 (–, *C*H$_2$), 22.7 (–, *C*H$_2$CH$_3$), 14.2 (+, *C*H$_3$) ppm. – **IR** (ATR, ṽ) = 2952, 2921, 2856, 1635, 1615, 1579, 1521, 1497, 1469, 1448, 1417, 1402, 1370, 1330, 1299, 1259, 1215, 1198, 1163, 1115, 1035, 989, 942, 891, 847, 824, 805, 796, 744, 727, 722, 693, 654, 642, 554, 517, 499, 467, 443, 395, 382 cm^{-1}. – **MS** (70 eV, EI, 70 °C) *m/z* (%) = 266 (39) [M]$^+$, 169 (18), 168 (100) [M-C$_6$H$_{14}$]$^+$, 167 (14) [M-C$_7$H$_{15}$]$^+$, 150 (25), 140 (18). – **HRMS** (C$_{15}$H$_{22}$O$_4$) = calc.: 266.1513; found: 266.1512.

Allyl (3,5-dimethoxyphenyl)carbamate (40)

To a solution of 3,5-dimethoxyaniline (**39**, 1.00 g, 6.53 mmol, 1.00 equiv.) and K$_2$CO$_3$ (1.81 g, 13.1 mmol, 2.00 equiv.) in abs. THF (20 ml) at 0 °C, allyl chloroformate (0.84 ml, 944 mg, 7.83 mmol, 1.20 equiv.) was added and the reaction mixture was stirred for 2 h at this temperature. The reaction was quenched by addition of H$_2$O and extracted with EtOAc. The combined organic layers were washed with 1 M aq. HCl and brine, dried over MgSO$_4$, filtrated and the volatiles were removed under reduced pressure. The title compound was obtained without any further purification as a white solid (1.51 g, 6.35 mmol, 97%).

^{1}H NMR (400 MHz, CDCl$_3$) δ = 6.65 (s, 1H, N*H*), 6.62 (d, ^4J = 2.2 Hz, 2H, 2 × C*H*$_{ar}$), 6.19 (t, ^4J = 2.2 Hz, 1H, C*H*$_{ar}$), 5.96 (ddt, ^3J = 17.2, 10.4, 5.7 Hz, 1H, C*H*), 5.36 (dq, $^{3,4}J$ = 17.2, 1.4 Hz, 1H, CHCH*H*^{a}), 5.26 (dq, $^{3,4}J$ = 10.4, 1.4 Hz, 1H, CHCH*H*^{b}), 4.66 (dt, $^{3,4}J$ = 5.7, 1.4 Hz, 2H, C*H*$_2$), 3.77 (s, 6H, 2 × OC*H*$_3$) ppm. – **^{13}C NMR** (101 MHz, CDCl$_3$) δ = 161.3 (C$_q$, 2 × *C*$_{ar}$OCH$_3$), 153.1 (C$_q$, *C*=O), 140.0 (C$_q$, *C*$_{ar}$N), 132.5 (+, *C*H), 118.4 (–, CH*C*H$_2$), 97.0 (+, *C*$_{ar}$H), 96.0 (+, 2 × *C*$_{ar}$H), 66.0 (–, O*C*H$_2$), 55.5 (+, 2 × O*C*H$_3$) ppm. – **IR** (ATR, ṽ) = 3363, 1727, 1620, 1550, 1485, 1454, 1421, 1336, 1261, 1210, 1160, 1084, 1073, 1038, 993, 969, 925, 828, 817, 765, 678, 650, 605, 541, 496 cm^{-1}. – **MS** (70 eV, EI, 60 °C) *m/z* (%) = 238 (14) [M+H]$^+$, 237 (100)

[M]⁺, 192 (10), 181 (36), 179 (14), 178 (17), 169 (11), 166 (10), 162 (11), 153 (23), 152 (16) [M–C$_4$H$_6$O$_2$]⁺, 125 (43). – **HRMS** (EI, C$_{12}$H$_{15}$NO$_4$) = calc.: 237.1001; found: 237.0999.

Allyl (3,5-dimethoxyphenyl)(methyl)carbamate (41)

In a pre-dried flask NaH (60% dispersion in mineral oil, 38.0 mg, 1.58 mmol, 1.50 equiv.) was suspended in abs. THF (2.10 ml) and allyl (3,5-dimethoxyphenyl)carbamate (**40**, 250 mg, 1.05 mmol, 1.00 equiv.) was added in portions at 0 °C. After stirring for 30 min, MeI (0.13 ml, 299 mg, 2.11 mmol, 2.00 equiv.) was added and the reaction mixture was stirred for 90 min at rt. By addition of H$_2$O the reaction was quenched and extracted with EtOAc. The combined organic layers were washed with brine, dried over Na$_2$SO$_4$ and the volatiles were removed under reduced pressure. The title compound was obtained without any further purification as a yellow oil (227 mg, 1.00 mmol, 95%)

^{1}H NMR (400 MHz, CDCl$_3$) δ = 6.42 (d, ^4J = 2.3 Hz, 2H, 2 × CH_{ar}), 6.33 (t, ^4J = 2.3 Hz, 1H, CH_{ar}), 5.91 (ddt, J = 17.3, 10.5, 5.3 Hz, 1H, CH), 5.25 (d, ^3J = 17.3 Hz, 1H, CHCHH^a), 5.18 (d, $^{3,4}J$ = 10.5, 1.4 Hz, 1H, CHCHH^b), 4.74 – 4.57 (m, 2H, CH_2), 3.77 (s, 6H, 2 × OCH_3), 3.28 (s, 3H, NCH_3) ppm. – **^{13}C NMR** (101 MHz, CDCl$_3$) δ = 160.9 (C$_q$, 2 × C_{ar}OCH$_3$), 155.3 (C$_q$, C=O), 145.0 (C$_q$, C_{ar}N), 132.9 (+, CH), 117.5 (–, CHCH$_2$), 104.4 (+, 2 × C_{ar}H), 98.5 (+, C_{ar}H), 66.4 (–, OCH$_2$), 55.5 (+, 2 × OCH$_3$), 37.9 (+, NCH$_3$) ppm. – **IR** (ATR, ṽ) = 2936, 1704, 1595, 1457, 1382, 1325, 1246, 1204, 1152, 1062, 962, 927, 831, 766, 695 cm^{-1}. – **MS** (70 eV, EI, 40 °C) m/z (%) = 252 (10) [M+H]⁺, 251 (72) [M]⁺, 194 (18) [M–C$_3$H$_5$O]⁺, 192 (14), 167 (20), 166 (63) [M–C$_4$H$_5$O$_2$]⁺, 162 (11), 138 (14), 119 (15), 100 (12). – **HRMS** (EI, C$_{13}$H$_{17}$NO$_4$) = calc.: 251.1158; found: 251.1158.

Allyl (3,5-dimethoxy-4-formylphenyl)(methyl)carbamate (42)

To a solution of allyl (3,5-dimethoxyphenyl)(methyl)carbamate (**41**, 600 mg, 2.39 mmol, 1.00 equiv.) in DMF (0.28 ml, 262 mg, 3.58 mmol, 1.50 equiv.), at 0 °C POCl$_3$ (0.44 ml, 732 mg, 4.78 mmol, 2.00 equiv.) was added. The mixture was stirred at this temperature for 1 h and additional 4 h at rt. The reaction was quenched by the addition of ice, stirred for further 30 min and followed by neutralization (~pH 6) of the very acidic aqueous solution with 20% aq. NaOH (w/w). After extraction with EtOAc, the combined organic layers were dried over Na$_2$SO$_4$ and the volatiles were removed under reduced pressure. Purification by flash column chromatography (cHex/EtOAc 2:1) resulted in the title compound as a yellow oil (132 mg, 0.47 mmol, 20%).

R_f (cHex/EtOAc 2:1) = 0.15. – ^{1}H NMR (400 MHz, CDCl$_3$) δ = 10.42 (s, 1H, C*H*O), 6.54 (s, 2H, 2 × C*H*$_{ar}$), 5.95 (ddt, ^3J = 17.3, 10.7, 5.6 Hz, 1H, C*H*), 5.37 – 5.14 (m, 2H, CHC*H*$_2$), 4.66 (d, ^3J = 5.6 Hz, 2H, C*H*$_2$), 3.88 (s, 6H, 2 × OC*H*$_3$), 3.37 (s, 3H, NC*H*$_3$) ppm. – ^{13}C NMR (101 MHz, CDCl$_3$) δ = 162.6 (C$_q$, 2 × *C*$_{ar}$OCH$_3$), 154.7 (C$_q$, *C*=O), 150.1 (C$_q$, *C*$_{ar}$N), 132.4 (+, *C*H), 118.3 (–, CH*C*H$_2$), 111.9 (C$_q$, *C*$_{ar}$CHO), 100.9 (+, 2 × *C*$_{ar}$H), 66.7 (–, *O*CH$_2$), 56.3 (+, 2 × O*C*H$_3$), 37.3 (+, N*C*H$_3$) ppm. – IR (ATR, ṽ) = 2941, 1708, 1680, 1599, 1569, 1456, 1385, 1352, 1324, 1243, 1221, 1144, 1073, 1017, 965, 932, 838, 763, 743, 583, 517, 380 cm^{-1}. – MS (70 eV, EI, 70 °C) *m/z* (%) = 279 (52) [M]$^+$, 262 (19), 251 (41) 238 (16) [M–C$_3$H$_5$]$^+$, 222 (11) [M–C$_3$H$_5$O]$^+$, 195 (13), 194 (100) [M–C$_4$H$_5$O$_2$]$^+$, 169 (10), 166 (21), 119 (17), 100 (11), 58 (77). – HRMS (EI, C$_{14}$H$_{17}$NO$_5$) = calc.: 279.1107; found: 279.1109.

3,5-Dimethoxy-*N*-methylaniline (45) 3,5-Dimethoxy-*N*,*N*-dimethylaniline (112)

A) To a solution of 3,5-dimethoxyaniline (**39**, 100 mg, 0.65 mmol, 1.00 equiv.) and Et$_3$N (0.10 ml, 73.0 mg, 0.72 mmol, 1.10 equiv.) in abs. THF (1.31 ml), MeI (50.0 μl, 111 mg, 0.78 mmol, 1.20 equiv.) was added and the reaction mixture was stirred under reflux overnight. The reaction was quenched by addition of 2 M aq. NaOH and extracted with EtOAc. The combined organic layers were washed with brine, dried over Na$_2$SO$_4$ and the volatiles were removed under reduced pressure. Purification by flash column chromatography (cHex/EtOAc 4:1) resulted in the title compound **45** (30.3 mg, 0.18 mmol, 28%) as green oil and the title compound **112** (51.8 mg, 0.29 mmol, 48%) as a green solid.

B) Under inert atmosphere a round-bottom flask was charged with allyl (3,5-dimethoxyphenyl)(methyl)carbamate (**XX**, 400 mg, 1.59 mmol, 1.00 equiv.), Pd(Ph$_3$P)$_4$ (92.0 mg, 0.08 mmol, 0.05 equiv.) and 1,3-dimethylbarbituric acid (497 mg, 3.18 mmol, 2.00 equiv.) in THF (0.50 M, 3.18 ml) and the reaction mixture was stirred overnight at 50 °C. The reaction was quenched by addition of Na$_2$CO$_3$ sol. and extracted with EtOAc. The combined organic layers were washed with brine, dried over Na$_2$SO$_4$ and the volatiles were removed under reduced pressure. Purification by flash column chromatography (cHex/EtOAc 3:1) resulted in the title compound **45** as green oil (240 mg, 0.14 mmol, 90%).

45) R_f (cHex/EtOAc 4:1) = 0.32. – ^{1}H NMR (400 MHz, CDCl$_3$) δ = 5.89 (t, ^4J = 2.2 Hz, 1H, C*H*$_{ar}$), 5.81 (d, ^4J = 2.2 Hz, 2H, 2 × C*H*$_{ar}$), 3.76 (s, 6H, 2 × OC*H*$_3$), 2.81 (s, 3H, NC*H*$_3$) ppm. – ^{13}C NMR (101 MHz, CDCl$_3$) δ = 162.6 (C$_q$, 2 × *C*$_{ar}$OCH$_3$), 152.1 (C$_q$, *C*$_{ar}$N), 92.2 (+, 2 × *C*$_{ar}$H), 90.6 (+, *C*$_{ar}$H), 56.1 (+, 2 × O*C*H$_3$), 31.7 (+, NH*C*H$_3$) ppm. – IR (ATR, ṽ) = 3408, 2934, 2837, 1591, 1514, 1486, 1455, 1409, 1254, 1200, 1176, 1148, 1086, 1062, 987, 926, 808, 682,

538 cm^{-1}. – **MS** (70 eV, EI, 20 °C) m/z (%) = 168 (11) [M+H]$^+$, 167 (100) [M]$^+$, 166 (27) [M–H]$^+$, 138 (31) [M–CH$_3$NH]$^+$, 131 (11), 69 (15). – **HRMS** (EI, C$_9$H$_{13}$NO$_2$) = calc.: 167.0946; found: 167.0948.

112) R_f (cHex/EtOAc 4:1) = 0.55. – **^{1}H NMR** (400 MHz, CDCl$_3$) δ = 5.91 (s, 3H, 3 × CH$_{ar}$), 3.78 (s, 6H, 2 × OCH$_3$), 2.92 (s, 6H, N(CH$_3$)$_2$) ppm. – **^{13}C NMR** (101 MHz, CDCl$_3$) δ = 161.7 (C$_q$, 2 × C$_{ar}$OCH$_3$), 152.6 (C$_q$, C$_{ar}$N), 92.1 (+, 2 × C$_{ar}$H), 89.0 (+, C$_{ar}$H), 55.3 (+, 2 × OCH$_3$), 40.8 (+, N(CH$_3$)$_2$) ppm. – **IR** (ATR, ṽ) = 3003, 2960, 2886, 1595, 1501, 1463, 1439, 1374, 1281, 1241, 1194, 1143, 1061, 1016, 794, 678, 593, 455 cm^{-1}. – **MS** (70 eV, EI, 20 °C) m/z (%) = 182 (10) [M+H]$^+$, 181 (100) [M]$^+$, 180 (34) [M–H]$^+$, 166 (12) [M–CH$_3$]$^+$, 152 (24), 131 (12), 69 (22). – **HRMS** (EI, C$_{10}$H$_{15}$NO$_2$) = calc.: 181.1103; found: 181.1103.

3,5-Dimethoxy-N-ethyl-N-methylaniline (46a)

Under inert atmosphere a round-bottom flask was charged with 3,5-di-methoxy-N-methylaniline (**45**, 2.76 g, 16.5 mmol, 1.00 equiv.), Et$_3$N (6.86 ml, 5.01 g, 49.5 mmol, 3.00 equiv.), 1-bromoethane (6.15 ml, 8.99 g, 82.5 mmol, 5.00 equiv.) and abs. dioxane (33 ml), and stirred under reflux until no starting material was detected anymore. Subsequently the reaction was quenched by addition of H$_2$O and extracted with EtOAc. The combined organic layers were washed with brine, dried over Na$_2$SO4 and the volatiles were removed under reduced pressure. The product was obtained without any further purification as a yellow oil (2.36 g, 12.1 mmol, 73%).

^{1}H NMR (500 MHz, CDCl$_3$) δ = 5.97 – 5.81 (m, 3H, 3 × CH$_{ar}$), 3.78 (s, 6H, 2 × OCH$_3$), 3.37 (q, ^3J = 7.1 Hz, 2H, NCH$_2$), 2.89 (s, 3H, NCH$_3$), 1.12 (t, ^3J = 7.1 Hz, 3H, CH$_3$) ppm. – **^{13}C NMR** (126 MHz, CDCl$_3$) δ = 161.8 (C$_q$, 2 × C$_{ar}$OCH$_3$), 151.1 (C$_q$, C$_{ar}$N), 91.8 (+, 2 × C$_{ar}$H), 88.2 (+, C$_{ar}$H), 55.3 (+, 2 × OCH$_3$), 47.1 (–, NCH$_2$), 37.7 (+, NCH$_3$), 11.5 (+, CH$_3$) ppm. – **IR** (ATR, ṽ) = 2932, 2837, 1610, 1578, 1492, 1457, 1374, 1349, 1330, 1272, 1202, 1141, 1067, 1016, 928, 804, 681, 540, 446 cm^{-1}. – **HRMS** (ESI, C$_{11}$H$_{17}$O$_2$N) = calc.: 196.1332 [M+H]$^+$; found: 196.1325 [M+H]$^+$.

N-Butyl-3,5-dimethoxy-N-methylanilin (46b)

Under inert atmosphere a *vial* was charged with 3,5-dimethoxy-N-methylaniline (**45**, 190 mg, 1.14 mmol, 1.00 equiv.), Et$_3$N (0.79 ml, 575 mg, 3.41 mmol, 3.00 equiv.), 1-bromobutane (0.36 ml, 467 mg, 3.41 mmol, 3.00 equiv.) and abs. dioxane (2.27 ml), and stirred under reflux until no starting material was detected anymore. Subsequently the reaction was quenched by addition of H$_2$O and extracted with EtOAc. The combined organic layers were washed with brine, dried over Na$_2$SO$_4$ and the volatiles were removed under reduced pressure. The product was obtained without any further purification as a yellow oil (57.8 mg, 0.26 mmol, 85%).

^{1}H NMR (500 MHz, CDCl$_3$) δ = 5.87 (s, 3H, 3 × CH_{ar}), 3.78 (s, 6H, 2 × OCH_3), 3.29 – 3.25 (m, 2H, NCH_2), 2.91 (s, 3H, NCH_3), 1.61 – 1.51 (m, 2H, NCH$_2$CH_2), 1.33 (sex., ^3J = 7.4 Hz, 2H, CH_2CH$_3$), 0.94 (t, ^3J = 7.4 Hz, 3H, CH_3) ppm. – ^{13}C NMR (101 MHz, CDCl$_3$) δ = 161.8 (C$_q$, 2 × C$_{ar}$OCH$_3$) 151.3 (C$_q$, C$_{ar}$N), 91.5 (+, 2 × C$_{ar}$H), 88.0 (+, C$_{ar}$H), 55.3 (+, 2 × OCH$_3$), 52.7 (–, NCH$_2$), 38.6 (+, NCH$_3$), 29.1 (–, CH$_2$), 20.5 (–, CH$_2$), 14.2 (+, CH$_3$) ppm. – HRMS (ESI, C$_{13}$H$_{21}$O$_2$N) = calc.: 224.1645 [M+H]$^+$; found: 224.1643 [M+H]$^+$. – Analytical data are consistent with literature.[228]

N-Hexyl-3,5-dimethoxy-*N*-methylaniline (46c)

Under inert atmosphere a round-bottom flask was charged with 3,5-dimethoxy-*N*-methylaniline (**45**, 3.55 g, 21.2 mmol, 1.00 equiv.), Et$_3$N (8.82 ml, 6.44 g, 63.6 mmol, 3.00 equiv.), 1-bromoethane (8.90 ml, 10.5 g, 63.6 mmol, 3.00 equiv.) and abs. dioxane (43 ml), and stirred under reflux until no starting material was detected anymore. Subsequently the reaction was quenched by addition of H$_2$O and extracted with EtOAc. The combined organic layers were washed with brine, dried over Na$_2$SO$_4$ and the volatiles were removed under reduced pressure. The product was obtained without any further purification as a yellow oil (5.29 g, 21.0 mmol, 99%).

^{1}H NMR (500 MHz, CDCl$_3$) δ = 5.88 (s, 3H, 3 × CH_{ar}), 3.78 (s, 6H, 2 × OCH_3), 3.26 (t, ^3J = 7.4 Hz, 2H, NCH_2), 2.91 (s, 3H, NCH_3), 1.65 – 1.49 (m, 2H, CH_2), 1.37 – 1.25 (m, 6H, 3 × CH_2), 0.93 – 0.84 (m, 3H, CH_3) ppm. – ^{13}C NMR (126 MHz, CDCl$_3$) δ = 161.8 (C$_q$, 2 × C$_{ar}$OCH$_3$), 151.3 (C$_q$, C$_{ar}$N), 91.5 (+, 2 × C$_{ar}$H), 88.0 (+, C$_{ar}$H), 55.3 (+, 2 × OCH$_3$), 53.0 (–, NCH$_2$), 38.6 (+, NCH$_3$), 31.9 (–, CH$_2$), 27.0 (–, CH$_2$), 26.9 (–, CH$_2$), 22.8 (–, CH$_2$), 14.2 (+, CH$_3$) ppm. – IR (ATR, ṽ) = 2927, 2856, 1611, 1578, 1492, 1458, 1369, 1331, 1269, 1245, 1201, 1174, 1150, 1063, 929, 804, 726, 680, 540, 386 cm^{-1}. – HRMS (ESI, C$_{15}$H$_{25}$O$_2$N) = calc.: 252.1958 [M+H]$^+$; found: 252.1948 [M+H]$^+$.

2,6-Dimethoxy-4-(ethyl(methyl)amino)benzaldehyde (44a)

To a solution of *N*-ethyl-3,5-dimethoxy-*N*-methylaniline (**46a**, 2.36 g, 12.1 mmol, 1.00 equiv.) in DCM (81 ml) at 0 °C, POCl$_3$ (1.33 ml, 2.23 g, 14.5 mmol, 1.20 equiv.) and DMF (1.12 ml, 1.06 g, 14.5 mmol, 1.20 equiv.) were added. The mixture was stirred at this temperature for 1 h and additional 4 h at rt. The reaction was quenched by the addition of ice, stirred for further 30 min and followed by neutralization (~pH 6) of the very acidic aqueous solution with 20% aq. NaOH (w/w). After extraction with DCM, the combined organic layers were dried over Na_2SO_4 and the volatiles were removed under reduced pressure. Purification by flash column chromatography (cHex/EtOAc 1:1) resulted in the title compound as a pink solid (0.77 g, 3.45 mmol, 29%).

R_f (cHex/EtOAc 1:1) = 0.09. – ^{1}H NMR (500 MHz, CDCl$_3$) δ = 10.19 (s, 1H, CHO), 5.68 (s, 2H, 2 × CH$_{ar}$), 3.82 (s, 6H, 2 × OCH$_3$), 3.42 (q, ^3J = 7.1 Hz, 2H, NCH$_2$), 2.99 (s, 3H, NCH$_3$), 1.16 (t, ^3J = 7.1 Hz, 3H, CH$_3$) ppm. – ^{13}C NMR (126 MHz, CDCl$_3$) δ = 186.5 (+, CHO), 164.3 (C$_q$, 2 × C$_{ar}$OCH$_3$), 154.8 (C$_q$, C$_{ar}$N), 104.9 (C$_q$, C$_{ar}$CHO), 86.8 (+, 2 × C$_{ar}$H), 55.6 (+, 2 × OCH$_3$), 46.8 (–, NCH$_2$), 37.7 (+, NCH$_3$), 11.8 (+, CH$_3$) ppm. – IR (ATR, ṽ) = 2966, 2928, 2849, 2774, 1655, 1596, 1540, 1475, 1451, 1436, 1408, 1378, 1347, 1303, 1258, 1218, 1190, 1145, 1118, 1084, 1072, 1040, 1017, 959, 928, 846, 790, 721, 691, 649, 592, 564, 541, 509, 484, 405, 381 cm^{-1}. – HRMS (ESI, C$_{12}$H$_{17}$O$_3$N) = calc.: 224.1281 [M+H]$^+$; found: 224.1273 [M+H]$^+$.

4-(Butyl(methyl)amino)-2,6-dimethoxybenzaldehyde (44b)

To a solution of N-butyl-3,5-dimethoxy-N-methylaniline (46b, 140 mg, 0.63 mmol, 1.00 equiv.) in DCM (6.27 ml), at 0 °C POCl$_3$ (0.07 ml, 115 mg, 0.75 mmol, 1.20 equiv.) and DMF (0.06 ml, 55.0 mg, 0.75 mmol, 1.20 equiv.) were added. The mixture was stirred at this temperature for 1 h and additional 4 h at rt. The reaction was quenched by the addition of ice, stirred for further 30 min, followed by neutralization (~pH 6) of the very acidic aqueous solution with 20% aq. NaOH (w/w). After extraction with DCM, the combined organic layers were dried over Na_2SO_4 and the volatiles were removed under reduced pressure. Purification by flash column chromatography (cHex/EtOAc 1:1) resulted in the title compound as a yellow oil (116 mg, 0.46 mmol, 73%).

R_f (cHex/EtOAc 1:1) = 0.09 – ^{1}H NMR (400 MHz, CDCl$_3$) δ = 10.23 (s, 1H, CHO), 5.72 (s, 2H, 2 × CH$_{ar}$), 3.87 (s, 6H, 2 × OCH$_3$), 3.39 (t, ^3J = 7.5 Hz, 2H, NCH$_2$), 3.05 (s, 3H, NCH$_3$), 1.68 – 1.54 (m, 2H, NCH$_2$CH$_2$), 1.47 – 1.28 (m, 2H, CH$_2$CH$_3$), 1.03 – 0.93 (m, 3H, CH$_3$) ppm. – ^{13}C NMR (101 MHz, CDCl$_3$) δ = 186.6 (+, CHO), 164.4 (C$_q$, 2 × C$_{ar}$OCH$_3$), 155.1 (C$_q$, C$_{ar}$N), 105.3 (C$_q$, C$_{ar}$CHO), 87.2 (+, 2 × C$_{ar}$H), 55.8 (+, 2 × OCH$_3$), 52.4 (–, NCH$_2$), 38.7 (+, NCH$_3$), 29.4 (–, CH$_2$), 20.4 (–, CH$_2$), 14.0 (+, CH$_3$) ppm. – IR (ATR, ṽ) = 2929, 2774, 2870, 1657, 1592, 1543, 1448, 1411, 1379, 1255, 1220, 1201, 1120, 932, 795, 692, 650, 585, 509 cm^{-1}. – MS (70 eV, EI, 80 °C) m/z (%) = 251 (33) [M]$^+$, 209 (12), 208 (100) [M–C$_3$H$_7$]$^+$. – HRMS (EI, C$_{14}$H$_{21}$NO$_3$) = calc.: 251.1521; found: 251.1521.

4-(Hexyl(methyl)amino)-2,6-dimethoxybenzaldehyde (44c)

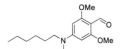

To a solution of *N*-hexyl-3,5-dimethoxy-*N*-methylaniline (**46c**, 5.28 g, 21.0 mmol, 1.00 equiv.) in DCM (140 ml) at 0 °C, POCl$_3$ (2.30 ml, 3.86 g, 25.2 mmol, 1.20 equiv.) and DMF (1.94 ml, 1.84 g, 25.2 mmol, 1.20 equiv.) were added. The mixture was stirred at this temperature for 1 h and additional 4 h at rt. The reaction was quenched by the addition of ice, stirred for further 30 min, followed by neutralization (~pH 6) of the very acidic aqueous solution with 20% aq. NaOH (w/w). After extraction with DCM, the combined organic layers were dried over Na$_2$SO$_4$ and the volatiles were removed under reduced pressure. Purification by flash column chromatography (*c*Hex/EtOAc 1:1) resulted in the title compound as an off-white solid (3.06 g, 10.9 mmol, 52%).

R_f (*c*Hex/EtOAc 1:1) = 0.12. – **^{1}H NMR** (500 MHz, CDCl$_3$) δ = 10.22 (s, 1H, C*H*O), 5.71 (s, 2H, 2 × C*H*$_{ar}$), 3.86 (s, 6H, 2 × OC*H*$_3$), 3.44 – 3.31 (m, 2H, NC*H*$_2$), 3.04 (s, 3H, NC*H*$_3$), 1.62 (quin., ^3J = 7.3 Hz, 2H, NCH$_2$C*H*$_2$), 1.40 – 1.28 (m, 6H, 3 × C*H*$_2$), 0.96 – 0.83 (m, 3H, C*H*$_3$) ppm. – **^{13}C NMR** (126 MHz, CDCl$_3$) δ = 186.6 (+, *C*HO), 164.3 (C$_q$, 2 × *C*$_{ar}$OCH$_3$), 155.0 (C$_q$, *C*$_{ar}$N), 105.1 (C$_q$, *C*$_{ar}$CHO), 87.0 (+, 2 × *C*$_{ar}$H), 55.7 (+, 2 × O*C*H$_3$), 52.7 (–, N*C*H$_2$), 38.7 (+, N*C*H$_3$), 31.7 (–, *C*H$_2$), 27.2 (–, *C*H$_2$), 26.8 (–, *C*H$_2$), 22.7 (–, *C*H$_2$), 14.1 (+, *C*H$_3$) ppm. – **IR** (ATR, ṽ) = 3002, 2955, 2924, 2853, 2774, 1655, 1598, 1537, 1476, 1466, 1446, 1409, 1380, 1367, 1339, 1313, 1281, 1255, 1220, 1193, 1171, 1140, 1120, 1094, 1038, 932, 901, 875, 810, 792, 758, 721, 693, 650, 613, 588, 567, 543, 530, 506, 496, 458, 443, 422, 409, 385, 378 cm^{-1}. – **HRMS** (ESI, C$_{16}$H$_{25}$O$_3$N) = calc.: 280.1907 [M+H]$^+$; found: 280.1895 [M+H]$^+$.

4-(Ethyl(methyl)amino)-2-hydroxy-6-methoxybenzaldehyde (19h)

According to **GP3**, to a solution of 4-(ethyl(methyl)amino)-2,6-dimethoxybenzaldehyde (**44a**, 700 mg, 3.14 mmol, 1.00 equiv.), dissolved in a mixture of abs. ACN and abs. DCM (2:1, 30.0 ml, 0.10 M), AlCl$_3$ (836 mg, 6.27 mmol, 2.00 equiv.) and NaI (940 g, 6.27 mmol, 2.00 equiv.) were added in portions at 0 °C and the reaction was stirred for 1 h at rt. Purification by flash column chromatography (*c*Hex/EtOAc 2:1 to 1:1) resulted in the title compound as a grey solid (466 mg, 2.23 mmol, 71%).

R_f (*c*Hex/EtOAc 2:1) = 0.47. – **^{1}H NMR** (500 MHz, CDCl$_3$) δ = 12.61 (s, 1H, O*H*), 9.88 (s, 1H, C*H*O), 5.70 (d, ^4J = 2.2 Hz, 1H, C*H*$_{ar}$), 5.58 (d, ^4J = 2.2 Hz, 1H, C*H*$_{ar}$), 3.84 (s, 3H, OC*H*$_3$), 3.43 (q, ^3J = 7.1 Hz, 2H, NC*H*$_2$), 3.01 (s, 3H, NC*H*$_3$), 1.18 (t, ^3J = 7.1 Hz, 3H, C*H*$_3$) ppm. – **^{13}C NMR** (126 MHz, CDCl$_3$) δ = 189.4 (+, *C*HO), 165.8 (C$_q$, *C*$_{ar}$OH), 163.9 (C$_q$, *C*$_{ar}$OCH$_3$), 156.5

(C_q, C_{ar}N), 103.1 (C_q, C_{ar}CHO), 90.7 (+, C_{ar}H), 85.7 (+, C_{ar}H), 55.4 (+, OCH_3), 47.0 (–, NCH_2), 37.9 (+, NCH_3), 12.0 (+, CH_3) ppm. – **IR** (ATR, $\tilde{v}$) = 2921, 1642, 1593, 1552, 1514, 1466, 1426, 1390, 1349, 1319, 1289, 1238, 1139, 1108, 1085, 954, 848, 799, 782, 724, 698, 659, 556, 514, 464, 407 cm^{-1}. – **HRMS** (ESI, $C_{11}H_{15}O_3N$) = calc.: 210.1125 [M+H]$^+$; found: 210.1127 [M+H]$^+$.

4-(Butyl(methyl)amino)-2-hydroxy-6-methoxybenzaldehyde (19i)

According to **GP3**, to a solution of 4-(butyl(methyl)amino)-2,6-dimethoxybenzaldehyde (**44b**, 920 mg, 3.66 mmol, 1.00 equiv.), dissolved in a mixture of abs. ACN and DCM (2:1, 26.0 ml), AlCl$_3$ (976 mg, 7.32 mmol, 2.00 equiv.) and NaI (1.10 g, 7.32 mmol, 2.00 equiv.) were added in portions at 0 °C and the reaction was stirred for 1 h at rt. Purification by flash column chromatography (cHex/EtOAc 4:1) resulted in the title compound as an off-white solid (694 mg, 2.92 mmol, 80%).

R_f (cHex/EtOAc 4:1) = 0.22. – **^{1}H NMR** (500 MHz, CDCl$_3$) δ = 12.63 (s, 1H, OH), 9.89 (s, 1H, CHO), 5.70 (d, ^4J = 2.2 Hz, 1H, CH_{ar}), 5.58 (d, ^4J = 2.2 Hz, 1H, CH_{ar}), 3.84 (s, 3H, OCH_3), 3.35 (t, ^3J = 7.5 Hz, 2H, NCH_2), 3.02 (s, 3H, NCH_3), 1.66 – 1.54 (m, 2H, NCH$_2$CH_2), 1.35 (sex., ^3J = 7.5 Hz, 2H, CH_2CH$_3$), 0.94 (t, ^3J = 7.4 Hz, 3H, CH_3) ppm. – **^{13}C NMR** (126 MHz, CDCl$_3$) δ = 189.4 (+, CHO), 165.8 (C_q, C_{ar}OH), 163.8 (C_q, C_{ar}OCH$_3$), 156.7 (C_q, C_{ar}N), 103.1 (C_q, C_{ar}CHO), 90.8 (+, C_{ar}H), 85.7 (+, C_{ar}H), 55.4 (+, OCH_3), 52.5 (–, NCH_2), 38.8 (+, NCH_3), 29.4 (–, CH_2), 20.3 (–, CH_2), 14.1 (+, CH_3) ppm. – **IR** (ATR, $\tilde{v}$) = 2956, 2928, 2871, 2863, 1650, 1591, 1553, 1517, 1492, 1477, 1465, 1456, 1429, 1390, 1370, 1322, 1290, 1244, 1228, 1191, 1146, 1111, 1085, 833, 809, 783, 727, 703, 660, 558, 516, 484, 477, 421 cm^{-1}. – **HRMS** (ESI, $C_{13}H_{19}O_3N$) = calc.: 238.1438 [M+H]$^+$; found: 238.1431 [M+H]$^+$.

4-(Hexyl(methyl)amino)-2-hydroxy-6-methoxybenzaldehyde (19j)

According to **GP3**, to a solution of 4-(hexyl(methyl)amino)-2,6-dimethoxybenzaldehyde (**44c**, 1.20 g, 4.30 mmol, 1.00 equiv.), dissolved in a mixture of abs. ACN and abs. DCM (2:1, 29.0 ml), AlCl$_3$ (1.15 g, 8.59 mmol, 2.00 equiv.) and NaI (1.29 g, 8.59 mmol, 2.00 equiv.) were added in portions at 0 °C and the reaction was stirred for 1 h at rt. Filtration over a small silica plug (cHex/EtOAc 5:1) resulted in the title compound as a yellow oil (655 mg, 2.47 mmol, 57%).

R_f (cHex/EtOAc 5:1) = 0.28. – **^{1}H NMR** (500 MHz, CDCl$_3$) δ = 12.63 (s, 1H, OH), 9.89 (s, 1H, CHO), 5.69 (d, ^4J = 2.2 Hz, 1H, CH_{ar}), 5.58 (d, ^4J = 2.2 Hz, 1H, CH_{ar}), 3.84 (s, 3H, OCH_3),

3.37 – 3.30 (m, 2H, NCH$_2$), 3.02 (s, 3H, NCH$_3$), 1.62 – 1.55 (m, 2H, NCH$_2$CH$_2$), 1.36 – 1.28 (m, 6H, 3 × CH$_2$), 0.93 – 0.86 (m, 3H, CH$_3$) ppm. – ^{13}C NMR (126 MHz, CDCl$_3$) δ = 189.4 (+, CHO), 165.8 (C$_q$, C$_{ar}$OH), 163.8 (C$_q$, C$_{ar}$OCH$_3$), 156.6 (C$_q$, C$_{ar}$N), 103.1 (C$_q$, C$_{ar}$CHO), 90.8 (+, C$_{ar}$H), 85.7 (+, C$_{ar}$H), 55.4 (+, OCH$_3$), 52.8 (–, NCH$_2$), 38.8 (+, NCH$_3$), 31.7 (–, CH$_2$), 27.3 (–, CH$_2$), 26.8 (–, CH$_2$), 22.7 (–, CH$_2$), 14.1 (+, CH$_3$) ppm. – IR (ATR, ṽ) = 2961, 2928, 2885, 2873, 2859, 1652, 1591, 1553, 1516, 1468, 1435, 1425, 1390, 1371, 1349, 1319, 1299, 1241, 1187, 1150, 1112, 1086, 1026, 999, 972, 895, 875, 816, 805, 781, 724, 703, 657, 558, 516, 496, 475, 425 cm^{-1}. – HRMS (ESI, C$_{15}$H$_{23}$O$_3$N) = calc.: 266.1751 [M+H]$^+$; found: 266.1743 [M+H]$^+$.

1-(3,5-Dimethoxyphenyl)ethan-1-one (48a)

According to GP4, to a solution of 3,5-dimethoxy benzoic acid (47, 1.01 g, 5.55 mmol, 1.00 equiv.) in abs. THF (22.2 ml), under an inert atmosphere methyl lithium (1.6 M in Et$_2$O, 7.28 ml, 256 mg, 11.7 mmol, 2.10 equiv.) was added at –78 °C and the reaction was stirred for 2 h at 0 °C. After extraction and concentration under reduced pressure, the product was obtained without any further purification as yellow oil (942 mg, 5.22 mmol, 94%).

^{1}H NMR (400 MHz, CDCl$_3$) δ = 7.06 (d, ^4J = 2.3 Hz, 2H, 2 × CH$_{ar}$), 6.62 (t, ^4J = 2.3 Hz, 1H, CH$_{ar}$), 3.80 (s, 6H, 2 × OCH$_3$), 2.55 (s, 3H, CH$_3$) ppm. – ^{13}C NMR (101 MHz, CDCl$_3$) δ = 197.8 (C$_q$, C=O), 160.9 (C$_q$, 2 × C$_{ar}$OCH$_3$), 139.1 (C$_q$, C$_{ar}$CO), 106.0 (+, 2 × C$_{ar}$H), 105.4 (+, C$_{ar}$H), 55.6 (+, 2 × OCH$_3$), 26.8 (+, CH$_3$) ppm. – IR (ATR, ṽ) = 3003, 2965, 2939, 2837, 1680, 1591, 1455, 1424, 1356, 1316, 1298, 1252, 1220, 1203, 1153, 1064, 1043, 992, 970, 938, 922, 843, 681, 643, 633, 606, 578, 540, 467, 450, 441, 419, 409, 388 cm^{-1}. – MS (70 eV, EI, 20 °C) m/z (%) = 180 (72) [M]$^+$, 165 (100) [M–CH$_3$]$^+$, 137 (25) [M–C$_2$H$_3$O]$^+$, 122 (23), 107 (11). – HRMS (EI, C$_{10}$H$_{12}$O$_3$) = calc.: 180.0786; found: 180.0786.

1-(3,5-Dimethoxyphenyl)pentan-1-one (48b)

According to GP4, to a solution of 3,5-dimethoxy benzoic acid (47, 1.50 g, 8.23 mmol, 1.00 equiv.) in abs. THF (66.0 ml), under an inert atmosphere n-butyllithium (2.5 M in hexane, 6.92 ml, 1.11 g, 17.3 mmol, 2.10 equiv.) was added at –78 °C and the reaction was stirred for 2 h at 0 °C. After extraction and concentration under reduced pressure, the product was purified by flash column chromatography (cHex/EtOAc 10:1 to 5:1) to yield the title compound as a white solid (1.56 g, 7.01 mmol, 85%).

R_f (cHex/EtOAc 10:1) = 0.33. – **^{1}H NMR** (500 MHz, CDCl$_3$) δ = 7.09 (d, ^4J = 2.3 Hz, 2H, 2 ×

CH_{ar}), 6.63 (t, ^4J = 2.3 Hz, 1H, CH_{ar}), 3.83 (s, 6H, 2 × OCH_3), 2.91 (t, ^3J = 7.4 Hz, 2H, COCH_2),

1.70 (quin., J = 7.4 Hz, 2H, CH_2), 1.39 (sex, ^3J = 7.4 Hz, 2H, CH_2CH$_3$), 0.94 (t, J = 7.4 Hz, 3H,

CH_3) ppm. – **^{13}C NMR** (126 MHz, CDCl$_3$) δ = 200.4 (C$_q$, C=O), 160.9 (C$_q$, 2 × C_{ar}OCH$_3$),

139.2 (C$_q$, C_{ar}CO), 106.0 (+, 2 × C_{ar}H), 105.1 (+, C_{ar}H), 55.7 (+, 2 × OCH$_3$), 38.6 (–, CH$_2$), 26.7

(–, CH$_2$), 22.6 (–, CH$_2$), 14.1 (+, CH$_3$) ppm. – **IR** (ATR, ṽ) = 2956, 2933, 2871, 2838, 1683,

1591, 1456, 1425, 1348, 1293, 1252, 1204, 1153, 1067, 1028, 1001, 925, 845, 754, 714, 679,

623, 542, 457 cm^{-1}. – **HRMS** (ESI, C$_{13}$H$_{18}$O$_3$) = calc.: 223.1329 [M+H]$^+$; found: 223.1324

[M+H]$^+$.

1-(3,5-Dimethoxyphenyl)heptan-1-one (48c)

According to **GP4**, to a solution of 3,5-dimethoxy benzoic acid (**47**, 1.09 g, 5.99 mmol, 1.00 equiv.) in abs. THF (24.0 ml), under an inert atmosphere n-hexyllithium (2.3 M in hexane, 5.47 ml, 1.16 g, 12.6 mmol, 2.10 equiv.) was added at –78 °C and the reaction was stirred for 2 h at 0 °C. Extraction and concentration under reduced pressure, gave the product as a colourless oil (1.47 g, 5.87 mmol, 98%).

^{1}H NMR (400 MHz, CDCl$_3$) δ = 7.09 (s, 2H, 2 × CH_{ar}), 6.64 (s, 1H, CH_{ar}), 3.83 (s, 6H, 2 ×

OCH_3), 2.91 (t, ^3J = 7.4 Hz, 2H, COCH_2), 1.70 (quin., ^3J = 7.4 Hz, 2H, COCH$_2$CH_2), 1.44 –

1.25 (m, 6H, 3 × CH_2), 0.89 (t, ^3J = 6.7 Hz, 3H, CH_3) ppm. – **^{13}C NMR** (101 MHz, CDCl$_3$) δ

= 200.4 (C$_q$, C=O), 161.0 (C$_q$, 2 × C_{ar}OCH$_3$), 139.2 (C$_q$, C_{ar}CO), 106.0 (+, 2 × C_{ar}H), 105.2 (+,

C_{ar}H), 55.7 (+, 2 × OCH$_3$), 38.9 (–, CH$_2$), 31.8 (–, CH$_2$), 29.2 (–, CH$_2$), 24.6 (–, CH$_2$), 22.7 (–,

CH$_2$), 14.2 (+, CH$_3$) ppm. – **IR** (ATR, ṽ) = 3093, 3001, 2971, 2955, 2945, 2916, 2894, 2873,

2848, 1681, 1592, 1466, 1453, 1426, 1405, 1355, 1290, 1250, 1206, 1154, 1112, 1059, 1024,

1001, 940, 925, 896, 888, 864, 852, 845, 831, 760, 724, 680, 632, 542, 520, 483, 459, 452, 408

cm^{-1}. – **MS** (70 eV, EI, 50 °C) m/z (%) = 251 (11) [M+H]$^+$, 250 (66) [M]$^+$, 181 (21), 180 (100),

166 (20), 165 (100) [M–C$_6$H$_{13}$]$^+$, 152 (35), 138 (18), 137 (26), 122 (18), 69 (13). – **HRMS** (EI,

C$_{15}$H$_{22}$O$_3$) = calc.: 250.1563; found: 250.1563.

2-(3,5-Dimethoxyphenyl)propan-2-ol (49a)

According to **GP5**, to a solution of 1-(3,5-dimethoxyphenyl)ethan-1-one (**48a**, 670 mg, 3.72 mmol, 1.00 equiv.) in abs. THF (14.9 ml), MeMgBr (3.00 M in Et$_2$O, 1.49 ml, 532 mg, 4.46 mmol, 1.20 equiv.) was added at –78 °C and the reaction was stirred for 2 h at rt. Purification by flash column chromatography (cHex/EtOAc 3:1) resulted in the title compound as a cloudy oil (543 mg, 2.77 mmol, 74%).

R_f (cHex/EtOAc 3:1) = 0.28. – ^{1}H NMR (300 MHz, CDCl$_3$) δ = 6.65 (d, ^4J = 2.3 Hz, 2H, 2 × CH$_{ar}$), 6.36 (t, ^4J = 2.3 Hz, 1H, CH$_{ar}$), 3.81 (s, 6H, 2 × OCH$_3$), 1.56 (s, 6H, 2 × CH$_3$) ppm. – Analytical data are consistent with literature.[151]

2-(3,5-Dimethoxyphenyl)hexan-2-ol (49b)

According to **GP5**, to a solution of 1-(3,5-dimethoxyphenyl)pentan-1-one (**48b**, 3.79 g, 17.1 mmol, 1.00 equiv.) in abs. THF (68.0 ml), MeMgBr (3.00 M in Et$_2$O, 11.4 ml, 4.07 g, 34.1 mmol, 2.00 equiv.) was added at −78 °C and the reaction was stirred for 2 h at rt. Purification by flash column chromatography (cHex/EtOAc 10:1) resulted in the title compound as a colourless oil (2.94 g, 12.4 mmol, 72%).

R_f (cHex/EtOAc 10:1) = 0.38. – ^{1}H NMR (500 MHz, CDCl$_3$) δ = 6.59 (d, ^4J = 2.3 Hz, 2H, 2 × CH$_{ar}$), 6.34 (t, ^4J = 2.3 Hz, 1H, CH$_{ar}$), 3.80 (s, 6H, 2 × OCH$_3$), 1.83 – 1.69 (m, 2H, CCH$_2$), 1.52 (s, 3H, CCH$_3$), 1.31 – 1.07 (m, 4H, 2 × CH$_2$), 0.85 (t, ^3J = 7.0 Hz, 3H, CH$_3$) ppm. – ^{13}C NMR (126 MHz, CDCl$_3$) δ = 160.7 (C$_q$, 2 × C$_{ar}$OCH$_3$), 151.1 (C$_q$, C$_{ar}$CO), 103.4 (+, 2 × C$_{ar}$H), 98.1 (+, C$_{ar}$H), 75.0 (C$_q$, COH), 55.4 (+, 2 × OCH$_3$), 43.9 (–, CCH$_2$), 30.3 (+, CCH$_3$), 26.2 (–, CH$_2$), 23.2 (–, CH$_2$), 14.2 (+, CH$_3$) ppm. – IR (ATR, ṽ) = 3463, 2955, 2934, 2862, 2836, 2119, 2018, 1593, 1455, 1423, 1338, 1289, 1253, 1203, 1151, 1046, 994, 923, 844, 828, 731, 701, 629, 591, 543 cm^{-1}. – HRMS (ESI, C$_{14}$H$_{22}$O$_3$) = calc.: 239.1642 [M+H]$^+$; found. 239.1629 [M+H]$^+$.

2-(3,5-Dimethoxyphenyl)octan-2-ol (49c)

According to **GP5**, to a solution of 1-(3,5-dimethoxyphenyl)heptan-1-one (**48c**, 235 mg, 0.94 mmol, 1.00 equiv.) in abs. THF (3.75 ml), MeMgBr (3 M in Et$_2$O, 0.63 ml, 224 mg, 1.88 mmol, 2.00 equiv.) was added at −78 °C and the reaction was stirred for 2 h at rt. Purification by flash column chromatography (cHex/EtOAc 5:1) resulted in the title compound as a colourless oil (205 mg, 0.77 mmol, 82%).

R_f (cHex/EtOAc 5:1) = 0.30. – ^{1}H NMR (400 MHz, CDCl$_3$) δ = 6.59 (d, ^4J = 2.3 Hz, 2H, 2 × CH$_{ar}$), 6.34 (t, ^4J = 2.3 Hz, 1H, CH$_{ar}$), 3.80 (s, 6H, 2 × OCH$_3$), 1.82 – 1.67 (m, 2H, CCH$_2$), 1.51 (s, 3H, CCH$_3$), 1.28 – 1.10 (m, 8H, 4 × CH$_2$), 0.88 – 0.81 (m, 3H, CH$_3$) ppm. – ^{13}C NMR (101 MHz, CDCl$_3$) δ = 160.7 (C$_q$, 2 × C$_{ar}$OCH$_3$), 151.1 (C$_q$, C$_{ar}$COH), 103.4 (+, 2 × C$_{ar}$H), 98.2 (+,

C_{ar}H), 75.0 (C_q, COH), 55.4 (+, 2 × OCH$_3$), 44.2 (–, CCH$_2$), 31.9 (–, CH$_2$), 30.3 (+, CCH$_3$), 29.7 (–, CH$_2$), 24.0 (–, CH$_2$), 22.7 (–, CH$_2$), 14.2 (+, CH$_3$) ppm. – **IR** (ATR, ṽ) = 2953, 2931, 2859, 1595, 1456, 1422, 1290, 1204, 1153, 1048, 925, 844, 827, 700 cm^{-1}. – **MS** (70 eV, EI, 30 °C) m/z (%) = 266 (30) [M]$^+$, 182 (47), 181 (100) [M–C$_6$H$_{13}$]$^+$, 180 (24), 165 (31), 139 (51). – **HRMS** (EI, C$_{16}$H$_{26}$O$_3$) = calc.: 266.1882; found: 266.1883.

1-(*tert*-Butyl)-3,5-dimethoxybenzene (50a)

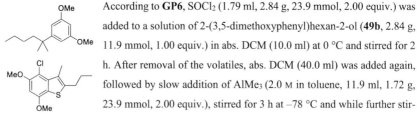

According to **GP6**, SOCl$_2$ (0.11 ml, 184 mg, 1.54 mmol, 2.00 equiv.) was added to a solution of 2-(3,5-dimethoxyphenyl)propan-2-ol (**49a**, 152 mg, 0.77 mmol, 1.00 equiv.) in abs. DCM (0.50 ml) at 0 °C and stirred for 2 h. After removal of the volatiles, abs. DCM (2.56 ml) was added again, followed by slow addition of AlMe$_3$ (2.00 M in toluene, 0.77 ml, 111 mg, 1.54 mmol, 2.00 equiv.), stirred for 3 h at –78 °C and while further stirring overnight the reaction mixture was allowed to warm up to rt. Subsequently extraction and purification by flash column chromatography (*n*Hex/DCM 4:1) resulted in the title compound as a colourless oil (82.0 mg, 0.42 mmol, 55%).

R_f (*n*Hex/DCM 5:1) = 0.20. – **^{1}H NMR** (400 MHz, CDCl$_3$) δ = 6.55 (d, ^4J = 2.3 Hz, 2H, 2 × CH_{ar}), 6.31 (t, ^4J = 2.3 Hz, 1H, CH_{ar}), 3.80 (s, 6H, 2 × OCH_3), 1.30 (s, 9H, 3 × CH_3) ppm. – **^{13}C NMR** (101 MHz, CDCl$_3$) δ = 160.6 (C_q, 2 × C_{ar}OCH$_3$), 154.0 (C_q, C_{ar}C), 104.2 (+, 2 × C_{ar}H), 96.9 (+, C_{ar}H), 55.4 (+, 2 × OCH$_3$), 35.1 (C_q, C(CH$_3$)$_3$), 31.4 (+, C(CH$_3$)$_3$) ppm. – **IR** (ATR, ṽ) = 2955, 2904, 2867, 2834, 1594, 1455, 1421, 1363, 1341, 1312, 1293, 1259, 1198, 1153, 1122, 1052, 1024, 925, 844, 830, 697 cm^{-1}. – **MS** (70 eV, EI, 20 °C) m/z (%) = 194 (65) [M]$^+$, 180 (13), 179 (100) [M–CH$_3$]$^+$, 152 (27), 151 (14), 139 (19), 91 (10). – **HRMS** (EI, C$_{12}$H$_{18}$O$_2$) = calc.: 194.1307; found: 194.1308.

1,3-Dimethoxy-5-(2-methylhexan-2-yl)benzene (50b), 4-Chloro-5,7-dimethoxy-3-methyl-2-propylbenzo[b]thiophene (52a)

According to **GP6**, SOCl$_2$ (1.79 ml, 2.84 g, 23.9 mmol, 2.00 equiv.) was added to a solution of 2-(3,5-dimethoxyphenyl)hexan-2-ol (**49b**, 2.84 g, 11.9 mmol, 1.00 equiv.) in abs. DCM (10.0 ml) at 0 °C and stirred for 2 h. After removal of the volatiles, abs. DCM (40.0 ml) was added again, followed by slow addition of AlMe$_3$ (2.0 M in toluene, 11.9 ml, 1.72 g, 23.9 mmol, 2.00 equiv.), stirred for 3 h at –78 °C and while further stirring overnight the reaction mixture was allowed to warm up to rt. Subsequently extraction and purification by flash column chromatography (*n*Pen/EtOAc 40:1) resulted in the title compound

50b as a cloudy oil (1.98 g, 8.38 mmol, 70%) and the side product **52a** as an off-white solid (284 mg, 1.00 mmol, 8.4%).[151]

50b) R_f(nPen/EtOAc 10:1) = 0.86. – **^{1}H NMR** (500 MHz, CDCl$_3$) δ = 6.50 (d, ^4J = 2.1 Hz, 2H, 2 × CH_{ar}), 6.31 (d, ^4J = 2.1 Hz, 1H, CH_{ar}), 3.80 (s, 6H, 2 × OCH_3), 1.60 – 1.54 (m, 2H, CCH_2), 1.27 (s, 6H, 2 × CCH_3), 1.25 – 1.18 (m, 2H, CH_2), 1.10 – 1.01 (m, 2H, CH_2), 0.84 (t, ^3J = 7.4 Hz, 3H, CH_3) ppm. – **^{13}C NMR** (126 MHz, CDCl$_3$) δ = 160.6 (C$_q$, 2 × C$_{ar}$OCH$_3$), 152.7 (C$_q$, C$_{ar}$C), 104.8 (+, 2 × C$_{ar}$H), 96.7 (+, C$_{ar}$H), 55.3 (+, 2 × OCH$_3$), 44.4 (–, CCH$_2$), 38.1 (C$_q$, C(CH$_3$)$_2$), 29.1 (+, C(CH$_3$)$_2$), 27.1 (–, CH$_2$), 23.5 (–, CH$_2$), 14.2 (+, CH$_3$) ppm. – **IR** (ATR, ṽ) = 2955, 2929, 2860, 2835, 1593, 1455, 1421, 1364, 1337, 1322, 1290, 1260, 1203, 1153, 1052, 927, 845, 831, 730, 698, 629, 540, 438 cm^{-1}. – **HRMS** (ESI, C$_{15}$H$_{24}$O$_2$) = calc.: 237.1849 [M+H]$^+$; found: 237.1842 [M+H]$^+$.

52a) R_f(nPen/EtOAc 10:1) = 0.74. – **^{1}H NMR** (500 MHz, CDCl$_3$) δ = 6.49 (s, 1H, CH_{ar}), 3.97 (s, 3H, OCH_3), 3.95 (s, 3H, OCH_3), 2.79 (t, ^3J = 7.4 Hz, 2H, CCH_2), 2.59 (s, 3H, CCH_3), 1.70 (quin., ^3J = 7.4 Hz, 2H, CH_2), 1.00 (t, ^3J = 7.4 Hz, 3H, CH_3) ppm. – **^{13}C NMR** (126 MHz, CDCl$_3$) δ = 153.7 (C$_q$, C$_{ar}$OCH$_3$), 152.8 (C$_q$, C$_{ar}$OCH$_3$), 142.2 (C$_q$, C$_{ar}$CH$_2$), 138.4 (C$_q$, 3a-C$_{ar}$), 128.3 (C$_q$, C$_{ar}$CH$_3$), 121.4 (C$_q$, 7a-C$_{ar}$), 109.1 (C$_q$, C$_{ar}$Cl), 92.9 (+, C$_{ar}$H), 57.8 (+, OCH$_3$), 56.1 (+, OCH$_3$), 30.9 (–, CH$_2$), 24.4 (–, CH$_2$), 15.5 (+, 3-CCH$_3$), 13.9 (+, CH$_3$) ppm. – **IR** (ATR, ṽ) = 3003, 2958, 2927, 2868, 2844, 1592, 1565, 1536, 1460, 1451, 1426, 1384, 1374, 1341, 1322, 1276, 1238, 1207, 1145, 1118, 1086, 1043, 1001, 959, 912, 887, 870, 786, 772, 744, 701, 629, 620, 579, 540, 520, 497, 401 cm^{-1}. – **HRMS** (ESI, C$_{14}$H$_{17}$O$_2$SCl) = calc.: 285.0711/287.0681 [M+H]$^+$; found: 285.0701/287.0670 [M+H]$^+$.

1,3-Dimethoxy-5-(2-methyloctan-2-yl)benzene (50c), 4-Chloro-5,7-dimethoxy-3-methyl-2-pentylbenzo[b]thiophene (52b)

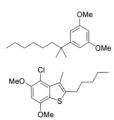

According to **GP6**, SOCl$_2$ (2.18 ml, 3.57 g, 30.0 mmol, 2.00 equiv.) was added to a solution of 2-(3,5-dimethoxyphenyl)octan-2-ol (49c, 4.00 g, 15.0 mmol, 1.00 equiv.) in abs. DCM (10.0 ml) at 0 °C and stirred for 2 h. After removal of the volatiles, abs. DCM (50.0 ml) was added again, followed by slow addition of AlMe$_3$ (2.0 M in toluene, 15.0 ml, 2.17 g, 30.0 mmol, 2.00 equiv.), stirred for 3 h at –78 °C and while further stirring overnight the reaction mixture was allowed to warm up to rt. Subsequently extraction and purification by flash column chromatography (nPen/EtOAc 100:1) resulted in the title compound **50c** as a colourless oil (2.14 g, 8.10 mmol, 54%) and the side product **52b** as a white solid (987 mg, 3.15 mmol, 21%).[151]

50c) R_f(nPen/EtOAc 50:1) = 0.41. – **^{1}H NMR** (500 MHz, CDCl$_3$) δ = 6.49 (d, ^4J = 2.3 Hz, 2H, 2 × CH_{ar}), 6.31 (d, ^4J = 2.3 Hz, 1H, CH_{ar}), 3.80 (s, 6H, 2 × OCH_3), 1.59 – 1.52 (m, 2H, CCH_2), 1.26 (s, 6H, 2 × CCH_3), 1.24 – 1.17 (m, 6H, 3 × CH_2), 1.10 – 1.02 (m, 2H, CH_2), 0.84 (t, ^3J = 7.0 Hz, 3H, CH_3) ppm. – **^{13}C NMR** (126 MHz, CDCl$_3$) δ = 160.5 (C$_q$, 2 × C_{ar}OCH$_3$), 152.7 (C$_q$, C_{ar}C), 104.8 (+, 2 × C_{ar}H), 96.7 (+, C_{ar}H), 55.3 (+, 2 × OCH$_3$), 44.7 (–, CCH$_2$), 38.1 (C$_q$, C(CH$_3$)$_2$), 31.9 (–, CH$_2$), 30.2 (–, CH$_2$), 29.1 (+, C(CH$_3$)$_2$), 24.8 (–, CH$_2$), 22.8 (–, CH$_2$), 14.2 (+, CH$_3$) ppm. – **IR** (ATR, $\tilde{v}$) = 2955, 2928, 2856, 1593, 1455, 1421, 1364, 1323, 1293, 1203, 1153, 1054, 927, 846, 831, 724, 699, 629, 541, 444 cm^{-1}. – **HRMS** (ESI, C$_{17}$H$_{28}$O$_2$) = calc.: 265.2162 [M+H]$^+$; found: 265.2150 [M+H]$^+$.

52b) R_f(nPen/EtOAc 50:1) = 0.27. – **^{1}H NMR** (500 MHz, CDCl$_3$) δ = 6.49 (s, 1H, CH_{ar}), 3.97 (s, 3H, OCH_3), 3.95 (s, 3H, OCH_3), 2.80 (t, ^3J = 7.6 Hz, 2H, CCH_2), 2.59 (s, 3H, CCH_3), 1.70 – 1.60 (m, 2H, CH_2), 1.41 – 1.28 (m, 4H, 2 × CH_2), 0.94 – 0.84 (m, 3H, CH_3) ppm. – **^{13}C NMR** (126 MHz, CDCl$_3$) δ = 153.7 (C$_q$, C_{ar}OCH$_3$), 152.8 (C$_q$, C_{ar}OCH$_3$), 142.5 (C$_q$, C_{ar}CH$_2$), 138.4 (C$_q$, 3a-C_{ar}), 128.2 (C$_q$, C_{ar}CH$_3$), 121.4 (C$_q$, 7a-C_{ar}), 109.1 (C$_q$, C_{ar}Cl), 92.9 (+, C_{ar}H), 57.8 (+, OCH$_3$), 56.1 (+, OCH$_3$), 31.5 (–, CH$_2$), 30.8 (–, CH$_2$), 28.8 (–, CH$_2$), 22.6 (–, CH$_2$), 15.4 (+, 3-CCH$_3$), 14.1 (+, CH$_3$) ppm. – **IR** (ATR, $\tilde{v}$) = 2953, 2925, 2854, 2846, 1585, 1567, 1543, 1451, 1426, 1385, 1373, 1341, 1317, 1272, 1241, 1232, 1208, 1154, 1145, 1113, 1086, 1045, 1001, 992, 962, 950, 897, 795, 773, 730, 707, 693, 577, 555, 398 cm^{-1}. – **HRMS** (ESI, C$_{16}$H$_{21}$O$_2$SCl) = calc.: 313.1024/315.0994 [M+H]$^+$; found: 313.1017/315.0985 [M+H]$^+$.

4-(*tert*-Butyl)-2,6-dimethoxybenzaldehyde (51a)

Similar to **GP2**, to a solution of 1-(*tert*-butyl)-3,5-dimethoxybenzene (**50a**, 1.85 g, 9.52 mmol, 1.00 equiv.) and TMEDA (2.16 ml, 1.66 g, 14.3 mmol, 1.50 equiv.) in abs. THF (34.0 ml), first *n*-butyllithium (2.5 M in hexane, 5.71 ml, 915 mg, 14.3 mmol, 1.50 equiv.) was added and the reaction was stirred at rt for 2 h. Subsequently followed by addition of DMF (2.22 ml, 2.09 g, 28.6 mmol, 3.00 equiv.) at 0 °C and stirred for another 3 h at rt. Purification by flash column chromatography (*c*Hex/EtOAc 3:1) resulted in the title compound as a yellow oil (1.28 g, 5.76 mmol, 60%).

R_f (*c*Hex/EtOAc 3:1) = 0.30. – **^{1}H NMR** (400 MHz, CDCl$_3$) δ = 10.46 (s, 1H, CHO), 6.58 (s, 2H, 2 × CH_{ar}), 3.91 (s, 6H, 2 × OCH_3), 1.34 (s, 9H, 3 × CH_3) ppm. – **^{13}C NMR** (101 MHz, CDCl$_3$) δ = 189.1 (+, CHO), 162.2 (C$_q$, 2 × C_{ar}OCH$_3$), 160.7 (C$_q$, C_{ar}C), 112.4 (C$_q$, C_{ar}CHO), 101.4 (+, 2 × C_{ar}H), 56.1 (+, 2 × OCH$_3$), 36.1 (C$_q$, C(CH$_3$)$_3$), 31.1 (+, C(CH$_3$)$_3$) ppm. – **IR** (ATR, $\tilde{v}$) = 3007, 2958, 2939, 2867, 2853, 2772, 1744, 1679, 1647, 1601, 1561, 1507, 1462, 1402, 1364, 1317, 1309, 1238, 1201, 1128, 1109, 1031, 1009, 969, 935, 833, 819, 781, 728,

667, 581, 561, 507, 458, 418, 402, 375 cm^{-1}. – **MS** (70 eV, EI, 50 °C) *m/z* (%) = 223 (21) [M+H]$^+$, 222 (100) [M]$^+$, 221 (27) [M–H]$^+$, 207 (69) [M–CH$_3$]$^+$, 205 (17), 204 (32), 189 (11), 180 (11), 177 (12), 176 (14), 165 (10) [M–C$_4$H$_9$]$^+$, 139 (13), 91 (14). – **HRMS** (EI, C$_{13}$H$_{18}$O$_3$) = calc.: 222.1256 found: 222.1258.

2,6-Dimethoxy-4-(2-methylhexan-2-yl)benzaldehyde (51b)

According to **GP2**, to a solution of 1,3-dimethoxy-5-(2-methylhexan-2-yl)benzene (**50b**, 1.92 g, 8.12 mmol, 1.00 equiv.) and TMEDA (1.84 ml, 1.42 g, 12.2 mmol, 1.50 equiv.) in abs. Et$_2$O (29.0 ml), first *n*-butyllithium (2.5 M in hexane, 4.87 ml, 7.80 g, 12.2 mmol, 1.50 equiv.) was added and the reaction was stirred at rt for 2 h. Subsequently followed by addition of DMF (1.89 ml, 1.78 g, 24.4 mmol, 3.00 equiv.) at 0 °C and stirred for another 4 h at rt. Purification by flash column chromatography (*c*Hex/EtOAc 5:1 to 2:1) resulted in the title compound as a yellow oil (1.77 g, 6.88 mmol, 82%).

R$_f$(*c*Hex/EtOAc 5:1) = 0.29. – **^{1}H NMR** (500 MHz, CDCl$_3$) δ = 10.44 (s, 1H, C*H*O), 6.51 (s, 2H, 2 × C*H*$_{ar}$), 3.88 (s, 6H, 2 × OC*H*$_3$), 1.62 – 1.55 (m, 2H, CC*H*$_2$), 1.28 (s, 6H, 2 × CC*H*$_3$), 1.25 – 1.17 (m, 2H, C*H*$_2$), 1.07 – 0.98 (m, 2H, C*H*$_2$CH$_3$), 0.85 – 0.80 (m, 3H, C*H*$_3$) ppm. – **^{13}C NMR** (126 MHz, CDCl$_3$) δ = 189.1 (+, CHO), 162.1 (C$_q$, 2 × *C*$_{ar}$OCH$_3$), 159.6 (C$_q$, *C*$_{ar}$C), 112.2 (C$_q$, *C*$_{ar}$CHO), 101.9 (+, 2 × *C*$_{ar}$H), 56.0 (+, 2 × O*C*H$_3$), 44.1 (–, *C*H$_2$), 39.2 (C$_q$, *C*(CH$_3$)$_2$), 28.8 (+, C(*C*H$_3$)$_2$), 27.0 (–, *C*H$_2$), 23.4 (–, *C*H$_2$), 14.1 (+, *C*H$_3$) ppm. – **IR** (ATR, ṽ) = 2955, 2930, 2860, 2776, 1733, 1682, 1602, 1563, 1454, 1403, 1322, 1240, 1203, 1129, 1112, 930, 832, 791, 730, 674, 619, 584, 510 cm^{-1}. – **HRMS** (ESI, C$_{16}$H$_{24}$O$_3$) = calc.: 265.1798 [M+H]$^+$; found: 265.1790 [M+H]$^+$.

2,6-Dimethoxy-4-(2-methyloctan-2-yl)benzaldehyde (51c)

According to **GP2**, to a solution of 1,3-dimethoxy-5-(2-methyloctan-2-yl)benzene (**50c**, 2.00 g, 7.56 mmol, 1.00 equiv.) and TMEDA (1.71 ml, 1.32 g, 11.4 mmol, 1.50 equiv.) in abs. Et$_2$O (27.0 ml), first *n*-butyl lithium (2.5 M in hexane, 4.54 ml, 727 mg, 11.4 mmol, 1.50 equiv.) was added and the reaction was stirred at rt for 2 h. Subsequently followed by addition of DMF (1.76 ml, 1.66 g, 22.7 mmol, 3.00 equiv.) at 0 °C and stirred for another 4 h at rt. Purification by flash column chromatography (*c*Hex/EtOAc 5:1) resulted in the title compound as a pale-yellow oil (1.94 g, 6.64 mmol, 88%).

R_f (cHex/EtOAc 5:1) = 0.27. – **^{1}H NMR** (500 MHz, CDCl$_3$) δ = 10.45 (s, 1H, CHO), 6.51 (s, 2H, 2 × CH_{ar}), 3.89 (s, 6H, 2 × OCH_3), 1.62 – 1.55 (m, 2H, CH_2), 1.29 (s, 6H, C(CH_3)$_2$), 1.25 – 1.17 (m, 6H, 3 × CH_2), 1.09 – 1.01 (m, 2H, CH_2), 0.84 (t, ^3J = 6.9 Hz, 3H, CH_3) ppm. – **^{13}C NMR** (126 MHz, CDCl$_3$) δ = 189.2 (+, CHO), 162.1 (C$_q$, 2 × C_{ar}OCH$_3$), 159.7 (C$_q$, C_{ar}C), 112.2 (C$_q$, C_{ar}CHO), 101.9 (+, 2 × C_{ar}H), 56.0 (+, 2 × OCH$_3$), 44.4 (–, CCH$_2$), 39.2 (C$_q$, C(CH$_3$)$_2$), 31.8 (–, CH$_2$), 30.0 (–, CH$_2$), 28.8 (+, C(CH$_3$)$_2$), 24.7 (–, CH$_2$), 22.7 (–, CH$_2$), 14.2 (+, CH$_3$) ppm. – **IR** (ATR, ṽ) = 2955, 2928, 2857, 2775, 1683, 1602, 1564, 1455, 1404, 1366, 1323, 1241, 1203, 1130, 1113, 1034, 948, 932, 891, 832, 788, 725, 673, 622, 584, 543, 510, 475 cm^{-1}. – **HRMS** (ESI, C$_{18}$H$_{28}$O$_3$) = calc.: 293.2111 [M+H]$^+$; found: 293.2102 [M+H]$^+$.

4-(*tert*-Butyl)-2-hydroxy-6-methoxybenzaldehyde (19k)

According to **GP3**, to a solution of 4-(*tert*-butyl)-2,6-dimethoxybenzaldehyde (**51a**, 1.18 g, 5.31 mmol, 1.00 equiv.), dissolved in a mixture of abs. ACN and abs. DCM (2:1, 53.0 ml, 0.10 M), AlCl$_3$ (1.77 mg, 13.3 mmol, 2.50 equiv.) and NaI (1.99 g, 13.3 mmol, 2.50 equiv.) were added in portions at 0 °C and the reaction was stirred for 1 h at rt. Purification by flash column chromatography (cHex/EtOAc 4:1) resulted in the title compound as an off-white solid (992 mg, 4.76 mmol, 90%).

R_f (cHex/EtOAc 4:1) = 0.67 – **^{1}H NMR** (400 MHz, CDCl$_3$) δ = 11.93 (s, 1H, OH), 10.26 (s, 1H, CHO), 6.55 (d, ^4J = 1.5 Hz, 1H, CH_{ar}), 6.39 (d, ^4J = 1.5 Hz, 1H, CH_{ar}), 3.99 (s, 3H, OCH_3), 1.30 (s, 9H, 3 × CH_3) ppm. – **^{13}C NMR** (101 MHz, CDCl$_3$) δ = 193.7 (+, CHO), 163.6 (C$_q$, C_{ar}), 163.5 (C$_q$, C_{ar}), 162.2 (C$_q$, C_{ar}), 109.1 (C$_q$, C_{ar}CHO), 107.1 (+, C_{ar}H), 98.8 (+, C_{ar}H), 55.7 (+, OCH$_3$), 36.1 (C$_q$, C(CH$_3$)$_3$), 30.9 (+, C(CH$_3$)$_3$) ppm. – **IR** (ATR, ṽ) = 2961, 2904, 2870, 1636, 1625, 1568, 1533, 1499, 1462, 1412, 1394, 1363, 1346, 1313, 1228, 1201, 1113, 1101, 1027, 984, 925, 853, 826, 782, 730, 687, 674, 626, 565, 547, 514, 456, 439, 419, 385 cm^{-1}. – **MS** (70 eV, EI, 20 °C) *m/z* (%) = 209 (13) [M+H]$^+$, 208 (98) [M]$^+$, 194 (12), 193 (100) [M–CH$_3$]$^+$, 190 (25), 181 (58), 175 (12), 166 (23) 165 (16), 162 (16), 147 (15), 137 (19), 133 (11), 125 (11), 119 (15), 100 (17), 93 (13), 91 (17), 79 (11), 77 (16), 69 (72). – **HRMS** (EI, C$_{12}$H$_{16}$O$_3$) = calc.: 208.1099; found: 208.1100.

2-Hydroxy-6-methoxy-4-(2-methylhexan-2-yl)benzaldehyde (19l)

According to **GP3**, to a solution of 2,6-dimethoxy-4-(2-methylhexan-2-yl)benzaldehyde (**51b**, 1.70 g, 6.43 mmol, 1.00 equiv.), dissolved in a mixture of abs. ACN and abs. DCM (2:1, 64.0 ml), AlCl$_3$ (2.14 g, 16.1 mmol, 2.50 equiv.) and NaI (2.41 g, 16.1 mmol, 2.50 equiv.) were added in portions at 0 °C and the reaction was stirred for 1 h at rt. Purification by flash column chromatography (cHex/EtOAc 5:1) resulted in the title compound as a dark blue oil (1.54 g, 6.15 mmol, 96%).

R_f (cHex/EtOAc 5:1) = 0.16. – **^{1}H NMR** (500 MHz, CDCl$_3$) δ = 11.93 (s, 1H, O*H*), 10.25 (s, 1H, C*H*O), 6.50 (d, ^4J = 1.4 Hz, 1H, C*H*$_{ar}$), 6.33 (d, ^4J = 1.4 Hz, 1H, C*H*$_{ar}$), 3.89 (s, 3H, OC*H*$_3$), 1.60 – 1.53 (m, 2H, CC*H*$_2$), 1.26 (s, 6H, 2 × CC*H*$_3$), 1.24 – 1.18 (m, 2H, C*H*$_2$), 1.08 – 0.99 (m, 2H, C*H*$_2$CH$_3$), 0.83 (t, ^3J = 7.3 Hz, 3H, C*H*$_3$) ppm. – **^{13}C NMR** (126 MHz, CDCl$_3$) δ = 193.7 (+, *C*HO), 163.4 (C$_q$, *C*$_{ar}$), 162.6 (C$_q$, *C*$_{ar}$), 162.2 (C$_q$, *C*$_{ar}$), 109.0 (C$_q$, *C*$_{ar}$CHO), 107.7 (+, *C*$_{ar}$H), 99.1 (+, *C*$_{ar}$H), 55.7 (+, O*C*H$_3$), 43.9 (–, *C*H$_2$), 39.2 (C$_q$, *C*(CH$_3$)$_2$), 28.7 (+, 2 × C(*C*H$_3$)$_2$), 27.0 (–, *C*H$_2$), 23.3 (–, *C*H$_2$), 14.1 (+, *C*H$_3$) ppm. – **IR** (ATR, ṽ) = 2957, 2929, 2860, 1628, 1567, 1462, 1412, 1395, 1346, 1315, 1228, 1211, 1114, 986, 921, 854, 829, 780, 729, 679, 549, 514 cm^{-1}. – **HRMS** (ESI, C$_{15}$H$_{22}$O$_3$) = calc.: 251.1642 [M+H]$^+$; found. 251.1632 [M+H]$^+$.

2-Hydroxy-6-methoxy-4-(2-methyloctan-2-yl)benzaldehyde (19m)

 According to **GP3**, to a solution of 2,6-dimethoxy-4-(2-methyloctan-2-yl)benzaldehyde (**51c**, 1.70 g, 5.81 mmol, 1.00 equiv.), dissolved in a mixture of abs. ACN and abs. DCM (2:1, 39.0 ml), AlCl$_3$ (1.55 g, 11.6 mmol, 2.00 equiv.) and NaI (1.74 g, 11.6 mmol, 2.00 equiv.) were added in portions at 0 °C and the reaction was stirred for 1 h at rt. Filtration over a small silica plug (cHex/EtOAc 4:1) resulted in the title compound as a yellow oil (1.53 g, 5.47 mmol, 94%).

R_f (cHex/EtOAc 4:1) = 0.49. – **^{1}H NMR** (500 MHz, CDCl$_3$) δ = 11.93 (s, 1H, O*H*), 10.26 (s, 1H, C*H*O), 6.50 (d, ^4J = 1.5 Hz, 1H, C*H*$_{ar}$), 6.33 (d, ^4J = 1.5 Hz, 1H, C*H*$_{ar}$), 3.89 (s, 3H, OC*H*$_3$), 1.60 – 1.53 (m, 2H, CC*H*$_2$), 1.26 (s, 6H, C(C*H*$_3$)$_2$), 1.24 – 1.16 (m, 6H, 3 × C*H*$_2$), 1.10 – 1.01 (m, 2H, C*H*$_2$), 0.85 (t, ^3J = 7.0 Hz, 3H, C*H*$_3$) ppm. – **^{13}C NMR** (126 MHz, CDCl$_3$) δ = 193.7 (+, *C*HO), 163.4 (C$_q$, *C*$_{ar}$), 162.7 (C$_q$, *C*$_{ar}$), 162.2 (C$_q$, *C*$_{ar}$), 109.0 (C$_q$, *C*$_{ar}$CHO), 107.8 (+, *C*$_{ar}$H), 99.2 (+, *C*$_{ar}$H), 55.7 (+, O*C*H$_3$), 44.2 (–, C*C*H$_2$), 39.2 (C$_q$, *C*(CH$_3$)$_2$), 31.9 (–, *C*H$_2$), 30.1 (–, *C*H$_2$), 28.7 (+, C(*C*H$_3$)$_2$), 24.7 (–, *C*H$_2$), 22.8 (–, *C*H$_2$), 14.2 (+, *C*H$_3$) ppm. – **IR** (ATR, ṽ) = 2956, 2928, 2857, 1628, 1567, 1534, 1502, 1462, 1412, 1395, 1346, 1315, 1228, 1210, 1113, 986, 918, 854, 829, 779, 728, 679, 625, 548, 514 cm^{-1}. – **HRMS** (ESI, C$_{17}$H$_{26}$O$_3$) = calc.: 279.1955 [M+H]$^+$; found. 279.1949 [M+H]$^+$.

2-Hydroxy-3,6-dimethylbenzaldehyde (19n)

Under inert atmosphere, to a solution of 2-dimethylphenol (2.00 g, 16.4 mmol, 1.00 equiv.), paraformaldehyde (3.32 g, 110 mmol, 6.75 equiv.) and MgCl$_2$ (2.34 g, 24.6 mmol, 1.50 equiv.) in abs ACN (80.0 mL), Et$_3$N (8.56 ml, 6.21 g, 61.4 mmol, 3.75

equiv.) was added and the reaction mixture was stirred under reflux for 4 h. After cooling to rt, the reaction was quenched by adding 5% aq. HCl and stirring for another 30 min. After extraction with Et$_2$O, the combined organic layers were dried over Na$_2$SO$_4$, filtrated and the volatiles were removed under reduced pressure. Purification by flash column chromatography (cHex/EtOAc 40:1) resulted in the title compound as yellow solid (1.09 mg, 7.27 mmol, 44%).

R_f (cHex/EtOAc 40:1) = 0.20 – ^{1}H NMR (300 MHz, CDCl$_3$) δ = 12.18 (s, 1H, OH), 10.30 (s, 1H, CHO), 7.24 (d, ^3J = 7.5 Hz, 1H, CH$_{ar}$), 6.62 (d, ^3J = 7.5 Hz, 1H, CH$_{ar}$), 2.57 (s, 3H, CH$_3$), 2.21 (s, 3H, CH$_3$) ppm. – The analytical data are consistent with literature.[141]

2,5-Dihydroxy-3,4,6-trimethylbenzaldhyd (55)

A) Under inert atmosphere, to a solution of 2,3,5-trimethyl-1,4-hydroquinone (54, 1.20 g, 7.88 mmol, 1.00 equiv.), paraformaldehyde (1.60 g, 53.2 mmol, 6.75 equiv.) and MgCl$_2$ (1.13 g, 11.8 mmol, 1.50 equiv.) in abs ACN (40.0 mL), Et$_3$N (4.12 ml, 2.99 g, 29.6 mmol, 3.75 equiv.) was added and the reaction mixture was stirred under reflux for 4 h. After cooling to rt, the reaction was quenched by adding 5% aq. HCl and stirring for another 30 min. After extraction with Et$_2$O, the combined organic layers were dried over Na$_2$SO$_4$, filtrated and the volatiles were removed under reduced pressure. Purification by flash column chromatography (cHex/EtOAc 5:1) resulted in the title compound as yellow solid (143 mg, 0.79 mmol, 10%).

B) Under inert atmosphere, to a solution of 2,3,5-trimethyl-1,4-hydroquinone (54, 7.00 g, 46.0 mmol, 1.00 equiv.) in abs. DCM (130 mL), TiCl$_4$ (25.2 ml, 43.6 g, 230 mmol, 5.00 equiv.) was added slowly over a period of 15 min, at 0 °C. The reaction mixture was stirred for 1 h at this temperature, then dichloromethyl methyl ether (4.29 ml, 5.55 g, 48.3 mmol, 1.05 equiv.) was added over a period of 15 min and the reaction mixture was stirred for another 2 d at rt. By addition of NH$_4$Cl sol. the reaction was quenched and extracted with DCM. The combined organic layers were washed with 2% aq. HCl, NaHCO$_3$ sol. and brine, dried over Na$_2$SO$_4$ and the volatiles were removed under reduced pressure. Purification by flash column chromatography (cHex/EtOAc 5:1) resulted in the title compound as yellow solid (3.79 g, 21.0 mmol, 46%).[107]

R_f (cHex/EtOAc 5:1) = 0.17 – ^{1}H NMR (400 MHz, CDCl$_3$) δ = 12.08 (s, 1H, OH), 10.24 (s, 1H, CHO), 4.40 (s, 1H, OH), 2.45 (s, 3H, CH$_3$), 2.24 (s, 3H, CH$_3$), 2.17 (s, 3H, CH$_3$) ppm. – The analytical data are consistent with literature.[107]

2,5-Dimethoxy-3,4,6-trimethylbenzaldehyde (56)

To a solution of 2,5-dihydroxy-3,4,6-trimethylbenzaldehyde (**55**, 3.79 g, 23.1 mmol, 1.00 equiv.) and K_2CO_3 (8.98 g, 65.0 mmol, 3.00 equiv.) in abs. acetone (45 mL), dimethyl sulphate (6.16 ml, 8.20 g, 65.0 mmol, 3.00 equiv.) was added and the reaction mixture was stirred under reflux for 4 h. The reaction was quenched by the addition of H_2O (60 mL), extracted with EtOAc, the combined organic layers were dried over Na_2SO_4, filtrated and the volatiles were removed under reduced pressure. Purification by flash column chromatography (cHex/EtOAc 10:1) resulted in the title compound as yellow solid (3.35 g, 16.1 mmol, 76%).[107]

R_f (cHex/EtOAc 10:1) = 0.35. – ^1H NMR (300 MHz, $CDCl_3$) δ = 10.48 (s, 1H, C*H*O), 3.77 (s, 3H, OC*H*$_3$), 3.64 (s, 3H, OC*H*$_3$), 2.49 (s, 3H, C*H*$_3$), 2.26 (s, 3H, C*H*$_3$), 2.20 (s, 3H, C*H*$_3$) ppm. – The analytical data are consistent with literature.[141]

2-Hydroxy-5-methoxy-3,4,6-trimethylbenzaldehyde (19o)

According to **GP3**, to a solution of 2,5-dimethoxy-3,4,6-trimethylbenzalde- hyde (**56**, 3.10 g, 14.9 mmol, 1.00 equiv.), dissolved in a mixture of abs. ACN and abs. DCM (2:1, 175 ml), $AlCl_3$ (4.96 g, 37.2 mmol, 2.50 equiv.) and NaI (5.58 g, 37.2 mmol, 2.50 equiv.) were added in portions at 0 °C and the reaction was stirred for 1 h at rt. Purification by flash column chromatography (cHex/EtOAc 10:1) resulted in the title compound as a yellow solid (2.68 g, 13.8 mmol, 92%).

R_f (cHex/EtOAc 10:1) = 0.42. – ^1H NMR (300 MHz, $CDCl_3$) δ = 12.20 (s, 3H, O*H*), 10.24 (s, 1H, C*H*O), 3.64 (s, 3H, OC*H*$_3$), 2.51 (s, 3H, C*H*$_3$), 2.27 (s, 3H, C*H*$_3$), 2.15 (s, 3H, C*H*$_3$) ppm. – The analytical data are consistent with literature.[141]

5.1.5   Synthesis and characterisation of cinnamaldehydes

(E)-3-(o-Tolyl)acrylaldehyde (20b)[157]

According to **GP7**, 1-iodo-2-methylbenzene (3.49 ml, 5.97 g, 27.4 mmol, 1.00 equiv.), acrolein (2.74 ml, 2.30 g, 41.0 mmol, 1.50 equiv.), $NaHCO_3$ (5.75 g, 68.4 mmol, 2.50 equiv.), TBAI (10.1 g, 27.4 mmol, 1.00 equiv.) and $Pd(OAc)_2$ (614 mg, 2.74 mmol, 0.10 equiv.) in abs. DMF (90.0 ml) were stirred for 24 h at 60 °C. Purification by flash column chromatography (cHex/EtOAc 20:1 to 10:1) resulted in the title compound as a yellow oil (2.93 g, 20.0 mmol, 73%).

R_f (cHex/EtOAc 15:1) = 0.21. – **^{1}H NMR** (300 MHz, CDCl$_3$) δ = 9.65 (d, ^3J = 7.7 Hz, 1H, C*H*O), 7.70 (d, ^3J = 15.8 Hz, 1H, C$_{ar}$C*H*), 7.52 (dd, $^{3,4}J$ = 8.1, 1.5 Hz, 1H, p-C*H*$_{ar}$), 7.32 – 7.06 (m, 3H, 3 × C*H*$_{ar}$), 6.59 (dd, ^3J = 15.8, 7.7 Hz, 1H, C*H*CHO), 2.41 (s, 3H, C*H*$_3$) ppm. – The analytical data are consistent with literature.[141]

(*E*)-3-(2-Fluorophenyl)acrylaldehyde (20e)

According to **GP7**, 1-fluoro-2-iodobenzene (1.55 ml, 2.96 g, 13.3 mmol, 1.00 equiv.), acrolein (1.33 ml, 1.12 g, 20.0 mmol, 1.50 equiv.), NaHCO$_3$ (2.80 g, 33.3 mmol, 2.50 equiv.), TBAI (4.92 g, 13.3 mmol, 1.00 equiv.) and Pd(OAc)$_2$ (299 mg, 1.33 mmol, 0.10 equiv.) in abs. DMF (53.5 ml) were stirred for 24 h at 60 °C. Purification by flash column chromatography (cHex/EtOAc 20:1 to 10:1) resulted in the title compound as an orange oil (1.32 g, 8.80 mmol, 66%).

R_f (cHex/EtOAc 10:1) = 0.21. – **^{1}H NMR** (500 MHz, CDCl$_3$) δ = 9.72 (d, ^3J = 7.7 Hz, 1H, C*H*O), 7.66 (d, ^3J = 16.2 Hz, 1H, C$_{ar}$C*H*), 7.59 (td, $^{3,4}J$ = 7.7, 1.7 Hz, 1H, p-C*H*$_{ar}$), 7.43 (qd, $^{3,4}J$ = 7.7, 1.7 Hz, 1H, m-C*H*$_{ar}$), 7.21 (d, ^3J = 7.5 Hz, 1H, m-C*H*$_{ar}$), 7.18 – 7.10 (m, 1H, o-C*H*$_{ar}$), 6.79 (dd, ^3J = 16.2, 7.7 Hz, 1H, C*H*CHO) ppm. – **^{13}C NMR** (126 MHz, CDCl$_3$) δ = 194.0 (+, CHO), 161.3 (C$_q$, d, ^1J = 255 Hz, C$_{ar}$F), 144.9 (+, d, ^3J = 3.6 Hz C$_{ar}$CH), 133.0 (+, d, ^3J = 8.9 Hz, m-C$_{ar}$H), 130.6 (+, d, ^4J = 5.4 Hz, CHCHO), 128.9 (+, d, ^4J = 2.6 Hz, p-C$_{ar}$H), 124.8 (+, d, ^3J = 3.6 Hz, m-C$_{ar}$H), 122.3 (C$_q$, d, ^2J = 11.6 Hz, o-C$_{ar}$), 116.5 (+, d, ^2J = 21.8 Hz, o-C$_{ar}$H) ppm. – **IR** (ATR, ṽ) = 2821, 2743, 1678, 1627, 1610, 1580, 1485, 1459, 1319, 1288, 1228, 1197, 1125, 1094, 1007, 974, 856, 793, 756, 597, 536, 481, 467 cm^{-1}. – **HRMS** (ESI, C$_9$H$_7$OF) = calc.: 151.0554 [M+H]$^+$; found: 151.0541 [M+H]$^+$.

(*E*)-3-(2-(Trifluoromethyl)phenyl)acrylaldehyde (20g)

According to **GP7**, 1-iodo-2-(trifluoromethyl)benzene (1.42 ml, 2.72 g, 9.99 mmol, 1.00 equiv.), acrolein (1.00 ml, 840 g, 15.0 mmol, 1.50 equiv.), NaHCO$_3$ (2.10 g, 25.0 mmol, 2.50 equiv.), TBAI (3.69 g, 9.99 mmol, 1.00 equiv.) and Pd(OAc)$_2$ (224 mg, 1.00 mmol, 0.10 equiv.) in abs. DMF (40.0 ml) were stirred for 24 h at 60 °C. Purification by flash column chromatography (cHex/EtOAc 20:1 to 10:1) resulted in the title compound as a brown solid (1.28 g, 6.37 mmol, 64%).

R_f (cHex/EtOAc 15:1) = 0.23. – **^{1}H NMR** (500 MHz, CDCl$_3$) δ = 9.76 (d, ^3J = 7.7 Hz, 1H, C*H*O), 7.87 (dd, $^{3,4}J$ = 15.8, 2.2 Hz, 1H, C$_{ar}$C*H*), 7.75 (dd, J = 7.9, 3.8 Hz, 2H, m- and p-C*H*$_{ar}$), 7.62 (t, ^3J = 7.6 Hz, 1H, o-C*H*$_{ar}$), 7.54 (t, ^3J = 7.7 Hz, 1H, m-C*H*$_{ar}$), 6.70 (dd, ^3J = 15.8, 7.7 Hz, 1H, C*H*CHO) ppm. – **^{13}C NMR** (126 MHz, CDCl$_3$) δ = 193.5 (+, CHO), 147.6 (+, t, ^4J = 2.2

Hz, C$_{ar}$CH), 132.8 (C$_q$, q, ^3J = 1.7 Hz, o-C$_{ar}$), 132.5 (+, m-C$_{ar}$H), 132.1 (+, m-C$_{ar}$H), 130.6 (+, p-C$_{ar}$H), 129.0 (C$_q$, q, ^2J = 30.5 Hz, C$_{ar}$CF$_3$), 128.1 (+, CHCHO), 126.5 (+, q, ^3J = 5.6 Hz, o–C$_{ar}$H), 124.0 (C$_q$, q, ^1J = 274.0 Hz, CF$_3$) ppm. – **IR** (ATR, ṽ) = 3068, 3051, 2840, 2768, 2731, 1677, 1628, 1604, 1574, 1547, 1487, 1453, 1398, 1312, 1288, 1249, 1208, 1164, 1109, 1060, 1031, 1000, 975, 891, 844, 768, 741, 656, 606, 594, 568, 520, 473, 439 cm^{-1}. – **HRMS** (ESI, C$_{10}$H$_7$OF$_3$) = calc.: 201.0522 [M+H]$^+$; found: 201.0516 [M+H]$^+$.

5.1.6   Synthesis and characterisation of 3-benzyl-7-alkoxycoumarins

3-Benzyl-5,7-dimethoxy-2H-chromen-2-one (21aa)

According to **GP8**, 2-hydroxy-4,6-dimethoxybenzaldehyde (**19a**, 91.0 mg, 0.50 mmol, 1.00 equiv.), cinnamaldehyde (0.16 ml, 165 mg, 1.25 mmol, 2.50 equiv.), K$_2$CO$_3$ (83.0 mg, 0.60 mmol, 1.20 equiv.) and 1,3-dimethylimidazolium dimethyl phosphate (0.10 ml, 133 mg, 0.60 mmol, 1.20 equiv.) in toluene (1.65 ml) were heated at 110 °C for 50 min *via* microwave irradiation. Purification by flash column chromatography (cHex/EtOAc 5:1) resulted in a mixture of the desired product and a methyl ester as impurity. Stirring in 2 M aq. NaOH (25 ml) for several hours, followed by extraction with EtOAc, drying over Na$_2$SO$_4$ and removal of the volatiles, gave the title compound as a yellow solid (26.0 mg, 0.08 mmol, 17%).

R_f (cHex/EtOAc 5:1) = 0.19. – ^{1}**H NMR** (400 MHz, CDCl$_3$) δ = 7.62 (s, 1H, CH), 7.32 – 7.23 (m, 4H, 4 × CH_{ar}), 7.23 – 7.16 (m, 1H, CH_{ar}), 6.35 (d, ^4J = 2.1 Hz, 1H, CH_{ar}), 6.20 (d, ^4J = 2.1 Hz, 1H, CH_{ar}), 3.80 – 3.79 (m, 2H, CH_2), 3.78 (s, 6H, 2 × OCH_3) ppm. – ^{13}**C NMR** (101 MHz, CDCl$_3$) δ = 163.0 (C$_q$, COO), 162.4 (C$_q$, 8a-C$_{ar}$), 156.6 (C$_q$, 7-C$_{ar}$OCH$_3$), 155.8 (C$_q$, 5-C$_{ar}$OCH$_3$), 138.7 (C$_q$, CH$_2$C$_{ar}$), 135.0 (+, CH), 129.3 (+, 2 × C$_{ar}$H), 128.7 (+, 2 × C$_{ar}$H), 126.6 (+, C$_{ar}$H), 123.7 (C$_q$, 3-C), 104.5 (C$_q$, 4a-C$_{ar}$), 94.9 (+, C$_{ar}$H), 92.5 (+, C$_{ar}$H), 56.0 (+, OCH$_3$), 55.9 (+, OCH$_3$), 36.8 (–, CH$_2$) ppm. – **IR** (ATR, ṽ) = 3050, 3029, 2998, 2950, 2917, 2838, 1698, 1606, 1495, 1467, 1450, 1426, 1366, 1315, 1296, 1243, 1229, 1205, 1165, 1158, 1144, 1109, 1079, 1049, 1031, 998, 981, 971, 949, 937, 896, 877, 830, 796, 759, 721, 701, 666, 637, 620, 609, 586, 531, 486, 456, 402 cm^{-1}. – **MS** (70 eV, EI, 120 °C) m/z (%) = 297 (23) [M+H]$^+$, 296 (100) [M]$^+$, 295 (15) [M–H]$^+$, 267 (13) [M–C$_2$H$_5$]$^+$. – **HRMS** (EI, C$_{18}$H$_{16}$O$_4$) = calc.: 296.1043; found: 296.1043.

5,7-Dimethoxy-3-(2-methylbenzyl)-2*H*-chromen-2-one (21ab)

According to **GP8**, 2-hydroxy-4,6-dimethoxybenzaldehyde (**19a**, 100 mg, 0.55 mmol, 1.00 equiv.), 2-methylcinnamaldehyde (201 mg, 1.37 mmol, 2.50 equiv.), K_2CO_3 (91.0 mg, 0.66 mmol, 1.20 equiv.) and 1,3-dimethylimidazolium dimethyl phosphate (0.11 ml, 146 mg, 0.66 mmol, 1.20 equiv.) in toluene (1.81 ml) were heated at 110 °C for 50 min *via* microwave irradiation. Filtration over a small silica plug (*c*Hex/EtOAc 1:1) and purification by prepHPLC (5–95_STD) resulted in the title compound as an off-white solid (19.0 mg, 0.06 mmol, 11%).

R_t (5%–95% in 27 min, STD) = 28.0 min. – **^{1}H NMR** (500 MHz, CDCl$_3$) δ = 7.38 (d, ^4J = 1.5 Hz, 1H, C*H*), 7.22 – 7.18 (m, 4H, 4 × C*H*$_{ar}$), 6.43 (d, ^4J = 2.2 Hz, 1H, C*H*$_{ar}$), 6.24 (d, ^4J = 2.2 Hz, 1H, C*H*$_{ar}$), 3.86 – 3.83 (m, 5H, C*H*$_2$ and OC*H*$_3$), 3.79 (s, 3H, OC*H*$_3$), 2.28 (s, 3H, C*H*$_3$) ppm. – **^{13}C NMR** (126 MHz, CDCl$_3$) δ = 162.9 (C$_q$, *C*OO), 162.5 (C$_q$, 7-*C*$_{ar}$OCH$_3$), 156.6 (C$_q$, 5-*C*$_{ar}$OCH$_3$), 155.7 (C$_q$, 8a-*C*$_{ar}$), 137.0 (C$_q$, *C*$_{ar}$), 136.4 (C$_q$, *C*$_{ar}$), 134.5 (+, *C*H), 130.6 (+, *C*$_{ar}$H), 130.3 (+, *C*$_{ar}$H), 127.1 (+, *C*$_{ar}$H), 126.3 (+, *C*$_{ar}$H), 122.9 (C$_q$, 3-*C*), 104.5 (C$_q$, 4a-*C*$_{ar}$), 94.9 (+, *C*$_{ar}$H), 92.5 (+, *C*$_{ar}$H), 55.9 (+, 2 × O*C*H$_3$), 34.0 (–, *C*H$_2$), 19.7 (+, *C*H$_3$) ppm. – **IR** (ATR, ṽ) = 3075, 2997, 2965, 2945, 2919, 2840, 1701, 1672, 1608, 1496, 1468, 1452, 1441, 1424, 1377, 1366, 1316, 1298, 1241, 1231, 1205, 1194, 1164, 1143, 1109, 1052, 1043, 999, 975, 950, 936, 898, 880, 864, 837, 798, 771, 761, 748, 728, 691, 663, 637, 613, 599, 578, 560, 534, 499, 469, 443, 424, 404 cm^{-1}. – **MS** (70 eV, EI, 100 °C) *m/z* (%) = 310 (58) [M]$^+$, 293 (22), 292 (34), 219 (14) [M–C$_7$H$_7$]$^+$, 145 (11), 131 (18), 129 (11), 128 (11), 119 (73), 118 (100), 117 (28), 115 (18), 105 (22), 104 (18), 91 (28), 69 (22). – **HRMS** (EI, C$_{19}$H$_{18}$O$_4$) = calc.: 310.1205 found: 310.1204.

5,7-Dimethoxy-3-(2-methoxybenzyl)-2*H*-chromen-2-one (21ac)

According to **GP8**, 2-hydroxy-4,6-dimethoxybenzaldehyde (**19a**, 75.0 mg, 0.41 mmol, 1.00 equiv.), 2-methoxycinnamaldehyde (167 mg, 1.03 mmol, 2.50 equiv.), K_2CO_3 (68.0 mg, 0.49 mmol, 1.20 equiv.) and 1,3-dimethylimidazolium dimethyl phosphate (0.09 ml, 110 mg, 0.49 mmol, 1.20 equiv.) in toluene (1.36 ml) were heated at 110 °C for 50 min *via* microwave irradiation. Purification by flash column chromatography (*c*Hex/EtOAc 3:1) resulted in the title compound as amber crystals (37.0 mg, 0.11 mmol, 28%).

R_f (*c*Hex/EtOAc 3:1) = 0.32. – **^{1}H NMR** (400 MHz, CDCl$_3$) δ = 7.60 (s, 1H, C*H*), 7.29 – 7.20 (m, 2H, 2 × C*H*$_{ar}$), 6.96 – 6.86 (m, 2H, 2 × C*H*$_{ar}$), 6.40 (d, ^4J = 2.2 Hz, 1H, C*H*$_{ar}$), 6.23 (d, ^4J =

2.2 Hz, 1H, CH_{ar}), 3.84 (s, 2H, CH_2), 3.82 (s, 3H, OCH_3), 3.81 (s, 6H, 2 × OCH_3) ppm. – ^{13}C

NMR (101 MHz, CDCl$_3$) δ = 162.7 (C$_q$, COO), 162.6 (C$_q$, 7-C$_{ar}$OCH$_3$), 157.7 (C$_q$, C$_{ar}$OCH$_3$),

156.6 (C$_q$, 5-C$_{ar}$OCH$_3$), 155.7 (C$_q$, 8a-C$_{ar}$), 134.7 (+, CH), 131.0 (+, C$_{ar}$H), 128.1 (+, C$_{ar}$H),

126.8 (C$_q$, CH$_2$C$_{ar}$), 123.5 (C$_q$, 3-C), 120.7 (+, C$_{ar}$H), 110.7 (+, C$_{ar}$H), 104.7 (C$_q$, 4a-C$_{ar}$), 94.8

(+, C$_{ar}$H), 92.5 (+, C$_{ar}$H), 55.9 (+, OCH$_3$), 55.8 (+, OCH$_3$), 55.5 (+, OCH$_3$), 30.9 (–, CH$_2$) ppm.

– **IR** (ATR, ṽ) = 2997, 2973, 2963, 2938, 2917, 2833, 1701, 1605, 1492, 1460, 1424, 1417,

1381, 1366, 1323, 1315, 1298, 1285, 1242, 1228, 1207, 1194, 1160, 1145, 1109, 1048, 1037,

1023, 1000, 949, 935, 898, 877, 854, 830, 805, 769, 752, 725, 688, 664, 642, 613, 596, 565,

547, 521, 489, 476, 453, 408, 375 cm^{-1}. – **MS** (70 eV, EI, 110 °C) m/z (%) = 327 (20) [M+H]$^+$,

326 (100) [M]$^+$, 324 (27), 295 (22) [M–CH$_3$O]$^+$, 231 (11), 219 (46) [M–C$_7$H$_7$O]$^+$, 181 (44), 154

(14), 135 (10), 134 (90), 131 (42), 119 (37), 91 (24), 69 (68), 58 (13). – **HRMS** (EI, C$_{19}$H$_{18}$O$_5$)

= calc.: 326.1154; found: 326.1155.

5,7-Dimethoxy-3-(4-methoxybenzyl)-2H-chromen-2-one (21ad)

According to **GP8**, 2-hydroxy-4,6-dimethoxybenzaldehyde (**19a**, 75.0 mg, 0.41 mmol, 1.00 equiv.), 4-methoxycinnamalde-hyde (167 mg, 1.03 mmol, 2.50 equiv.), K$_2$CO$_3$ (68.0 mg, 0.49 mmol, 1.20 equiv.) and 1,3-dimethylimidazolium dimethyl phosphate (0.09 ml, 110 mg, 0.49 mmol, 1.20 equiv.) in toluene (1.36 ml) were heated at 110 °C for 50 min *via* microwave irra-diation. Purification by flash column chromatography (cHex/EtOAc 4:1) resulted in the title compound as a white solid (40.0 mg, 0.12 mmol, 30%).

R_f (cHex/EtOAc 3:1) = 0.42. – **^{1}H NMR** (400 MHz, CDCl$_3$) δ = 7.64 (s, 1H, CH), 7.24 – 7.19 (m, 2H, 2 × CH_{ar}), 6.89 – 6.84 (m, 2H, 2 × CH_{ar}), 6.41 (d, ^4J = 2.2 Hz, 1H, CH_{ar}), 6.25 (d, ^4J = 2.2 Hz, 1H, CH_{ar}), 3.84 (s, 3H, OCH_3), 3.83 (s, 3H, OCH_3), 3.80 (s, 3H, OCH_3), 3.78 (s, 2H, CH_2) ppm. – **^{13}C NMR** (101 MHz, CDCl$_3$) δ = 162.9 (C$_q$, COO), 162.4 (C$_q$, 7-C$_{ar}$OCH$_3$), 158.4 (C$_q$, C$_{ar}$OCH$_3$), 156.7 (C$_q$, 5-C$_{ar}$OCH$_3$), 155.8 (C$_q$, 8a-C$_{ar}$), 134.7 (+, CH), 130.7 (C$_q$, CH$_2$C$_{ar}$), 130.3 (+, 2 × C$_{ar}$H), 124.2 (C$_q$, 3-C), 114.2 (+, 2 × C$_{ar}$H), 104.5 (C$_q$, 4a-C$_{ar}$), 94.9 (+, C$_{ar}$H), 92.5 (+, C$_{ar}$H), 56.0 (+, OCH$_3$), 55.9 (+, OCH$_3$), 55.4 (+, OCH$_3$), 35.9 (–, CH$_2$) ppm. – **IR** (ATR, ṽ) = 2997, 2973, 2963, 2938, 2917, 2833, 1701, 1605, 1492, 1460, 1434, 1424, 1417, 1381, 1366, 1315, 1298, 1285, 1242, 1228, 1207, 1194, 1160, 1145, 1109, 1048, 1037, 1023, 1000, 949, 935, 898, 877, 854, 830, 805, 769, 752, 725, 688, 664, 642, 613, 596, 565, 547, 521, 489, 476, 453, 408, 375 cm^{-1}. – **MS** (70 eV, EI, 120 °C) m/z (%) = 327 (21) [M+H]$^+$, 326 (100) [M]$^+$, 325 (15) [M–H]$^+$, 324 (22), 297 (11), 295 (18) [M–CH$_3$O]$^+$, 219 (16) [M–C$_7$H$_7$O]$^+$, 181

(13), 163 (10), 135 (11), 134 (98), 131 (12), 121 (16), 119 (12), 69 (24), 58 (20). – **HRMS** (EI, $C_{19}H_{18}O_5$) = calc.: 326.1154; found: 326.1153.

3-(4-Fluorobenzyl)-5,7-dimethoxy-2*H*-chromen-2-one (21af)

According to **GP8**, 2-hydroxy-4,6-dimethoxybenzaldehyde (**19a**, 75.0 mg, 0.41 mmol, 1.00 equiv.), 4-fluorocinnamaldehyde (0.13 ml, 155 mg, 1.03 mmol, 2.50 equiv.), K_2CO_3 (68.0 mg, 0.49 mmol, 1.20 equiv.) and 1,3-dimethylimidazolium dimethyl phosphate (0.09 ml, 110 mg, 0.49 mmol, 1.20 equiv.) in toluene (1.36 ml) were heated at 110 °C for 50 min *via* microwave irradiation. Purification by flash column chromatography (*c*Hex/EtOAc 3:1) resulted in the title compound as an orange solid (23.8 mg, 0.08 mmol, 19%, purity 91%).

R_f (*c*Hex/EtOAc 3:1) = 0.38. – **^{1}H NMR** (400 MHz, CDCl$_3$) δ = 7.65 (s, 1H, C*H*), 7.29 – 7.22 (m, 2H, 2 × C*H*$_{ar}$), 7.05 – 6.92 (m, 2H, 2 × C*H*$_{ar}$), 6.41 (d, ^4J = 2.2 Hz, 1H, C*H*$_{ar}$), 6.26 (d, ^4J = 2.2 Hz, 1H, C*H*$_{ar}$), 3.84 (s, 3H, OC*H*$_3$), 3.83 (s, 3H, OC*H*$_3$), 3.80 (s, 2H, C*H*$_2$) ppm. – **^{13}C NMR** (101 MHz, CDCl$_3$) δ = 163.1 (C$_q$, *C*OO), 162.3 (C$_q$, 7-*C*$_{ar}$OCH$_3$), 161.8 (C$_q$, d, ^1J = 244.4 Hz, *C*$_{ar}$F), 156.7 (C$_q$, 5-*C*$_{ar}$OCH$_3$), 155.9 (C$_q$, 8a-*C*$_{ar}$), 135.0 (+, *C*H), 134.3 (d, ^4J = 3.3 Hz, CH$_2$*C*$_{ar}$), 130.7 (+, d, ^3J = 7.9 Hz, 2 × *m*-*C*$_{ar}$H), 123.5 (C$_q$, 3-*C*), 115.5 (+, d, ^2J = 21.3 Hz, 2 × *o*-*C*$_{ar}$H), 104.4 (C$_q$, 4a-*C*$_{ar}$), 95.0 (+, *C*$_{ar}$H), 92.6 (+, *C*$_{ar}$H), 56.0 (+, O*C*H$_3$), 55.9 (+, O*C*H$_3$), 36.1 (–, *C*H$_2$) ppm. – **^{19}F NMR** (376 MHz, CDCl$_3$) δ = –121.0 (C*F*) ppm. – **IR** (ATR, ṽ) = 3089, 3070, 3004, 2989, 2965, 2939, 2927, 2907, 2873, 2840, 1701, 1611, 1579, 1509, 1497, 1466, 1455, 1436, 1424, 1380, 1367, 1303, 1245, 1222, 1211, 1191, 1150, 1113, 1092, 1052, 1040, 1014, 1000, 977, 960, 950, 932, 902, 885, 841, 819, 771, 761, 722, 683, 654, 643, 608, 599, 562, 530, 486, 463, 421, 412 cm^{-1}. – **MS** (70 eV, EI, 80 °C) *m/z* (%) = 314 (13) [M]$^+$, 194 (14), 182 (11), 154 (31), 125 (21), 122 (100), 109 (11), 69 (15). – **HRMS** (EI, $C_{18}H_{15}O_4F$) = calc.: 314.0954; found: 314.0956.

3-(4-Chlorobenzyl)-5,7-dimethoxy-2*H*-chromen-2-one (21ah)

According to **GP8**, 2-hydroxy-4,6-dimethoxybenzaldehyde (**XX**, 75.0 mg, 0.41 mmol, 1.00 equiv.), 4-chlorocinnamaldehyde (171 mg, 1.03 mmol, 2.50 equiv.), K_2CO_3 (68.0 mg, 0.49 mmol, 1.20 equiv.) and 1,3-dimethylimidazolium dimethyl phosphate (0.09 ml, 110 mg, 0.49 mmol, 1.20 equiv.) in toluene (1.36 ml) were heated at 110 °C for 50 min *via* microwave irradiation. Purification by flash column chromatography (*c*Hex/EtOAc 3:1) resulted in the title compound as a white solid (41.0 mg, 0.12 mmol, 30%, purity 93%).

R_f (cHex/EtOAc 3:1) = 0.36. – **^{1}H NMR** (400 MHz, CDCl$_3$) δ = 7.67 (s, 1H, C*H*), 7.30 – 7.21 (m, 4H, 4 × C*H*$_{ar}$), 6.41 (d, ^4J = 2.2 Hz, 1H, C*H*$_{ar}$), 6.26 (d, ^4J = 2.2 Hz, 1H, C*H*$_{ar}$), 3.85 (s, 3H, OC*H*$_3$), 3.84 (s, 3H, OC*H*$_3$), 3.80 (s, 2H, C*H*$_2$) ppm. – **^{13}C NMR** (101 MHz, CDCl$_3$) δ = 163.1 (C$_q$, *C*OO), 162.3 (C$_q$, 7-*C*$_{ar}$OCH$_3$), 156.7 (C$_q$, 5-*C*$_{ar}$OCH$_3$), 155.9 (C$_q$, 8a-*C*$_{ar}$), 137.2 (C$_q$, *C*$_{ar}$Cl), 135.2 (+, *C*H), 132.5 (C$_q$, CH$_2$*C*$_{ar}$), 130.6 (+, 2 × *C*$_{ar}$H), 128.8 (+, 2 × *C*$_{ar}$H), 123.1 (C$_q$, 3-*C*), 104.4 (C$_q$, 4a-*C*$_{ar}$), 95.0 (+, *C*$_{ar}$H), 92.6 (+, *C*$_{ar}$H), 56.0 (+, O*C*H$_3$), 55.9 (+, O*C*H$_3$), 36.2 (–, *C*H$_2$) ppm. – **IR** (ATR, ṽ) = 2921, 1703, 1612, 1490, 1465, 1423, 1366, 1303, 1244, 1211, 1193, 1153, 1114, 1086, 1054, 1013, 1000, 931, 903, 839, 807, 760, 725, 696, 644, 529, 505, 467, 405 cm^{-1}. – **MS** (70 eV, EI, 110 °C) *m/z* (%) = 331 (24) [M+H]$^+$, 332/330 (40/100) [M]$^+$, 301 (12), 295 (45) [M–Cl]$^+$, 219 (11), 191 (11), 181 (24), 165 (15), 140 (13), 138 (41), 131 (20), 125 (16), 69 (44). – **HRMS** (EI, C$_{18}$H$_{15}$O$_4{}^{35}$Cl) = calc.: 330.0659; found: 330.0660.

3-(4-(Dimethylamino)benzyl)-5,7-dimethoxy-2*H*-chromen-2-one (21ai)

According to **GP8**, 2-hydroxy-4,6-dimethoxybenzaldehyde (**19a**, 75.0 mg, 0.41 mmol, 1.00 equiv.), 4-(dimethylamino)cinnamaldehyde (180 mg, 1.03 mmol, 2.50 equiv.), K$_2$CO$_3$ (68.0 mg, 0.49 mmol, 1.20 equiv.) and 1,3-dimethylimidazolium dimethyl phosphate (0.09 ml, 110 mg, 0.49 mmol, 1.20 equiv.) in toluene (1.36 ml) were heated at 110 °C for 50 min via microwave irradiation. Purification by flash column chromatography (cHex/EtOAc 3:1) resulted in the title compound as a black to dark red solid (25.0 mg, 0.07 mmol, 18%).

R_f (cHex/EtOAc 3:1) = 0.31. – **^{1}H NMR** (400 MHz, CDCl$_3$) δ = 7.66 (s, 1H, C*H*), 7.19 (d, ^3J = 8.6 Hz, 2H, 2 × C*H*$_{ar}$), 6.77 (d, ^3J = 8.1 Hz, 2H, 2 × C*H*$_{ar}$), 6.40 (d, ^4J = 2.2 Hz, 1H, C*H*$_{ar}$), 6.25 (d, ^4J = 2.2 Hz, 1H, C*H*$_{ar}$), 3.83 (s, 3H, OC*H*$_3$), 3.83 (s, 3H, OC*H*$_3$), 3.75 (s, 2H, C*H*$_2$), 2.94 (s, 6H, N(C*H*$_3$)$_2$) ppm. – **^{13}C NMR** (101 MHz, CDCl$_3$) δ = 162.8 (C$_q$, *C*OO), 162.5 (C$_q$, 7-*C*$_{ar}$OCH$_3$), 156.6 (C$_q$, 5-*C*$_{ar}$OCH$_3$), 155.7 (C$_q$, 8a-*C*$_{ar}$), 149.2 (C$_q$, *C*$_{ar}$N(CH$_3$)$_3$), 134.5 (+, *C*H), 130.1 (+, 2 × *C*$_{ar}$H), 124.5 (C$_q$, 3-*C*), 113.6 (+, 2 × *C*$_{ar}$H), 104.6 (C$_q$, 4a-*C*$_{ar}$), 94.8 (+, *C*$_{ar}$H), 92.5 (+, *C*$_{ar}$H), 56.0 (+, O*C*H$_3$), 55.9 (+, O*C*H$_3$), 41.2 (+, N(*C*H$_3$)$_2$), 35.8 (–, *C*H$_2$) ppm. – **IR** (ATR, ṽ) = 2915, 2895, 2853, 2815, 1701, 1605, 1521, 1497, 1482, 1465, 1421, 1380, 1364, 1347, 1313, 1293, 1239, 1228, 1207, 1191, 1163, 1143, 1108, 1048, 999, 945, 895, 832, 805, 762, 738, 724, 691, 643, 564, 534, 480 cm^{-1}. – **MS** (70 eV, EI, 130 °C) *m/z* (%) = 340 (23) [M+H]$^+$, 339 (100) [M]$^+$, 338 (13) [M–H]$^+$, 322 (16), 147 (46), 146 (13), 134 (26), 84 (15), 71 (12), 69 (12), 58 (20), 57 (18). – **HRMS** (EI, C$_{20}$H$_{21}$O$_4$N) = calc.: 339.1471; found: 339.1472.

3-Benzyl-7-ethoxy-5-methoxy-2*H*-chromen-2-one (21ba)

According to **GP8**, 4-ethoxy-2-hydroxy-6-methoxybenzaldehyde (**19b**, 75.0 mg, 0.38 mmol, 1.00 equiv.), cinnamaldehyde (0.12 ml, 120 mg, 0.96 mmol, 2.50 equiv.), K_2CO_3 (63.0 mg, 0.46 mmol, 1.20 equiv.) and 1,3-dimethylimidazolium dimethyl phosphate (0.08 ml, 102 mg, 0.46 mmol, 1.20 equiv.) in toluene (1.26 ml) were heated at 110 °C for 50 min *via* microwave irradiation. Purification by flash column chromatography (*c*Hex/EtOAc 5:1) resulted in a mixture of product, starting material and methyl ester as impurity. Another purification by prepTLC (*c*Hex/EtOAc 3:1) resulted in the title compound as an off-white solid (18.0 mg, 0.06 mmol, 15%, purity 90%).

R_f (*c*Hex/EtOAc 5:1) = 0.47. – **¹H NMR** (400 MHz, CDCl$_3$) δ = 7.67 (s, 1H, C*H*), 7.33 – 7.29 (m, 5H, 5 × C*H*$_{ar}$), 6.39 (d, ^4J = 2.2 Hz, 1H, C*H*$_{ar}$), 6.25 (d, ^4J = 2.2 Hz, 1H, C*H*$_{ar}$), 4.05 (q, ^3J = 7.0 Hz, 2H, OC*H*$_2$), 3.85 – 3.84 (m, 2H, CC*H*$_2$), 3.83 (s, 3H, OC*H*$_3$), 1.43 (t, ^3J = 7.0 Hz, 3H, C*H*$_3$) ppm. – **¹³C NMR** (101 MHz, CDCl$_3$) δ = 162.5 (C$_q$, *C*OO), 162.3 (C$_q$, 7-C$_{ar}$OCH$_2$), 156.6 (C$_q$, 5-C$_{ar}$OCH$_3$), 155.9 (C$_q$, 8a-*C*$_{ar}$), 138.7 (C$_q$, CH$_2$*C*$_{ar}$), 135.1 (+, *C*H), 129.3 (+, 2 × *C*$_{ar}$H), 128.7 (+, 2 × *C*$_{ar}$H), 126.7 (+, *C*$_{ar}$H), 123.6 (C$_q$, 3-*C*), 104.4 (C$_q$, 4a-*C*$_{ar}$), 95.2 (+, *C*$_{ar}$H), 93.0 (+, *C*$_{ar}$H), 64.2 (–, O*C*H$_2$), 56.0 (+, O*C*H$_3$), 36.8 (–, *C*CH$_2$), 14.7 (+, *C*H$_3$) ppm. – **IR** (ATR, ṽ) = 3089, 3047, 3030, 2990, 2963, 2919, 2877, 2851, 1782, 1725, 1660, 1613, 1572, 1496, 1455, 1419, 1394, 1378, 1366, 1303, 1251, 1224, 1194, 1180, 1164, 1152, 1115, 1072, 1043, 1010, 1003, 970, 949, 929, 925, 884, 856, 822, 799, 788, 773, 747, 731, 693, 664, 639, 611, 589, 551, 535, 511, 484, 439 cm^{-1}. – **MS** (70 eV, EI, 80 °C) *m/z* (%) = 310 (13) [M]$^+$, 264 (24), 256 (12), 196 (12), 131 (13), 105 (12), 104 (100), 103 (14), 97 (10), 83 (11), 78 (10), 71 (11), 69 (17), 57 (16), 55 (14). – **HRMS** (EI, C$_{19}$H$_{18}$O$_4$) = calc.: 310.1205; found: 310.1204.

7-Ethoxy-5-methoxy-3-(2-methylbenzyl)-2*H*-chromen-2-one (21bb)

According to **GP8**, 4-ethoxy-2-hydroxy-6-methoxybenzaldehyde (**19b**, 38.0 mg, 0.19 mmol, 1.00 equiv.), 2-methylcinnamaldehyde (71.0 mg, 0.48 mmol, 2.50 equiv.), K_2CO_3 (32.0 mg, 0.23 mmol, 1.20 equiv.) and 1,3-dimethylimidazolium dimethyl phosphate (0.04 ml, 52.0 mg, 0.23 mmol, 1.20 equiv.) in toluene (0.64 ml) were heated at 110 °C for 50 min *via* microwave irradiation. Filtration over a small silica plug (*c*Hex/EtOAc 2:1) and purification by prepHPLC (5–95_STD) resulted in the title compound as colourless to pale-green crystals (14.0 mg, 0.04 mmol, 23%).

R_t (5%–95% in 27 min, STD) = 30.4 min. – **^{1}H NMR** (500 MHz, CDCl$_3$) δ = 7.38 (d, ^4J = 1.5 Hz, 1H, CH), 7.23 – 7.18 (m, 4H, 4 × CH_{ar}), 6.41 (d, ^4J = 2.2 Hz, 1H, CH_{ar}), 6.24 (d, ^4J = 2.2 Hz, 1H, CH_{ar}), 4.05 (q, ^3J = 7.0 Hz, 2H, OCH_2), 3.84 (d, ^4J = 1.5 Hz, 2H, CCH_2), 3.78 (s, 3H, OCH_3), 2.28 (s, 3H, C$_{ar}$CH_3), 1.43 (t, ^3J = 7.0 Hz, 3H, CH_3) ppm. – **^{13}C NMR** (126 MHz, CDCl$_3$) δ = 162.6 (C$_q$, COO), 162.2 (C$_q$, 7-C$_{ar}$OCH$_2$), 156.6 (C$_q$, 5-C$_{ar}$OCH$_3$), 155.6 (C$_q$, 8a-C_{ar}), 137.0 (C$_q$, C_{ar}), 136.4 (C$_q$, C_{ar}), 134.5 (+, CH), 130.6 (+, C_{ar}H), 130.3 (+, C_{ar}H), 127.0 (+, C_{ar}H), 126.3 (+, C_{ar}H), 122.7 (C$_q$, 3-C), 104.4 (C$_q$, 4a-C_{ar}), 95.2 (+, C_{ar}H), 93.0 (+, C_{ar}H), 64.2 (–, OCH$_2$), 55.9 (+, OCH$_3$), 33.9 (–, CCH$_2$), 19.7 (+, C$_{ar}$$CH_3$), 14.7 (+, CH$_3$) ppm. – **IR** (ATR, ṽ) = 3063, 3047, 3020, 3003, 2990, 2961, 2953, 2941, 2836, 1718, 1612, 1575, 1497, 1465, 1451, 1425, 1377, 1370, 1358, 1302, 1283, 1244, 1227, 1200, 1169, 1150, 1109, 1040, 1013, 975, 950, 938, 908, 888, 873, 841, 822, 809, 771, 755, 741, 727, 721, 688, 666, 642, 599, 545, 527, 500, 465, 432, 402 cm^{-1}. – **MS** (70 eV, EI, 100 °C) m/z (%) = 325 (24) [M+H]$^+$, 324 (100) [M]$^+$, 307 (19), 191 (17), 181 (28), 131 (15), 119 (12), 118 (24), 105 (20), 104 (21), 69 (30). – **HRMS** (EI, C$_{20}$H$_{20}$O$_4$) = calc.: 324.1362; found: 324.1363.

7-Ethoxy-5-methoxy-3-(2-methoxybenzyl)-2H-chromen-2-one (21bc)

According to **GP8**, 4-ethoxy-2-hydroxy-6-methoxybenzaldehyde (**19b**, 75.0 mg, 0.38 mmol, 1.00 equiv.), 2-methoxycinnamaldehyde (155 mg, 0.96 mmol, 2.50 equiv.), K$_2$CO$_3$ (63.0 mg, 0.46 mmol, 1.20 equiv.) and 1,3-dimethylimidazolium dimethyl phosphate (0.08 ml, 102 mg, 0.46 mmol, 1.20 equiv.) in toluene (1.26 ml) were heated at 110 °C for 50 min *via* microwave irradiation. Purification by flash column chromatography (cHex/EtOAc 5:1) resulted in the title compound as an orange solid (34.0 mg, 0.10 mmol, 26%).

R_f (cHex/EtOAc 5:1) = 0.36. – **^{1}H NMR** (400 MHz, CDCl$_3$) δ = 7.59 (s, 1H, CH), 7.30 – 7.20 (m, 2H, 2 × CH_{ar}), 6.98 – 6.85 (m, 2H, 2 × CH_{ar}), 6.39 (d, ^4J = 2.1 Hz, 1H, CH_{ar}), 6.24 (d, ^4J = 2.1 Hz, 1H, CH_{ar}), 4.04 (q, ^3J = 7.0 Hz, 2H, OCH_2), 3.84 (s, 2H, CCH_2), 3.81 (s, 6H, 2 × OCH_3), 1.43 (t, ^3J = 7.0 Hz, 3H, CH_3) ppm. – **^{13}C NMR** (101 MHz, CDCl$_3$) δ = 162.6 (C$_q$, COO), 162.0 (C$_q$, 7-C$_{ar}$OCH$_2$), 157.8 (C$_q$, C_{ar}OCH$_3$), 156.6 (C$_q$, 5-C$_{ar}$OCH$_3$), 155.7 (C$_q$, 8a-C_{ar}), 134.7 (+, CH), 131.1 (+, C_{ar}H), 128.1 (+, C_{ar}H), 126.9 (C$_q$, CH$_2$$C_{ar}$), 122.9 (C$_q$, 3-$C$), 120.7 (+, C_{ar}H), 110.7 (+, C_{ar}H), 104.6 (C$_q$, 4a-C_{ar}), 95.1 (+, C_{ar}H), 93.0 (+, C_{ar}H), 64.2 (–, OCH$_2$), 55.9 (+, OCH$_3$), 55.5 (+, OCH$_3$), 30.9 (–, CCH$_2$), 14.7 (+, CH$_3$), ppm. – **IR** (ATR, ṽ) = 2976, 2963, 2918, 2880, 2851, 2839, 1718, 1615, 1578, 1496, 1462, 1436, 1424, 1380, 1371, 1361, 1313, 1298, 1285, 1241, 1207, 1191, 1171, 1150, 1112, 1044, 1038, 1027, 1010, 977, 948, 932, 894, 871, 849, 837, 819, 798, 771, 752, 722, 696, 687, 666, 643, 598, 568, 554, 524, 473 cm^{-1}. – **MS**

(70 eV, EI, 110 °C) m/z (%) = 341 (22) [M+H]$^+$, 340 (100) [M]$^+$, 309 (14) [M–CH$_3$O]$^+$, 233 (34) [M–C$_7$H$_7$O]$^+$, 219 (16), 205 (12), 69 (13), 58 (11), 57 (14). – **HRMS** (EI, C$_{20}$H$_{20}$O$_5$) = calc.: 340.1311; found: 340.1309.

7-Ethoxy-5-methoxy-3-(4-methoxybenzyl)-2*H*-chromen-2-one (21bd)

According to **GP8**, 4-ethoxy-2-hydroxy-6-methoxybenzalde-hyde (**19b**, 75.0 mg, 0.38 mmol, 1.00 equiv.), 4-methox-ycinnamaldehyde (155 mg, 0.96 mmol, 2.50 equiv.), K$_2$CO$_3$ (63.0 mg, 0.46 mmol, 1.20 equiv.) and 1,3-dimethylimidazolium dimethyl phosphate (0.08 ml, 102 mg, 0.46 mmol, 1.20 equiv.) in toluene (1.26 ml) were heated at 110 °C for 50 min *via* microwave irradiation. Purification by flash column chromatography (*c*Hex/EtOAc 3:1) resulted in the title compound as a red solid (30.0 mg, 0.09 mmol, 23%).

R_f (*c*Hex/EtOAc 3:1) = 0.40. – **^{1}H NMR** (400 MHz, CDCl$_3$) δ = 7.64 (s, 1H, C*H*), 7.22 (d, ^3J = 8.6 Hz, 2H, 2 × C*H*$_{ar}$), 6.86 (d, ^3J = 8.6 Hz, 2H, 2 × C*H*$_{ar}$), 6.38 (d, ^4J = 2.1 Hz, 1H, C*H*$_{ar}$), 6.25 (d, ^4J = 2.1 Hz, 1H, C*H*$_{ar}$), 4.04 (q, ^3J = 7.0 Hz, 2H, OC*H*$_2$), 3.83 (s, 3H, OC*H*$_3$), 3.80 (s, 3H, OC*H*$_3$), 3.78 (d, ^4J = 1.1 Hz, 2H, CC*H*$_2$), 1.43 (t, ^3J = 7.0 Hz, 3H, C*H*$_3$) ppm. – **^{13}C NMR** (101 MHz, CDCl$_3$) δ = 162.5 (C$_q$, *C*OO), 162.2 (C$_q$, 7-*C*$_{ar}$OCH$_2$), 158.4 (C$_q$, *C*$_{ar}$OCH$_3$), 156.6 (C$_q$, 5-*C*$_{ar}$OCH$_3$), 155.8 (C$_q$, 8a-*C*$_{ar}$), 134.8 (+, *C*H), 130.7 (C$_q$, CH$_2$*C*$_{ar}$), 130.3 (+, 2 × *C*$_{ar}$H), 124.0 (C$_q$, 3-*C*), 114.1 (+, 2 × *C*$_{ar}$H), 104.4 (C$_q$, 4a-*C*$_{ar}$), 95.2 (+, *C*$_{ar}$H), 93.0 (+, *C*$_{ar}$H), 64.2 (–, O*C*H$_2$), 56.0 (+, O*C*H$_3$), 55.4 (+, O*C*H$_3$), 35.9 (–, C*C*H$_2$), 14.7 (+, *C*H$_3$), ppm. – **IR** (ATR, ṽ) = 3003, 2983, 2955, 2924, 2905, 2874, 2851, 2832, 1715, 1657, 1629, 1606, 1582, 1510, 1497, 1463, 1451, 1425, 1397, 1375, 1350, 1310, 1292, 1239, 1230, 1200, 1180, 1170, 1159, 1142, 1109, 1094, 1048, 1033, 1004, 956, 949, 938, 898, 887, 863, 837, 820, 810, 759, 725, 690, 653, 645, 568, 538, 528, 507, 452, 422 cm^{-1}. – **MS** (70 eV, EI) m/z (%) = 296 (30), 135 (10), 134 (100), 119 (14), 91 (11). – **HRMS** (EI, C$_{20}$H$_{20}$O$_5$) = calc.: 340.1311; found: 340.1310.

7-Ethoxy-3-(4-fluorobenzyl)-5-methoxy-2*H*-chromen-2-one (21bf)

According to **GP8**, 4-ethoxy-2-hydroxy-6-methoxybenzaldehyde (**19b**, 75.0 mg, 0.38 mmol, 1.00 equiv.), 4-fluorocinnamaldehyde (0.12 ml, 143 mg, 0.96 mmol, 2.50 equiv.), K$_2$CO$_3$ (63.0 mg, 0.46 mmol, 1.20 equiv.) and 1,3-dimethylimidazolium dimethyl phosphate (0.08 ml, 102 mg, 0.46 mmol, 1.20 equiv.) in toluene (1.26 ml) were heated at 110 °C for 50 min *via* microwave irra-diation. Purification by flash column chromatography (*c*Hex/EtOAc 5:1) resulted in a mixture of product, starting material and unidentified impurities. Another purification by prepTLC (cHex/EtOAc 3:1) gave the title compound as a yellow solid (31.0 mg, 0.10 mmol, 25%).

R_f (cHex/EtOAc 3:1) = 0.46. – **^{1}H NMR** (400 MHz, CDCl$_3$) δ = 7.65 (s, 1H, CH), 7.29 – 7.24 (m, 2H, 2 × CH_{ar}), 7.03 – 6.97 (m, 2H, 2 × CH_{ar}), 6.39 (d, ^4J = 2.1 Hz, 1H, CH_{ar}), 6.26 (d, ^4J = 2.1 Hz, 1H, CH_{ar}), 4.05 (q, ^3J = 7.0 Hz, 2H, OCH_2), 3.84 (s, 3H, OCH_3), 3.80 (s, 2H, CCH_2), 1.43 (t, ^3J = 7.0 Hz, 3H, CH_3) ppm. – **^{13}C NMR** (101 MHz, CDCl$_3$) δ = 162.4 (C$_q$, COO), 162.3 (C$_q$, 7-C$_{ar}$OCH$_2$), 161.8 (C$_q$, d, ^1J = 244.3 Hz, C$_{ar}$F), 156.7 (C$_q$, 5-C$_{ar}$OCH$_3$), 155.9 (C$_q$, 8a-C$_{ar}$), 135.1 (+, CH), 134.4 (C$_q$, d, ^4J = 3.1 Hz, CH$_2$C$_{ar}$), 130.7 (+, d, ^3J = 8.0 Hz, 2 × m-C$_{ar}$H), 123.4 (C$_q$, 3-C), 115.5 (+, d, ^2J = 21.3 Hz, 2 × o-C$_{ar}$H), 104.3 (C$_q$, 4a-C$_{ar}$), 95.3 (+, C$_{ar}$H), 93.1 (+, C$_{ar}$H), 64.3 (–, OCH$_2$), 56.0 (+, OCH$_3$), 36.1 (–, CCH$_2$), 14.7 (+, CH$_3$), ppm. – **^{19}F NMR** (376 MHz, CDCl$_3$) δ = –121.1 (s, CF) ppm – **IR** (ATR, ṽ) = 2980, 2936, 2919, 1707, 1618, 1575, 1510, 1499, 1463, 1425, 1402, 1366, 1306, 1245, 1225, 1201, 1171, 1156, 1115, 1051, 1014, 973, 959, 935, 904, 895, 871, 853, 833, 820, 792, 771, 761, 724, 690, 660, 646, 565, 543, 521, 486, 460, 422, 414 cm^{-1}. – **MS** (70 eV, EI, 80 °C) m/z (%) = 328 (11) [M–H]$^+$, 123 (15), 122 (100), 121 (10), 109 (19), 97 (11), 58 (14). – **HRMS** (EI, C$_{19}$H$_{17}$O$_4$F) = calc.: 329.1189; found: 329.1188.

3-(4-Chlorobenzyl)-7-ethoxy-5-methoxy-2H-chromen-2-one (21bh)

According to **GP8**, 4-ethoxy-2-hydroxy-6-methoxybenzaldehyde (**19b**, 75.0 mg, 0.38 mmol, 1.00 equiv.), 4-chlorocinnamaldehyde (159 mg, 0.96 mmol, 2.50 equiv.), K$_2$CO$_3$ (63.0 mg, 0.46 mmol, 1.20 equiv.) and 1,3-dimethylimidazolium dimethyl phosphate (0.08 ml, 102 mg, 0.46 mmol, 1.20 equiv.) in toluene (1.26 ml) were heated at 110 °C for 50 min via microwave irradiation. Purification by flash column chromatography (cHex/EtOAc 5:1) resulted in a mixture of product, starting material and unidentified impurities. After another purification by prepTLC (cHex/EtOAc 3:1) resulted in the title compound as an off-white solid (30.0 mg, 0.09 mmol, 23%).

R_f (cHex/EtOAc 3:1) = 0.54. – **^{1}H NMR** (400 MHz, CDCl$_3$) δ = 7.66 (d, ^4J = 1.0 Hz, 1H, CH), 7.32 – 7.20 (m, 4H, 4 × CH_{ar}), 6.39 (d, ^4J = 2.1 Hz, 1H, CH_{ar}), 6.26 (d, ^4J = 2.1 Hz, 1H, CH_{ar}), 4.05 (q, ^3J = 7.0 Hz, 2H, OCH_2), 3.85 (s, 3H, OCH_3), 3.80 (s, 2H, CCH_2), 1.43 (t, ^3J = 7.0 Hz, 3H, CH_3) ppm. – **^{13}C NMR** (101 MHz, CDCl$_3$) δ = 162.5 (C$_q$, COO), 162.3 (C$_q$, 7-C$_{ar}$OCH$_2$), 156.7 (C$_q$, 5-C$_{ar}$OCH$_3$), 155.9 (C$_q$, 8a-C$_{ar}$), 137.2 (C$_q$, C$_{ar}$Cl), 135.2 (+, CH), 132.5 (C$_q$, CH$_2$C$_{ar}$), 130.6 (+, 2 × C$_{ar}$H), 128.8 (+, 2 × C$_{ar}$H), 123.0 (C$_q$, 3-C), 104.3 (C$_q$, 4a-C$_{ar}$), 95.3 (+, C$_{ar}$H), 93.1 (+, C$_{ar}$H), 64.3 (–, OCH$_2$), 56.0 (+, OCH$_3$), 36.2 (–, CCH$_2$), 14.7 (+, CH$_3$) ppm. – **IR** (ATR, ṽ) = 3012, 2986, 2965, 2935, 1720, 1710, 1629, 1606, 1578, 1490, 1463, 1453, 1425, 1397, 1375, 1351, 1312, 1293, 1244, 1230, 1194, 1179, 1160, 1142, 1113, 1091, 1050, 1041, 1013, 1004, 962, 953, 904, 866, 837, 819, 809, 761, 727, 694, 679, 645, 551, 530, 511, 487, 432, 401 cm^{-1}. – **MS** (70 eV, EI, 110 °C) m/z (%) = 345 (14) [M+H]$^+$, 346/344 (21/60) [M]$^+$, 309 (15) [M–Cl]$^+$, 304 (11), 281 (13), 196 (14), 167 (11), 140 (35), 139 (13), 138 (100), 125 (13), 103 (19), 69 (13). – **HRMS** (EI, C$_{19}$H$_{17}$O$_4$^{35}Cl) = calc.: 344.0815; found: 344.0815.

3-(4-(Dimethylamino)benzyl)-7-ethoxy-5-methoxy-2*H*-chromen-2-one (21bi)

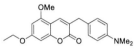

Similar to **GP8**, 4-ethoxy-2-hydroxy-6-methoxybenzaldehyde (**19b**, 100 mg, 0.51 mmol, 1.00 equiv.), 4-(dimethylamino)cinnamaldehyde (223 mg, 1.27 mmol, 2.50 equiv.), K$_2$CO$_3$ (85.0 mg, 0.61 mmol, 1.20 equiv.) and 1,3-dimethylimidazolium dimethyl phosphate (0.09 ml, 110 mg, 0.49 mmol, 1.20 equiv.) in toluene (1.68 ml) were heated at 110 °C for 65 min *via* microwave irradiation. Purification by flash column chromatography (*c*Hex/EtOAc 3:1) resulted in the title compound as dark red crystals (45.0 mg, 0.13 mmol, 25%).

R_f (*c*Hex/EtOAc 3:1) = 0.25. – **^{1}H NMR** (500 MHz, CDCl$_3$) δ = 7.65 (s, 1H, C*H*), 7.22 – 7.17 (m, 2H, 2 × C*H*$_{ar}$), 6.81 – 6.73 (m, 2H, 2 × C*H*$_{ar}$), 6.38 (d, ^4J = 2.1 Hz, 1H, C*H*$_{ar}$), 6.24 (d, ^4J = 2.1 Hz, 1H, C*H*$_{ar}$), 4.04 (q, ^3J = 7.0 Hz, 2H, OC*H*$_2$), 3.83 (s, 3H, OC*H*$_3$), 3.75 (s, 2H, CC*H*$_2$), 2.94 (s, 6H, N(C*H*$_3$)$_2$), 1.43 (t, ^3J = 7.0 Hz, 3H, C*H*$_3$) ppm. – **^{13}C NMR** (126 MHz, CDCl$_3$) δ = 162.6 (C$_q$, *C*OO), 162.1 (C$_q$, 7-*C*$_{ar}$OCH$_2$), 156.5 (C$_q$, 5-*C*$_{ar}$OCH$_3$), 155.7 (C$_q$, 8a-*C*$_{ar}$), 134.6 (+, *C*H), 130.0 (+, 2 × *C*$_{ar}$H), 124.3 (C$_q$, 3-*C*), 113.5 (+, 2 × *C*$_{ar}$H), 104.4 (C$_q$, 4a-*C*$_{ar}$), 95.1 (+, *C*$_{ar}$H), 92.9 (+, *C*$_{ar}$H), 64.2 (–, O*C*H$_2$), 55.9 (+, O*C*H$_3$), 41.2 (+, N(*C*H$_3$)$_2$), 35.7 (–, *C*H$_2$) 14.7 (+, *C*H$_3$) ppm. – **IR** (ATR, ṽ) = 3088, 2969, 2888, 2799, 1708, 1613, 1575, 1521, 1496, 1465, 1453, 1424, 1380, 1363, 1344, 1309, 1244, 1200, 1171, 1153, 1112, 1048, 1010, 948, 851, 824, 805, 762, 727, 713, 543, 472, 412, 391 cm^{-1}. – **MS** (70 eV, EI, 120 °C) m/z (%) = 353 (39) [M]$^+$, 323 (11), 322 (50) [M–CH$_3$O]$^+$, 148 (13), 147 (100), 146 (31), 134 (30), 58 (12). – **HRMS** (EI, C$_{21}$H$_{23}$O$_4$N) = calc.: 353.1627; found: 353.1627.

3-Benzyl-5-methoxy-7-propoxy-2*H*-chromen-2-one (21ca)

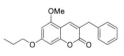

According to **GP8**, 4-propoxy-2-hydroxy-6-methoxybenzaldehyde (**19c**, 75.0 mg, 0.36 mmol, 1.00 equiv.), cinnamaldehyde (0.11 ml, 118 mg, 0.89 mmol, 2.50 equiv.), K$_2$CO$_3$ (59.0 mg, 0.43 mmol, 1.20 equiv.) and 1,3-dimethylimidazolium dimethyl phosphate (0.07 ml, 95.0 mg, 0.43

mmol, 1.20 equiv.) in toluene (1.18 ml) were heated at 110 °C for 50 min *via* microwave irra-diation. Purification by flash column chromatography (*c*Hex/EtO$_2$ 3:1) resulted in the title com-pound as a yellow solid (31.0 mg, 0.10 mmol, 27%).

R_f (*c*Hex/EtO$_2$ 3:1) = 0.28. – **^{1}H NMR** (400 MHz, CDCl$_3$) δ = 7.67 (d, ^4J = 1.3 Hz, 1H, C*H*), 7.35 – 7.29 (m, 3H, 3 × C*H*$_{ar}$), 7.28 – 7.21 (m, 2H, 2 × C*H*$_{ar}$), 6.39 (d, ^4J = 2.1 Hz, 1H, C*H*$_{ar}$), 6.26 (d, ^4J = 2.1 Hz, 1H, C*H*$_{ar}$), 3.93 (t, ^3J = 6.5 Hz, 2H, OC*H*$_2$), 3.85 – 3.84 (m, 2H, CC*H*$_2$), 3.83 (s, 3H, OC*H*$_3$), 1.88 – 1.77 (m, 2H, C*H*$_2$CH$_3$), 1.05 (t, ^3J = 7.4 Hz, 3H, C*H*$_3$) ppm. – **^{13}C NMR** (101 MHz, CDCl$_3$) δ = 162.5 (C$_q$, *C*OO), 162.5 (C$_q$, 7-*C*$_{ar}$OCH$_2$), 156.6 (C$_q$, 5-*C*$_{ar}$OCH$_3$), 155.9 (C$_q$, 8a-*C*$_{ar}$), 138.8 (C$_q$, CH$_2$*C*$_{ar}$), 135.1 (+, *C*H), 129.3 (+, 2 × *C*$_{ar}$H), 128.7 (+, 2 × *C*$_{ar}$H), 126.6 (+, *C*$_{ar}$H), 123.5 (C$_q$, 3-*C*), 104.3 (C$_q$, 4a-*C*$_{ar}$), 95.2 (+, *C*$_{ar}$H), 93.1 (+, *C*$_{ar}$H), 70.2 (–, O*C*H$_2$), 56.0 (+, O*C*H$_3$), 36.8 (–, C*C*H$_2$), 22.5 (–, *C*H$_2$CH$_3$), 10.6 (+, *C*H$_3$) ppm. – **IR** (ATR, ṽ) = 3082, 3065, 3029, 2975, 2961, 2935, 2918, 2876, 2851, 1710, 1622, 1574, 1496, 1455, 1422, 1366, 1309, 1295, 1242, 1231, 1218, 1196, 1156, 1113, 1050, 1018, 1001, 986, 950, 931, 915, 899, 833, 807, 761, 755, 727, 698, 642, 585, 545, 517, 487, 463, 443, 395 cm^{-1}. – **MS** (70 eV, EI, 100 °C) *m/z* (%) = 325 (24) [M+H]$^+$, 324 (100) [M]$^+$, 282 (45) 281 (20) [M–C$_3$H$_7$]$^+$, 254 (11), 253 (18), 236 (31), 210 (17), 177 (12), 168 (26), 167 (13), 150 (15), 140 (10), 105 (13), 104 (86), 103 (11), 91 (11), 69 (14). – **HRMS** (EI, C$_{20}$H$_{20}$O$_4$) = calc.: 324.1362; found: 324.1361.

5-Methoxy-3-(2-methylbenzyl)-7-propoxy-2*H*-chromen-2-one (21cb)

According to **GP8**, 4-propoxy-2-hydroxy-6-methoxybenzalde-hyde (**19c**, 75.0 mg, 0.36 mmol, 1.00 equiv.), 2-methylcinnamal-dehyde (130 mg, 0.89 mmol, 2.50 equiv.), K$_2$CO$_3$ (59.0 mg, 0.43 mmol, 1.20 equiv.) and 1,3-dimethylimidazolium dimethyl phosphate (0.07 ml, 95.0 mg, 0.43 mmol, 1.20 equiv.) in toluene (1.18 ml) were heated at 110 °C for 50 min *via* microwave irra-diation. Filtration over a small silica plug (*c*Hex/EtOAc 2:1) and purification by prepHPLC (5–95_STD) resulted in the title compound as an off-white solid (18.0 mg, 0.05 mmol, 15%).

R_t (5%–95% in 27 min, STD) = 31.6 min. – **^{1}H NMR** (500 MHz, CDCl$_3$) δ = 7.38 (s, 1H, C*H*), 7.23 – 7.18 (m, 4H, 4 × C*H*$_{ar}$), 6.41 (d, ^4J = 2.2 Hz, 1H, C*H*$_{ar}$), 6.24 (d, ^4J = 2.2 Hz, 1H, C*H*$_{ar}$), 3.94 (t, ^3J = 6.5 Hz, 2H, OC*H*$_2$), 3.84 (d, ^4J = 1.5 Hz, 2H, CC*H*$_2$), 3.79 (s, 3H, OC*H*$_3$), 2.28 (s, 3H, C$_{ar}$C*H*$_3$), 1.83 (sex., ^3J = 7.1 Hz, 2H, C*H*$_2$CH$_3$), 1.05 (t, ^3J = 7.4 Hz, 3H, C*H*$_3$) ppm. – **^{13}C NMR** (126 MHz, CDCl$_3$) δ = 162.6 (C$_q$, *C*OO), 162.4 (C$_q$, 7-*C*$_{ar}$OCH$_2$), 156.6 (C$_q$, 5-*C*$_{ar}$OCH$_3$), 155.7 (C$_q$, 8a-*C*$_{ar}$), 137.0 (C$_q$, *C*$_{ar}$), 136.4 (C$_q$, *C*$_{ar}$), 134.5 (+, *C*H), 130.6 (+, *C*$_{ar}$H), 130.3 (+, *C*$_{ar}$H), 127.0 (+, *C*$_{ar}$H), 126.3 (+, *C*$_{ar}$H), 122.7 (C$_q$, 3-*C*), 104.3 (C$_q$, 4a-*C*$_{ar}$), 95.2 (+, *C*$_{ar}$H), 93.0

(+, C_{ar}H), 70.2 (−, OCH$_2$), 55.9 (+, OCH$_3$), 34.0 (−, CCH$_2$), 22.5 (−, CH$_2$CH$_3$), 19.7 (+, $C_{ar}C$H$_3$), 10.6 (+, CH$_3$) ppm. − **IR** (ATR, ṽ) = 2975, 2962, 2939, 2876, 1710, 1613, 1575, 1496, 1460, 1453, 1425, 1404, 1364, 1343, 1309, 1242, 1200, 1169, 1153, 1115, 1048, 989, 955, 932, 916, 908, 881, 834, 819, 759, 742, 725, 690, 666, 646, 599, 547, 499, 470, 458, 438, 377 cm^{-1}. − **MS** (70 eV, EI, 110 °C) m/z (%) = 339 (25) [M+H]$^+$, 338 (100) [M]$^+$, 321 (16), 205 (13), 104 (20), 69 (11). − **HRMS** (EI, $C_{21}H_{22}O_4$) = calc.: 338.1518; found: 338.1517.

5-Methoxy-3-(2-methoxybenzyl)-7-propoxy-2*H*-chromen-2-one (21cc)

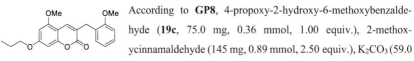

According to **GP8**, 4-propoxy-2-hydroxy-6-methoxybenzaldehyde (**19c**, 75.0 mg, 0.36 mmol, 1.00 equiv.), 2-methoxycinnamaldehyde (145 mg, 0.89 mmol, 2.50 equiv.), K$_2$CO$_3$ (59.0 mg, 0.43 mmol, 1.20 equiv.) and 1,3-dimethylimidazolium dimethyl phosphate (0.07 ml, 95.0 mg, 0.43 mmol, 1.20 equiv.) in toluene (1.18 ml) were heated at 110 °C for 50 min *via* microwave irradiation. Purification by flash column chromatography (*c*Hex/EtO$_2$ 3:1) resulted in the title compound as an off-white solid (28.0 mg, 0.08 mmol, 22%).

R_f (*c*Hex/EtO$_2$ 3:1) = 0.36. − **^{1}H NMR** (400 MHz, CDCl$_3$) δ = 7.59 (d, ^4J = 1.1 Hz, 1H, CH), 7.29 − 7.21 (m, 2H, 2 × CH_{ar}), 6.97 − 6.87 (m, 2H, 2 × CH_{ar}), 6.39 (d, ^4J = 2.1 Hz, 1H, CH_{ar}), 6.24 (d, ^4J = 2.1 Hz, 1H, CH_{ar}), 3.93 (t, ^3J = 6.5 Hz, 2H, OCH_2), 3.84 (d, ^4J = 1.1 Hz, 2H, CCH_2), 3.81 (s, 6H, 2 × OCH_3), 1.82 (sex., ^3J = 7.3 Hz, 2H, CH_2CH$_3$), 1.04 (t, ^3J = 7.3 Hz, 3H, CH_3) ppm. − **^{13}C NMR** (101 MHz, CDCl$_3$) δ = 162.6 (C$_q$, COO), 162.2 (C$_q$, 7-C_{ar}OCH$_2$), 157.8 (C$_q$, C_{ar}OCH$_3$), 156.6 (C$_q$, 5-C_{ar}OCH$_3$), 155.7 (C$_q$, 8a-C_{ar}), 134.7 (+, CH), 131.1 (+, C_{ar}H), 128.1 (+, C_{ar}H), 126.9 (C$_q$, CH$_2C_{ar}$), 122.9 (C$_q$, 3-C), 120.7 (+, C_{ar}H), 110.7 (+, C_{ar}H), 104.5 (C$_q$, 4a-C_{ar}), 95.1 (+, C_{ar}H), 93.0 (+, C_{ar}H), 70.1 (−, OCH$_2$), 56.0 (+, OCH$_3$), 55.5 (+, OCH$_3$), 30.9 (−, CCH$_2$), 22.5 (−, CH$_2$CH$_3$), 10.6 (+, CH$_3$), ppm. − **IR** (ATR, ṽ) = 2961, 2927, 2874, 2853, 2837, 1713, 1613, 1496, 1455, 1426, 1407, 1366, 1300, 1288, 1244, 1201, 1170, 1157, 1112, 1050, 1031, 1001, 987, 950, 935, 912, 898, 836, 819, 788, 768, 748, 727, 690, 664, 646, 613, 596, 568, 552, 533, 524, 475, 458, 435, 375 cm^{-1}. − **MS** (70 eV, EI, 160 °C) m/z (%) = 355 (27) [M+H]$^+$, 354 (100) [M]$^+$, 281 (12), 247 (18), 205 (26). − **HRMS** (EI, $C_{21}H_{22}O_5$) = calc.: 354.1467; found: 354.1467.

5-Methoxy-3-(4-methoxybenzyl)-7-propoxy-2*H*-chromen-2-one (21cd)

According to **GP8**, 4-propoxy-2-hydroxy-6-methoxybenzaldehyde (**19c**, 75.0 mg, 0.36 mmol, 1.00 equiv.), 4-methoxycinnamaldehyde (145 mg, 0.89 mmol, 2.50 equiv.), K$_2$CO$_3$

(59.0 mg, 0.43 mmol, 1.20 equiv.) and 1,3-dimethylimidazolium dimethyl phosphate (0.07 ml, 95.0 mg, 0.43 mmol, 1.20 equiv.) in toluene (1.18 ml) were heated at 110 °C for 50 min *via* microwave irradiation. Purification by flash column chromatography (*c*Hex/EtO$_2$ 3:1) resulted in the title compound as an orange solid (19.0 mg, 0.05 mmol, 15%).

R_f (*c*Hex/EtO$_2$ 3:1) = 0.38. – ^{1}H NMR (400 MHz, CDCl$_3$) δ = 7.63 (d, ^4J = 1.2 Hz, 1H, C*H*), 7.25 – 7.19 (m, 2H, 2 × C*H*$_{ar}$), 6.88 – 6.84 (m, 2H, 2 × C*H*$_{ar}$), 6.38 (d, ^4J = 2.2 Hz, 1H, C*H*$_{ar}$), 6.26 (d, ^4J = 2.2 Hz, 1H, C*H*$_{ar}$), 3.91 (t, ^3J = 6.5 Hz, 2H, OC*H*$_2$), 3.83 (s, 3H, OC*H*$_3$), 3.80 (s, 3H, OC*H*$_3$), 3.78 – 3.77 (m, 2H, CC*H*$_2$), 1.82 (sex., ^3J = 7.2 Hz, 2H, C*H*$_2$CH$_3$), 1.04 (t, ^3J = 7.4 Hz, 3H, C*H*$_3$) ppm. – ^{13}C NMR (101 MHz, CDCl$_3$) δ = 162.5 (C$_q$, *C*OO), 162.4 (C$_q$, 7-*C*$_{ar}$OCH$_2$), 158.4 (C$_q$, *C*$_{ar}$OCH$_3$), 156.6 (C$_q$, 5-*C*$_{ar}$OCH$_3$), 155.8 (C$_q$, 8a-*C*$_{ar}$), 134.8 (+, *C*H), 130.7 (C$_q$, CH$_2$*C*$_{ar}$), 130.3 (+, 2 × *C*$_{ar}$H), 123.9 (C$_q$, 3-*C*), 114.1 (+, 2 × *C*$_{ar}$H), 104.4 (C$_q$, 4a-*C*$_{ar}$), 95.2 (+, *C*$_{ar}$H), 93.0 (+, *C*$_{ar}$H), 70.2 (–, O*C*H$_2$), 56.0 (+, O*C*H$_3$), 55.4 (+, O*C*H$_3$), 35.9 (–, C*C*H$_2$), 22.5 (–, *C*H$_2$CH$_3$), 10.6 (+, *C*H$_3$) ppm. – IR (ATR, ṽ) = 3082, 3061, 3013, 2962, 2931, 2902, 2876, 2836, 1711, 1609, 1577, 1511, 1499, 1466, 1452, 1425, 1363, 1299, 1242, 1201, 1170, 1153, 1129, 1108, 1048, 1023, 999, 949, 932, 908, 897, 880, 839, 830, 819, 807, 759, 724, 693, 683, 660, 643, 629, 567, 545, 517, 489, 465, 436, 387 cm^{-1}. – MS (70 eV, EI, 120 °C) *m/z* (%) = 354 (100) [M]$^+$, 312 (13), 281 (13), 230 (10), 205 (16), 140 (11), 134 (20), 121 (21), 111 (15), 97 (15), 85 (14), 83 (14), 81 (10), 71 (19), 69 (22), 57 (29), 55 (13). – HRMS (EI, C$_{21}$H$_{22}$O$_5$) = calc.: 354.1467; found: 354.1469.

3-(4-Fluorobenzyl)-5-methoxy-7-propoxy-2*H*-chromen-2-one (21cf)

According to **GP8**, 4-propoxy-2-hydroxy-6-methoxybenzalde-hyde (**19c**, 75.0 mg, 0.36 mmol, 1.00 equiv.), 4-fluorocinnamal-dehyde (0.11 ml, 134 mg, 0.89 mmol, 2.50 equiv.), K$_2$CO$_3$ (59.0 mg, 0.43 mmol, 1.20 equiv.) and 1,3-dimethylimidazolium dimethyl phosphate (0.07 ml, 95.0 mg, 0.43 mmol, 1.20 equiv.) in toluene (1.18 ml) were heated at 110 °C for 50 min *via* micro-wave irradiation. Purification by flash column chromatography (*c*Hex/EtO$_2$ 3:1) resulted in the title compound as a yellow solid (17.7 mg, 0.05 mmol, 15%, purity 93%).

R_f (*c*Hex/ EtO$_2$ 3:1) = 0.41. – ^{1}H NMR (400 MHz, CDCl$_3$) δ = 7.65 (d, ^4J = 1.5 Hz, 1H, C*H*), 7.29 – 7.23 (m, 2H, 2 × C*H*$_{ar}$), 7.03 – 6.97 (m, 2H, 2 × C*H*$_{ar}$), 6.39 (d, ^4J = 2.1 Hz, 1H, C*H*$_{ar}$), 6.27 (d, ^4J = 2.1 Hz, 1H, C*H*$_{ar}$), 3.93 (t, ^3J = 6.6 Hz, 2H, OC*H*$_2$), 3.85 (s, 3H, OC*H*$_3$), 3.80 (s, 2H, CC*H*$_2$), 1.82 (sex., ^3J = 7.4 Hz, 2H, C*H*$_2$CH$_3$), 1.05 (t, ^3J = 7.4 Hz, 3H, C*H*$_3$) ppm. – ^{13}C NMR (101 MHz, CDCl$_3$) δ = 162.6 (C$_q$, *C*OO), 162.4 (C$_q$, 7-*C*$_{ar}$OCH$_2$), 161.8 (C$_q$, d, ^1J = 244.3 Hz, *C*$_{ar}$F), 156.6 (C$_q$, 5-*C*$_{ar}$OCH$_3$), 155.9 (C$_q$, 8a-*C*$_{ar}$), 135.1 (+, *C*H), 134.4 (C$_q$, d, ^4J = 3.3 Hz,

CH_2C_{ar}), 130.7 (+, d, ^3J = 7.8 Hz, 2 × m-C_{ar}H), 123.3 (C_q, 3-C), 115.5 (+, d, ^2J = 21.3 Hz, 2 × o-C_{ar}H), 104.2 (C_q, 4a-C_{ar}), 95.3 (+, C_{ar}H), 93.1 (+, C_{ar}H), 70.2 (–, OCH_2), 56.0 (+, OCH_3), 36.1 (–, CCH_2), 22.5 (–, CH_2CH_3), 10.6 (+, CH_3) ppm. – ^{19}F NMR (376 MHz, CDCl$_3$) δ = – 121.1 (s, CF) ppm. – IR (ATR, ṽ) = 3088, 3068, 3007, 2968, 2944, 2919, 2878, 2853, 1708, 1613, 1575, 1509, 1462, 1453, 1425, 1404, 1366, 1309, 1298, 1247, 1222, 1200, 1170, 1154, 1129, 1113, 1092, 1065, 1047, 1030, 1017, 1001, 989, 960, 952, 933, 915, 901, 885, 853, 834, 819, 761, 725, 691, 684, 662, 646, 565, 544, 523, 482, 469, 452, 421, 390, 375 cm^{-1}. – MS (70 eV, EI, 100 °C) m/z (%) = 342 (100) [M]$^+$, 330 (14), 301 (11), 300 (58), 299 (14) [M–C$_3$H$_7$]$^+$, 272 (16), 271 (19), 210 (17), 177 (13), 168 (15), 167 (14), 150 (14), 149 (13), 135 (10), 125 (12), 123 (23), 122 (13), 111 (15), 109 (32), 97 (19), 95 (15), 85 (20), 83 (18), 81 (12), 71 (28), 69 (32), 59 (17), 57 (37), 55 (16). – HRMS (EI, C$_{20}$H$_{19}$O$_4$F) = calc.: 342.1267; found: 342.1266.

3-(4-Chlorobenzyl)-5-methoxy-7-propoxy-2H-chromen-2-one (21ch)

According to GP8, 4-propoxy-2-hydroxy-6-methoxybenzalde-hyde (19c, 75.0 mg, 0.36 mmol, 1.00 equiv.), 4-chlorocin-namaldehyde (149 mg, 0.89 mmol, 2.50 equiv.), K$_2$CO$_3$ (59.0 mg, 0.43 mmol, 1.20 equiv.) and 1,3-dimethylimidazolium dimethyl phosphate (0.07 ml, 95.0 mg, 0.43 mmol, 1.20 equiv.) in toluene (1.18 ml) were heated at 110 °C for 50 min via micro-wave irradiation. Purification by flash column chromatography (cHex/ EtO$_2$ 3:1) resulted in the title compound as a white solid (15.0 mg, 0.05 mmol, 15%).

R_f (cHex/EtO$_2$ 3:1) = 0.26. – ^{1}H NMR (400 MHz, CDCl$_3$) δ = 7.66 (s, 1H, CH), 7.30 –7.20 (m, 4H, 4 × CH_{ar}), 6.38 (d, ^4J = 2.1 Hz, 1H, CH_{ar}), 6.26 (d, ^4J = 2.1 Hz, 1H, CH_{ar}), 3.93 (t, ^3J = 6.5 Hz, 2H, OCH_2), 3.84 (s, 3H, OCH_3), 3.79 (s, 2H, CCH_2), 1.81 (sex., ^3J = 7.0 Hz, 2H, CH_2CH_3), 1.04 (t, ^3J = 7.4 Hz, 3H, CH_3) ppm. – ^{13}C NMR (101 MHz, CDCl$_3$) δ = 162.7 (C_q, COO), 162.3 (C_q, 7-$C_{ar}OCH_2$), 156.7 (C_q, 5-$C_{ar}OCH_3$), 155.9 (C_q, 8a-C_{ar}), 137.3 (C_q, $C_{ar}Cl$), 135.2 (+, CH), 132.5 (C_q, CH_2C_{ar}), 130.6 (+, 2 × C_{ar}H), 128.8 (+, 2 × C_{ar}H), 122.9 (C_q, 3-C), 104.2 (C_q, 4a-C_{ar}), 95.3 (+, C_{ar}H), 93.1 (+, C_{ar}H), 70.2 (–, OCH_2), 56.0 (+, OCH_3), 36.2 (–, CCH_2), 22.5 (–, CH_2CH_3), 10.6 (+, CH_3), ppm. – IR (ATR, ṽ) = 3088, 3067, 2976, 2959, 2939, 2919, 2874, 1717, 1613, 1575, 1489, 1459, 1451, 1426, 1404, 1366, 1307, 1247, 1201, 1169, 1156, 1115, 1088, 1064, 1048, 1016, 989, 953, 932, 915, 902, 887, 841, 817, 812, 778, 761, 725, 694, 684, 645, 613, 557, 544, 507, 470, 453, 411, 395, 375 cm^{-1}. – MS (70 eV, EI, 110 °C) m/z (%) = 359 (22) [M+H]$^+$, 360/358 (35/100) [M]$^+$, 318/316 (11/30), 287 (14), 282 (11), 281 (54), 177 (11), 125 (15), 69 (13), 57 (13). – HRMS (EI, C$_{20}$H$_{19}$O$_4$^{35}Cl) = calc.: 358.0972; found: 358.0971.

3-(4-(Dimethylamino)benzyl)-5-methoxy-7-propoxy-2*H*-chromen-2-one (21ci)

According to **GP8**, 4-propoxy-2-hydroxy-6-methoxybenzaldehyde (**19c**, 75.0 mg, 0.36 mmol, 1.00 equiv.), 4-(dimethylamino)cinnamaldehyde (156 mg, 0.89 mmol, 2.50 equiv.), K_2CO_3 (59.0 mg, 0.43 mmol, 1.20 equiv.) and 1,3-dimethylimidazolium dimethyl phosphate (0.07 ml, 95.0 mg, 0.43 mmol, 1.20 equiv.) in toluene (1.18 ml) were heated at 110 °C for 50 min *via* microwave irradiation. Purification by flash column chromatography (*c*Hex/EtO$_2$ 3:1 to 2:1) resulted in the title compound as a brown solid (36.0 mg, 0.10 mmol, 27%, purity 91%).

R_f (*c*Hex/EtO$_2$ 3:1) = 0.31. – **^{1}H NMR** (500 MHz, CDCl$_3$) δ = 7.66 (s, 1H, C*H*), 7.20 (d, ^3J = 8.1 Hz, 2H, 2 × C*H*$_{ar}$), 6.88 – 6.73 (m, 2H, 2 × C*H*$_{ar}$), 6.38 (d, ^4J = 2.1 Hz, 1H, C*H*$_{ar}$), 6.25 (d, ^4J = 2.1 Hz, 1H, C*H*$_{ar}$), 3.93 (t, ^3J = 6.5 Hz, 2H, OC*H*$_2$), 3.83 (s, 3H, OC*H*$_3$), 3.75 (s, 2H, CC*H*$_2$), 2.94 (s, 6H, N(C*H*$_3$)$_2$), 1.82 (sex., ^3J = 7.0 Hz, 2H, C*H*$_2$CH$_3$), 1.05 (t, ^3J = 7.4 Hz, 3H, C*H*$_3$) ppm. – **^{13}C NMR** (126 MHz, CDCl$_3$) δ = 162.6 (C$_q$, COO), 162.3 (C$_q$, 7-C$_{ar}$OCH$_2$), 156.5 (C$_q$, 5-C$_{ar}$OCH$_3$), 155.7 (C$_q$, 8a-C$_{ar}$), 134.6 (+, CH), 130.1 (+, 2 × C$_{ar}$H), 124.2 (C$_q$, 3-C), 113.5 (+, 2 × C$_{ar}$H), 104.4 (C$_q$, 4a-C$_{ar}$), 95.1 (+, C$_{ar}$H), 92.9 (+, C$_{ar}$H), 70.1 (–, OCH$_2$), 55.9 (+, OCH$_3$), 41.2 (+, N(CH$_3$)$_2$), 35.8 (–, CCH$_2$), 22.5 (–, CH$_2$CH$_3$), 10.6 (+, CH$_3$) ppm. – **IR** (ATR, ṽ) = 3088, 3071, 2962, 2927, 2890, 2874, 2851, 2802, 2737, 1708, 1612, 1575, 1521, 1497, 1465, 1451, 1426, 1380, 1364, 1347, 1313, 1305, 1293, 1242, 1221, 1201, 1171, 1154, 1129, 1109, 1050, 1031, 999, 948, 932, 909, 897, 840, 826, 806, 762, 741, 727, 715, 693, 656, 643, 629, 595, 564, 544, 527, 514, 480, 465, 407, 385 cm^{-1}. – **MS** (70 eV, EI, 150 °C) *m/z* (%) = 368 (25) [M+H]$^+$, 367 (100) [M]$^+$, 364 (10), 355 (14), 341 (24), 160 (11), 147 (48), 146 (11), 134 (59), 111 (13), 97 (16), 95 (11), 85 (12), 83 (13), 71 (17), 69 (25), 57 (27), 55 (11). – **HRMS** (EI, C$_{22}$H$_{25}$O$_4$N) = calc.: 367.1784; found: 367.1784.

3-Benzyl-7-butoxy-5-methoxy-2*H*-chromen-2-one (21da)

According to **GP8**, 4-butoxy-2-hydroxy-6-methoxybenzaldehyde (**19d**, 75.0 mg, 0.33 mmol, 1.00 equiv.), cinnamaldehyde (0.11 ml, 110 mg, 0.84 mmol, 2.50 equiv.), K_2CO_3 (55.0 mg, 0.40 mmol, 1.20 equiv.) and 1,3-dimethylimidazolium dimethyl phosphate (0.07 ml, 89.0 mg, 0.40 mmol, 1.20 equiv.) in toluene (1.10 ml) were heated at 110 °C for 50 min *via* microwave irradiation. Purification by flash column chromatography (*c*Hex/EtO$_2$ 5:1) resulted in a mixture of product and starting material. Another purification by prepHPLC (5–95_STD) resulted in the title compound as a yellow solid (17.0 mg, 0.05 mmol, 15%).

R_f (cHex/EtO$_2$ 5:1) = 0.28. – **^{1}H NMR** (500 MHz, CDCl$_3$) δ = 7.67 (s, 1H, C*H*), 7.35 – 7.29 (m, 4H, 4 × C*H*$_{ar}$), 7.27 – 7.22 (m, 1H, C*H*$_{ar}$), 6.39 (d, ^4J = 2.1 Hz, 1H, C*H*$_{ar}$), 6.25 (d, ^4J = 2.1 Hz, 1H, C*H*$_{ar}$), 3.97 (t, ^3J = 6.5 Hz, 2H, OC*H*$_2$), 3.84 (s, 2H, CC*H*$_2$), 3.83 (s, 3H, OC*H*$_3$), 1.78 (quin., ^3J = 7.4, 6.5 Hz, 2H, OCH$_2$C*H*$_2$), 1.50 (quin., ^3J = 7.4 Hz, 2H, C*H*$_2$CH$_3$), 0.98 (t, ^3J = 7.4 Hz, 3H, C*H*$_3$) ppm. – **^{13}C NMR** (126 MHz, CDCl$_3$) δ = 162.5 (C$_q$, *C*OO), 162.5 (C$_q$, 7-*C*$_{ar}$OCH$_2$), 156.6 (C$_q$, 5-*C*$_{ar}$OCH$_3$), 155.8 (C$_q$, 8a-*C*$_{ar}$), 138.7 (C$_q$, CH$_2$*C*$_{ar}$), 135.1 (+, *C*H), 129.3 (+, 2 × *C*$_{ar}$H), 128.7 (+, 2 × *C*$_{ar}$H), 126.6 (+, *C*$_{ar}$H), 123.5 (C$_q$, 3-*C*), 104.3 (C$_q$, 4a-*C*$_{ar}$), 95.2 (+, *C*$_{ar}$H), 93.0 (+, *C*$_{ar}$H), 68.4 (–, O*C*H$_2$), 56.0 (+, O*C*H$_3$), 36.8 (–, *C*CH$_2$), 31.2 (–, O*C*H$_2$CH$_2$), 19.3 (–, *C*H$_2$CH$_3$), 14.0 (+, *C*H$_3$) ppm. – **IR** (ATR, ṽ) = 3088, 3063, 2952, 2927, 2871, 1715, 1616, 1496, 1466, 1455, 1425, 1366, 1247, 1231, 1204, 1174, 1154, 1116, 1048, 1001, 950, 921, 899, 823, 761, 727, 704, 645, 585, 544, 497, 445, 402, 381 cm^{-1}. – **MS** (70 eV, EI, 120 °C) m/z (%) = 339 (23) [M+H]$^+$, 338 (100) [M]$^+$, 283 (11), 282 (59), 281 (22) [M–C$_4$H$_9$]$^+$, 254 (13), 253 (18), 181 (14), 177 (11), 131 (12), 69 (19). – **HRMS** (EI, C$_{21}$H$_{22}$O$_4$) = calc.: 338.1518; found: 338.1517.

7-Butoxy-5-methoxy-3-(2-methylbenzyl)-2*H*-chromen-2-one (21db)

According to **GP8**, 4-butoxy-2-hydroxy-6-methoxybenzalde-hyde (**19d**, 75.0 mg, 0.33 mmol, 1.00 equiv.), 2-methylcin-namaldehyde (122 mg, 0.84 mmol, 2.50 equiv.), K$_2$CO$_3$ (55.0 mg, 0.40 mmol, 1.20 equiv.) and 1,3-dimethylimidazolium dimethyl phosphate (0.07 ml, 89.0 mg, 0.40 mmol, 1.20 equiv.) in toluene (1.10 ml) were heated at 110 °C for 50 min *via* micro-wave irradiation. Filtration over a small silica plug (cHex/EtOAc 1:1) and purification by prepHPLC (5–95_STD) resulted in the title compound as an off-white solid (25.0 mg, 0.07 mmol, 21%).

R_t (5%–95% in 27 min, STD) = 33.2 min. – **^{1}H NMR** (500 MHz, CDCl$_3$) δ = 7.38 (s, 1H, C*H*), 7.23 – 7.18 (m, 4H, 4 × C*H*$_{ar}$), 6.41 (d, ^4J = 2.2 Hz, 1H, C*H*$_{ar}$), 6.24 (d, ^4J = 2.2 Hz, 1H, C*H*$_{ar}$), 3.98 (t, ^3J = 6.5 Hz, 2H, OC*H*$_2$), 3.84 (d, ^4J = 1.4 Hz, 2H, CC*H*$_2$), 3.78 (s, 3H, OC*H*$_3$), 2.28 (s, 3H, C$_{ar}$C*H*$_3$), 1.83 – 1.73 (m, 2H, OCH$_2$C*H*$_2$), 1.50 (quin., ^3J = 7.5 Hz, 2H, C*H*$_2$CH$_3$), 0.98 (t, ^3J = 7.5 Hz, 3H, C*H*$_3$) ppm. – **^{13}C NMR** (126 MHz, CDCl$_3$) δ = 162.6 (C$_q$, *C*OO), 162.4 (C$_q$, 7-*C*$_{ar}$OCH$_2$), 156.6 (C$_q$, 5-*C*$_{ar}$OCH$_3$), 155.6 (C$_q$, 8a-*C*$_{ar}$), 137.0 (C$_q$, *C*$_{ar}$), 136.4 (C$_q$, *C*$_{ar}$), 134.5 (+, *C*H), 130.6 (+, *C*$_{ar}$H), 130.3 (+, *C*$_{ar}$H), 127.0 (+, *C*$_{ar}$H), 126.3 (+, *C*$_{ar}$H), 122.7 (C$_q$, 3-*C*), 104.3 (C$_q$, 4a-*C*$_{ar}$), 95.2 (+, *C*$_{ar}$H), 93.0 (+, *C*$_{ar}$H), 68.4 (–, O*C*H$_2$), 55.9 (+, O*C*H$_3$), 33.9 (–, *C*CH$_2$), 31.2 (–, O*C*H$_2$CH$_2$), 19.7 (+, C$_{ar}$*C*H$_3$), 19.3 (–, *C*H$_2$CH$_3$), 14.0 (+, *C*H$_3$) ppm. – **IR** (ATR, ṽ) = 3067, 3013, 2951, 2931, 2870, 2840, 1711, 1613, 1575, 1494, 1463, 1456, 1425,

1405, 1363, 1303, 1244, 1232, 1200, 1169, 1154, 1113, 1069, 1052, 1031, 1016, 1003, 986, 956, 939, 905, 884, 866, 844, 822, 769, 758, 744, 725, 690, 667, 647, 637, 599, 567, 547, 530, 499, 455, 414, 399 cm^{-1}. – **MS** (70 eV, EI, 110 °C) m/z (%) = 353 (25) [M+H]$^+$, 352 (100) [M]$^+$, 335 (11), 205 (16), 104 (18), 69 (12). – **HRMS** (EI, C$_{22}$H$_{24}$O$_4$) = calc.: 352.1675; found: 352.1673.

7-Butoxy-5-methoxy-3-(2-methoxybenzyl)-2H-chromen-2-one (21dc)

According to **GP8**, 4-butoxy-2-hydroxy-6-methoxybenzalde-hyde (**19d**, 75.0 mg, 0.33 mmol, 1.00 equiv.), 2-methoxycinnamaldehyde (136 mg, 0.84 mmol, 2.50 equiv.), K$_2$CO$_3$ (55.0 mg, 0.40 mmol, 1.20 equiv.) and 1,3-dimethylimidazolium dimethyl phosphate (0.07 ml, 89.0 mg, 0.40 mmol, 1.20 equiv.) in toluene (1.10 ml) were heated at 110 °C for 50 min via microwave irradiation. Purification by flash column chromatography (cHex/EtO$_2$ 5:1) resulted in a mixture of product and starting material. Another purification by prepHPLC (5–95_STD) resulted in the title compound as tan crystals (37.4 mg, 0.10 mmol, 30%).

R_f (cHex/EtO$_2$ 6:1) = 0.29. – **^{1}H NMR** (500 MHz, CDCl$_3$) δ = 7.59 (d, ^4J = 1.3 Hz, 1H, CH), 7.28 – 7.22 (m, 2H, 2 × CH_{ar}), 6.95 – 6.87 (m, 2H, 2 × CH_{ar}), 6.39 (d, ^4J = 2.1 Hz, 1H, CH_{ar}), 6.24 (d, ^4J = 2.1 Hz, 1H, CH_{ar}), 3.97 (t, ^3J = 6.6 Hz, 2H, OCH_2), 3.84 (s, 2H, CCH_2), 3.83 (s, 6H, 2 × OCH_3), 1.77 (quin., ^3J = 6.6 Hz, 2H, OCH$_2$CH_2), 1.49 (sex., ^3J = 7.4, Hz, 2H, CH_2CH$_3$), 0.98 (t, ^3J = 7.4 Hz, 3H, CH_3) ppm. – **^{13}C NMR** (126 MHz, CDCl$_3$) δ = 162.6 (C$_q$, COO), 162.2 (C$_q$, 7-C_{ar}OCH$_2$), 157.7 (C$_q$, C_{ar}OCH$_3$), 156.6 (C$_q$, 5-C_{ar}OCH$_3$), 155.7 (C$_q$, 8a-C_{ar}), 134.7 (+, CH), 131.1 (+, C_{ar}H), 128.1 (+, C_{ar}H), 126.9 (C$_q$, CH$_2$$C_{ar}$), 122.8 (C$_q$, 3-$C$), 120.7 (+, C_{ar}H), 110.7 (+, C_{ar}H), 104.5 (C$_q$, 4a-C_{ar}), 95.1 (+, C_{ar}H), 93.0 (+, C_{ar}H), 68.3 (–, OCH$_2$), 55.9 (+, OCH$_3$), 55.5 (+, OCH$_3$), 31.2 (–, OCH$_2$CH$_2$), 30.9 (–, CCH$_2$), 19.3 (–, CH$_2$CH$_3$), 14.0 (+, CH$_3$), ppm. – **IR** (ATR, ṽ) = 2944, 2867, 1711, 1615, 1494, 1465, 1456, 1426, 1405, 1367, 1363, 1312, 1286, 1245, 1201, 1170, 1159, 1113, 1048, 1031, 984, 950, 899, 849, 836, 819, 768, 748, 725, 687, 555, 526, 492, 470, 462, 449, 436, 398 cm^{-1}. – **MS** (70 eV, EI, 120 °C) m/z (%) = 369 (27) [M+H]$^+$, 368 (100) [M]$^+$, 312 (10), 281 (16) [M–C$_7$H$_7$O]$^+$, 269 (11), 261 (15), 231 (14), 205 (37), 181 (35), 169 (14), 131 (33), 119 (18), 69 (48). – **HRMS** (EI, C$_{22}$H$_{24}$O$_5$) = calc.: 368.1624; found: 368.1622.

7-Butoxy-5-methoxy-3-(4-methoxybenzyl)-2H-chromen-2-one (21dd)

According to **GP8**, 4-butoxy-2-hydroxy-6-methoxyben-zaldehyde (**19d**, 75.0 mg, 0.33 mmol, 1.00 equiv.), 4-meth-oxycinnamaldehyde (136 mg, 0.84 mmol, 2.50 equiv.), K_2CO_3 (55.0 mg, 0.40 mmol, 1.20 equiv.) and 1,3-dimethylimidazolium dimethyl phosphate (0.07 ml, 89.0 mg, 0.40 mmol, 1.20 equiv.) in toluene (1.10 ml) were heated at 110 °C for 50 min *via* microwave irradiation. Purification by prepHPLC (5–95_STD) resulted in the title compound as a yellow solid (29.0 mg, 0.07 mmol, 22%).

R_t (5%–95% in 27 min, STD) = 32.0 min – **^{1}H NMR** (400 MHz, CDCl$_3$) δ = 7.63 (d, ^4J = 1.4 Hz, 1H, CH), 7.22 (d, ^3J = 8.6 Hz, 2H, 2 × CH_{ar}), 6.86 (d, ^3J = 8.6 Hz, 2H, 2 × CH_{ar}), 6.39 (d, ^4J = 2.2 Hz, 1H, CH_{ar}), 6.25 (d, ^4J = 2.2 Hz, 1H, CH_{ar}), 3.97 (t, ^3J = 6.5 Hz, 2H, OCH_2), 3.83 (s, 3H, OCH_3), 3.80 (s, 3H, OCH_3), 3.78 (s, 2H, CCH_2), 1.82 – 1.73 (m, 2H, OCH$_2$CH_2), 1.49 (sex., ^3J = 7.4 Hz, 2H, CH_2CH$_3$), 0.98 (t, ^3J = 7.4 Hz, 3H, CH_3) ppm. – **^{13}C NMR** (101 MHz, CDCl$_3$) δ = 162.5 (C$_q$, COO), 162.4 (C$_q$, 7-C_{ar}OCH$_2$), 158.4 (C$_q$, C_{ar}OCH$_3$), 156.6 (C$_q$, 5-C_{ar}OCH$_3$), 155.8 (C$_q$, 8a-C_{ar}), 134.8 (+, CH), 130.7 (C$_q$, CH$_2$$C_{ar}$), 130.3 (+, 2 × C_{ar}H), 123.9 (C$_q$, 3-C), 114.1 (+, 2 × C_{ar}H), 104.3 (C$_q$, 4a-C_{ar}), 95.2 (+, C_{ar}H), 93.0 (+, C_{ar}H), 68.4 (–, OCH$_2$), 55.9 (+, OCH$_3$), 55.4 (+, OCH$_3$), 35.9 (–, CCH$_2$), 31.2 (–, OCH$_2$$CH_2$), 19.3 (–, CH$_2CH_3$), 14.0 (+, CH$_3$) ppm. – **IR** (ATR, ṽ) = 3009, 2958, 2922, 2853, 1711, 1612, 1577, 1513, 1500, 1466, 1455, 1429, 1408, 1375, 1366, 1313, 1298, 1242, 1201, 1170, 1154, 1126, 1113, 1052, 1044, 1027, 1003, 956, 933, 898, 880, 843, 824, 810, 761, 739, 724, 694, 660, 645, 632, 568, 544, 513, 496, 465, 443, 414 cm^{-1}. – **MS** (70 eV, EI, 120 °C) m/z (%) = 369 (26) [M+H]$^+$, 368 (100) [M]$^+$, 361 (11), 331 (11), 312 (21), 311 (11) [M–C$_4$H$_9$]$^+$, 283 (11), 281 (25) [M–C$_7$H$_7$O]$^+$, 257 (20), 256 (19), 243 (11), 231 (18), 205 (19), 181 (53), 169 (15), 167 (14), 149 (26), 131 (49), 121 (13), 119 (19), 111 (13), 109 (11), 100 (11), 97 (18), 95 (14), 85 (15), 83 (17), 81 (13), 71 (20), 69 (89), 57 (28), 55 (15). – **HRMS** (EI, C$_{22}$H$_{24}$O$_5$) = calc.: 368.1624 found: 368.1624.

3-(4-Fluorobenzyl)-5-methoxy-7-butoxy-2H-chromen-2-one (21df)

According to **GP8**, 4-butoxy-2-hydroxy-6-methoxybenzalde-hyde (**19d**, 75.0 mg, 0.33 mmol, 1.00 equiv.), 4-fluorocin-namaldehyde (0.11 ml, 126 mg, 0.84 mmol, 2.50 equiv.), K_2CO_3 (55.0 mg, 0.40 mmol, 1.20 equiv.) and 1,3-dimethylimidazolium dimethyl phosphate (0.07 ml, 89.0 mg, 0.40 mmol, 1.20 equiv.) in toluene (1.10 ml) were heated at 110 °C for 50 min *via* microwave irradiation. Purification by prepHPLC (5–95_STD) resulted in the title compound as a yellow solid (18.0 mg, 0.05 mmol, 15%).

R_t (5%–95% in 27 min, STD) = 32.5 min. – ^{1}**H NMR** (400 MHz, CDCl$_3$) δ = 7.66 (d, ^4J = 1.4 Hz, 1H, C*H*), 7.31 – 7.22 (m, 2H, 2 × C*H*$_{ar}$), 7.04 – 6.96 (m, 2H, 2 × C*H*$_{ar}$), 6.39 (d, ^4J = 2.2 Hz, 1H, C*H*$_{ar}$), 6.26 (d, ^4J = 2.2 Hz, 1H, C*H*$_{ar}$), 3.98 (t, ^3J = 6.5 Hz, 2H, OC*H*$_2$), 3.85 (s, 3H, OC*H*$_3$), 3.80 (s, 2H, CC*H*$_2$), 1.84 – 1.72 (m, 2H, OCH$_2$C*H*$_2$), 1.50 (sex., ^3J = 7.4 Hz, 2H, C*H*$_2$CH$_3$), 0.98 (t, ^3J = 7.4 Hz, 3H, C*H*$_3$) ppm. – ^{13}**C NMR** (101 MHz, CDCl$_3$) δ = 162.6 (C$_q$, *C*OO), 162.3 (C$_q$, 7-C$_{ar}$OCH$_2$), 161.8 (C$_q$, d, ^1J = 244.2 Hz, *C*$_{ar}$F), 156.6 (C$_q$, 5-C$_{ar}$OCH$_3$), 155.9 (C$_q$, 8a-*C*$_{ar}$), 135.0 (+, *C*H), 134.4 (C$_q$, d, ^4J = 3.3 Hz, CH$_2$*C*$_{ar}$), 130.7 (+, d, ^3J = 7.9 Hz, 2 × *m*-*C*$_{ar}$H), 123.3 (C$_q$, 3-*C*), 115.5 (+, d, ^2J = 21.2 Hz, 2 × *o*-*C*$_{ar}$H), 104.2 (C$_q$, 4a-*C*$_{ar}$), 95.3 (+, *C*$_{ar}$H), 93.1 (+, *C*$_{ar}$H), 68.4 (–, O*C*H$_2$), 56.0 (+, O*C*H$_3$), 36.1 (–, *C*CH$_2$), 31.1 (–, OCH$_2$*C*H$_2$), 19.3 (–, *C*H$_2$CH$_3$), 13.9 (+, *C*H$_3$) ppm. – ^{19}**F NMR** (376 MHz, CDCl$_3$) δ = –121.1 (s, C*F*) ppm. – **IR** (ATR, ṽ) = 2968, 2952, 2927, 2870, 1711, 1615, 1577, 1509, 1466, 1455, 1426, 1407, 1366, 1309, 1293, 1247, 1231, 1218, 1203, 1173, 1153, 1115, 1094, 1074, 1048, 1016, 1003, 953, 902, 884, 851, 832, 822, 810, 768, 761, 741, 725, 686, 663, 646, 611, 585, 565, 544, 524, 499, 479, 456, 418, 399, 388, 381 cm^{-1}. – **MS** (70 eV, EI, 120 °C) *m/z* (%) = 357 (24) [M+H]$^+$, 356 (100) [M]$^+$, 301 (10), 300 (62), 299 (19) [M–C$_4$H$_9$]$^+$, 272 (14), 271 (22), 183 (14), 181 (32), 177 (11), 131 (33), 109 (14), 69 (44), 57 (11). – **HRMS** (EI, C$_{21}$H$_{21}$O$_4$F) = calc.: 356.1424; found: 356.1423.

3-(4-Chlorobenzyl)-5-methoxy-7-butoxy-2*H*-chromen-2-one (21dh)

According to **GP8**, 4-butoxy-2-hydroxy-6-methoxybenzaldehyde (**19d**, 75.0 mg, 0.33 mmol, 1.00 equiv.), 4-chlorocinnamaldehyde (139 mg, 0.84 mmol, 2.50 equiv.), K$_2$CO$_3$ (55.0 mg, 0.40 mmol, 1.20 equiv.) and 1,3-dimethylimidazolium dimethyl phosphate (0.07 ml, 89.0 mg, 0.40 mmol, 1.20 equiv.) in toluene (1.10 ml) were heated at 110 °C for 50 min *via* microwave irradiation. Filtration over a small silica plug (*c*Hex/EtOAc 2:1) and purification by prepHPLC (5–95_STD) resulted in the title compound as a yellow solid (12.2 mg, 0.03 mmol, 10%).

R_t (5%–95% in 27 min, STD) = 32.5 min. – ^{1}**H NMR** (400 MHz, CDCl$_3$) δ = 7.66 (s, 1H, C*H*), 7.30 – 7.20 (m, 4H, 4 × C*H*$_{ar}$), 6.39 (d, ^4J = 2.1 Hz, 1H, C*H*$_{ar}$), 6.26 (d, ^4J = 2.1 Hz, 1H, C*H*$_{ar}$), 3.98 (t, ^3J = 6.5 Hz, 2H, OC*H*$_2$), 3.85 (s, 3H, OC*H*$_3$), 3.80 (s, 2H, CC*H*$_2$), 1.84 – 1.72 (m, 2H, OCH$_2$C*H*$_2$), 1.49 (sex., ^3J = 7.4 Hz, 2H, C*H*$_2$CH$_3$), 0.98 (t, ^3J = 7.4 Hz, 3H, C*H*$_3$) ppm. – ^{13}**C NMR** (101 MHz, CDCl$_3$) δ = 162.7 (C$_q$, *C*OO), 162.3 (C$_q$, 7-C$_{ar}$OCH$_2$), 156.6 (C$_q$, 5-C$_{ar}$OCH$_3$), 155.9 (C$_q$, 8a-*C*$_{ar}$), 137.2 (C$_q$, *C*$_{ar}$Cl), 135.2 (+, *C*H), 132.5 (C$_q$, CH$_2$*C*$_{ar}$), 130.6 (+, 2 × *C*$_{ar}$H), 128.8 (+, 2 × *C*$_{ar}$H), 122.9 (C$_q$, 3-*C*), 104.2 (C$_q$, 4a-*C*$_{ar}$), 95.3 (+, *C*$_{ar}$H), 93.1 (+, *C*$_{ar}$H), 68.4 (–,

OCH_2), 56.0 (+, OCH_3), 36.2 (–, CCH_2), 31.1 (–, OCH$_2$$CH_2$), 19.3 (–, CH_2CH$_3$), 13.9 (+, CH_3) ppm. – **IR** (ATR, $\tilde{v}$) = 3087, 3067, 3009, 2963, 2951, 2925, 2867, 1706, 1616, 1575, 1492, 1462, 1425, 1405, 1366, 1309, 1293, 1247, 1232, 1201, 1171, 1154, 1115, 1089, 1050, 1040, 1016, 1001, 962, 952, 901, 884, 837, 820, 809, 761, 725, 696, 683, 643, 557, 543, 511, 489, 455, 422, 398 cm^{-1}. – **MS** (70 eV, EI, 130 °C) m/z (%) = 373 (26) [M+H]$^+$, 374/372 (36/100) [M]$^+$, 318/316 (13/34), 317/315 (12/16) [M–C$_4$H$_9$]$^+$, 287 (16), 281 (58), 243 (11), 181 (39), 177 (13), 165 (13), 162 (12), 131 (46), 125 (13), 100 (12), 69 (57), 57 (16). – **HRMS** (EI, C$_{21}$H$_{21}$O$_4$^{35}Cl) = calc.: 372.1128; found: 372.1129.

3-(4-(Dimethylamino)benzyl)-5-methoxy-7-butoxy-2H-chromen-2-one (21di)

According to **GP8**, 4-butoxy-2-hydroxy-6-methoxyben-zaldehyde (**19d**, 75.0 mg, 0.33 mmol, 1.00 equiv.), 4-(di-methylamino)cinnamaldehyde (147 mg, 0.84 mmol, 2.50 equiv.), K$_2$CO$_3$ (55.0 mg, 0.40 mmol, 1.20 equiv.) and 1,3-dimethylimidazolium dimethyl phosphate (0.07 ml, 89.0 mg, 0.40 mmol, 1.20 equiv.) in toluene (1.10 ml) were heated at 110 °C for 50 min via microwave irradiation. Filtration over a small silica plug (cHex/EtOAc 2:1) and purification by prepHPLC (5–95_STD) resulted in the title compound as a brown solid (27.0 mg, 0.07 mmol, 21%).

R_t (5%–95% in 27 min, STD) = 20.2 min – **^{1}H NMR** (400 MHz, CDCl$_3$) δ = 7.66 (d, ^4J = 1.3 Hz, 1H, CH), 7.21 (d, ^3J = 8.6 Hz, 2H, 2 × CH_{ar}), 6.82 (d, ^3J = 8.6 Hz, 2H, 2 × CH_{ar}), 6.38 (d, ^4J = 2.2 Hz, 1H, CH_{ar}), 6.25 (d, ^4J = 2.2 Hz, 1H, CH_{ar}), 3.97 (t, ^3J = 6.5 Hz, 2H, OCH_2), 3.83 (s, 3H, OCH_3), 3.76 (s, 2H, CCH_2), 2.95 (s, 6H, N(CH_3)$_2$), 1.83 – 1.71 (m, 2H, OCH$_2$$CH_2$), 1.49 (sex., ^3J = 7.4 Hz, 2H, CH_2CH$_3$), 0.98 (t, ^3J = 7.4 Hz, 3H, CH_3) ppm. – **^{13}C NMR** (101 MHz, CDCl$_3$) δ = 162.6 (C$_q$, COO), 162.4 (C$_q$, 7-C_{ar}OCH$_2$), 156.6 (C$_q$, 5-C_{ar}OCH$_3$), 155.8 (C$_q$, 8a-C_{ar}), 148.7 (C$_q$, C_{ar}N(CH$_3$)$_2$), 134.7 (+, CH), 130.1 (+, 2 × C_{ar}H), 124.1 (C$_q$, 3-C), 114.1 (+, 2 × C_{ar}H), 104.4 (C$_q$, 4a-C_{ar}), 95.2 (+, C_{ar}H), 93.0 (+, C_{ar}H), 68.4 (–, OCH_2), 55.9 (+, OCH_3), 41.6 (+, N(CH_3)$_2$), 35.8 (–, CCH_2), 31.2 (–, OCH$_2$$CH_2$), 19.3 (–, CH_2CH$_3$), 13.9 (+, CH_3) ppm. – **IR** (ATR, $\tilde{v}$) = 3087, 3009, 2958, 2927, 2898, 2868, 2854, 2803, 1720, 1711, 1611, 1574, 1523, 1497, 1465, 1451, 1426, 1407, 1374, 1364, 1350, 1332, 1312, 1292, 1266, 1242, 1232, 1201, 1171, 1150, 1123, 1108, 1062, 1048, 1041, 1021, 997, 948, 928, 895, 877, 836, 824, 800, 761, 735, 724, 714, 693, 654, 643, 630, 564, 543, 527, 513, 494, 479, 462, 398 cm^{-1}. – **MS** (70 eV, EI, 150 °C) m/z (%) = 382 (25) [M+H]$^+$, 381 (100) [M]$^+$, 324 (11) [M–C$_4$H$_9$]$^+$, 181 (25), 134 (14), 131 (28), 69 (37). – **HRMS** (EI, C$_{23}$H$_{27}$O$_4$N) = calc.: 381.1940; found: 381.1942.

3-Benzyl-5-methoxy-7-(pentyloxy)-2*H*-chromen-2-one (21ea)

According to **GP8**, 4-(pentyloxy)-2-hydroxy-6-methoxyben-zaldehyde (**19e**, 75.0 mg, 0.31 mmol, 1.00 equiv.), cinnamal-dehyde (0.10 ml, 104 mg, 0.79 mmol, 2.50 equiv.), K$_2$CO$_3$ (52.0 mg, 0.38 mmol, 1.20 equiv.) and 1,3-dimethylimidazolium dimethyl phosphate (0.07 ml, 84.0 mg, 0.38 mmol, 1.20 equiv.) in toluene (1.04 ml) were heated at 110 °C for 50 min *via* microwave irradiation. Filtration over a small silica plug (*c*Hex/EtOAc 1:1) and purification by prepHPLC (5–95_STD) resulted in the title compound as yellow crystals (18.0 mg, 0.05 mmol, 16%).

R_t (5%–95% in 27 min, STD) = 32.5 min – **^{1}H NMR** (500 MHz, CDCl$_3$) δ = 7.67 (s, 1H, C*H*), 7.35 – 7.29 (m, 4H, 4 × C*H*$_{ar}$), 7.26 – 7.21 (m, 1H, C*H*$_{ar}$), 6.39 (d, ^4J = 2.1 Hz, 1H, C*H*$_{ar}$), 6.25 (d, ^4J = 2.1 Hz, 1H, C*H*$_{ar}$), 3.97 (t, ^3J = 6.5 Hz, 2H, OC*H*$_2$), 3.84 (s, 2H, CC*H*$_2$), 3.83 (s, 3H, OC*H*$_3$), 1.84 – 1.75 (m, 2H, OCH$_2$C*H*$_2$), 1.48 – 1.34 (m, 4H, 2 × C*H*$_2$), 0.94 (t, ^3J = 7.1 Hz, 3H, C*H*$_3$) ppm. – **^{13}C NMR** (126 MHz, CDCl$_3$) δ = 162.5 (C$_q$, *C*OO and 7-*C*$_{ar}$OCH$_2$), 156.6 (C$_q$, 5-*C*$_{ar}$OCH$_3$), 155.8 (C$_q$, 8a-*C*$_{ar}$), 138.7 (C$_q$, CH$_2$*C*$_{ar}$), 135.1 (+, *C*H), 129.3 (+, 2 × *C*$_{ar}$H), 128.7 (+, 2 × *C*$_{ar}$H), 126.6 (+, *C*$_{ar}$H), 123.5 (C$_q$, 3-*C*), 104.3 (C$_q$, 4a-*C*$_{ar}$), 95.2 (+, *C*$_{ar}$H), 93.0 (+, *C*$_{ar}$H), 68.7 (–, O*C*H$_2$), 56.0 (+, O*C*H$_3$), 36.8 (–, *C*CH$_2$), 28.8 (–, *C*H$_2$), 28.3 (–, *C*H$_2$), 22.6 (–, *C*H$_2$CH$_3$), 14.1 (+, *C*H$_3$) ppm. – **IR** (ATR, ṽ) = 3087, 3063, 3026, 3003, 2958, 2925, 2870, 2859, 1711, 1613, 1577, 1496, 1453, 1425, 1405, 1364, 1340, 1309, 1293, 1244, 1234, 1201, 1170, 1156, 1113, 1074, 1051, 1028, 1003, 966, 952, 916, 899, 873, 854, 829, 822, 779, 758, 745, 727, 701, 670, 645, 635, 585, 544, 501, 482, 426, 378 cm^{-1}. – **MS** (70 eV, EI, 100 °C) *m/z* (%) = 353 (16) [M+H]$^+$, 352 (68) [M]$^+$, 282 (16), 281 (15) [M–C$_5$H$_{11}$]$^+$, 253 (11), 231 (15), 181 (51), 131 (47), 119 (15), 100 (12), 69 (100), 58 (15). – **HRMS** (EI, C$_{22}$H$_{24}$O$_4$) = calc.: 352.1675; found: 352.1673.

5-Methoxy-3-(2-methylbenzyl)-7-(pentyloxy)-2*H*-chromen-2-one (21eb)

According to **GP8**, 4-(pentyloxy)-2-hydroxy-6-methoxyben-zaldehyde (**19e**, 75.0 mg, 0.31 mmol, 1.00 equiv.), 2-methylcinnamaldehyde (115 mg, 0.79 mmol, 2.50 equiv.), K$_2$CO$_3$ (52.0 mg, 0.38 mmol, 1.20 equiv.) and 1,3-dimethylimidazolium dimethyl phosphate (0.07 ml, 84.0 mg, 0.38 mmol, 1.20 equiv.) in toluene (1.04 ml) were heated at 110 °C for 50 min *via* microwave irradiation. Filtration over a small silica plug (*c*Hex/EtOAc 1:1) and purifi-cation by prepHPLC (5–95_STD) resulted in the title compound as an off-white solid (12.0 mg, 0.03 mmol, 10%).

R_t (5%–95% in 27 min, STD) = 34.7 min. – ^{1}H NMR (400 MHz, CDCl$_3$) δ = 7.38 (s, 1H, CH), 7.22 – 7.18 (m, 4H, 4 × CH_{ar}), 6.41 (d, ^4J = 2.2 Hz, 1H, CH_{ar}), 6.24 (d, ^4J = 2.2 Hz, 1H, CH_{ar}), 3.97 (t, ^3J = 6.5 Hz, 2H, OCH_2), 3.84 (s, 2H, CCH_2), 3.78 (s, 3H, OCH_3), 2.28 (s, 3H, C$_{ar}$CH_3), 1.85 –1.75 (m, 2H, OCH$_2$CH_2), 1.49 – 1.34 (m, 4H, 2 × CH_2), 0.94 (t, ^3J = 7.0 Hz, 3H, CH_3) ppm. – ^{13}C NMR (101 MHz, CDCl$_3$) δ = 162.6 (C$_q$, COO), 162.5 (C$_q$, 7-C$_{ar}$OCH$_2$), 156.6 (C$_q$, 5-C$_{ar}$OCH$_3$), 155.7 (C$_q$, 8a-C$_{ar}$), 137.0 (C$_q$, C$_{ar}$), 136.4 (C$_q$, C$_{ar}$), 134.5 (+, CH), 130.6 (+, C$_{ar}$H), 130.3 (+, C$_{ar}$H), 127.0 (+, C$_{ar}$H), 126.3 (+, C$_{ar}$H), 122.7 (C$_q$, 3-C), 104.3 (C$_q$, 4a-C$_{ar}$), 95.2 (+, C$_{ar}$H), 93.0 (+, C$_{ar}$H), 68.7 (–, OCH$_2$), 55.9 (+, OCH$_3$), 34.0 (–, CCH$_2$), 28.9 (–, CH$_2$), 28.3 (–, CH$_2$), 22.6 (–, CH$_2$CH$_3$), 19.7 (+, C$_{ar}$CH$_3$), 14.1 (+, CH$_3$) ppm. – IR (ATR, ṽ) = 3085, 3065, 3017, 2953, 2928, 2918, 2870, 2856, 1710, 1613, 1575, 1494, 1456, 1424, 1401, 1374, 1363, 1303, 1242, 1198, 1167, 1153, 1112, 1074, 1051, 1018, 1003, 992, 956, 941, 908, 894, 877, 819, 771, 758, 744, 725, 690, 666, 646, 599, 567, 548, 528, 500, 456, 426, 401, 394, 375 cm^{-1}. – MS (70 eV, EI, 110 °C) m/z (%) = 367 (27) [M+H]$^+$, 366 (100) [M]$^+$, 349 (15), 296 (18), 281 (11), 279 (14) [M–C$_5$H$_{11}$O]$^+$, 205 (22), 157 (10), 153 (10), 119 (12), 104 (21). – HRMS (EI, C$_{23}$H$_{26}$O$_4$) = calc.: 366.1831; found: 366.1829.

5-Methoxy-3-(2-methoxybenzyl)-7-(pentyloxy)-2H-chromen-2-one (21ec)

According to GP8, 4-(pentyloxy)-2-hydroxy-6-methoxyben-zaldehyde (19e, 75.0 mg, 0.31 mmol, 1.00 equiv.), 2-methox-ycinnamaldehyde (128 mg, 0.79 mmol, 2.50 equiv.), K$_2$CO$_3$ (52.0 mg, 0.38 mmol, 1.20 equiv.) and 1,3-dimethylimidazolium dimethyl phosphate (0.07 ml, 84.0 mg, 0.38 mmol, 1.20 equiv.) in toluene (1.04 ml) were heated at 110 °C for 50 min via microwave irradiation. Filtration over a small silica plug (cHex/EtOAc 1:1) and purification by prepHPLC (5–95_STD) resulted in the title compound as an off-white solid (30.0 mg, 0.08 mmol, 25%).

R_t (5%–95% in 27 min, STD) = 33.8 min. – ^{1}H NMR (400 MHz, CDCl$_3$) δ = 7.60 (d, ^4J = 1.6 Hz, 1H, CH), 7.29 – 7.21 (m, 2H, 2 × CH_{ar}), 6.96 – 6.86 (m, 2H, 2 × CH_{ar}), 6.39 (d, ^4J = 2.1 Hz, 1H, CH_{ar}), 6.24 (d, ^4J = 2.1 Hz, 1H, CH_{ar}), 3.96 (t, ^3J = 6.6 Hz, 2H, OCH_2), 3.84 (s, 2H, CCH_2), 3.81 (s, 6H, 2 × OCH_3), 1.84 – 1.75 (m, 2H, OCH$_2$CH_2), 1.49 – 1.32 (m, 4H, 2 × CH_2), 0.94 (t, ^3J = 7.0 Hz, 3H, CH_3) ppm. – ^{13}C NMR (101 MHz, CDCl$_3$) δ = 162.6 (C$_q$, COO), 162.2 (C$_q$, 7-C$_{ar}$OCH$_2$), 157.8 (C$_q$, C$_{ar}$OCH$_3$), 156.6 (C$_q$, 5-C$_{ar}$OCH$_3$), 155.7 (C$_q$, 8a-C$_{ar}$), 134.7 (+, CH), 131.1 (+, C$_{ar}$H), 128.1 (+, C$_{ar}$H), 126.9 (C$_q$, CH$_2$C$_{ar}$), 122.8 (C$_q$, 3-C), 120.7 (+, C$_{ar}$H), 110.7 (+, C$_{ar}$H), 104.5 (C$_q$, 4a-C$_{ar}$), 95.1 (+, C$_{ar}$H), 93.0 (+, C$_{ar}$H), 68.7 (–, OCH$_2$), 55.9 (+, OCH$_3$), 55.5 (+, OCH$_3$), 30.9 (–, CCH$_2$), 28.9 (–, CH$_2$), 28.3 (–, CH$_2$), 22.6 (–, CH$_2$CH$_3$), 14.1

(+, C*H*₃) ppm. – **IR** (ATR, ṽ) = 3087, 3068, 3004, 2959, 2935, 2868, 2861, 2836, 1708, 1615, 1587, 1577, 1494, 1458, 1425, 1402, 1364, 1312, 1296, 1288, 1242, 1200, 1169, 1157, 1112, 1048, 1031, 1004, 962, 950, 932, 898, 877, 854, 834, 820, 769, 747, 725, 688, 667, 645, 636, 596, 569, 554, 524, 499, 469, 426, 377 cm⁻¹. – **MS** (70 eV, EI, 40 °C) *m/z* (%) = 382 (28) [M]⁺, 231 (14), 205 (13), 181 (50), 131 (47), 119 (17), 100 (11), 71 (11), 69 (100), 58 (36), 57 (19), 55 (12). – **HRMS** (EI, C₂₃H₂₆O₅) = calc.: 382.1780; found: 382.1781.

5-Methoxy-3-(4-methoxybenzyl)-7-(pentyloxy)-2*H*-chromen-2-one (21ed)

According to **GP8**, 4-(pentyloxy)-2-hydroxy-6-methox-ybenzaldehyde (**19e**, 75.0 mg, 0.31 mmol, 1.00 equiv.), 4-methoxycinnam-aldehyde (128 mg, 0.79 mmol, 2.50 equiv.), K₂CO₃ (52.0 mg, 0.38 mmol, 1.20 equiv.) and 1,3-dimethylimidazolium dimethyl phosphate (0.07 ml, 84.0 mg, 0.38 mmol, 1.20 equiv.) in toluene (1.04 ml) were heated at 110 °C for 50 min *via* microwave irradiation. Filtration over a small silica plug (*c*Hex/EtOAc 1:1) and purification by prepHPLC (5–95_STD) resulted in the title compound as a white solid (8.00 mg, 0.02 mmol, 6.5%).

*R*ₜ (5%–95% in 27 min, STD) = 32.4 min. – **¹H NMR** (400 MHz, CDCl₃) δ = 7.64 (s, 1H, C*H*), 7.25 – 7.19 (m, 2H, 2 × C*H*ₐᵣ), 6.89 – 6.84 (m, 2H, 2 × C*H*ₐᵣ), 6.38 (d, ⁴*J* = 2.1 Hz, 1H, C*H*ₐᵣ), 6.25 (d, ⁴*J* = 2.1 Hz, 1H, C*H*ₐᵣ), 3.96 (t, ³*J* = 6.6 Hz, 2H, OC*H*₂), 3.83 (s, 3H, OC*H*₃), 3.80 (s, 3H, OC*H*₃), 3.78 (s, 2H, CC*H*₂), 1.85 – 1.74 (m, 2H, OCH₂C*H*₂), 1.50 – 1.34 (m, 4H, 2 × C*H*₂), 0.94 (t, ³*J* = 7.0 Hz, 3H, C*H*₃) ppm. – **¹³C NMR** (101 MHz, CDCl₃) δ = 162.5 (Cq, *C*OO), 162.4 (Cq, 7-*C*ₐᵣOCH₂), 158.4 (Cq, *C*ₐᵣOCH₃), 156.6 (Cq, 5-*C*ₐᵣOCH₃), 155.8 (Cq, 8a-*C*ₐᵣ), 134.8 (+, *C*H), 130.7 (Cq, CH₂*C*ₐᵣ), 130.3 (+, 2 × *C*ₐᵣH), 123.9 (Cq, 3-*C*), 114.1 (+, 2 × *C*ₐᵣH), 104.4 (Cq, 4a-*C*ₐᵣ), 95.2 (+, *C*ₐᵣH), 93.0 (+, *C*ₐᵣH), 68.7 (–, O*C*H₂), 56.0 (+, O*C*H₃), 55.4 (+, O*C*H₃), 35.9 (–, C*C*H₂), 28.8 (–, *C*H₂), 28.3 (–, *C*H₂), 22.6 (–, *C*H₂CH₃), 14.1 (+, *C*H₃) ppm. – **IR** (ATR, ṽ) = 3009, 2953, 2929, 2897, 2866, 2837, 1711, 1613, 1577, 1514, 1497, 1465, 1455, 1425, 1405, 1366, 1319, 1300, 1245, 1230, 1203, 1173, 1152, 1115, 1045, 1031, 1001, 960, 950, 899, 880, 844, 824, 809, 759, 725, 694, 662, 646, 632, 569, 547, 510, 490, 480, 445, 418, 392, 382 cm⁻¹. – **MS** (70 eV, EI, 150 °C) *m/z* (%) = 383 (28) [M+H]⁺, 382 (100) [M]⁺, 312 (23), 281 (12), 205 (13). – **HRMS** (EI, C₂₃H₂₆O₅) = calc.: 382.1780; found: 382.1778.

3-(4-Fluorobenzyl)-5-methoxy-7-(pentyloxy)-2*H*-chromen-2-one (21ef)

According to **GP8**, 4-(pentyloxy)-2-hydroxy-6-methoxybenzaldehyde (**19e**, 75.0 mg, 0.31 mmol, 1.00 equiv.), 4-fluorocinnamaldehyde (0.10 ml, 118 mg, 0.79 mmol, 2.50 equiv.), K_2CO_3 (52.0 mg, 0.38 mmol, 1.20 equiv.) and 1,3-dimethylimidazolium dimethyl phosphate (0.07 ml, 84.0 mg, 0.38 mmol, 1.20 equiv.) in toluene (1.04 ml) were heated at 110 °C for 50 min *via* microwave irradiation. Filtration over a small silica plug (*c*Hex/EtOAc 1:1) and purification by prepHPLC (5–95_STD) resulted in the title compound as a yellow solid (11.0 mg, 0.03 mmol, 9.7%).

R_t (5%–95% in 27 min, STD) = 33.5 min. – **^{1}H NMR** (400 MHz, CDCl$_3$) δ = 7.65 (s, 1H, C*H*), 7.33 – 7.23 (m, 2H, 2 × C*H*$_{ar}$), 7.05 – 6.96 (m, 2H, 2 × C*H*$_{ar}$), 6.38 (d, ^4J = 2.2 Hz, 1H, C*H*$_{ar}$), 6.26 (d, ^4J = 2.2 Hz, 1H, C*H*$_{ar}$), 3.97 (t, ^3J = 6.5 Hz, 2H, OC*H*$_2$), 3.84 (s, 3H, OC*H*$_3$), 3.80 (s, 2H, CC*H*$_2$), 1.79 (quin., ^3J = 6.7 Hz, 2H, OCH$_2$C*H*$_2$), 1.49 – 1.33 (m, 4H, 2 × C*H*$_2$), 0.94 (t, ^3J = 7.0 Hz, 3H, C*H*$_3$) ppm. – **^{13}C NMR** (101 MHz, CDCl$_3$) δ = 162.6 (C$_q$, *C*OO), 162.3 (C$_q$, 7-*C*$_{ar}$OCH$_2$), 161.8 (C$_q$, d, ^1J = 244.4 Hz, *C*$_{ar}$F), 156.6 (C$_q$, 5-*C*$_{ar}$OCH$_3$), 155.9 (C$_q$, 8a-*C*$_{ar}$), 135.0 (+, *C*H), 134.4 (C$_q$, d, ^4J = 3.3 Hz, CH$_2$*C*$_{ar}$), 130.7 (+, d, ^3J = 7.8 Hz, 2 × *m*-*C*$_{ar}$H), 123.3 (C$_q$, 3-*C*), 115.5 (d, ^2J = 21.3 Hz, 2 × *o*-*C*$_{ar}$H), 104.2 (C$_q$, 4a-*C*$_{ar}$), 95.3 (+, *C*$_{ar}$H), 93.1 (+, *C*$_{ar}$H), 68.7 (–, O*C*H$_2$), 56.0 (+, O*C*H$_3$), 36.1 (–, C*C*H$_2$), 28.8 (–, *C*H$_2$), 28.3 (–, *C*H$_2$), 22.6 (–, *C*H$_2$CH$_3$), 14.1 (+, *C*H$_3$) ppm. – **^{19}F NMR** (376 MHz, CDCl$_3$) δ = –121.0 (s, C*F*) ppm. – **IR** (ATR, ṽ) = 3088, 3003, 2958, 2934, 2873, 2859, 1708, 1613, 1577, 1509, 1463, 1455, 1425, 1405, 1364, 1310, 1293, 1244, 1224, 1201, 1170, 1156, 1112, 1094, 1051, 1031, 1017, 1003, 956, 952, 901, 882, 875, 853, 830, 822, 772, 761, 745, 725, 691, 662, 646, 630, 567, 544, 524, 501, 480, 459, 441, 419, 382 cm^{-1}. – **MS** (70 eV, EI, 130 °C) *m/z* (%) = 371 (25) [M+H]$^+$, 370 (100) [M]$^+$, 301 (12), 300 (60), 299 (11) [M–C$_5$H$_{11}$]$^+$, 271 (12). – **HRMS** (EI, C$_{22}$H$_{23}$O$_4$F) = calc.: 370.1580; found: 370.1580.

3-(4-Chlorobenzyl)-5-methoxy-7-(pentyloxy)-2*H*-chromen-2-one (21eh)

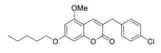

According to **GP8**, 4-(pentyloxy)-2-hydroxy-6-methoxybenzaldehyde (**19e**, 75.0 mg, 0.31 mmol, 1.00 equiv.), 4-chlorocinnamaldehyde (131 mg, 0.79 mmol, 2.50 equiv.), K_2CO_3 (52.0 mg, 0.38 mmol, 1.20 equiv.) and 1,3-dimethylimidazolium dimethyl phosphate (0.07 ml, 84.0 mg, 0.38 mmol, 1.20 equiv.) in toluene (1.04 ml) were heated at 110 °C for 50 min *via* microwave irradiation. Filtration over a small silica plug (*c*Hex/EtOAc 1:1) and purification by prepHPLC (5–95_STD) resulted in the title compound as a yellow solid (18.0 mg, 0.05 mmol, 15%).

R_t (5%–95% in 27 min, STD) = 35.2 min. – **^{1}H NMR** (400 MHz, CDCl$_3$) δ = 7.67 (d, ^4J = 1.5 Hz, 1H, CH), 7.31 – 7.21 (m, 4H, 4 × CH_{ar}), 6.39 (d, ^4J = 2.1 Hz, 1H, CH_{ar}), 6.26 (d, ^4J = 2.1 Hz, 1H, CH_{ar}), 3.97 (t, ^3J = 6.6 Hz, 2H, OCH_2), 3.85 (s, 3H, OCH_3), 3.80 (s, 2H, CCH_2), 1.86 – 1.74 (m, 2H, OCH$_2$CH_2), 1.50 – 1.35 (m, 4H, 2 × CH_2), 0.94 (t, ^3J = 7.0 Hz, 3H, CH_3) ppm. – **^{13}C NMR** (101 MHz, CDCl$_3$) δ = 162.7 (C$_q$, COO), 162.3 (C$_q$, 7-C_{ar}OCH$_2$), 156.7 (C$_q$, 5-C_{ar}OCH$_3$), 155.9 (C$_q$, 8a-C_{ar}), 137.3 (C$_q$, C_{ar}Cl), 135.2 (+, CH), 132.5 (C$_q$, CH$_2C_{ar}$), 130.6 (+, 2 × C_{ar}H), 128.8 (+, 2 × C_{ar}H), 122.9 (C$_q$, 3-C), 104.2 (C$_q$, 4a-C_{ar}), 95.3 (+, C_{ar}H), 93.1 (+, C_{ar}H), 68.8 (–, OCH$_2$), 56.0 (+, OCH$_3$), 36.2 (–, CCH$_2$), 28.8 (–, CH$_2$), 28.3 (–, CH$_2$), 22.6 (–, CH$_2$CH$_3$), 14.1 (+, CH$_3$) ppm. – **IR** (ATR, ṽ) = 2956, 2932, 2922, 2870, 2859, 1710, 1615, 1575, 1492, 1456, 1425, 1405, 1378, 1364, 1309, 1245, 1201, 1171, 1156, 1113, 1091, 1051, 1031, 1016, 1004, 953, 901, 877, 854, 830, 822, 810, 779, 761, 747, 725, 696, 684, 643, 543, 511, 407, 380 cm^{-1}. – **MS** (70 eV, EI, 140 °C) m/z (%) = 387 (25) [M+H]$^+$, 386/388 (100/34) [M]$^+$, 318 (18), 317 (14), 316 (49), 315 (10) [M–C$_5$H$_{11}$]$^+$, 282 (12), 281 (60), 205 (12), 69 (17). – **HRMS** (EI, C$_{22}$H$_{23}$O$_4$^{35}Cl) = calc.: 386.1285; found: 386.1283.

3-(4-(Dimethylamino)benzyl)-5-methoxy-7-(pentyloxy)-2H-chromen-2-one (21ei)

According to **GP8**, 4-(pentyloxy)-2-hydroxy-6-methoxybenzaldehyde (**19e**, 75.0 mg, 0.31 mmol, 1.00 equiv.), 4-(dimethylamino)cinnamaldehyde (138 mg, 0.79 mmol, 2.50 equiv.), K$_2$CO$_3$ (52.0 mg, 0.38 mmol, 1.20 equiv.) and 1,3-dimethylimidazolium dimethyl phosphate (0.07 ml, 84.0 mg, 0.38 mmol, 1.20 equiv.) in toluene (1.04 ml) were heated at 110 °C for 50 min *via* microwave irradiation. Filtration over a small silica plug (*c*Hex/EtOAc 1:1) and purification by prepHPLC (5–95_STD) resulted in the title compound as a purple solid (16.0 mg, 0.04 mmol, 13%).

R_t (5%–95% in 27 min, STD) = 21.7 min. – **^{1}H NMR** (400 MHz, CDCl$_3$) δ = 7.66 (d, ^4J = 1.3 Hz, 1H, CH), 7.20 (d, ^3J = 8.6 Hz, 2H, 2 × CH_{ar}), 6.79 (d, ^3J = 8.6 Hz, 2H, 2 × CH_{ar}), 6.38 (d, ^4J = 2.1 Hz, 1H, CH_{ar}), 6.25 (d, ^4J = 2.1 Hz, 1H, CH_{ar}), 3.96 (t, ^3J = 6.5 Hz, 2H, OCH_2), 3.83 (s, 3H, OCH_3), 3.75 (s, 2H, CCH_2), 2.95 (s, 6H, N(CH_3)$_2$), 1.84 – 1.75 (m, 2H, OCH$_2$CH_2), 1.49 – 1.34 (m, 4H, 2 × CH_2), 0.94 (t, ^3J = 7.0 Hz, 3H, CH_3) ppm. – **^{13}C NMR** (101 MHz, CDCl$_3$) δ = 162.6 (C$_q$, COO), 162.4 (C$_q$, 7-C_{ar}OCH$_2$), 156.6 (C$_q$, 5-C_{ar}OCH$_3$), 155.8 (C$_q$, 8a-C_{ar}), 148.9 (C$_q$, C_{ar}N(CH$_3$)$_2$), 134.6 (+, CH), 130.1 (+, 2 × C_{ar}H), 127.9 (C$_q$, CH$_2C_{ar}$), 124.2 (C$_q$, 3-C), 113.9 (+, 2 × C_{ar}H), 104.4 (C$_q$, 4a-C_{ar}), 95.2 (+, C_{ar}H), 93.0 (+, C_{ar}H), 68.7 (–, OCH$_2$), 55.9 (+, OCH$_3$), 41.4 (+, N(CH$_3$)$_2$), 35.8 (–, CCH$_2$), 28.9 (–, CH$_2$), 28.3 (–, CH$_2$), 22.6 (–, CH$_2$CH$_3$), 14.1 (+, CH$_3$) ppm. – **IR** (ATR, ṽ) = 2955, 2944, 2931, 2898, 2885, 2866, 2851, 2798, 1723, 1615,

1575, 1523, 1494, 1466, 1453, 1425, 1401, 1377, 1366, 1344, 1309, 1242, 1227, 1200, 1170, 1150, 1126, 1108, 1045, 1031, 997, 946, 895, 822, 806, 759, 722, 693, 656, 640, 564, 545, 526, 489, 479, 391 cm^{-1}. – **MS** (70 eV, EI, 150 °C) m/z (%) = 396 (29) [M+H]$^+$, 395 (100) [M]$^+$, 69 (14). – **HRMS** (EI, C$_{24}$H$_{29}$O$_4$N) = calc.: 395.2097; found: 395.2098.

3-Benzyl-7-(hexyloxy)-5-methoxy-2*H*-chromen-2-one (21fa)

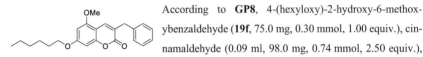

According to **GP8**, 4-(hexyloxy)-2-hydroxy-6-methox-ybenzaldehyde (**19f**, 75.0 mg, 0.30 mmol, 1.00 equiv.), cin-namaldehyde (0.09 ml, 98.0 mg, 0.74 mmol, 2.50 equiv.), K$_2$CO$_3$ (49.0 mg, 0.36 mmol, 1.20 equiv.) and 1,3-dimethylimidazolium dimethyl phosphate (0.06 ml, 79.0 mg, 0.36 mmol, 1.20 equiv.) in toluene (0.98 ml) were heated at 110 °C for 50 min *via* microwave irradiation. Filtration over a small silica plug (*c*Hex/EtOAc 1:1) and purifi-cation by prepHPLC (5–95_MOD) resulted in the title compound as a yellow solid (11.0 mg, 0.03 mmol, 10%).

R_t (5%–95% in 22 min, MOD) = 35.2 min. – **^{1}H NMR** (400 MHz, CDCl$_3$) δ = 7.67 (d, ^4J = 1.1 Hz, 1H, C*H*), 7.36 – 7.29 (m, 4H, 4 × C*H*$_{ar}$), 7.27 – 7.21 (m, 1H, C*H*$_{ar}$), 6.39 (d, ^4J = 2.1 Hz, 1H, C*H*$_{ar}$), 6.25 (d, ^4J = 2.1 Hz, 1H, C*H*$_{ar}$), 3.97 (t, ^3J = 6.5 Hz, 2H, OC*H*$_2$), 3.84 (d, ^4J = 1.1 Hz, 2H, CC*H*$_2$), 3.83 (s, 3H, OC*H*$_3$), 1.84 – 1.74 (m, 2H, OCH$_2$C*H*$_2$), 1.52 – 1.41 (m, 2H, C*H*$_2$), 1.38 – 1.31 (m, 4H, 2 × C*H*$_2$), 0.91 (t, ^3J = 7.0 Hz, 3H, C*H*$_3$) ppm. – **^{13}C NMR** (101 MHz, CDCl$_3$) δ = 162.5 (C$_q$, *C*OO and 7-*C*$_{ar}$OCH$_2$), 156.6 (C$_q$, 5-*C*$_{ar}$OCH$_3$), 155.8 (C$_q$, 8a-*C*$_{ar}$), 138.8 (C$_q$, CH$_2$*C*$_{ar}$), 135.1 (+, *C*H), 129.3 (+, 2 × *C*$_{ar}$H), 128.7 (+, 2 × *C*$_{ar}$H), 126.6 (+, *C*$_{ar}$H), 123.5 (C$_q$, 3-*C*), 104.3 (C$_q$, 4a-*C*$_{ar}$), 95.2 (+, *C*$_{ar}$H), 93.0 (+, *C*$_{ar}$H), 68.7 (–, O*C*H$_2$), 56.0 (+, O*C*H$_3$), 36.8 (–, C*C*H$_2$), 31.1 (–, *C*H$_2$), 29.1 (–, *C*H$_2$), 25.8 (–, *C*H$_2$), 22.7 (–, *C*H$_2$CH$_3$), 14.2 (+, *C*H$_3$) ppm. – **IR** (ATR, ṽ) = 2956, 2928, 2915, 2870, 2854, 1711, 1615, 1577, 1496, 1465, 1453, 1425, 1364, 1309, 1245, 1201, 1171, 1153, 1115, 1075, 1047, 1033, 1001, 953, 943, 919, 901, 820, 809, 759, 725, 700, 671, 646, 585, 547, 504, 487, 456, 394, 378 cm^{-1}. – **MS** (70 eV, EI, 130 °C) m/z (%) = 367 (25) [M+H]$^+$, 366 (100) [M]$^+$, 283 (14), 282 (65), 281 (19) [M–C$_6$H$_{13}$]$^+$, 254 (10), 253 (13). – **HRMS** (EI, C$_{23}$H$_{26}$O$_4$) = calc.: 366.1831; found: 366.1832.

7-(Hexyloxy)-5-methoxy-3-(2-methylbenzyl)-2*H*-chromen-2-one (21fb)

According to **GP8**, 4-(hexyloxy)-2-hydroxy-6-methoxybenzaldehyde (**19f**, 75.0 mg, 0.30 mmol, 1.00 equiv.), 2-methylcinnamaldehyde (109 mg, 0.74 mmol, 2.50 equiv.), K₂CO₃ (49.0 mg, 0.36 mmol, 1.20 equiv.) and 1,3-dimethylimidazolium dimethyl phosphate (0.06 ml, 79.0 mg, 0.36 mmol, 1.20 equiv.) in toluene (0.98 ml) were heated at 110 °C for 50 min *via* microwave irradiation. Filtration over a small silica plug (*c*Hex/EtOAc 1:1) and purification by prepHPLC (5–95_STD) resulted in the title compound as a yellow solid (14.0 mg, 0.04 mmol, 12%).

R_t (5%–95% in 27 min, STD) = 36.5 min. – **¹H NMR** (500 MHz, CDCl₃) δ = 7.38 (d, ^4J = 1.5 Hz, 1H, C*H*), 7.23 – 7.17 (m, 4H, 4 × C*H*$_{ar}$), 6.41 (d, ^4J = 2.1 Hz, 1H, C*H*$_{ar}$), 6.24 (d, ^4J = 2.1 Hz, 1H, C*H*$_{ar}$), 3.97 (t, ^3J = 6.6 Hz, 2H, OC*H*₂), 3.84 (d, ^4J = 1.5 Hz, 2H, CC*H*₂), 3.78 (s, 3H, OC*H*₃), 2.28 (s, 3H, C$_{ar}$C*H*₃), 1.79 (quin., ^3J = 7.1, 6.6 Hz, 2H, OCH₂C*H*₂), 1.46 (quin., ^3J = 7.1 Hz, 2H, C*H*₂), 1.38 – 1.32 (m, 4H, 2 × C*H*₂), 0.91 (t, ^3J = 6.9 Hz, 3H, C*H*₃) ppm. – **¹³C NMR** (101 MHz, CDCl₃) δ = 162.6 (C$_q$, *C*OO), 162.4 (C$_q$, 7-*C*$_{ar}$OCH₂), 156.6 (C$_q$, 5-*C*$_{ar}$OCH₃), 155.7 (C$_q$, 8a-*C*$_{ar}$), 136.9 (C$_q$, *C*$_{ar}$), 136.4 (C$_q$, *C*$_{ar}$), 134.5 (+, *C*H), 130.6 (+, *C*$_{ar}$H), 130.3 (+, *C*$_{ar}$H), 127.0 (+, *C*$_{ar}$H), 126.3 (+, *C*$_{ar}$H), 122.7 (C$_q$, 3-*C*), 104.3 (C$_q$, 4a-*C*$_{ar}$), 95.2 (+, *C*$_{ar}$H), 93.0 (+, *C*$_{ar}$H), 68.7 (–, O*C*H₂), 55.9 (+, O*C*H₃), 33.9 (–, C*C*H₂), 31.7 (–, *C*H₂), 29.1 (–, *C*H₂), 25.8 (–, *C*H₂), 22.7 (–, *C*H₂CH₃), 19.7 (+, C$_{ar}$*C*H₃), 14.2 (+, *C*H₃) ppm. – **IR** (ATR, ṽ) = 3019, 2953, 2921, 2857, 1720, 1612, 1575, 1494, 1453, 1422, 1375, 1364, 1303, 1242, 1222, 1197, 1167, 1149, 1112, 1050, 1041, 1016, 1006, 973, 956, 941, 907, 881, 843, 819, 771, 758, 745, 725, 690, 667, 645, 637, 599, 568, 548, 531, 500, 462, 438, 402, 391 cm⁻¹. – **MS** (70 eV, EI, 110 °C) *m/z* (%) = 381 (35) [M+H]⁺, 380 (85) [M]⁺, 296 (19) 281 (12), 279 (13) [M–C₆H₁₃O]⁺, 251 (13), 231 (17), 205 (20), 153 (11), 119 (19), 104 (18), 100 (12), 58 (13), 57 (13), 55 (11). – **HRMS** (EI, C₂₄H₂₈O₄) = calc.: 380.1988; found: 380.1989.

7-(Hexyloxy)-5-methoxy-3-(2-methoxybenzyl)-2*H*-chromen-2-one (21fc)

According to **GP8**, 4-(hexyloxy)-2-hydroxy-6-methoxybenzaldehyde (**19f**, 75.0 mg, 0.30 mmol, 1.00 equiv.), 2-methoxycinnamaldehyde (121 mg, 0.74 mmol, 2.50 equiv.), K₂CO₃ (49.0 mg, 0.36 mmol, 1.20 equiv.) and 1,3-dimethylimidazolium dimethyl phosphate (0.06 ml, 79.0 mg, 0.36 mmol, 1.20 equiv.) in toluene (0.98 ml) were heated at 110 °C for 50 min *via* microwave irradiation. Filtration over a small silica plug (*c*Hex/EtOAc 1:1) and purification by prepHPLC (5–95_MOD) resulted in the title compound as a yellow solid (27.0 mg, 0.07 mmol, 23%).

R_t (5%–95% in 22 min, MOD) = 31.0 min. – **^{1}H NMR** (400 MHz, CDCl$_3$) δ = 7.59 (s, 1H, C*H*), 7.29 – 7.21 (m, 2H, 2 × C*H*$_{ar}$), 6.96 – 6.87 (m, 2H, 2 × C*H*$_{ar}$), 6.39 (d, ^4J = 2.1 Hz, 1H, C*H*$_{ar}$), 6.24 (d, ^4J = 2.1 Hz, 1H, C*H*$_{ar}$), 3.96 (t, ^3J = 6.6 Hz, 2H, OC*H*$_2$), 3.84 (s, 2H, CC*H*$_2$), 3.81 (s, 6H, 2 × OC*H*$_3$), 1.78 (quin., ^3J = 6.7 Hz, 2H, OCH$_2$C*H*$_2$), 1.51 – 1.41 (m, 2H, C*H*$_2$), 1.37 – 1.30 (m, 4H, 2 × C*H*$_2$), 0.91 (t, ^3J = 6.9 Hz, 3H, C*H*$_3$) ppm. – **^{13}C NMR** (101 MHz, CDCl$_3$) δ = 162.6 (C$_q$, *C*OO), 162.2 (C$_q$, 7-*C*$_{ar}$OCH$_2$), 157.8 (C$_q$, *C*$_{ar}$OCH$_3$), 156.6 (C$_q$, 5-*C*$_{ar}$OCH$_3$), 155.7 (C$_q$, 8a-*C*$_{ar}$), 134.7 (+, *C*H), 131.1 (+, *C*$_{ar}$H), 128.1 (+, *C*$_{ar}$H), 126.9 (C$_q$, CH$_2$*C*$_{ar}$), 122.9 (C$_q$, 3-*C*), 120.7 (+, *C*$_{ar}$H), 110.7 (+, *C*$_{ar}$H), 104.5 (C$_q$, 4a-*C*$_{ar}$), 95.1 (+, *C*$_{ar}$H), 93.0 (+, *C*$_{ar}$H), 68.7 (–, O*C*H$_2$), 55.9 (+, O*C*H$_3$), 55.5 (+, O*C*H$_3$), 31.7 (–, C*C*H$_2$), 30.9 (–, *C*H$_2$), 29.1 (–, *C*H$_2$), 25.8 (–, *C*H$_2$), 22.7 (–, *C*H$_2$CH$_3$), 14.2 (+, *C*H$_3$) ppm. – **IR** (ATR, ṽ) = 3004, 2952, 2929, 2866, 2859, 2836, 1715, 1615, 1588, 1577, 1496, 1465, 1456, 1435, 1425, 1402, 1366, 1310, 1296, 1289, 1245, 1200, 1169, 1154, 1112, 1047, 1030, 1006, 976, 952, 938, 915, 898, 885, 834, 820, 788, 769, 749, 725, 690, 666, 645, 616, 596, 569, 554, 521, 475, 439, 424, 390 cm^{-1}. – **HRMS** (ESI, C$_{24}$H$_{28}$O$_5$) = calc.: 397.2010 [M+H]$^+$; found: 397.2006 [M+H]$^+$.

7-(Hexyloxy)-5-methoxy-3-(4-methoxybenzyl)-2*H*-chromen-2-one (21fd)

According to **GP8**, 4-(hexyloxy)-2-hydroxy-6-meth-oxybenzaldehyde (**19f**, 75.0 mg, 0.30 mmol, 1.00 equiv.), 4-methoxycinnamaldehyde (121 mg, 0.74 mmol, 2.50 equiv.), K$_2$CO$_3$ (49.0 mg, 0.36 mmol, 1.20 equiv.) and 1,3-dimethylimidazolium dimethyl phosphate (0.06 ml, 79.0 mg, 0.36 mmol, 1.20 equiv.) in toluene (0.98 ml) were heated at 110 °C for 50 min *via* microwave irradiation. Filtration over a small silica plug (*c*Hex/EtOAc 1:1) and purification by prepHPLC (5–95_MOD) resulted in the title compound as an orange solid (18.0 mg, 0.05 mmol, 15%).

R_t (5%–95% in 22 min, MOD) = 30.4 min. – **^{1}H NMR** (400 MHz, CDCl$_3$) δ = 7.63 (d, ^4J = 1.3 Hz, 1H, C*H*), 7.22 (d, ^3J = 8.7 Hz, 2H, 2 × C*H*$_{ar}$), 6.86 (d, ^3J = 8.7 Hz, 2H, 2 × C*H*$_{ar}$), 6.38 (d, ^4J = 2.1 Hz, 1H, C*H*$_{ar}$), 6.25 (d, ^4J = 2.1 Hz, 1H, C*H*$_{ar}$), 3.96 (t, ^3J = 6.5 Hz, 2H, OC*H*$_2$), 3.83 (s, 3H, OC*H*$_3$), 3.80 (s, 3H, OC*H*$_3$), 3.77 (s, 2H, CC*H*$_2$), 1.83 – 1.74 (m, 2H, OCH$_2$C*H*$_2$), 1.51 – 1.41 (m, 2H, C*H*$_2$), 1.38 – 1.30 (m, 4H, 2 × C*H*$_2$), 0.91 (t, ^3J = 6.8 Hz, 3H, C*H*$_3$) ppm. – **^{13}C NMR** (101 MHz, CDCl$_3$) δ = 162.5 (C$_q$, *C*OO), 162.4 (C$_q$, 7-*C*$_{ar}$OCH$_2$), 158.4 (C$_q$, *C*$_{ar}$OCH$_3$), 156.6 (C$_q$, 5-*C*$_{ar}$OCH$_3$), 155.8 (C$_q$, 8a-*C*$_{ar}$), 134.8 (+, *C*H), 130.7 (C$_q$, CH$_2$*C*$_{ar}$), 130.3 (+, 2 × *C*$_{ar}$H), 123.9 (C$_q$, 3-*C*), 114.1 (+, 2 × *C*$_{ar}$H), 104.4 (C$_q$, 4a-*C*$_{ar}$), 95.2 (+, *C*$_{ar}$H), 93.0 (+, *C*$_{ar}$H), 68.7 (–, O*C*H$_2$), 55.9 (+, O*C*H$_3$), 55.4 (+, O*C*H$_3$), 35.9 (–, C*C*H$_2$), 31.7 (–, *C*H$_2$), 29.1 (–, *C*H$_2$), 25.8 (–, *C*H$_2$), 22.7 (–, *C*H$_2$CH$_3$), 14.2 (+, *C*H$_3$) ppm. – **IR** (ATR, ṽ) = 3009, 2958, 2932, 2911,

2900, 2876, 2851, 2836, 1706, 1606, 1578, 1513, 1497, 1455, 1424, 1404, 1366, 1300, 1244, 1201, 1173, 1152, 1129, 1113, 1045, 1033, 1011, 1000, 959, 953, 941, 916, 901, 881, 844, 822, 806, 759, 725, 694, 687, 662, 643, 632, 569, 545, 511, 490, 477, 443, 431, 405, 395, 377 cm^{-1}. – **MS** (70 eV, EI, 150 °C) m/z (%) = 397 (26) [M+H]$^+$, 396 (100) [M]$^+$, 312 (15), 181 (10), 69 (22). – **HRMS** (EI, C$_{24}$H$_{28}$O$_5$) = calc.: 396.1937; found: 396.1938.

3-(4-Fluorobenzyl)-7-(hexyloxy)-5-methoxy-2*H*-chromen-2-one (21ff)

According to **GP8**, 4-(hexyloxy)-2-hydroxy-6-methoxybenzaldehyde (**19f**, 75.0 mg, 0.30 mmol, 1.00 equiv.), 4-fluorocinnamaldehyde (0.09 ml, 112 mg, 0.74 mmol, 2.50 equiv.), K$_2$CO$_3$ (49.0 mg, 0.36 mmol, 1.20 equiv.) and 1,3-dimethylimidazolium dimethyl phosphate (0.06 ml, 79.0 mg, 0.36 mmol, 1.20 equiv.) in toluene (0.98 ml) were heated at 110 °C for 50 min *via* microwave irradiation. Filtration over a small silica plug (*c*Hex/EtOAc 1:1) and purification by prepHPLC (5–95_MOD) resulted in the title compound as an orange solid (17.0 mg, 0.05 mmol, 15%).

R_t (5%–95% in 22 min, MOD) = 30.6 min. – **^{1}H NMR** (400 MHz, CDCl$_3$) δ = 7.65 (s, 1H, C*H*), 7.30 – 7.22 (m, 2H, 2 × C*H*$_{ar}$), 7.04 – 6.96 (m, 2H, 2 × C*H*$_{ar}$), 6.39 (d, ^4J = 2.1 Hz, 1H, C*H*$_{ar}$), 6.26 (d, ^4J = 2.1 Hz, 1H, C*H*$_{ar}$), 3.97 (t, ^3J = 6.5 Hz, 2H, OC*H*$_2$), 3.85 (s, 3H, OC*H*$_3$), 3.80 (s, 2H, CC*H*$_2$), 1.84 – 1.74 (m, 2H, OCH$_2$C*H*$_2$), 1.52 – 1.42 (m, 2H, C*H*$_2$), 1.38 – 1.31 (m, 4H, 2 × C*H*$_2$), 0.91 (t, ^3J = 6.9 Hz, 3H, C*H*$_3$) ppm. – **^{13}C NMR** (101 MHz, CDCl$_3$) δ = 162.6 (C$_q$, *C*OO), 162.4 (C$_q$, 7-*C*$_{ar}$OCH$_2$), 161.8 (C$_q$, d, ^1J = 244.0 Hz, *C*$_{ar}$F), 156.6 (C$_q$, 5-*C*$_{ar}$OCH$_3$), 155.9 (C$_q$, 8a-*C*$_{ar}$), 135.1 (+, *C*H), 134.4 (C$_q$, d, ^4J = 3.2 Hz, CH$_2$*C*$_{ar}$), 130.7 (+, d, ^3J = 8.1 Hz, 2 × *m*-*C*$_{ar}$H), 123.3 (C$_q$, 3-*C*), 115.5 (+, d, ^2J = 21.2 Hz, 2 × *o*-*C*$_{ar}$H), 104.2 (C$_q$, 4a-*C*$_{ar}$), 95.3 (+, *C*$_{ar}$H), 93.1 (+, *C*$_{ar}$H), 68.7 (–, O*C*H$_2$), 56.0 (+, O*C*H$_3$), 36.1 (–, C*C*H$_2$), 31.7 (–, *C*H$_2$), 29.1 (–, *C*H$_2$), 25.8 (–, *C*H$_2$), 22.7 (–, *C*H$_2$CH$_3$), 14.2 (+, *C*H$_3$) ppm. – **^{19}F NMR** (376 MHz, CDCl$_3$) δ = –121.0 (s, C*F*) ppm – **IR** (ATR, ṽ) = 3091, 3016, 2953, 2934, 2919, 2871, 2854, 1708, 1629, 1602, 1575, 1504, 1499, 1458, 1425, 1400, 1377, 1366, 1303, 1251, 1221, 1200, 1166, 1156, 1145, 1112, 1094, 1051, 1017, 999, 986, 972, 943, 929, 914, 853, 832, 817, 790, 751, 728, 717, 686, 666, 645, 616, 571, 550, 523, 503, 486, 450, 421, 391 cm^{-1}. – **MS** (70 eV, EI, 110 °C) m/z (%) = 384 (100) [M]$^+$, 301 (11), 300 (51), 299 (11) [M–C$_6$H$_{13}$]$^+$, 284 (25), 271 (11), 256 (14), 252 (20), 181 (27), 169 (16), 168 (47), 150 (14), 131 (21), 119 (11), 109 (16), 97 (12), 85 (12), 83 (15), 73 (15), 71 (15), 69 (74), 60 (13), 57 (31), 55 (32). – **HRMS** (EI, C$_{23}$H$_{25}$O$_4$F) = calc.: 384.1737; found: 384.1738.

3-(4-Chlorobenzyl)-7-(hexyloxy)-5-methoxy-2*H*-chromen-2-one (21fh)

According to **GP8**, 4-(hexyloxy)-2-hydroxy-6-methoxybenzaldehyde (**19f**, 75.0 mg, 0.30 mmol, 1.00 equiv.), 4-chlorocinnamaldehyde (124 mg, 0.74 mmol, 2.50 equiv.), K_2CO_3 (49.0 mg, 0.36 mmol, 1.20 equiv.) and 1,3-dimethylimidazolium dimethyl phosphate (0.06 ml, 79.0 mg, 0.36 mmol, 1.20 equiv.) in toluene (0.98 ml) were heated at 110 °C for 50 min *via* microwave irradiation. Filtration over a small silica plug (*c*Hex/EtOAc 1:1) and purification by prepHPLC (5–95_MOD) resulted in the title compound as a yellow solid (22.0 mg, 0.05 mmol, 18%).

R_t (5%–95% in 22 min, MOD) = 32.5 min. – **^{1}H NMR** (400 MHz, CDCl$_3$) δ = 7.67 (d, ^4J = 1.6 Hz, 1H, C*H*), 7.31 – 7.21 (m, 4H, 4 × C*H*$_{ar}$), 6.39 (d, ^4J = 2.1 Hz, 1H, C*H*$_{ar}$), 6.26 (d, ^4J = 2.1 Hz, 1H, C*H*$_{ar}$), 3.97 (t, ^3J = 6.5 Hz, 2H, OC*H*$_2$), 3.85 (s, 3H, OC*H*$_3$), 3.80 (s, 2H, CC*H*$_2$), 1.85 – 1.75 (m, 2H, OCH$_2$C*H*$_2$), 1.51 – 1.41 (m, 2H, C*H*$_2$), 1.38 – 1.30 (m, 4H, 2 × C*H*$_2$), 0.91 (t, ^3J = 6.7 Hz, 3H, C*H*$_3$) ppm. – **^{13}C NMR** (101 MHz, CDCl$_3$) δ = 162.7 (C$_q$, *C*OO), 162.3 (C$_q$, 7-*C*$_{ar}$OCH$_2$), 156.7 (C$_q$, 5-*C*$_{ar}$OCH$_3$), 155.9 (C$_q$, 8a-*C*$_{ar}$), 137.3 (C$_q$, *C*$_{ar}$Cl), 135.2 (+, *C*H), 132.5 (C$_q$, CH$_2$*C*$_{ar}$), 130.6 (+, 2 × *C*$_{ar}$H), 128.8 (+, 2 × *C*$_{ar}$H), 122.9 (C$_q$, 3-*C*), 104.2 (C$_q$, 4a-*C*$_{ar}$), 95.3 (+, *C*$_{ar}$H), 93.1 (+, *C*$_{ar}$H), 68.8 (–, O*C*H$_2$), 56.0 (+, O*C*H$_3$), 36.2 (–, *C*CH$_2$), 31.7 (–, *C*H$_2$), 29.1 (–, *C*H$_2$), 25.8 (–, *C*H$_2$), 22.7 (–, *C*H$_2$CH$_3$), 14.2 (+, *C*H$_3$) ppm. – **IR** (ATR, ṽ) = 2953, 2924, 2856, 1713, 1615, 1575, 1490, 1458, 1425, 1405, 1377, 1366, 1305, 1247, 1201, 1169, 1153, 1113, 1089, 1048, 1016, 1001, 819, 803, 776, 761, 725, 715, 694, 652, 645, 615, 564, 545, 507 cm^{-1}. – **HRMS** (ESI, C$_{23}$H$_{25}$O$_4$^{35}Cl) = calc.: 401.1514 [M+H]$^+$; found: 401.1510 [M+H]$^+$.

3-(4-(Dimethylamino)benzyl)-7-(hexyloxy)-5-methoxy-2*H*-chromen-2-one (21fi)

According to **GP8**, 4-(hexyloxy)-2-hydroxy-6-methoxybenzaldehyde (**19f**, 75.0 mg, 0.30 mmol, 1.00 equiv.), 4-(dimethylamino)cinnamaldehyde (130 mg, 0.74 mmol, 2.50 equiv.), K_2CO_3 (49.0 mg, 0.36 mmol, 1.20 equiv.) and 1,3-dimethylimidazolium dimethyl phosphate (0.06 ml, 79.0 mg, 0.36 mmol, 1.20 equiv.) in toluene (0.98 ml) were heated at 110 °C for 50 min *via* microwave irradiation. Filtration over a small silica plug (*c*Hex/EtOAc 1:1) and purification by prepHPLC (5–95_STD) resulted in the title compound as a purple solid (12.0 mg, 0.03 mmol, 10%).

R_t (5%–95% in 27 min, STD) = 22.8 min. – **^{1}H NMR** (400 MHz, CDCl$_3$) δ = 7.96 (s, 1H, C*H*), 7.31 – 7.24 (m, 2H, 2 × C*H*$_{ar}$), 7.00 (d, ^3J = 8.2 Hz, 2H, 2 × C*H*$_{ar}$), 6.38 (d, ^4J = 2.1 Hz, 1H, C*H*$_{ar}$), 6.25 (d, ^4J = 2.1 Hz, 1H, C*H*$_{ar}$), 3.96 (t, ^3J = 6.5 Hz, 2H, OC*H*$_2$), 3.84 (s, 3H, OC*H*$_3$), 3.78 (s, 2H, CC*H*$_2$), 3.01 (s, 6H, N(C*H*$_3$)$_2$), 1.84 – 1.74 (m, 2H, OCH$_2$C*H*$_2$), 1.50 – 1.41 (m, 2H,

CH_2), 1.38 – 1.30 (m, 4H, 2 × CH_2), 0.91 (t, ^3J = 6.8 Hz, 3H, CH_3) ppm. – ^{13}C NMR (101 MHz, CDCl$_3$) δ = 162.5 (C$_q$, COO and 7-C_{ar}OCH$_2$), 156.6 (C$_q$, 5-C_{ar}OCH$_3$), 155.8 (C$_q$, 8a-C_{ar}), 146.9 (C$_q$, C_{ar}N(CH$_3$)$_2$), 135.0 (+, CH), 130.4 (+, 2 × C_{ar}H), 123.5 (C$_q$, 3-C), 115.9 (+, 2 × C_{ar}H), 104.3 (C$_q$, 4a-C_{ar}), 95.2 (+, C_{ar}H), 93.0 (+, C_{ar}H), 68.7 (–, OCH$_2$), 56.0 (+, OCH$_3$), 42.9 (+, N(CH$_3$)$_2$), 36.0 (–, CCH$_2$), 31.7 (–, CH$_2$), 29.1 (–, CH$_2$), 25.8 (–, CH$_2$), 22.7 (–, CH$_2$CH$_3$), 14.2 (+, CH$_3$) ppm. – **IR** (ATR, ṽ) = 2951, 2928, 2919, 2867, 2854, 1724, 1609, 1575, 1524, 1497, 1466, 1455, 1425, 1378, 1366, 1353, 1310, 1290, 1241, 1228, 1200, 1167, 1150, 1126, 1109, 1047, 1033, 1010, 993, 945, 894, 875, 841, 826, 803, 761, 741, 721, 693, 656, 642, 564, 545, 527, 514, 476, 418, 390 cm^{-1}. – **HRMS** (ESI, C$_{25}$H$_{31}$O$_4$N) = calc.: 410.2326 [M+H]$^+$; found: 410.2319 [M+H]$^+$.

3-Benzyl-7-(heptyloxy)-5-methoxy-2*H*-chromen-2-one (21ga)

According to **GP8**, 4-(heptyloxy)-2-hydroxy-6-methox-ybenzaldehyde (**19g**, 75.0 mg, 0.28 mmol, 1.00 equiv.), cinnamaldehyde (0.09 ml, 93.0 mg, 0.70 mmol, 2.50 equiv.), K$_2$CO$_3$ (47.0 mg, 0.34 mmol, 1.20 equiv.) and 1,3-dimethylimidazolium dimethyl phosphate (0.06 ml, 75.0 mg, 0.34 mmol, 1.20 equiv.) in toluene (0.93 ml) were heated at 110 °C for 50 min *via* microwave irradiation. Filtration over a small silica plug (*c*Hex/EtOAc 1:1) and purification by prepHPLC (5–95_MOD) resulted in the title compound as a yellow solid (14.0 mg, 0.04 mmol, 13%).

R_t (5%–95% in 22 min, MOD) = 32.7 min. – ^{1}H NMR (400 MHz, CDCl$_3$) δ = 7.66 (s, 1H, CH), 7.35 – 7.28 (m, 4H, 4 × CH_{ar}), 7.25 – 7.21 (m, 1H, CH_{ar}), 6.39 (d, ^4J = 2.1 Hz, 1H, CH_{ar}), 6.25 (d, ^4J = 2.1 Hz, 1H, CH_{ar}), 3.96 (t, ^3J = 6.5 Hz, 2H, OCH_2), 3.84 (s, 2H, CCH_2), 3.83 (s, 3H, OCH_3), 1.84 – 1.74 (m, 2H, OCH$_2$CH_2), 1.50 – 1.41 (m, 2H, CH_2), 1.40 – 1.28 (m, 6H, 3 × CH_2), 0.90 (t, ^3J = 6.6 Hz, 3H, CH_3) ppm. – ^{13}C NMR (101 MHz, CDCl$_3$) δ = 162.5 (C$_q$, COO and 7-C_{ar}OCH$_2$), 156.6 (C$_q$, 5-C_{ar}OCH$_3$), 155.8 (C$_q$, 8a-C_{ar}), 138.7 (C$_q$, CH$_2$$C_{ar}$), 135.1 (+, CH), 129.3 (+, 2 × C_{ar}H), 128.7 (+, 2 × C_{ar}H), 126.6 (+, C_{ar}H), 123.5 (C$_q$, 3-C), 104.3 (C$_q$, 4a-C_{ar}), 95.2 (+, C_{ar}H), 93.0 (+, C_{ar}H), 68.7 (–, OCH$_2$), 56.0 (+, OCH$_3$), 36.8 (–, CCH$_2$), 31.9 (–, CH$_2$), 29.2 (–, 2 × CH$_2$), 26.1 (–, CH$_2$), 22.7 (–, CH$_2$CH$_3$), 14.2 (+, CH$_3$) ppm. – **IR** (ATR, ṽ) = 3088, 3063, 3043, 3021, 2955, 2917, 2868, 2853, 2798, 1708, 1673, 1602, 1575, 1493, 1452, 1424, 1398, 1377, 1364, 1302, 1249, 1222, 1200, 1166, 1145, 1111, 1074, 1065, 1050, 1031, 1023, 1003, 997, 938, 932, 909, 878, 870, 856, 834, 819, 772, 754, 732, 696, 666, 645, 619, 589, 550, 506, 483, 450, 435, 424, 402, 378 cm^{-1}. – **MS** (70 eV, EI, 130 °C) *m/z* (%) =

381 (29) $[M+H]^+$, 380 (100) $[M]^+$, 283 (13), 282 (58), 281 (14) $[M–C_7H_{15}]^+$, 253 (11), 69 (12).
– **HRMS** (EI, $C_{24}H_{28}O_4$) = calc.: 380.1988; found: 380.1990.

7-(Heptyloxy)-5-methoxy-3-(2-methylbenzyl)-2*H*-chromen-2-one (21gb)

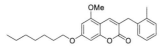

According to **GP8**, 4-(heptyloxy)-2-hydroxy-6-methoxybenzaldehyde (**19g**, 75.0 mg, 0.28 mmol, 1.00 equiv.), 2-methylcinnamaldehyde (103 mg, 0.70 mmol, 2.50 equiv.), K_2CO_3 (47.0 mg, 0.34 mmol, 1.20 equiv.) and 1,3-dimethylimidazolium dimethyl phosphate (0.06 ml, 75.0 mg, 0.34 mmol, 1.20 equiv.) in toluene (0.93 ml) were heated at 110 °C for 50 min *via* microwave irradiation. Filtration over a small silica plug (*c*Hex/EtOAc 1:1) and purification by prepHPLC (5–95_STD) resulted in the title compound as an off-white solid (18.0 mg, 0.05 mmol, 16%).

R_t (5%–95% in 27 min, STD) = 38.6 min. – **¹H NMR** (400 MHz, CDCl₃) δ = 7.38 (s, 1H, C*H*), 7.23 – 7.17 (m, 4H, 4 × C*H*ₐᵣ), 6.41 (d, ^4J = 2.1 Hz, 1H, C*H*ₐᵣ), 6.24 (d, ^4J = 2.1 Hz, 1H, C*H*ₐᵣ), 3.97 (t, ^3J = 6.5 Hz, 2H, OC*H₂*), 3.84 (d, ^4J = 1.4 Hz, 2H, CC*H₂*), 3.78 (s, 3H, OC*H₃*), 2.28 (s, 3H, CₐᵣC*H₃*), 1.85 – 1.73 (m, 2H, OCH₂C*H₂*), 1.51 – 1.41 (m, 2H, C*H₂*), 1.40 – 1.28 (m, 6H, 3 × C*H₂*), 0.90 (t, ^3J = 6.7 Hz, 3H, C*H₃*) ppm. – **¹³C NMR** (101 MHz, CDCl₃) δ = 162.6 (C_q, *C*OO), 162.5 (C_q, 7-*C*ₐᵣOCH₂), 156.6 (C_q, 5-*C*ₐᵣOCH₃), 155.5 (C_q, 8a-*C*ₐᵣ), 137.0 (C_q, *C*ₐᵣ), 136.4 (C_q, *C*ₐᵣ), 134.5 (+, *C*H), 130.6 (+, *C*ₐᵣH), 130.3 (+, *C*ₐᵣH), 127.0 (+, *C*ₐᵣH), 126.3 (+, *C*ₐᵣH), 122.7 (C_q, 3-*C*), 104.3 (C_q, 4a-*C*ₐᵣ), 95.2 (+, *C*ₐᵣH), 93.0 (+, *C*ₐᵣH), 68.7 (–, O*C*H₂), 55.9 (+, O*C*H₃), 33.9 (–, C*C*H₂), 31.9 (–, *C*H₂), 29.2 (–, 2 × *C*H₂), 26.1 (–, *C*H₂), 22.7 (–, *C*H₂CH₃), 19.7 (+, Cₐᵣ*C*H₃), 14.2 (+, *C*H₃) ppm. – **IR** (ATR, ṽ) = 3016, 2919, 2853, 1703, 1602, 1497, 1456, 1424, 1363, 1308, 1241, 1199, 1151, 1112, 1045, 956, 908, 819, 772, 758, 744, 726, 689, 646, 600, 548, 499, 460 cm⁻¹. – **MS** (70 eV, EI, 100 °C) *m/z* (%) = 394 (33) $[M]^+$, 169 (11), 119 (16), 100 (11), 97 (14), 85 (12), 83 (16), 71 (19), 70 (10). – **HRMS** (EI, $C_{25}H_{30}O_4$) = calc.: 394.2144; found: 394.2142.

7-(Heptyloxy)-5-methoxy-3-(2-methoxybenzyl)-2*H*-chromen-2-one (21gc)

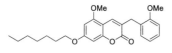

According to **GP8**, 4-(heptyloxy)-2-hydroxy-6-methoxybenzaldehyde (**19g**, 75.0 mg, 0.28 mmol, 1.00 equiv.), 2-methoxycinnamaldehyde (114 mg, 0.70 mmol, 2.50 equiv.), K_2CO_3 (47.0 mg, 0.34 mmol, 1.20 equiv.) and 1,3-dimethylimidazolium dimethyl phosphate (0.06 ml, 75.0 mg, 0.34 mmol, 1.20 equiv.) in toluene (0.93 ml) were heated at 110 °C for 50 min *via* microwave irradiation. Filtration over a small silica plug (*c*Hex/EtOAc 1:1) and purification by prepHPLC (5–95_MOD) resulted in the title compound as a yellow solid (25.0 mg, 0.06 mmol, 22%).

R_t (5%–95% in 22 min, MOD) = 32.9 min. – **^{1}H NMR** (400 MHz, CDCl$_3$) δ = 7.59 (s, 1H, C*H*), 7.29 – 7.21 (m, 2H, 2 × C*H*$_{ar}$), 6.97 – 6.86 (m, 2H, 2 × C*H*$_{ar}$), 6.39 (d, ^4J = 2.1 Hz, 1H, C*H*$_{ar}$), 6.24 (d, ^4J = 2.1 Hz, 1H, C*H*$_{ar}$), 3.96 (t, ^3J = 6.5 Hz, 2H, OC*H*$_2$), 3.84 (s, 2H, CC*H*$_2$), 3.81 (s, 6H, 2 × OC*H*$_3$), 1.84 – 1.74 (m, 2H, OCH$_2$C*H*$_2$), 1.50 – 1.41 (m, 2H, C*H*$_2$), 1.40 – 1.28 (m, 6H, 3 × C*H*$_2$), 0.89 (t, ^3J = 6.8 Hz, 3H, C*H*$_3$) ppm. – **^{13}C NMR** (101 MHz, CDCl$_3$) δ = 162.6 (C$_q$, *C*OO), 162.3 (C$_q$, 7-*C*$_{ar}$OCH$_2$), 157.8 (C$_q$, *C*$_{ar}$OCH$_3$), 156.6 (C$_q$, 5-*C*$_{ar}$OCH$_3$), 155.7 (C$_q$, 8a-*C*$_{ar}$), 134.7 (+, *C*H), 131.1 (+, *C*$_{ar}$H), 128.1 (+, *C*$_{ar}$H), 126.9 (C$_q$, CH$_2$*C*$_{ar}$), 122.8 (C$_q$, 3-*C*), 120.7 (+, *C*$_{ar}$H), 110.7 (+, *C*$_{ar}$H), 104.5 (C$_q$, 4a-*C*$_{ar}$), 95.1 (+, *C*$_{ar}$H), 93.0 (+, *C*$_{ar}$H), 68.7 (–, O*C*H$_2$), 55.9 (+, O*C*H$_3$), 55.5 (+, O*C*H$_3$), 31.9 (–, C*C*H$_2$), 30.9 (–, *C*H$_2$), 29.2 (–, 2 × *C*H$_2$), 26.1 (–, *C*H$_2$), 22.7 (–, *C*H$_2$CH$_3$), 14.2 (+, *C*H$_3$) ppm. – **IR** (ATR, ṽ) = 2925, 2854, 1715, 1611, 1494, 1459, 1426, 1368, 1244, 1197, 1147, 1112, 1044, 1031, 948, 822, 805, 752, 725, 691, 666, 639, 554, 487, 475 cm^{-1}. – **HRMS** (ESI, C$_{25}$H$_{30}$O$_5$) = calc.: 411.2166 [M+H]$^+$; found: 411.2157 [M+H]$^+$.

7-(Heptyloxy)-5-methoxy-3-(4-methoxybenzyl)-2*H*-chromen-2-one (21gd)

According to **GP8**, 4-(heptyloxy)-2-hydroxy-6-methoxybenzaldehyde (**19g**, 75.0 mg, 0.28 mmol, 1.00 equiv.), 4-methoxycinnamaldehyde (114 mg, 0.70 mmol, 2.50 equiv.), K$_2$CO$_3$ (47.0 mg, 0.34 mmol, 1.20 equiv.) and 1,3-dimethylimidazolium dimethyl phosphate (0.06 ml, 75.0 mg, 0.34 mmol, 1.20 equiv.) in toluene (0.93 ml) were heated at 110 °C for 50 min *via* microwave irradiation. Filtration over a small silica plug (*c*Hex/EtOAc 1:1) and purification by prepHPLC (5–95_MOD) resulted in the title compound as a yellow solid (17.0 mg, 0.04 mmol, 15%).

R_t (5%–95% in 22 min, MOD) = 31.8 min. – **^{1}H NMR** (400 MHz, CDCl$_3$) δ = 7.63 (s, 1H, C*H*), 7.22 (d, ^3J = 8.6 Hz, 2H, 2 × C*H*$_{ar}$), 6.86 (d, ^3J = 8.6 Hz, 2H, 2 × C*H*$_{ar}$), 6.38 (d, ^4J = 2.1 Hz, 1H, C*H*$_{ar}$), 6.25 (d, ^4J = 2.1 Hz, 1H, C*H*$_{ar}$), 3.96 (t, ^3J = 6.5 Hz, 2H, OC*H*$_2$), 3.83 (s, 3H, OC*H*$_3$), 3.80 (s, 3H, OC*H*$_3$), 3.78 (s, 2H, CC*H*$_2$), 1.83 – 1.74 (m, 2H, OCH$_2$C*H*$_2$), 1.50 – 1.41 (m, 2H, C*H*$_2$), 1.40 – 1.27 (m, 6H, 3 × C*H*$_2$), 0.89 (t, ^3J = 6.7 Hz, 3H, C*H*$_3$) ppm. – **^{13}C NMR** (101 MHz, CDCl$_3$) δ = 162.5 (C$_q$, *C*OO), 162.4 (C$_q$, 7-*C*$_{ar}$OCH$_2$), 158.4 (C$_q$, *C*$_{ar}$OCH$_3$), 156.6 (C$_q$, 5-*C*$_{ar}$OCH$_3$), 155.8 (C$_q$, 8a-*C*$_{ar}$), 134.8 (+, *C*H), 130.7 (C$_q$, CH$_2$*C*$_{ar}$), 130.3 (+, 2 × *C*$_{ar}$H), 123.9 (C$_q$, 3-*C*), 114.1 (+, 2 × *C*$_{ar}$H), 104.3 (C$_q$, 4a-*C*$_{ar}$), 95.2 (+, *C*$_{ar}$H), 93.0 (+, *C*$_{ar}$H), 68.7 (–, O*C*H$_2$), 55.9 (+, O*C*H$_3$), 55.4 (+, O*C*H$_3$), 35.9 (–, C*C*H$_2$), 31.9 (–, *C*H$_2$), 29.2 (–, 2 × *C*H$_2$), 26.1

(–, CH_2), 22.7 (–, CH_2CH_3), 14.2 (+, CH_3) ppm. – **IR** (ATR, ṽ) = 2958, 2948, 2914, 2859, 2849, 1706, 1621, 1605, 1578, 1511, 1499, 1468, 1455, 1424, 1367, 1299, 1242, 1201, 1173, 1152, 1115, 1034, 1001, 960, 943, 902, 882, 844, 824, 805, 779, 759, 725, 687, 663, 645, 571, 547, 510, 482, 462, 433, 381 cm^{-1}. – **HRMS** (ESI, $C_{25}H_{30}O_5$) = calc.: 411.2166 [M+H]$^+$; found: 411.2163 [M+H]$^+$.

3-(4-Fluorobenzyl)-7-(heptyloxy)-5-methoxy-2*H*-chromen-2-one (21gf)

According to **GP8**, 4-(heptyloxy)-2-hydroxy-6-meth-oxybenzaldehyde (**19g**, 75.0 mg, 0.28 mmol, 1.00 equiv.), 4-fluorocinnamaldehyde (0.09 ml, 106 mg, 0.70 mmol, 2.50 equiv.), K_2CO_3 (47.0 mg, 0.34 mmol, 1.20 equiv.) and 1,3-dimethylimidazo-lium dimethyl phosphate (0.06 ml, 75.0 mg, 0.34 mmol, 1.20 equiv.) in toluene (0.93 ml) were heated at 110 °C for 50 min *via* microwave irradiation. Filtration over a small silica plug (*c*Hex/EtOAc 1:1) and purification by prepHPLC (5–95_MOD) resulted in the title compound as a yellow solid (8.00 mg, 0.02 mmol, 7.1%).

R_t (5%–95% in 22 min, MOD) = 32.8 min. – **^{1}H NMR** (400 MHz, CDCl$_3$) δ = 7.65 (s, 1H, C*H*), 7.31 – 7.22 (m, 2H, 2 × C*H*$_{ar}$), 7.08 – 6.93 (m, 2H, 2 × C*H*$_{ar}$), 6.39 (d, ^4J = 2.1 Hz, 1H, C*H*$_{ar}$), 6.26 (d, ^4J = 2.1 Hz, 1H, C*H*$_{ar}$), 3.97 (t, ^3J = 6.5 Hz, 2H, OC*H*$_2$), 3.85 (s, 3H, OC*H*$_3$), 3.80 (s, 2H, CC*H*$_2$), 1.84 – 1.74 (m, 2H, OCH$_2$C*H*$_2$), 1.50 – 1.40 (m, 2H, C*H*$_2$), 1.38 – 1.28 (m, 6H, 3 × C*H*$_2$), 0.89 (t, ^3J = 6.7 Hz, 3H, C*H*$_3$) ppm. – **^{13}C NMR** (101 MHz, CDCl$_3$) δ = 162.6 (C$_q$, *C*OO), 162.4 (C$_q$, 7-*C*$_{ar}$OCH$_2$), 161.8 (C$_q$, d, ^1J = 244.2 Hz, *C*$_{ar}$F), 156.6 (C$_q$, 5-*C*$_{ar}$OCH$_3$), 155.9 (C$_q$, 8a-*C*$_{ar}$), 135.1 (+, *C*H), 134.4 (C$_q$, d, ^4J = 3.2 Hz, CH$_2$*C*$_{ar}$), 130.7 (+, d, ^3J = 7.9 Hz, 2 × *m*-*C*$_{ar}$H), 123.3 (C$_q$, 3-*C*), 115.5 (+, d, ^2J = 21.2 Hz, 2 × *o*-*C*$_{ar}$H), 104.2 (C$_q$, 4a-*C*$_{ar}$), 95.3 (+, *C*$_{ar}$H), 93.1 (+, *C*$_{ar}$H), 68.8 (–, O*C*H$_2$), 56.0 (+, O*C*H$_3$), 36.1 (–, C*C*H$_2$), 31.9 (–, *C*H$_2$), 29.9 (–, *C*H$_2$), 29.2 (–, *C*H$_2$), 26.1 (–, *C*H$_2$), 22.7 (–, *C*H$_2$CH$_3$), 14.2 (+, *C*H$_3$) ppm. – **^{19}F NMR** (376 MHz, CDCl$_3$) δ = –121.1 (s, C*F*) ppm. – **IR** (ATR, ṽ) = 2956, 2918, 2871, 2851, 1710, 1629, 1618, 1604, 1507, 1459, 1425, 1377, 1366, 1302, 1251, 1224, 1200, 1166, 1145, 1112, 1092, 1051, 931, 851, 830, 817, 790, 751, 728, 686, 636, 612, 572, 551, 523, 489, 449 cm^{-1}. – **MS** (70 eV, EI, 150 °C) *m/z* (%) = 399 (27) [M+H]$^+$, 398 (100) [M]$^+$, 301 (14), 300 (66), 299 (17) [M–C$_7$H$_{15}$]$^+$, 271 (15), 183 (11), 109 (13), 69 (18), 57 (31), 55 (13). – **HRMS** (EI, C$_{24}$H$_{27}$O$_4$F) = calc.: 398.1893; found: 398.1894.

3-(4-Chlorobenzyl)-7-(heptyloxy)-5-methoxy-2*H*-chromen-2-one (21gh)

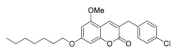

According to **GP8**, 4-(heptyloxy)-2-hydroxy-6-meth-oxybenzaldehyde (**19g**, 75.0 mg, 0.28 mmol, 1.00 equiv.), 4-chlorocinnamaldehyde (117 mg, 0.70 mmol, 2.50 equiv.), K_2CO_3 (47.0 mg, 0.34 mmol, 1.20 equiv.) and 1,3-dimethylimidazolium dimethyl phosphate (0.06 ml, 75.0 mg, 0.34 mmol, 1.20 equiv.) in toluene (0.93 ml) were heated at 110 °C for 50 min *via* microwave irradiation. Filtration over a small silica plug (*c*Hex/EtOAc 1:1) and purification by prepHPLC (5–95_MOD) resulted in the title compound as a yellow solid (18.0 mg, 0.04 mmol, 15%).

R_t (5%–95% in 22 min, MOD) = 35.2 min. – **^{1}H NMR** (400 MHz, CDCl$_3$) δ = 7.66 (s, 1H, C*H*), 7.31 – 7.20 (m, 4H, 4 × C*H*$_{ar}$), 6.39 (d, ^4J = 2.1 Hz, 1H, C*H*$_{ar}$), 6.26 (d, ^4J = 2.1 Hz, 1H, C*H*$_{ar}$), 3.97 (t, ^3J = 6.5 Hz, 2H, OC*H*$_2$), 3.85 (s, 3H, OC*H*$_3$), 3.80 (s, 2H, CC*H*$_2$), 1.84 – 1.74 (m, 2H, OCH$_2$C*H*$_2$), 1.50 – 1.41 (m, 2H, C*H*$_2$), 1.39 – 1.27 (m, 6H, 3 × C*H*$_2$), 0.89 (t, ^3J = 6.7 Hz, 3H, C*H*$_3$) ppm. – **^{13}C NMR** (101 MHz, CDCl$_3$) δ = 162.7 (C$_q$, COO), 162.3 (C$_q$, 7-C$_{ar}$-OCH$_2$), 156.6 (C$_q$, 5-C$_{ar}$OCH$_3$), 155.9 (C$_q$, 8a-C$_{ar}$), 137.2 (C$_q$, C$_{ar}$Cl), 135.2 (+, C*H*), 132.5 (C$_q$, CH$_2$C$_{ar}$), 130.6 (+, 2 × C$_{ar}$H), 128.8 (+, 2 × C$_{ar}$H), 122.9 (C$_q$, 3-C), 104.2 (C$_q$, 4a-C$_{ar}$), 95.3 (+, C$_{ar}$H), 93.1 (+, C$_{ar}$H), 68.8 (–, OCH$_2$), 56.0 (+, OCH$_3$), 36.2 (–, CCH$_2$), 31.9 (–, CH$_2$), 29.2 (–, CH$_2$), 29.1 (–, CH$_2$), 26.1 (–, CH$_2$), 22.7 (–, CH$_2$CH$_3$), 14.2 (+, CH$_3$) ppm. – **IR** (ATR, ṽ) = 3089, 2917, 2856, 1708, 1674, 1616, 1602, 1577, 1496, 1489, 1458, 1425, 1407, 1375, 1366, 1302, 1249, 1232, 1201, 1167, 1147, 1112, 1091, 1050, 1017, 1003, 983, 962, 955, 943, 932, 904, 877, 830, 819, 807, 799, 776, 761, 739, 727, 715, 693, 664, 650, 618, 565, 545, 506, 455, 426, 409, 385, 380 cm^{-1}. – **HRMS** (ESI, C$_{24}$H$_{27}$O$_4$^{35}Cl) = calc.: 415.1671 [M+H]$^+$; found: 415.1668 [M+H]$^+$.

3-(4-(Dimethylamino)benzyl)-7-(heptyloxy)-5-methoxy-2*H*-chromen-2-one (21gi)

According to **GP8**, 4-(heptyloxy)-2-hydroxy-6-methoxybenzaldehyde (**19g**, 75.0 mg, 0.28 mmol, 1.00 equiv.), 4-(dimethylamino)cinnamaldehyde (123 mg, 0.70 mmol, 2.50 equiv.), K_2CO_3 (47.0 mg, 0.34 mmol, 1.20 equiv.) and 1,3-dimethylimidazolium dimethyl phosphate (0.06 ml, 75.0 mg, 0.34 mmol, 1.20 equiv.) in toluene (0.93 ml) were heated at 110 °C for 50 min *via* microwave irradiation. Filtration over a small silica plug (*c*Hex/EtOAc 1:1) and purification by prepHPLC (5–95_STD) resulted in the title compound as a purple solid (14.0 mg, 0.03 mmol, 12%).

R_t (5%–95% in 27 min, STD) = 35.2 min. – **^{1}H NMR** (400 MHz, CDCl$_3$) δ = 7.79 (s, 1H, C*H*), 7.48 (s, 3H, 3 × C*H*$_{ar}$), 7.26 (s, 1H, C*H*$_{ar}$), 6.38 (d, ^4J = 2.1 Hz, 1H, C*H*$_{ar}$), 6.28 (d, ^4J = 2.1 Hz,

1H, CH_{ar}), 3.97 (t, ^3J = 6.5 Hz, 2H, OCH_2), 3.87 (s, 3H, OCH_3), 3.86 (s, 2H, CCH_2), 3.20 (s, 6H, N(CH_3)$_3$), 1.84 – 1.74 (m, 2H, OCH$_2$CH_2), 1.50 – 1.40 (m, 2H, CH_2), 1.39 – 1.28 (m, 6H, 3 × CH_2), 0.89 (t, ^3J = 6.7 Hz, 3H, CH_3) ppm. – ^{13}C NMR (101 MHz, CDCl$_3$) δ = 163.0 (C$_q$, COO), 162.3 (C$_q$, 7-C_{ar}OCH$_2$), 156.8 (C$_q$, 5-C_{ar}OCH$_3$), 156.1 (C$_q$, 8a-C_{ar}), 141.6 (C$_q$), 141.5 (C$_q$), 135.9 (+, CH), 131.2 (+, 2 × C_{ar}H), 121.8 (C$_q$, 3-C), 120.6 (+, 2 × C_{ar}H), 104.1 (C$_q$, 4a-C_{ar}), 95.4 (+, C_{ar}H), 93.2 (+, C_{ar}H), 68.8 (–, OCH$_2$), 56.1 (+, OCH$_3$), 46.6 (+, N(CH$_3$)$_2$), 36.6 (–, CCH$_2$), 31.9 (–, CH$_2$), 29.1 (–, 2 × CH$_2$), 26.1 (–, CH$_2$), 22.7 (–, CH$_2$CH$_3$), 14.2 (+, CH$_3$) ppm. – IR (ATR, ṽ) = 2928, 2856, 1776, 1713, 1672, 1609, 1577, 1510, 1499, 1465, 1428, 1368, 1344, 1322, 1306, 1254, 1197, 1146, 1115, 1050, 1020, 999, 943, 904, 824, 798, 786, 721, 705, 639, 588, 552, 520, 411, 388 cm^{-1}. – HRMS (ESI, C$_{26}$H$_{33}$O$_4$N) = calc.: 424.2482 [M+H]$^+$; found: 424.2481 [M+H]$^+$.

5-Hydroxy-3-(2-methylbenzyl)-7-propoxy-2H-chromen-2-one (58a)

According to GP9, to a solution of the 5-methoxycoumarin 21cb (8.60 mg, 25.4 μmol, 1.00 equiv.) in abs. DCM (0.51 ml), was added BBr$_3$ (1 M in DCM, 0.13 ml, 31.8 mg, 130 μmol, 5.00 equiv.) at –78 °C. After stirring for 30 min at this temperature, the reaction mixture was allowed to warm to rt and stirred overnight. Purification by flash column chromatography (cHex/EtOAc 5:1 to 3:1) resulted in the title compound as a colourless oil (2.2 mg, 6.86 μmol, 27%).

R_f (cHex/EtOAc 2:1) = 0.53. – ^{1}H NMR (500 MHz, acetone-D$_6$) δ = 9.49 (bs, 1H, OH), 7.46 (s, 1H, CH), 7.27 – 7.14 (m, 4H, 4 × CH_{ar}), 6.40 (d, ^4J = 2.2 Hz, 1H, CH_{ar}), 6.35 (d, ^4J = 2.2 Hz, 1H, CH_{ar}), 3.99 (t, ^3J = 6.5 Hz, 2H, OCH_2), 3.80 (d, ^4J = 1.4 Hz, 2H, CCH_2), 2.29 (s, 3H, C$_{ar}$CH_3), 1.85 – 1.72 (m, 2H, CH_2CH$_3$), 1.01 (t, ^3J = 7.4 Hz, 3H, CH_3) ppm. – ^{13}C NMR (126 MHz, acetone-D$_6$) δ = 163.2 (C$_q$, COO), 162.2 (C$_q$, 7-C_{ar}OCH$_2$), 156.7 (C$_q$, 5-C_{ar}OH), 155.6 (C$_q$, 8a-C_{ar}), 137.6 (C$_q$, C_{ar}), 137.5 (C$_q$, C_{ar}), 134.9 (+, CH), 131.2 (+, C_{ar}H), 130.9 (+, C_{ar}H), 127.7 (+, C_{ar}H), 127.0 (+, C_{ar}H), 123.1 (C$_q$, 3-C), 104.1 (C$_q$, 4a-C_{ar}), 98.7 (+, C_{ar}H), 93.5 (+, C_{ar}H), 70.6 (–, OCH$_2$), 34.3 (–, CCH$_2$), 23.0 (–, CH$_2$CH$_3$), 19.5 (+, C_{ar}CH$_3$), 10.7 (+, CH$_3$) ppm. – IR (ATR, ṽ) = 2961, 2918, 2876, 2850, 1666, 1618, 1520, 1490, 1443, 1401, 1363, 1278, 1248, 1218, 1166, 1108, 1086, 1061, 1048, 1018, 950, 899, 844, 798, 768, 747, 725, 683, 521, 504 cm^{-1}. – HRMS (ESI, C$_{20}$H$_{20}$O$_4$) = calc.: 325.1434 [M+H]$^+$; found: 325.1421 [M+H]$^+$.

5-Hydroxy-3-(2-hydroxybenzyl)-7-propoxy-2H-chromen-2-one (58b)

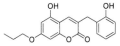 According to **GP9**, to a solution of the dimethoxycoumarin **21cc** (10.0 mg, 28.2 μmol, 1.00 equiv.) in abs. DCM (0.56 ml), was added BBr₃ (1 M in DCM, 0.28 ml, 71.0 mg, 280 μmol, 10.0 equiv.) at −78 °C. After stirring for 30 min at this temperature, the reaction mixture was allowed to warm to rt and stirred overnight. Purification by flash column chromatography (cHex/EtOAc 2:1) resulted in the title compound as a cloudy oil (1.2 mg, 3.66 μmol, 13%).

R_f (cHex/EtOAc 2:1) = 0.27. – **^{1}H NMR** (500 MHz, acetone-D₆) δ = 9.47 (bs, 1H, OH), 8.43 (bs, 1H, OH), 7.75 (d, ^4J = 1.3 Hz, 1H, CH), 7.25 (dd, $^{3,4}J$ = 7.5, 1.7 Hz, 1H, CH_{ar}), 7.10 (td, $^{3,4}J$ = 7.7, 1.7 Hz, 1H, CH_{ar}), 6.88 (dd, $^{3,4}J$ = 8.1, 1.2 Hz, 1H, CH_{ar}), 6.83 (td, $^{3,4}J$ = 7.4, 1.2 Hz, 1H, CH_{ar}), 6.40 (d, ^4J = 2.3 Hz, 1H, CH_{ar}), 6.36 (d, ^4J = 2.3 Hz, 1H, CH_{ar}), 3.98 (t, ^3J = 6.5 Hz, 2H, OCH_2), 3.80 (s, 2H, CCH_2), 1.84 – 1.73 (m, 2H, OCH₂CH_2), 1.01 (t, ^3J = 7.4 Hz, 3H, CH_3) ppm. – **^{13}C NMR** (126 MHz, acetone-D₆) δ = 163.3 (C$_q$, COO), 163.1 (C$_q$, 7-C$_{ar}$OCH₂), 156.8 (C$_q$, 5-C$_{ar}$OH), 155.8 (C$_q$, 8a-C$_{ar}$), 135.4 (+, CH), 132.1 (+, C$_{ar}$H), 128.9 (+, C$_{ar}$H), 123.4 (C$_q$, 3-C), 120.8 (+, C$_{ar}$H), 116.6 (+, C$_{ar}$H), 98.9 (+, C$_{ar}$H), 93.7 (+, C$_{ar}$H), 70.7 (−, OCH₂), 31.6 (−, CCH₂), 23.2 (−, CH₂CH₃), 10.8 (+, CH₃) ppm. – **HRMS** (ESI, C₁₉H₁₈O₅) = calc.: 327.1227 [M+H]⁺; found: 327.1218 [M+H]⁺.

5-Hydroxy-3-(2-methylbenzyl)-7-(pentyloxy)-2H-chromen-2-one (58d)

According to **GP9**, to a solution of the 5-methoxycoumarin **21eb** (6.30 mg, 17.2 μmol, 1.00 equiv.) in abs. DCM (0.34 ml), was added BBr₃ (1 M in DCM, 0.09 ml, 21.5 mg, 90.0 μmol, 5.00 equiv.) at −78 °C. After stirring for 30 min at this temperature, the reaction mixture was allowed to warm to rt and stirred overnight. Purification by flash column chromatography (cHex/EtOAc 6:1 to 4:1) resulted in the title compound as a white solid (1.2 mg, 3.44 μmol, 20%, purity 90%).

R_f (cHex/EtOAc 2:1) = 0.76. – **^{1}H NMR** (500 MHz, acetone-D₆) δ = 7.51 – 7.43 (m, 1H, CH), 7.26 – 7.15 (m, 4H, 4 × CH_{ar}), 6.39 (s, 2H, 2 × CH_{ar}), 4.03 (t, ^3J = 6.5 Hz, 2H, OCH_2), 3.80 (d, ^4J = 1.4 Hz, 2H, CCH_2), 2.29 (s, 3H, C$_{ar}$CH_3), 1.81 –1.73 (m, 2H, OCH₂CH_2), 1.48 – 1.37 (m, 4H, 2 × CH_2), 0.92 (t, ^3J = 7.1 Hz, 3H, CH_3) ppm. – **^{13}C NMR** (126 MHz, acetone-D₆) δ = 163.2 (C$_q$, COO), 162.2 (C$_q$, 7-C$_{ar}$OCH₂), 156.7 (C$_q$, 5-C$_{ar}$OH), 137.6 (C$_q$, C$_{ar}$), 137.5 (C$_q$, C$_{ar}$), 135.0 (+, CH), 131.2 (+, C$_{ar}$H), 130.9 (+, C$_{ar}$H), 127.7 (+, C$_{ar}$H), 127.0 (+, C$_{ar}$H), 123.0 (C$_q$, 3-C), 104.1 (C$_q$, 4a-C$_{ar}$), 98.7 (+, C$_{ar}$H), 93.5 (+, C$_{ar}$H), 69.1 (−, OCH₂), 34.3 (−, CCH₂), 28.8 (−, CH₂), 23.1 (−, CH₂CH₃), 19.5 (+, C$_{ar}$CH₃), 14.3 (+, CH₃) ppm. – **IR** (ATR, ṽ) = 2926, 1667,

1618, 1443, 1362, 1248, 1164, 1087, 842, 747, 726, 455 cm^{-1}. – **HRMS** (ESI, C$_{22}$H$_{24}$O$_4$) = calc.: 353.1747 [M+H]$^+$; found: 353.1733 [M+H]$^+$.

5-Hydroxy-3-(2-hydroxybenzyl)-7-(pentyloxy)-2*H*-chromen-2-one (58e)

According to GPXX, to a solution of the dimethoxycoumarin **21cc** (10.0 mg, 26.1 µmol, 1.00 equiv.) in abs. DCM (0.56 ml), was added BBr$_3$ (1 M in DCM, 0.26 ml, 66.0 mg, 260 µmol, 10.0 equiv.) at –78 °C. After stirring for 30 min at this temperature, the reaction mixture was allowed to warm to rt and stirred overnight. Purification by prepTLC (*c*Hex/EtOAc 2:1) resulted in the title compound as a white solid (2.4 mg, 6.79 µmol, 26%).

R_f (*c*Hex/EtOAc 2:1) = 0.27. – **^{1}H NMR** (500 MHz, acetone-D$_6$) δ = 7.75 (s, 1H, C*H*), 7.25 (dd, $^{3,4}J$ = 7.5, 1.7 Hz, 1H, C*H*$_{ar}$), 7.10 (td, $^{3,4}J$ = 7.7, 1.7 Hz, 1H, C*H*$_{ar}$), 6.88 (dd, $^{3,4}J$ = 8.1, 1.2 Hz, 1H, C*H*$_{ar}$), 6.82 (td, $^{3,4}J$ = 7.4, 1.2 Hz, 1H, C*H*$_{ar}$), 6.40 (d, ^4J = 2.2 Hz, 1H, C*H*$_{ar}$), 6.35 (d, ^4J = 2.2 Hz, 1H, C*H*$_{ar}$), 4.02 (t, ^3J = 6.5 Hz, 2H, OC*H*$_2$), 3.80 (d, ^4J = 1.1 Hz, 2H, CC*H*$_2$), 1.82 – 1.72 (m, 2H, OCH$_2$C*H*$_2$), 1.48 – 1.35 (m, 4H, 2 × C*H*$_2$), 0.92 (t, ^3J = 7.1 Hz, 3H, C*H*$_3$) ppm. – **^{13}C NMR** (126 MHz, acetone-D$_6$) δ = 163.3 (C$_q$, *C*OO), 163.1 (C$_q$, 7-*C*$_{ar}$OCH$_2$), 156.8 (C$_q$, 5-*C*$_{ar}$OH), 156.3 (C$_q$, *C*$_{ar}$OH), 155.8 (C$_q$, 8a-*C*$_{ar}$), 135.4 (+, *C*H), 132.1 (+, *C*$_{ar}$H), 128.9 (+, *C*$_{ar}$H), 126.0 (C$_q$, *C*$_{ar}$), 123.4 (C$_q$, 3-*C*), 120.8 (+, *C*$_{ar}$H), 116.5 (+, *C*$_{ar}$H), 104.4 (C$_q$, 4a-*C*$_{ar}$), 98.9 (+, *C*$_{ar}$H), 93.7 (+, *C*$_{ar}$H), 69.2 (–, O*C*H$_2$), 31.6 (–, C*C*H$_2$), 29.2 (–, *C*H$_2$), 29.0 (–, *C*H$_2$), 23.2 (–, *C*H$_2$), 14.4 (+, *C*H$_3$) ppm. – **IR** (ATR, ṽ) = 3305, 2956, 2927, 2871, 2854, 1690, 1656, 1621, 1490, 1449, 1397, 1380, 1367, 1286, 1261, 1222, 1193, 1157, 1088, 1027, 967, 895, 827, 805, 754, 730, 657, 640, 585, 526, 513, 466, 441, 415, 407 cm^{-1}. – **HRMS** (ESI, C$_{21}$H$_{22}$O$_5$) = calc.: 355.1540 [M+H]$^+$; found: 355.1528 [M+H]$^+$.

7-(Heptyloxy)-5-hydroxy-3-(2-methylbenzyl)-2*H*-chromen-2-one (58g)

According to **GP9**, to a solution of the 5-methoxycoumarin **21fb** (8.00 mg, 20.3 µmol, 1.00 equiv.) in abs. DCM (0.41 ml), was added BBr$_3$ (1.0 M in DCM, 0.10 ml, 25.0 mg, 100 µmol, 5.00 equiv.) at –78 °C. After stirring for 30 min at this temperature, the reaction mixture was allowed to warm to rt and stirred overnight. Purification by flash column chromatography (*c*Hex/EtOAc 9:1 to 5:1) resulted in the title compound as a white solid (3.5 mg, 9.14 µmol, 45%).

R_f (*c*Hex/EtOAc 2:1) = 0.67. – **^{1}H NMR** (500 MHz, acetone-D$_6$) δ = 9.59 (bs, 1H, O*H*), 7.46 (d, ^4J = 1.6 Hz, 1H, C*H*), 7.26 – 7.20 (m, 2H, 2 × C*H*$_{ar}$), 7.20 – 7.15 (m, 2H, 2 × C*H*$_{ar}$), 6.40 (d,

^4J = 2.3 Hz, 1H, CH_{ar}), 6.37 (d, ^4J = 2.3 Hz, 1H, CH_{ar}), 4.03 (t, ^3J = 6.5 Hz, 2H, OCH_2), 3.80 (d, ^4J = 1.4 Hz, 2H, CCH_2), 2.29 (s, 3H, C$_{ar}$CH_3), 1.85 – 1.70 (m, 2H, OCH$_2$CH_2), 1.50 – 1.42 (m, 2H, CH_2), 1.39 – 1.26 (m, 6H, 3 × CH_2), 0.94 – 0.84 (m, 3H, CH_3) ppm. – ^{13}C NMR (126 MHz, acetone-D$_6$) δ = 163.2 (C$_q$, COO), 162.2 (C$_q$, 7-C$_{ar}$OCH$_2$), 156.7 (C$_q$, 5-C$_{ar}$OH), 155.7 (C$_q$, 8a-C$_{ar}$), 137.6 (C$_q$, C$_{ar}$), 137.5 (C$_q$, C$_{ar}$), 134.9 (+, CH), 131.2 (+, C$_{ar}$H), 130.9 (+, C$_{ar}$H), 127.7 (+, C$_{ar}$H), 127.0 (+, C$_{ar}$H), 123.0 (C$_q$, 3-C), 104.1 (C$_q$, 4a-C$_{ar}$), 98.7 (+, C$_{ar}$H), 93.5 (+, C$_{ar}$H), 69.1 (–, OCH$_2$), 34.3 (–, CCH$_2$), 32.5 (–, CH$_2$), 29.2 (–, 2 × CH$_2$), 26.6 (–, CH$_2$), 23.3 (–, CH$_2$CH$_3$), 19.5 (+, C$_{ar}$CH$_3$), 14.3 (+, CH$_3$) ppm. – IR (ATR, ṽ) = 2918, 2850, 1664, 1618, 1443, 1363, 1248, 1163, 1086, 844, 798, 768, 747, 725, 679, 572, 543, 534, 517, 458 cm^{-1}. – HRMS (ESI, C$_{24}$H$_{28}$O$_4$) = calc.: 381.2060 [M+H]$^+$; found: 381.2044 [M+H]$^+$.

7-(Heptyloxy)-5-hydroxy-3-(2-hydroxybenzyl)-2*H*-chromen-2-one (58h)

According to **GP9**, to a solution of the dimethoxycoumarin **21fc** (10.0 mg, 24.4 μmol, 1.00 equiv.) in abs. DCM (0.49 ml), was added BBr$_3$ (1 M in DCM, 0.24 ml, 61.0 mg, 240 μmol, 10.0 equiv.) at –78 °C. After stirring for 30 min at this temperature, the reaction mixture was allowed to warm to rt and stirred overnight. Purification by prepTLC (*c*Hex/EtOAc 2:1) resulted in the title compound as an off-white solid (1.8 mg, 4.64 μmol, 19%, purity 90 – 95%).

R_f (*c*Hex/EtOAc 2:1) = 0.27. – ^{1}H NMR (500 MHz, acetone-D$_6$) δ = 9.43 (bs, 1H, OH), 8.44 (bs, 1H, OH), 7.75 (d, ^4J = 1.5 Hz, 1H, CH), 7.25 (dd, $^{3,4}J$ = 7.5, 1.7 Hz, 1H, CH_{ar}), 7.10 (td, $^{3,4}J$ = 7.7, 1.8 Hz, 1H, CH_{ar}), 6.88 (dd, $^{3,4}J$ = 8.1, 1.2 Hz, 1H, CH_{ar}), 6.83 (td, $^{3,4}J$ = 7.4, 1.2 Hz, 1H, CH_{ar}), 6.41 (d, ^4J = 2.2 Hz, 1H, CH_{ar}), 6.35 (d, ^4J = 2.2 Hz, 1H, CH_{ar}), 4.03 (t, ^3J = 6.6 Hz, 2H, OCH_2), 3.80 (d, ^4J = 1.2 Hz, 2H, CCH_2), 1.81 – 1.73 (m, 2H, OCH$_2$CH_2), 1.49 – 1.43 (m, 2H, CH_2), 1.40 – 1.35 (m, 2H, CH_2), 1.33 – 1.25 (m, 4H, 2 × CH_2), 0.91 – 0.84 (m, 3H, CH_3) ppm. – ^{13}C NMR (126 MHz, acetone-D$_6$) δ = 163.3 (C$_q$, COO), 163.0 (C$_q$, 7-C$_{ar}$OCH$_2$), 156.8 (C$_q$, 5-C$_{ar}$OH), 156.3 (C$_q$, C$_{ar}$OH), 155.7 (C$_q$, 8a-C$_{ar}$), 135.4 (+, CH), 132.1 (+, C$_{ar}$H), 128.9 (+, C$_{ar}$H), 126.0 (C$_q$, C$_{ar}$), 123.4 (C$_q$, 3-C), 120.8 (+, C$_{ar}$H), 116.5 (+, C$_{ar}$H), 104.4 (C$_q$, 4a-C$_{ar}$), 98.9 (+, C$_{ar}$H), 93.7 (+, C$_{ar}$H), 69.2 (–, OCH$_2$), 32.6 (–, CH$_2$), 31.6 (–, CCH$_2$), 26.8 (–, CH$_2$), 23.4 (–, CH$_2$CH$_3$), 14.4 (+, CH$_3$) ppm. – HRMS (ESI, C$_{23}$H$_{26}$O$_5$) = calc.: 383.1853 [M+H]$^+$; found: 383.1840 [M+H]$^+$.

3-(4-(Dimethylamino)benzyl)-7-(heptyloxy)-5-hydroxy-2*H*-chromen-2-one (58i)

According to **GP9**, to a solution of the 5-methoxycoumarin **21gi** (7.0 mg, 16.5 μmol, 1.00 equiv.) in abs. DCM (0.33 ml), was added BBr$_3$ (1.0 M in DCM, 0.08 ml, 21.0 mg, 80.0 μmol, 5.00 equiv.) at –78 °C. After stirring for 30 min at this temperature, the reaction mixture was allowed to warm to rt and stirred overnight. Purification by prepTLC (*c*Hex/EtOAc 2:1) resulted in the title compound as a purple solid (0.7 mg, 1.16 μmol, 10%, purity <90%).

R_f (*c*Hex/EtOAc 2:1) = 0.30. – **^{1}H NMR** (500 MHz, acetone-D$_6$) δ = 7.69 (d, ^4J = 1.2 Hz, 1H, C*H*), 7.18 – 7.13 (m, 2H, 2 × C*H*$_{ar}$), 6.74 – 6.69 (m, 2H, 2 × C*H*$_{ar}$), 6.39 (d, ^4J = 2.3 Hz, 1H, C*H*$_{ar}$), 6.35 (d, ^4J = 2.3 Hz, 1H, C*H*$_{ar}$), 4.02 (t, ^3J = 6.5 Hz, 2H, OC*H*$_2$), 3.68 (d, ^4J = 1.1 Hz, 2H, CC*H*$_2$), 1.81 – 1.72 (m, 2H, OCH$_2$C*H*$_2$), 1.50 – 1.44 (m, 2H, C*H*$_2$), 1.40 – 1.35 (m, 2H, C*H*$_2$), 1.33 – 1.27 (m, 4H, 2 × C*H*$_2$), 0.92 – 0.85 (m, 3H, C*H*$_3$) ppm. – **^{13}C NMR** insufficient quantities available. – **HRMS** (ESI, C$_{25}$H$_{31}$O$_4$N) = calc.: 410.2326 [M+H]$^+$; found: 410.2312 [M+H]$^+$.

5.1.7   Synthesis and characterisation of 3-benzyl-7-dialkylaminocoumarins

7-(Ethyl(methyl)amino)-5-methoxy-3-(2-methylbenzyl)-2*H*-chromen-2-one (21hb)

According to **GP10**, 4-(ethyl(methyl)amino)-2-hydroxy-6-methoxybenzaldehyde (**19h**, 13.0 mg, 62.1 μmol, 1.00 equiv.), 2-methylcinnamaldehyde (23.0 mg, 0.16 mmol, 2.50 equiv.), KOAc (21.0 mg, 0.22 mmol, 3.50 equiv.), glacial acetic acid (1.00 μl, 0.93 mg, 15.5 μmol, 0.25 equiv.), 1,3-dimethylimidazolium dimethyl phosphate (12.9 μl, 17.0 mg, 0.75 mmol, 1.20 equiv.) and MS (powder, 3Å) in toluene (0.21 ml) were heated at 110 °C for 75 min *via* microwave irradiation. Filtration over a small silica plug (*c*Hex/EtOAc 1:1) and purification by prepHPLC (5–95_amino) resulted in the title compound as an off-white solid (1.00 mg, 3.11 μmol, 4.8%).

R_t (5%–95% in 27 min, amino) = 29.3 min. – **^{1}H NMR** (500 MHz, CDCl$_3$) δ = 7.34 (s, 1H, C*H*), 7.21 – 7.16 (m, 4H, 4 × C*H*$_{ar}$), 6.16 (d, ^4J = 2.2 Hz, 1H, C*H*$_{ar}$), 5.96 (d, ^4J = 2.2 Hz, 1H, C*H*$_{ar}$), 3.81 (d, ^4J = 1.3 Hz, 2H, CC*H*$_2$), 3.79 (s, 3H, OC*H*$_3$), 3.43 (q, ^3J = 7.1 Hz, 2H, NC*H*$_2$), 2.98 (s, 3H, NC*H*$_3$), 2.28 (s, 3H, C$_{ar}$C*H*$_3$), 1.16 (t, ^3J = 7.1 Hz, 3H, C*H*$_3$) ppm. – **^{13}C NMR** (126 MHz, CDCl$_3$) δ = 163.1 (C$_q$, *C*OO), 156.8 (C$_q$, 5-C$_{ar}$OCH$_3$), 156.3 (C$_q$, 8a-C$_{ar}$), 151.7 (C$_q$, 7-C$_{ar}$N), 137.1 (C$_q$, *C*$_{ar}$), 137.0 (C$_q$, *C*$_{ar}$), 135.0 (+, *C*H), 130.5 (+, *C*$_{ar}$H), 130.3 (+, *C*$_{ar}$H), 126.8 (+, *C*$_{ar}$H), 126.2 (+, *C*$_{ar}$H), 119.3 (C$_q$, 3-*C*), 100.9 (C$_q$, 4a-C$_{ar}$), 91.2 (+, *C*$_{ar}$H), 90.4 (+, *C*$_{ar}$H), 55.6 (+, OCH$_3$), 47.1 (–, NCH$_2$), 37.9 (+, NCH$_3$), 33.9 (–, *C*CH$_2$), 19.7 (+, C$_{ar}$CH$_3$), 11.7 (+,

CH_3) ppm – **IR** (ATR, ṽ) = 3437, 2962, 2922, 2870, 2851, 1708, 1595, 1517, 1489, 1459, 1431, 1397, 1377, 1368, 1336, 1299, 1256, 1235, 1217, 1174, 1157, 1133, 1106, 1065, 1050, 1031, 1014, 955, 799, 744, 730 cm^{-1}. – **MS** (+FAB, 3-NBA) = 338 (46) [M+H]$^+$, 337 (69) [M]$^+$. – **HRMS** (FAB, 3-NBA, $C_{21}H_{23}O_3N$) = calc.: 337.1678; found: 337.1678.

7-(Ethyl(methyl)amino)-5-methoxy-3-(2-methoxybenzyl)-2*H*-chromen-2-one (21hc)

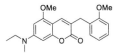 According to **GP10**, 4-(ethyl(methyl)amino)-2-hydroxy-6-methox-ybenzaldehyde (**19h**, 30.0 mg, 0.14 mmol, 1.00 equiv.), 2-methox-ycinnamaldehyde (58.0 mg, 0.36 mmol, 2.50 equiv.), KOAc (35.0 mg, 0.36 mmol, 2.50 equiv.), glacial acetic acid (2.00 µl, 2.15 mg, 0.04 mmol, 0.25 equiv.), 1,3-dimethylimidazolium dimethyl phosphate (0.03 ml, 38.0 mg, 0.17 mmol, 1.20 equiv.) and MS (powder, 3Å) in toluene (0.47 ml) were heated at 110 °C for 75 min *via* microwave irradi-ation. Filtration over a small silica plug (*c*Hex/EtOAc 1:1) and purification by prepHPLC (5–95_amino) resulted in the title compound as an off-white solid (9.0 mg, 25.2 µmol, 18%).

R_t (5%–95% in 27 min, amino) = 27.9 min. – **^{1}H NMR** (500 MHz, CDCl$_3$) δ = 7.56 (s, 1H, C*H*), 7.27 – 7.19 (m, 2H, 2 × C*H*$_{ar}$), 6.94 – 6.86 (m, 2H, 2 × C*H*$_{ar}$), 6.15 (d, ^4J = 2.2 Hz, 1H, C*H*$_{ar}$), 5.97 (d, ^4J = 2.2 Hz, 1H, C*H*$_{ar}$), 3.83 (s, 3H, OC*H*$_3$), 3.82 – 3.80 (m, 5H, OC*H*$_3$ and CC*H*$_2$), 3.43 (q, ^3J = 7.1 Hz, 2H, NC*H*$_2$), 2.98 (s, 3H, NC*H*$_3$), 1.16 (t, ^3J = 7.1 Hz, 2H, C*H*$_3$) ppm. – **^{13}C NMR** (126 MHz, CDCl$_3$) δ = 163.2 (C_q, *C*OO), 157.8 (C_q, C_{ar}OCH$_3$), 156.8 (C_q, 5-C_{ar}OCH$_3$), 156.4 (C_q, 8a-C_{ar}), 151.5 (C_q, 7-C_{ar}N), 135.3 (+, *C*H), 130.9 (+, C_{ar}H), 127.8 (+, C_{ar}H), 127.6 (C_q, CH$_2$$C_{ar}$), 120.7 (+, C_{ar}H), 119.6 (C_q, 3-*C*), 110.6 (+, C_{ar}H), 101.1 (C_q, 4a-C_{ar}), 91.2 (+, C_{ar}H), 90.5 (+, C_{ar}H), 55.7 (+, O*C*H$_3$), 55.5 (+, O*C*H$_3$), 47.1 (−, N*C*H$_2$), 37.9 (+, N*C*H$_3$), 30.7 (−, C*C*H$_2$), 11.6 (+, *C*H$_3$) ppm. – **HRMS** (ESI, $C_{21}H_{23}O_4N$) = calc.: 354.1700 [M+H]$^+$; found: 354.1687 [M+H]$^+$.

7-(Ethyl(methyl)amino)-3-(2-fluorobenzyl)-5-methoxy-2*H*-chromen-2-one (21he)

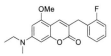 **A)** According to **GP8**, 4-(ethyl(methyl)amino)-2-hydroxy-6-methox-ybenzaldehyde (**19h**, 33.0 mg, 0.16 mmol, 1.00 equiv.), 2-fluorocin-namaldehyde (54.0 mg, 0.36 mmol, 2.50 equiv.), K$_2$CO$_3$ (24.0 mg, 0.17 mmol, 1.20 equiv.) and 1,3-dimethylimidazolium dimethyl phosphate (0.03 ml, 38.0 mg, 0.17 mmol, 1.20 equiv.) in toluene (0.47 ml) were heated at 110 °C for 50 min *via* microwave irradiation. Filtration over a small silica plug (*c*Hex/EtOAc 1:1) and purification by prepHPLC (5–95_amino) resulted in the title compound as an orange solid (3.00 mg, 0.01 mmol, 6.3%).

B) Similar to **GP8**, 4-(ethyl(methyl)amino)-2-hydroxy-6-methoxybenzaldehyde (**21h**, 30.0 mg, 0.14 mmol, 1.00 equiv.), 2-fluorocinnamaldehyde (54.0 mg, 0.36 mmol, 2.50 equiv.), K_2CO_3 (24.0 mg, 0.17 mmol, 1.20 equiv.), 1,3-dimethylimidazolium dimethyl phosphate (0.03 ml, 38.0 mg, 0.17 mmol, 1.20 equiv.) and additionally MS (powder, 3Å) in toluene (0.47 ml) were heated at 110 °C for 50 min *via* microwave irradiation. Filtration over a small silica plug (*c*Hex/EtOAc 1:1) and purification by prepHPLC (5–95_amino) resulted in the title compound as a yellow solid (2.6 mg, 7.00 μmol, 5.0%).

R_t (5%–95% in 27 min, amino) = 27.9 min. – **^{1}H NMR** (500 MHz, CDCl$_3$) δ = 7.65 (s, 1H, C*H*), 7.35 (td, $^{3,4}J$ = 7.6, 1.9 Hz, 1H, *m*-C*H*$_{ar}$), 7.24 – 7.18 (m, 1H, *m*-C*H*$_{ar}$), 7.10 – 7.01 (m, 2H, *o*– and *p*-C*H*$_{ar}$), 6.14 (d, ^4J = 2.2 Hz, 1H, C*H*$_{ar}$), 5.97 (d, ^4J = 2.2 Hz, 1H, C*H*$_{ar}$), 3.86 – 3.83 (m, 5H, OC*H*$_3$ and CC*H*$_2$), 3.43 (q, ^3J = 7.1 Hz, 2H, NC*H*$_2$), 2.98 (s, 3H, NC*H*$_3$), 1.16 (t, ^3J = 7.1 Hz, 3H, C*H*$_3$) ppm. – **^{13}C NMR** (126 MHz, CDCl$_3$) δ = 162.9 (C$_q$, COO), 161.4 (C$_q$, d, ^1J = 245.7 Hz, C$_{ar}$F), 156.9 (C$_q$, 5-C$_{ar}$OCH$_3$), 156.5 (C$_q$, 8a-C$_{ar}$), 151.8 (C$_q$, 7-C$_{ar}$N), 135.8 (+, *C*H), 131.7 (+, d, ^3J = 4.5 Hz, *m*-C$_{ar}$H), 128.3 (+, d, ^3J = 7.9 Hz, *m*-C$_{ar}$H), 126.2 (C$_q$, d, ^2J = 15.4 Hz, CH$_2$C$_{ar}$), 124.2 (+, d, ^4J = 3.5 Hz, *p*-C$_{ar}$H), 118.5 (C$_q$, 3-*C*), 115.4 (+, d, ^2J = 21.9 Hz, *o*-C$_{ar}$H), 100.9 (C$_q$, 4a-C$_{ar}$), 91.2 (+, C$_{ar}$H), 90.4 (+, C$_{ar}$H), 55.7 (+, OCH$_3$), 47.1 (–, NCH$_2$), 37.9 (+, NCH$_3$), 29.8 (–, CCH$_2$), 11.7 (+, CH$_3$) ppm – **^{19}F NMR** (471 MHz, CDCl$_3$) δ = –117.8 (s, C*F*) ppm. – **IR** (ATR, ṽ) = 2968, 2929, 2873, 2850, 1704, 1598, 1560, 1517, 1489, 1473, 1453, 1438, 1398, 1368, 1344, 1299, 1254, 1228, 1215, 1173, 1136, 1111, 1086, 1068, 1045, 1035, 990, 945, 922, 866, 847, 800, 755, 722, 697, 645, 606, 581, 554, 543, 528, 520, 482, 450, 433, 407, 394 cm^{-1}. – **HRMS** (ESI, C$_{20}$H$_{20}$O$_3$NF) = calc.: 342.1500 [M+H]$^+$; found: 342.1490 [M+H]$^+$.

7-(Ethyl(methyl)amino)-3-(4-fluorobenzyl)-5-methoxy-2*H*-chromen-2-one (21hf)

Similar to **GP8**, 4-(ethyl(methyl)amino)-2-hydroxy-6-methox-ybenzaldehyde (**19h**, 30.0 mg, 0.14 mmol, 1.00 equiv.), 4-fluo-rocinnamaldehyde (54.0 mg, 0.36 mmol, 2.50 equiv.), K_2CO_3 (24.0 mg, 0.17 mmol, 1.20 equiv.), 1,3-dimethylimidazolium dimethyl phosphate (0.03 ml, 38.0 mg, 0.17 mmol, 1.20 equiv.) and additionally 18-crown-6 (3.80 mg, 14.0 μmol, 0.10 equiv.) in toluene (0.47 ml) were heated at 110 °C for 50 min *via* microwave irradiation. Filtration over a small silica plug (*c*Hex/EtOAc 1:1) and purification by prepHPLC (5–95_amino) resulted in the title compound as a yellow solid (2.5 mg, 5.6 μmol, 4.0%, purity <90%).

R_t (5%–95% in 27 min, amino) = 28.4 min. – **^{1}H NMR** (500 MHz, CDCl$_3$) δ = 7.60 (s, 1H, C*H*), 7.31 – 7.22 (m, 2H, 2 × C*H*$_{ar}$), 7.02 – 6.93 (m, 2H, 2 × C*H*$_{ar}$), 6.15 (d, ^4J = 2.2 Hz, 1H,

CH_{ar}), 5.98 (d, ^4J = 2.2 Hz, 1H, CH_{ar}), 3.86 (s, 3H, OCH_3), 3.78 (s, 2H, CCH_2), 3.43 (q, ^3J = 7.1 Hz, 2H, NCH_2), 2.99 (s, 3H, NCH_3), 1.17 (t, ^3J = 7.1 Hz, 3H, CH_3) ppm. – ^{13}C **NMR** insufficient purity. – ^{19}F **NMR** (471 MHz, CDCl$_3$) δ = –117.1 (s, CF) ppm. – **HRMS** (ESI, $C_{20}H_{20}O_3NF$) = calc.: 342.1500 [M+H]$^+$; found: 342.1482 [M+H]$^+$.

7-(Ethyl(methyl)amino)-5-methoxy-3-(2-(trifluoromethyl)benzyl)-2H-chromen-2-one (21hg)

According to **GP8**, 4-(ethyl(methyl)amino)-2-hydroxy-6-methoxybenzaldehyde (**19h**, 30.0 mg, 0.14 mmol, 1.00 equiv.), 2-(trifluoro)cinnamaldehyde (72.0 mg, 0.36 mmol, 2.50 equiv.), K$_2$CO$_3$ (24.0 mg, 0.17 mmol, 1.20 equiv.) and 1,3-dimethylimidazolium dimethyl phosphate (0.03 ml, 38.0 mg, 0.17 mmol, 1.20 equiv.) in toluene (0.47 ml) were heated at 110 °C for 50 min *via* microwave irradiation. Filtration over a small silica plug (*c*Hex/EtOAc 1:1) and purification by prepHPLC (5–95_amino) resulted in the title compound as a yellow oil (1.4 mg, 3.50 μmol, 2.5%).

R_t (5%–95% in 27 min, amino) = 29.9 min. – ^1H **NMR** (500 MHz, CDCl$_3$) δ = 7.70 – 7.66 (m, 1H, CH), 7.51 – 7.44 (m, 1H, CH_{ar}), 7.44 (s, 1H, CH_{ar}), 7.39 (d, ^3J = 7.7 Hz, 1H, CH_{ar}), 7.34 (t, ^3J = 7.7 Hz, 1H, CH_{ar}), 6.17 (d, ^4J = 2.2 Hz, 1H, CH_{ar}), 5.96 (d, ^4J = 2.2 Hz, 1H, CH_{ar}), 4.03 (s, 2H, CCH_2), 3.81 (s, 3H, OCH_3), 3.44 (q, ^3J = 7.1 Hz, 2H, NCH_2), 2.99 (s, 3H, NCH_3), 1.17 (t, ^3J = 7.1 Hz, 3H, NCH_2CH_3) ppm. – ^{13}C **NMR** (126 MHz, CDCl$_3$) δ = 163.1 (C$_q$, COO), 156.9 (C$_q$, 5-C_{ar}OCH$_3$), 156.5 (C$_q$, 8a-C_{ar}), 151.9 (C$_q$, 7-C_{ar}N), 137.9 (C$_q$, CH$_2C_{ar}$), 136.3 (+, CH), 132.1 (+, C_{ar}H), 131.7 (+, C_{ar}H), 129.0 (C$_q$, C_{ar}CF$_3$), 126.6 (+, C_{ar}H), 126.2 (+, C_{ar}H), 118.9 (C$_q$, 3-C), 100.8 (C$_q$, 4a-C_{ar}), 91.1 (+, C_{ar}H), 90.4 (+, C_{ar}H), 55.7 (+, OCH$_3$), 47.1 (–, NCH$_2$), 37.9 (+, NCH$_3$), 32.7 (–, CCH$_2$), 11.7 (+, CH$_3$) ppm – ^{19}F **NMR** (471 MHz, CDCl$_3$) δ = –59.8 (s, CF_3) ppm. – **HRMS** (ESI, $C_{21}H_{20}O_3NF_3$) = calc.: 392.1468 [M+H]$^+$; found: 392.1449 [M+H]$^+$.

3-Benzyl-7-(butyl(methyl)amino)-5-methoxy-2H-chromen-2-one (21ia)

According to **GP10**, 4-(butyl(methyl)amino)-2-hydroxy-6-methoxybenzaldehyde (**XX**, 50.0 mg, 0.21 mmol, 1.00 equiv.), cinnamaldehyde (0.07 ml, 70.0 mg, 0.53 mmol, 2.50 equiv.), KOAc (52.0 mg, 0.53 mmol, 2.50 equiv.), glacial acetic acid (3.00 μl, 3.16 mg, 0.05 mmol, 0.25 equiv.), 1,3-dimethylimidazolium dimethyl phosphate (0.04 ml, 56.0 mg, 0.25 mmol, 1.20

equiv.) and MS (powder, 3Å) in toluene (0.70 ml) were heated at 110 °C for 75 min *via* microwave irradiation. Filtration over a small silica plug (*c*Hex/EtOAc 1:1) and purification by prepHPLC (5–95_amino) resulted in the title compound as a cloudy oil (6.1 mg, 16.8 μmol, 8.0%).

R_t (5%–95% in 27 min, amino) = 31.7 min. – **¹H NMR** (500 MHz, CDCl₃) δ = 7.67 (d, ^4J = 1.3 Hz, 1H, C*H*), 7.34 – 7.28 (m, 4H, 4 × C*H*ₐᵣ), 7.25 – 7.22 (m, 1H, C*H*ₐᵣ), 6.34 (d, ^4J = 2.2 Hz, 1H, C*H*ₐᵣ), 6.30 (d, ^4J = 2.2 Hz, 1H, C*H*ₐᵣ), 3.87 (s, 3H, OC*H₃*), 3.84 (s, 2H, CC*H₂*), 3.42 – 3.34 (m, 2H, NC*H₂*), 3.06 (s, 3H, NC*H₃*), 1.61 – 1.53 (m, 2H, NCH₂C*H₂*), 1.35 (sex., ^3J = 7.4 Hz, 2H, C*H₂*CH₃), 0.94 (t, ^3J = 7.3 Hz, 3H, C*H₃*) ppm. – **¹³C NMR** (126 MHz, CDCl₃) δ = 162.9 (C_q, *C*OO), 157.0 (C_q, 5-*C*ₐᵣOCH₃), 156.0 (C_q, 8a-*C*ₐᵣ), 149.8 (C_q, 7-*C*ₐᵣN), 138.8 (C_q, CH₂*C*ₐᵣ), 135.4 (+, *C*H), 129.2 (+, 2 × *C*ₐᵣH), 128.7 (+, 2 × *C*ₐᵣH), 126.6 (+, *C*ₐᵣH), 122.4 (C_q, 3-*C*), 103.6 (C_q, 4a-*C*ₐᵣ), 93.6 (+, *C*ₐᵣH), 93.2 (+, *C*ₐᵣH), 56.0 (+, O*C*H₃), 54.5 (–, N*C*H₂), 40.7 (+, N*C*H₃), 36.7 (–, C*C*H₂), 28.6 (–, *C*H₂), 20.2 (–, *C*H₂), 13.9 (+, *C*H₃) ppm – **IR** (ATR, ṽ) = 2956, 2927, 2871, 1706, 1601, 1560, 1516, 1492, 1487, 1453, 1438, 1398, 1364, 1306, 1258, 1231, 1200, 1183, 1136, 1111, 1077, 1044, 1031, 1001, 984, 966, 946, 800, 772, 755, 728, 698 cm⁻¹. – **HRMS** (ESI, C₂₂H₂₅O₃N) = calc.: 352.1907 [M+H]⁺; found: 352.1896 [M+H]⁺.

7-(Butyl(methyl)amino)-5-methoxy-3-(2-methylbenzyl)-2*H*-chromen-2-one (21ib)

According to **GP10**, 4-(butyl(methyl)amino)-2-hydroxy-6-methoxybenzaldehyde (**19i**, 40.0 mg, 0.17 mmol, 1.00 equiv.), 2-methylcinnamaldehyde (62.0 mg, 0.42 mmol, 2.50 equiv.), KOAc (41.0 mg, 0.42 mmol, 2.50 equiv.), glacial acetic acid (2.00 μl, 2.53 mg, 0.04 mmol, 0.25 equiv.), 1,3-dimethylimidazolium dimethyl phosphate (0.04 ml, 45.0 mg, 0.20 mmol, 1.20 equiv.) and MS (powder, 3Å) in toluene (0.56 ml) were heated at 110 °C for 75 min *via* microwave irradiation. Filtration over a small silica plug (*c*Hex/EtOAc 1:1) and purification by prepHPLC (5–95_amino) resulted in the title compound as a red solid (5.1 mg, 13.6 μmol, 8.0%).

R_t (5%–95% in 27 min, amino) = 32.5 min. – **¹H NMR** (500 MHz, CDCl₃) δ = 7.35 (d, ^4J = 1.3 Hz, 1H, C*H*), 7.22 – 7.16 (m, 4H, 4 × C*H*ₐᵣ), 6.21 (d, ^4J = 2.2 Hz, 1H, C*H*ₐᵣ), 6.06 (d, ^4J = 2.2 Hz, 1H, C*H*ₐᵣ), 3.82 (d, ^4J = 1.3 Hz, 2H, CC*H₂*), 3.80 (s, 3H, OC*H₃*), 3.36 (t, ^3J = 7.6 Hz, 2H, NC*H₂*), 3.02 (s, 3H, NC*H₃*), 2.28 (s, 3H, CₐᵣC*H₃*), 1.62 – 1.53 (m, 2H, NCH₂C*H₂*), 1.41 – 1.30 (m, 2H, C*H₂*CH₃), 0.95 (t, ^3J = 7.3 Hz, 3H, C*H₃*) ppm. – **¹³C NMR** (126 MHz, CDCl₃) δ = 163.2 (C_q, *C*OO), 156.8 (C_q, 5-*C*ₐᵣOCH₃), 156.1 (C_q, 8a-*C*ₐᵣ), 151.3 (C_q, 7-*C*ₐᵣN), 137.0 (C_q, *C*ₐᵣ), 136.9 (C_q, *C*ₐᵣ), 135.1 (+, *C*H), 130.5 (+, *C*ₐᵣH), 130.3 (+, *C*ₐᵣH), 126.9 (+, *C*ₐᵣH), 126.3 (+,

$C_{ar}H$), 119.9 (C_q, 3-C), 101.7 (C_q, 4a-C_{ar}), 91.8 (+, $C_{ar}H$), 91.3 (+, $C_{ar}H$), 55.7 (+, OCH_3), 53.2
(–, NCH_2), 39.4 (+, NCH_3), 33.9 (–, CCH_2), 29.0 (–, CH_2), 20.3 (–, CH_2), 19.7 (+, $C_{ar}CH_3$),
14.0 (+, CH_3) ppm – **IR** (ATR, ṽ) = 2958, 2928, 2871, 2857, 1707, 1596, 1516, 1489, 1459,
1398, 1364, 1340, 1305, 1256, 1197, 1162, 1136, 1109, 1051, 1038, 1016, 946, 799, 768, 744,
730, 722, 707, 547, 453 cm^{-1}. – **HRMS** (ESI, $C_{23}H_{27}O_3N$) = calc.: 366.2064 [M+H]$^+$; found:
366.2053 [M+H]$^+$.

7-(Butyl(methyl)amino)-5-methoxy-3-(2-methoxybenzyl)-2*H*-chromen-2-one (21ic)

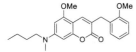 A microwave *vial* was charged with 4-(butyl(methyl)amino)-2-
hydroxy-6-methoxybenzaldehyde (**19h**, 40.0 mg, 0.17 mmol,
1.00 equiv.), 2-methoxycinnamaldehyde (68.0 mg, 0.42 mmol,
2.50 equiv.), KOAc (22.0 mg, 0.22 mmol, 1.30 equiv.), K_2CO_3 (28.0 mg, 0.20 mmol, 1.20
equiv.), glacial acetic acid (1.30 µl, 1.32 mg, 2.21 µmol, 0.13 equiv.), 1,3-dimethylimidazolium
dimethyl phosphate (0.04 ml, 45.0 mg, 0.20 mmol, 1.20 equiv.) and MS (powder, 3Å) in toluene
(0.56 ml). The reaction mixture was heated at 110 °C for 75 min, at max. 230 Watt microwave
irradiation and a maximum pressure of 7 bar. After cooling to rt, H_2O was added, followed by
extraction with EtOAc. The combined organic layers were dried over Na_2SO_4, filtrated and the
volatiles were removed under reduced pressure. The crude product was filtered over a small
pad of silica (*c*Hex/EtOAc 1:1) and concentrated again under reduced pressure. The filtrated
product was then purified by HPLC (5–95_amino), resulted in the title compound as a red solid
(8.10 mg, 22.1 µmol, 13%).

R_t (5%–95% in 27 min, amino) = 31.4 min. – **¹H NMR** (300 MHz^{10}, CDCl$_3$) δ = 7.64 (d, ^4J =
1.3 Hz, 1H, C*H*), 7.32 – 7.20 (m, 2H, 2 × C*H*$_{ar}$), 6.99 – 6.86 (m, 2H, 2 × C*H*$_{ar}$), 6.70 (d, ^4J =
2.1 Hz, 1H, C*H*$_{ar}$), 6.60 (d, ^4J = 2.1 Hz, 1H, C*H*$_{ar}$), 3.89 (s, 3H, OC*H*$_3$), 3.86 (d, ^4J = 1.3 Hz,
2H, CC*H*$_2$), 3.81 (s, 3H, OC*H*$_3$), 3.47 – 3.37 (m, 2H, NC*H*$_2$), 3.14 (s, 3H, NC*H*$_3$), 1.63 – 1.47
(m, 2H, NCH$_2$C*H*$_2$), 1.44 – 1.23 (m, 2H, C*H*$_2$CH$_3$), 0.91 (t, ^3J = 7.3 Hz, 3H, C*H*$_3$) ppm. – **¹³C
NMR** (126 MHz, CDCl$_3$) δ = 163.0 (C_q, *C*OO), 157.7 (C_q, C_{ar}OCH$_3$), 156.9 (C_q, 5-C_{ar}OCH$_3$),
155.9 (C_q, 8a-C_{ar}), 149.6 (C_q, 7-C_{ar}N), 135.1 (+, *C*H), 131.0 (+, C_{ar}H), 128.0 (+, C_{ar}H), 127.0
(C_q, CH$_2$$C_{ar}$), 121.7 ($C_q$, 3-$C$), 120.7 (+, C_{ar}H), 110.7 (+, C_{ar}H), 103.7 (C_q, 4a-C_{ar}), 93.5 (+,
C_{ar}H), 93.1 (+, C_{ar}H), 56.0 (+, O*C*H$_3$), 55.5 (+, O*C*H$_3$), 54.4 (–, N*C*H$_2$), 40.5 (+, N*C*H$_3$), 30.8
(–, C*C*H$_2$), 28.6 (–, *C*H$_2$), 20.2 (–, *C*H$_2$), 13.9 (+, *C*H$_3$) ppm. – **IR** (ATR, ṽ) = 2961, 2927, 2871,
2837, 1706, 1596, 1516, 1492, 1459, 1438, 1398, 1363, 1337, 1286, 1242, 1196, 1160, 1136,

---

^{10} Only low frequency H NMR available, compound started to decomposed during measurements.

1108, 1048, 1026, 946, 799, 749, 704 cm^{-1}. – **HRMS** (ESI, C$_{23}$H$_{27}$O$_4$N) = calc.: 382.2013 [M+H]$^+$; found: 382.2002 [M+H]$^+$.

7-(Butyl(methyl)amino)-5-methoxy-3-(4-methoxybenzyl)-2*H*-chromen-2-one (21id)

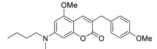

According to **GP10**, 4-(butyl(methyl)amino)-2-hydroxy-6-methoxybenzaldehyde (**19i**, 50.0 mg, 0.21 mmol, 1.00 equiv.), 4-methoxycinnamaldehyde (85.0 mg, 0.53 mmol, 2.50 equiv.), KOAc (52.0 mg, 0.53 mmol, 2.50 equiv.), glacial acetic acid (3.00 µl, 3.16 mg, 0.05 mmol, 0.25 equiv.), 1,3-dimethylimidazolium dimethyl phosphate (0.04 ml, 56.0 mg, 0.25 mmol, 1.20 equiv.) and MS (powder, 3Å) in toluene (0.70 ml) were heated at 110 °C for 75 min *via* microwave irradiation. Filtration over a small silica plug (*c*Hex/EtOAc 1:1) and purification by prepHPLC (5–95_amino) resulted in the title compound as a cloudy oil (13.8 mg, 35.7 µmol, 17%).

R$_t$ (5%–95% in 27 min, amino) = 31.3 min. – **^{1}H NMR** (500 MHz, CDCl$_3$) δ = 7.71 (d, ^4J = 1.3 Hz, 1H, C*H*), 7.20 (dd, $^{3,4}J$ = 8.6, 2.1 Hz, 2H, 2 × C*H*$_{ar}$), 6.96 (d, ^4J = 2.1 Hz, 1H, C*H*$_{ar}$), 6.88 (dd, $^{3,4}J$ = 8.6, 2.1 Hz, 2H, 2 × C*H*$_{ar}$), 6.79 (d, ^4J = 2.1 Hz, 1H, C*H*$_{ar}$), 3.93 (s, 3H, OC*H*$_3$), 3.82 (d, ^4J = 1.2 Hz, 2H, CC*H*$_2$), 3.81 (s, 3H, OC*H*$_3$), 3.50 – 3.42 (m, 2H, NC*H*$_2$), 3.21 (s, 3H, NC*H*$_3$), 1.59 – 1.50 (m, 2H, NCH$_2$C*H*$_2$), 1.34 (sex., ^3J = 7.4 Hz, 2H, C*H*$_2$CH$_3$), 0.90 (t, ^3J = 7.4 Hz, 3H, C*H*$_3$) ppm. – **^{13}C NMR** (126 MHz, CDCl$_3$) δ = 161.6 (C$_q$, *C*OO), 158.6 (C$_q$, *C*$_{ar}$OCH$_3$), 157.3 (C$_q$, 5-*C*$_{ar}$OCH$_3$), 154.9 (C$_q$, 8a-*C*$_{ar}$), 144.4 (C$_q$, 7-*C*$_{ar}$N), 133.9 (+, *C*H), 130.4 (+, 2 × *C*$_{ar}$H), 129.6 (C$_q$, *C*), 128.3 (C$_q$, *C*), 114.4 (+, 2 × *C*$_{ar}$H), 109.7 (C$_q$, 4a-*C*$_{ar}$), 99.0 (+, *C*$_{ar}$H), 98.9 (+, *C*$_{ar}$H), 58.8 (–, N*C*H$_2$), 56.7 (+, O*C*H$_3$), 55.5 (+, O*C*H$_3$), 44.8 (+, N*C*H$_3$), 36.1 (–, *C*CH$_2$), 27.4 (–, N*C*H$_2$*C*H$_2$), 19.8 (–, *C*H$_2$CH$_3$), 13.6 (+, *C*H$_3$) ppm. – **IR** (ATR, ṽ) = 2958, 2932, 2873, 2839, 1779, 1704, 1599, 1510, 1487, 1460, 1441, 1400, 1364, 1302, 1245, 1197, 1171, 1137, 1111, 1033, 945, 904, 830, 799, 747, 721, 703, 663, 646, 595, 564, 548, 535, 516, 482, 436, 387 cm$^-$^{1}. – **HRMS** (ESI, C$_{23}$H$_{27}$O$_4$N) = calc.: 382.2013 [M+H]$^+$; found: 382.2000 [M+H]$^+$.

7-(Butyl(methyl)amino)-3-(2-fluorobenzyl)-5-methoxy-2*H*-chromen-2-one (21ie)

According to **GP10**, 4-(butyl(methyl)amino)-2-hydroxy-6-methoxybenzaldehyde (**19i**, 50.0 mg, 0.21 mmol, 1.00 equiv.), 2-fluorocinnamaldehyde (79.0 mg, 0.53 mmol, 2.50 equiv.), KOAc (52.0 mg, 0.53 mmol, 2.50 equiv.), glacial acetic acid (3.00 µl, 3.16 mg, 0.05 mmol, 0.25 equiv.), 1,3-dimethylimidazolium dimethyl phosphate (0.04 ml, 56.0 mg, 0.25 mmol, 1.20

equiv.) and MS (powder, 3Å) in toluene (0.70 ml) were heated at 110 °C for 75 min *via* microwave irradiation. Filtration over a small silica plug (*c*Hex/EtOAc 1:1) and purification by prepHPLC (5–95_amino) resulted in the title compound as a yellow solid (4.1 mg, 10.5 μmol, 5.0%).

***R**$_t$* (5%–95% in 27 min, amino) = 33.1 min. – **^{1}H NMR** (300 MHz[11], CDCl$_3$) δ = 7.65 (s, 1H, C*H*), 7.34 (dd, $^{3,4}J$ = 7.6, 1.9 Hz, 1H, C*H*$_{ar}$), 7.20 (ddd, $^{3,3,4}J$ = 7.4, 5.3, 2.0 Hz, 1H, C*H*$_{ar}$), 7.12 – 6.99 (m, 2H, 2 × C*H*$_{ar}$), 6.16 (d, ^4J = 2.2 Hz, 1H, C*H*$_{ar}$), 6.01 (d, ^4J = 2.2 Hz, 1H, C*H*$_{ar}$), 3.85 (s, 5H, OC*H*$_3$ and CC*H*$_2$), 3.40 – 3.29 (m, 2H, NC*H*$_2$), 3.01 (s, 3H, NC*H*$_3$), 1.67 – 1.50 (m, 2H, NCH$_2$C*H*$_2$), 1.35 (sex., ^3J = 7.3 Hz, 2H, C*H*$_2$CH$_3$), 0.95 (t, ^3J = 7.3 Hz, 3H, C*H*$_3$) ppm. – **^{13}C NMR** due to degradation not measured. – **IR** (ATR, ṽ) = 2956, 2929, 2871, 1704, 1599, 1516, 1489, 1455, 1398, 1366, 1299, 1258, 1228, 1200, 1177, 1136, 1109, 1047, 1034, 943, 924, 800, 754, 724, 698, 548, 517, 433 cm^{-1}. – **HRMS** (ESI, C$_{22}$H$_{24}$O$_3$NF) = calc.: 370.1813 [M+H]$^+$; found: 370.1800 [M+H]$^+$.

7-(Butyl(methyl)amino)-3-(4-fluorobenzyl)-5-methoxy-2*H*-chromen-2-one (21if)

According to **GP10**, 4-(butyl(methyl)amino)-2-hydroxy-6-methoxybenzaldehyde (**19i**, 50.0 mg, 0.21 mmol, 1.00 equiv.), 4-fluorocinnamaldehyde (0.07 ml, 79.0 mg, 0.53 mmol, 2.50 equiv.), KOAc (52.0 mg, 0.53 mmol, 2.50 equiv.), glacial acetic acid (3.00 μl, 3.16 mg, 0.05 mmol, 0.25 equiv.), 1,3-dimethylimidazolium dimethyl phosphate (0.04 ml, 56.0 mg, 0.25 mmol, 1.20 equiv.) and MS (powder, 3Å) in toluene (0.70 ml) were heated at 110 °C for 75 min *via* microwave irradiation. Filtration over a small silica plug (*c*Hex/EtOAc 1:1) and purification by prepHPLC (5–95_amino) resulted in the title compound as a cloudy oil (6.2 mg, 16.8 μmol, 8.0%).

***R**$_t$* (5%–95% in 27 min, amino) = 31.6 min. – **^{1}H NMR** (500 MHz, CDCl$_3$) δ = 7.66 (d, ^4J = 1.2 Hz, 1H, C*H*), 7.29 – 7.22 (m, 2H, 2 × C*H*$_{ar}$), 7.00 (t, ^3J = 8.7 Hz, 2H, 2 × C*H*$_{ar}$), 6.38 (s, 2H, 2 × C*H*$_{ar}$), 3.89 (s, 3H, OC*H*$_3$), 3.81 (s, 2H, CC*H*$_2$), 3.42 –3.36 (m, 2H, NC*H*$_2$), 3.08 (s, 3H, NC*H*$_3$), 1.62 – 1.53 (m, 2H, NCH$_2$C*H*$_2$), 1.35 (sex., ^3J = 7.4 Hz, 2H, C*H*$_2$CH$_3$), 0.93 (t, ^3J = 7.4 Hz, 3H, C*H*$_3$) ppm. – **^{13}C NMR** (126 MHz, CDCl$_3$) δ = 162.7 (C$_q$, COO), 161.8 (C$_q$, d, ^1J = 244.4 Hz, C$_{ar}$F), 157.0 (C$_q$, 5-C$_{ar}$OCH$_3$), 155.9 (C$_q$, 8a-C$_{ar}$), 149.3 (C$_q$, 7-C$_{ar}$N), 135.3 (+, CH), 134.3 (C$_q$, d, ^4J = 3.2 Hz, CH$_2$C$_{ar}$), 130.7 (+, d, ^3J = 7.8 Hz, 2 × *m*-C$_{ar}$H), 122.7 (C$_q$, 3-C), 115.5 (+, d, ^2J = 21.2 Hz, 2 × *o*-C$_{ar}$H), 104.2 (C$_q$, 4a-C$_{ar}$), 94.1 (+, C$_{ar}$H), 93.9 (+, C$_{ar}$H), 56.1 (+,

---

[11] Only low frequency H NMR available, compound started to decomposed during measurements.

OCH$_3$), 55.0 (–, NCH$_2$), 41.1 (+, NCH$_3$), 36.7 (–, CCH$_2$), 28.4 (–, CH$_2$), 20.2 (–, CH$_2$), 13.9 (+, CH$_3$) ppm – **^{19}F NMR** (471 MHz, CDCl$_3$) δ = –75.8 (s, TFA), –115.2 (s, CF) ppm. – **IR** (ATR, ṽ) = 2959, 2935, 2876, 1703, 1598, 1560, 1507, 1487, 1460, 1436, 1400, 1364, 1258, 1220, 1198, 1156, 1139, 1112, 1095, 1047, 1016, 989, 945, 822, 795, 721, 704, 520, 487 cm^{-1}. – **HRMS** (ESI, C$_{22}$H$_{24}$O$_3$NF) = calc.: 370.1813 [M+H]$^+$; found: 370.1802 [M+H]$^+$.

7-(Butyl(methyl)amino)-5-methoxy-3-(2-(trifluoromethyl)benzyl)-2*H*-chromen-2-one (21ig)

According to **GP10**, 4-(butyl(methyl)amino)-2-hydroxy-6-methoxybenzaldehyde (**21i**, 50.0 mg, 0.21 mmol, 1.00 equiv.), 2-(trifluoromethyl)cinnamaldehyde (105 mg, 0.53 mmol, 2.50 equiv.), KOAc (52.0 mg, 0.53 mmol, 2.50 equiv.), glacial acetic acid (3.00 µl, 3.16 mg, 0.05 mmol, 0.25 equiv.), 1,3-dimethylimidazolium dimethyl phosphate (0.04 ml, 56.0 mg, 0.25 mmol, 1.20 equiv.) and MS (powder, 3Å) in toluene (0.70 ml) were heated at 110 °C for 75 min *via* microwave irradiation. Filtration over a small silica plug (*c*Hex/EtOAc 1:1) and purification by prepHPLC (5–95_amino) resulted in the title compound as a yellow solid (5.6 mg, 12.6 µmol, 6.0%).

R_t (5%–95% in 27 min, amino) = 33.1 min. – **^{1}H NMR** (300 MHz[12], CDCl$_3$) δ = 7.68 (d, ^3J = 7.9 Hz, 1H, CH$_{ar}$), 7.52 – 7.42 (m, 2H, CH and CH$_{ar}$), 7.42 – 7.30 (m, 2H, 2 × CH$_{ar}$), 6.16 (d, ^4J = 2.2 Hz, 1H, CH$_{ar}$), 5.96 (d, ^4J = 2.2 Hz, 1H, CH$_{ar}$), 4.03 (s, 2H, CCH$_2$), 3.81 (s, 3H, OCH$_3$), 3.41 – 3.31 (m, 2H, NCH$_2$), 3.01 (s, 3H, NCH$_3$), 1.65 – 1.52 (m, 2H, NCH$_2$CH$_2$), 1.43 – 1.24 (m, 2H, CH$_2$CH$_3$), 0.96 (t, ^3J = 7.3 Hz, 3H, CH$_3$) ppm. – **^{13}C NMR** due to degradation not measured. – **IR** (ATR, ṽ) = 2959, 2931, 2873, 1708, 1601, 1517, 1489, 1455, 1439, 1401, 1366, 1310, 1258, 1201, 1153, 1105, 1058, 1034, 958, 802, 765, 725, 652 cm^{-1}. – **HRMS** (ESI, C$_{23}$H$_{24}$O$_3$NF$_3$) = calc.: 420.1781 [M+H]$^+$; found: 420.1767 [M+H]$^+$.

3-Benzyl-7-(hexyl(methyl)amino)-5-methoxy-2*H*-chromen-2-one (21ja)

According to **GP10**, 4-(hexyl(methyl)amino)-2-hydroxy-6-methoxybenzaldehyde (**19j**, 50.0 mg, 0.19 mmol, 1.00 equiv.), cinnamaldehyde (0.06 ml, 62.0 mg, 0.47 mmol, 2.50 equiv.), KOAc (46.0 mg, 0.47 mmol, 2.50 equiv.), glacial acetic acid (3.00 µl, 2.83 mg, 0.05 mmol, 0.25 equiv.), 1,3-dimethylimidazolium dimethyl phosphate (0.04 ml, 50.0 mg, 0.23

---

[12] Only low frequency H NMR available, compound started to decomposed during measurements.

mmol, 1.20 equiv.) and MS (powder, 3Å) in toluene (0.62 ml) were heated at 110 °C for 75 min *via* microwave irradiation. Filtration over a small silica plug (*c*Hex/EtOAc 1:1) and purification by prepHPLC (5–95_amino) resulted in the title compound as a yellow oil (3.5 mg, 9.24 μmol, 4.9%).

R_t (5%–95% in 27 min, amino) = 34.5 min. – **¹H NMR** (500 MHz, CDCl₃) δ = 7.71 (s, 1H, C*H*), 7.36 – 7.27 (m, 4H, 4 × C*H*ₐᵣ), 7.27 – 7.25 (m, 1H, C*H*ₐᵣ), 6.66 (d, ^4J = 2.1 Hz, 1H, C*H*ₐᵣ), 6.59 (d, ^4J = 2.1 Hz, 1H, C*H*ₐᵣ), 3.90 (s, 3H, OC*H*₃), 3.86 (s, 2H, CC*H*₂), 3.45 – 3.37 (m, 2H, NC*H*₂), 3.13 (s, 3H, NC*H*₃), 1.62 – 1.52 (m, 2H, NCH₂C*H*₂), 1.35 – 1.22 (m, 6H, 3 × C*H*₂), 0.89 – 0.84 (m, 3H, C*H*₃) ppm. – **¹³C NMR** (126 MHz, CDCl₃) δ = 162.2 (C_q, COO), 157.2 (C_q, 5-C_arOCH₃), 155.4 (C_q, 8a-C_ar), 147.1 (C_q, 7-C_arN), 138.2 (C_q, CH₂C_ar), 134.8 (+, CH), 129.3 (+, 2 × C_arH), 128.8 (+, 2 × C_arH), 126.9 (+, C_arH), 106.8 (C_q, 4a-C_ar), 96.3 (+, C_arH), 57.0 (–, NCH₂), 56.4 (+, OCH₃), 42.8 (+, NCH₃), 36.8 (–, CCH₂), 31.4 (–, CH₂), 26.3 (–, CH₂), 25.9 (–, CH₂), 22.6 (–, CH₂), 14.0 (+, CH₃) ppm. – **IR** (ATR, ṽ) = 2952, 2922, 2851, 1710, 1605, 1516, 1455, 1397, 1367, 1259, 1180, 1137, 1112, 1044, 802, 700 cm⁻¹. – **HRMS** (ESI, C₂₄H₂₉O₃N) = calc.: 380.2220 [M+H]⁺; found: 380.2205 [M+H]⁺.

7-(Hexyl(methyl)amino)-5-methoxy-3-(2-methylbenzyl)-2*H*-chromen-2-one (21jb)

According to **GP10**, 4-(hexyl(methyl)amino)-2-hydroxy-6-methoxybenzaldehyde (**19j**, 50.0 mg, 0.19 mmol, 1.00 equiv.), 2-methylcinnamaldehyde (69.0 mg, 0.47 mmol, 2.50 equiv.), KOAc (46.0 mg, 0.47 mmol, 2.50 equiv.), glacial acetic acid (3.00 μl, 2.83 mg, 0.05 mmol, 0.25 equiv.), 1,3-dimethylimidazolium dimethyl phosphate (0.04 ml, 50.0 mg, 0.23 mmol, 1.20 equiv.) and MS (powder, 3Å) in toluene (0.62 ml) were heated at 110 °C for 75 min *via* microwave irradiation. Filtration over a small silica plug (*c*Hex/EtOAc 1:1) and purification by prepHPLC (5–95_amino) resulted in the title compound as a white solid (7.1 mg, 18.2 μmol, 10%).

R_t (5%–95% in 27 min, amino) = 35.2 min. – **¹H NMR** (500 MHz, CDCl₃) δ = 7.44 (d, ^4J = 1.6 Hz, 1H, C*H*), 7.25 – 7.22 (m, 2H, 2 × C*H*ₐᵣ), 7.22 – 7.16 (m, 2H, 2 × C*H*ₐᵣ), 6.86 (d, ^4J = 2.1 Hz, 1H, C*H*ₐᵣ), 6.73 (d, ^4J = 2.1 Hz, 1H, C*H*ₐᵣ), 3.89 – 3.86 (s, 5H, OC*H*₃ and CC*H*₂), 3.48 – 3.41 (m, 2H, NC*H*₂), 3.18 (s, 3H, NC*H*₃), 2.28 (s, 3H, C_arC*H*₃), 1.61 – 1.52 (m, 2H, NCH₂C*H*₂), 1.34 – 1.28 (m, 2H, C*H*₂), 1.28 – 1.23 (m, 4H, 2 × C*H*₂), 0.88 – 0.83 (m, 3H, C*H*₃) ppm. – **¹³C NMR** (126 MHz, CDCl₃) δ = 161.9 (C_q, COO), 157.2 (C_q, 5-C_arOCH₃), 154.9 (C_q, 8a-C_ar), 145.2 (C_q, 7-C_arN), 136.9 (C_q, C_ar), 135.5 (C_q, C_ar), 133.8 (+, CH), 130.8 (+, C_arH), 130.3 (+, C_arH), 127.4 (+, C_arH), 126.5 (+, C_arH), 126.4 (C_q, 3-C), 108.8 (C_q, 4a-C_ar), 98.2 (+, C_arH), 98.1

$(+, C_{ar}H)$, 58.5 $(-, NCH_2)$, 56.6 $(+, OCH_3)$, 44.2 $(+, NCH_3)$, 34.1 $(-, CCH_2)$, 31.3 $(-, CH_2)$, 26.2 $(-, CH_2)$, 25.5 $(-, CH_2)$, 22.5 $(-, CH_2)$, 19.6 $(+, CH_3)$, 13.9 $(+, CH_3)$ ppm. $-$ **IR** (ATR, $\tilde{v}$) = 2953, 2922, 2853, 1708, 1605, 1561, 1516, 1487, 1459, 1398, 1366, 1256, 1222, 1180, 1137, 1113, 1052, 802, 755, 742, 730 cm^{-1}. $-$ **HRMS** (ESI, $C_{25}H_{31}O_3N$) = calc.: 394.2377 [M+H]$^+$; found: 394.2366 [M+H]$^+$.

7-(Hexyl(methyl)amino)-5-methoxy-3-(2-methoxybenzyl)-2*H*-chromen-2-one (21jc)

According to **GP10**, 4-(hexyl(methyl)amino)-2-hydroxy-6-methoxybenzaldehyde (**21j**, 50.0 mg, 0.19 mmol, 1.00 equiv.), 2-methoxycinnamaldehyde (76.0 mg, 0.47 mmol, 2.50 equiv.), KOAc (46.0 mg, 0.47 mmol, 2.50 equiv.), glacial acetic acid (3.00 μl, 2.83 mg, 0.05 mmol, 0.25 equiv.), 1,3-dimethylimidazolium dimethyl phosphate (0.04 ml, 50.0 mg, 0.23 mmol, 1.20 equiv.) and MS (powder, 3Å) in toluene (0.62 ml) were heated at 110 °C for 75 min *via* microwave irradiation. Filtration over a small silica plug (*c*Hex/EtOAc 1:1) and purification by prepHPLC (5–95_amino) resulted in the title compound as a pale-yellow oil (8.0 mg, 19.7 μmol, 10%).

R_t (5%–95% in 27 min, amino) = 34.3 min. $-$ **^{1}H NMR** (500 MHz, CDCl$_3$) δ = 7.62 (d, ^4J = 1.3 Hz, 1H, C*H*), 7.28 – 7.22 (m, 2H, 2 × C*H*$_{ar}$), 6.93 (d, $^{3,4}J$ = 7.4, 1.1 Hz, 1H, C*H*$_{ar}$), 6.89 (dd, $^{3,4}J$ = 8.7, 1.1 Hz, 1H, C*H*$_{ar}$), 6.45 (d, ^4J = 2.1 Hz, 1H, C*H*$_{ar}$), 6.43 (d, ^4J = 2.1 Hz, 1H, C*H*$_{ar}$), 3.86 (s, 3H, OC*H*$_3$), 3.84 (d, ^4J = 1.3 Hz, 2H, CC*H*$_2$), 3.81 (s, 3H, OC*H*$_3$), 3.40 – 3.35 (m, 2H, NC*H*$_2$), 3.08 (s, 3H, NC*H*$_3$), 1.61 – 1.53 (m, 2H, NCH$_2$C*H*$_2$), 1.34 – 1.24 (m, 6H, 3 × C*H*$_2$), 0.90 – 0.85 (m, 3H, C*H*$_3$) ppm. $-$ **^{13}C NMR** (126 MHz, CDCl$_3$) δ = 162.8 (C$_q$, *C*OO), 157.7 (C$_q$, *C*$_{ar}$OCH$_3$), 157.0 (C$_q$, 5-*C*$_{ar}$OCH$_3$), 155.6 (C$_q$, 8a-*C*$_{ar}$), 148.4 (C$_q$, 7-*C*$_{ar}$N), 134.8 (+, *C*H), 131.1 (+, *C*$_{ar}$H), 128.2 (+, *C*$_{ar}$H), 126.7 (C$_q$, *C*H$_2$*C*$_{ar}$), 123.0 (C$_q$, 3-*C*), 120.8 (+, *C*$_{ar}$H), 110.7 (+, *C*$_{ar}$H), 105.2 (C$_q$, 4a-*C*$_{ar}$), 94.8 (+, *C*$_{ar}$H), 94.6 (+, *C*$_{ar}$H), 56.1 (+, O*C*H$_3$), 55.7 (-, N*C*H$_2$), 55.5 (+, O*C*H$_3$), 41.5 (+, N*C*H$_3$), 31.5 (-, *C*H$_2$), 30.9 (-, *C*CH$_2$), 26.5 (-, *C*H$_2$), 26.2 (-, *C*H$_2$), 22.6 (-, *C*H$_2$), 14.1 (+, *C*H$_3$) ppm. $-$ **IR** (ATR, $\tilde{v}$) = 2952, 2924, 2868, 2853, 1707, 1601, 1561, 1516, 1490, 1459, 1438, 1398, 1366, 1343, 1307, 1285, 1244, 1177, 1160, 1136, 1112, 1050, 1028, 983, 950, 921, 897, 882, 800, 751, 724, 645, 609, 552, 535 cm^{-1}. $-$ **HRMS** (ESI, $C_{25}H_{31}O_4N$) = calc.: 410.2326 [M+H]$^+$; found: 410.2314 [M+H]$^+$.

7-(Hexyl(methyl)amino)-5-methoxy-3-(4-methoxybenzyl)-2*H*-chromen-2-one (21jd)

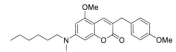

According to **GP10**, 4-(hexyl(methyl)amino)-2-hydroxy-6-methoxybenzaldehyde (**21j**, 50.0 mg, 0.19 mmol, 1.00 equiv.), 4-methoxycinnamaldehyde (76.0 mg, 0.47 mmol, 2.50 equiv.), KOAc (46.0 mg, 0.47 mmol, 2.50 equiv.), glacial acetic acid (3.00 µl, 2.83 mg, 0.05 mmol, 0.25 equiv.), 1,3-dimethylimidazolium dimethyl phosphate (0.04 ml, 50.0 mg, 0.23 mmol, 1.20 equiv.) and MS (powder, 3Å) in toluene (0.62 ml) were heated at 110 °C for 75 min *via* microwave irradiation. Filtration over a small silica plug (*c*Hex/EtOAc 1:1) and purification by prepHPLC (5–95_amino) resulted in the title compound as a pale-yellow oil (5.2 mg, 12.8 µmol, 6.7%).

R_t (5%–95% in 27 min, amino) = 33.9 min. – **^{1}H NMR** (500 MHz, CDCl$_3$) δ = 7.60 (d, ^4J = 1.1 Hz, 1H, C*H*), 7.21 (d, ^3J = 8.6 Hz, 2H, 2 × C*H*$_{ar}$), 6.85 (d, ^3J = 8.6 Hz, 2H, 2 × C*H*$_{ar}$), 6.16 (d, ^4J = 2.2 Hz, 1H, C*H*$_{ar}$), 6.02 (d, ^4J = 2.2 Hz, 1H, C*H*$_{ar}$), 3.85 (s, 3H, OC*H*$_3$), 3.79 (s, 3H, OC*H*$_3$), 3.76 (s, 2H, CC*H*$_2$), 3.34 (t, ^3J = 7.7 Hz, 2H, NC*H*$_2$), 3.00 (s, 3H, NC*H*$_3$), 1.58 (quin., ^3J = 7.4 Hz, 2H, NCH$_2$C*H*$_2$), 1.37 – 1.27 (m, 6H, 3 × C*H*$_2$), 0.89 (t, ^3J = 6.8 Hz, 3H, C*H*$_3$) ppm. – **^{13}C NMR** (126 MHz, CDCl$_3$) δ = 163.2 (C$_q$, *C*OO), 158.3 (C$_q$, *C*$_{ar}$OCH$_3$), 156.8 (C$_q$, 5-*C*$_{ar}$OCH$_3$), 156.3 (C$_q$, 8a-*C*$_{ar}$), 151.6 (C$_q$, 7-*C*$_{ar}$N), 135.4 (+, *C*H), 131.4 (C$_q$, CH$_2$*C*$_{ar}$), 130.2 (+, 2 × *C*$_{ar}$H), 120.9 (C$_q$, 3-*C*), 114.1 (+, 2 × *C*$_{ar}$H), 101.3 (C$_q$, 4a-*C*$_{ar}$), 91.5 (+, *C*$_{ar}$H), 90.9 (+, *C*$_{ar}$H), 55.7 (+, O*C*H$_3$), 55.4 (+, O*C*H$_3$), 53.2 (–, N*C*H$_2$), 39.1 (+, N*C*H$_3$), 35.8 (–, *C*CH$_2$), 31.8 (–, *C*H$_2$), 26.9 (–, *C*H$_2$), 26.8 (–, *C*H$_2$), 22.7 (–, *C*H$_2$), 14.2 (+, *C*H$_3$) ppm. – **IR** (ATR, ṽ) = 2953, 2918, 2870, 2851, 1704, 1663, 1626, 1604, 1558, 1514, 1486, 1465, 1456, 1435, 1398, 1364, 1317, 1303, 1288, 1245, 1205, 1176, 1154, 1136, 1109, 1027, 983, 953, 931, 905, 884, 871, 847, 834, 812, 790, 756, 721, 697, 643, 567, 560, 548, 526, 518, 439, 384 cm^{-1}. – **HRMS** (ESI, C$_{25}$H$_{31}$O$_4$N) = calc.: 410.2326 [M+H]$^+$; found: 410.2307 [M+H]$^+$.

3-(2-Fluorobenzyl)-7-(hexyl(methyl)amino)-5-methoxy-2*H*-chromen-2-one (21je)

According to **GP10**, 4-(hexyl(methyl)amino)-2-hydroxy-6-methoxybenzaldehyde (**21j**, 50.0 mg, 0.19 mmol, 1.00 equiv.), 2-fluorocinnamaldehyde (71.0 mg, 0.47 mmol, 2.50 equiv.), KOAc (46.0 mg, 0.47 mmol, 2.50 equiv.), glacial acetic acid (3.00 µl, 2.83 mg, 0.05 mmol, 0.25 equiv.), 1,3-dimethylimidazolium dimethyl phosphate (0.04 ml, 50.0 mg, 0.23 mmol, 1.20 equiv.) and MS (powder, 3Å) in toluene (0.62 ml) were heated at 110 °C for 75 min *via* microwave irradiation. Filtration over a small silica plug (*c*Hex/EtOAc 1:1) and purification by prepHPLC (5–95_amino) resulted in the title compound as a yellow oil (4.7 mg, 11.9 μmol, 6.3%).

R_t (5%–95% in 27 min, amino) = 34.5 min. – **^{1}H NMR** (500 MHz, CDCl$_3$) δ = 7.67 (d, ^4J = 1.3 Hz, 1H, CH), 7.34 (td, $^{3,4}J$ = 7.6, 1.8 Hz, 1H, *m*-CH_{ar}), 7.25 – 7.19 (m, 1H, *m*-CH_{ar}), 7.08 (td, $^{3,4}J$ = 7.5, 1.2 Hz, 1H *o*–CH_{ar}), 7.04 (ddd, $^{3,3,4}J$ = 9.7, 8.1, 1.2 Hz, *p*–CH_{ar}), 6.23 (d, ^4J = 2.2 Hz, 1H, CH_{ar}), 6.14 (d, ^4J = 2.2 Hz, 1H, CH_{ar}), 3.85 (s, 5H, OCH_3 and CCH_2), 3.38 – 3.32 (m, 2H, NCH_2), 3.02 (s, 3H, NCH_3), 2.50 (bs, 1H, NH), 1.57 (quin., ^3J = 7.3 Hz, 2H, NCH$_2$CH_2), 1.34 – 1.27 (m, 6H, 3 × CH_2), 0.91 – 0.85 (m, 3H, CH_3) ppm. – **^{13}C NMR** (126 MHz, CDCl$_3$) δ = 163.0 (C$_q$, COO), 161.4 (C$_q$, d, ^1J = 245.8 Hz, C$_{ar}$F), 157.0 (C$_q$, 5-C$_{ar}$OCH$_3$), 156.2 (C$_q$, 8a-C$_{ar}$), 150.9 (C$_q$, 7-C$_{ar}$N), 135.8 (+, CH), 131.7 (+, d, ^3J = 4.5 Hz, *m*-C$_{ar}$H), 128.4 (+, d, ^3J = 7.9 Hz, *m*-C$_{ar}$H), 125.9 (C$_q$, d, ^2J = 15.5 Hz, CH$_2$C$_{ar}$), 124.3 (+, d, ^4J = 3.5 Hz, *p*-C$_{ar}$H), 119.5 (C$_q$, 3-C), 115.5 (+, d, ^2J = 21.8 Hz, *o*-C$_{ar}$H), 102.2 (C$_q$, 4a-C$_{ar}$), 92.4 (+, C$_{ar}$H), 91.9 (+, C$_{ar}$H), 55.8 (+, OCH$_3$), 53.9 (–, NCH$_2$), 39.7 (+, NCH$_3$), 31.7 (–, CH$_2$), 29.9 (–, d, ^3J = 3.2 Hz, CCH$_2$), 26.7 (–, 2 × CH$_2$), 22.7 (–, CH$_2$), 14.1 (+, CH$_3$) ppm – **^{19}F NMR** (471 MHz, CDCl$_3$) δ = –75.9 (s, TFA)[13], –117.7 (s, CF) ppm. – **IR** (ATR, ṽ) = 2953, 2922, 2851, 1710, 1604, 1560, 1517, 1489, 1455, 1439, 1400, 1366, 1306, 1258, 1230, 1204, 1179, 1137, 1113, 1096, 1045, 1034, 982, 949, 924, 905, 881, 860, 846, 802, 755, 722, 700, 646, 608, 550 cm^{-1}. – **HRMS** (ESI, C$_{24}$H$_{28}$O$_3$NF) = calc.: 398.2126 [M+H]$^+$; found: 398.2115 [M+H]$^+$.

3-(4-Fluorobenzyl)-7-(hexyl(methyl)amino)-5-methoxy-2*H*-chromen-2-one (21jf)

According to **GP10**, 4-(hexyl(methyl)amino-2-hydroxy-6-methoxybenzaldehyde (**19j**, 50.0 mg, 0.19 mmol, 1.00 equiv.), 4-fluorocinnamaldehyde (71.0 mg, 0.47 mmol, 2.50 equiv.), KOAc (46.0 mg, 0.47 mmol, 2.50 equiv.), glacial acetic acid (3.00 μl, 2.83 mg, 0.05 mmol, 0.25 equiv.), 1,3-dimethylimidazolium dimethyl phosphate (0.04 ml, 50.0 mg, 0.23 mmol, 1.20 equiv.) and MS (powder, 3Å) in toluene (0.62 ml) were heated at 110 °C for 75 min *via* microwave irradiation. Filtration over a small silica plug (*c*Hex/EtOAc 1:1) and purification by prepHPLC (5–95_amino) resulted in the TFA salt of the title compound as a yellow oil (4.4 mg, 11.1 μmol, 5.8%).

R_t (5%–95% in 27 min, amino) = 34.3 min. – **^{1}H NMR** (500 MHz, CDCl$_3$) δ = 7.65 (d, ^4J = 1.2 Hz, 1H, CH), 7.29 – 7.22 (m, 2H, 2 × CH_{ar}), 7.02 – 6.97 (m, 2H, 2 × CH_{ar}), 6.30 (d, ^4J = 2.1 Hz,

---

[13] Only traces of the compound were identified as TFA salt.

1H, C*H*$_{ar}$), 6.26 (d, ^4J = 2.1 Hz, 1H, C*H*$_{ar}$), 3.88 (s, 3H, OC*H*$_3$), 3.80 (s, 2H, CC*H*$_2$), 3.41 – 3.33 (m, 2H, NC*H*$_2$), 3.05 (s, 3H, NC*H*$_3$), 1.63 – 1.53 (m, 2H, NCH$_2$C*H*$_2$), 1.36 – 1.24 (m, 6H, 3 × C*H*$_2$), 0.91 – 0.85 (m, 3H, C*H*$_3$) ppm. – ^{13}C NMR (126 MHz, CDCl$_3$) δ = 162.9 (C$_q$, COO), 161.8 (C$_q$, d, ^1J = 244.3 Hz, *C*$_{ar}$F), 157.0 (C$_q$, 5-*C*$_{ar}$OCH$_3$), 156.1 (C$_q$, 8a-*C*$_{ar}$), 150.2 (C$_q$, 7-*C*$_{ar}$N), 135.5 (+, *C*H), 134.6 (C$_q$, d, ^4J = 3.2 Hz, CH$_2$*C*$_{ar}$), 130.7 (+, d, ^3J = 7.8 Hz, 2 × *m*-*C*$_{ar}$H), 121.8 (C$_q$, 3-*C*), 115.5 (+, d, ^2J = 21.3 Hz, 2 × *o*-*C*$_{ar}$H), 103.1 (C$_q$, 4a-*C*$_{ar}$), 93.2 (+, *C*$_{ar}$H), 92.9 (+, *C*$_{ar}$H), 56.0 (+, OCH$_3$), 54.5 (–, NCH$_2$), 40.4 (+, NCH$_3$), 36.0 (–, CCH$_2$), 31.6 (–, CH$_2$), 26.7 (–, CH$_2$), 26.5 (–, CH$_2$), 22.7 (–, CH$_2$), 14.1 (+, CH$_3$) ppm. – ^{19}F NMR (471 MHz, CDCl$_3$) δ = –75.8 (s, TFA)[14], –116.8 (s, C*F*) ppm. – IR (ATR, ṽ) = 2953, 2922, 2853, 1707, 1602, 1560, 1509, 1486, 1460, 1436, 1398, 1364, 1312, 1258, 1221, 1180, 1156, 1136, 1112, 1094, 1045, 1016, 980, 972, 922, 894, 881, 849, 802, 748, 722, 693, 646, 594, 565, 548, 521, 487 cm^{-1}. – HRMS (ESI, C$_{24}$H$_{28}$O$_3$NF) = calc.: 398.2126 [M+H]$^+$; found: 398.2114 [M+H]$^+$.

7-(Hexyl(methyl)amino)-5-methoxy-3-(2-(trifluoromethyl)benzyl)-2*H*-chromen-2-one (21jg)

According to GP10, 4-(hexyl(methyl)amino)-2-hydroxy-6-methoxybenzaldehyde (19j, 50.0 mg, 0.19 mmol, 1.00 equiv.), 2-(trifluoro)cinnamaldehyde (94.0 mg, 0.47 mmol, 2.50 equiv.), KOAc (46.0 mg, 0.47 mmol, 2.50 equiv.), glacial acetic acid (3.00 µl, 2.83 mg, 0.05 mmol, 0.25 equiv.), 1,3-dimethylimidazolium dimethyl phosphate (0.04 ml, 50.0 mg, 0.23 mmol, 1.20 equiv.) and MS (powder, 3Å) in toluene (0.62 ml) were heated at 110 °C for 75 min *via* microwave irradiation. Filtration over a small silica plug (*c*Hex/EtOAc 1:1) and purification by prepHPLC (5–95_amino) resulted in the title compound as a yellow oil (2.8 mg, 6.33 µmol, 3.3%).

R$_t$ (5%–95% in 27 min, amino) = 35.3 min. – ^{1}H NMR (500 MHz, CDCl$_3$) δ = 7.70 – 7.67 (m, 1H, C*H*), 7.38 (td, $^{3,4}J$ = 7.6, 1.3 Hz, 1H, *m*-C*H*$_{ar}$), 7.45 (s, 1H, C*H*$_{ar}$), 7.40 – 7.32 (m, 2H, 2 × C*H*$_{ar}$), 6.20 (d, ^4J = 2.2 Hz, 1H, C*H*$_{ar}$), 6.03 (d, ^4J = 2.2 Hz, 1H, C*H*$_{ar}$), 4.04 (s, 2H, CC*H*$_2$), 3.82 (s, 3H, OC*H*$_3$), 3.38 – 3.32 (m, 2H, NC*H*$_2$), 3.02 (s, 3H, NC*H*$_3$), 1.58 (quin., ^3J = 7.3 Hz, 2H, NCH$_2$C*H*$_2$), 1.35 – 1.24 (m, 6H, 3 × C*H*$_2$), 0.91 – 0.85 (m, 3H, C*H*$_3$) ppm. – ^{13}C NMR (126 MHz, CDCl$_3$) δ = 163.1 (C$_q$, COO), 156.9 (C$_q$, 5-*C*$_{ar}$OCH$_3$), 156.3 (C$_q$, 8a-*C*$_{ar}$), 151.7 (C$_q$, 7-*C*$_{ar}$N), 137.8 (C$_q$, CH$_2$*C*$_{ar}$), 136.4 (+, *C*H), 132.1 (+, *C*$_{ar}$H), 131.7 (+, *C*$_{ar}$H), 129.1 (C$_q$, d, ^2J = 29.8 Hz, *C*$_{ar}$CF$_3$), 126.7 (+, *C*$_{ar}$H), 126.2 (+, d, ^3J = 5.9 Hz, *o*-*C*$_{ar}$H), 119.3 (C$_q$, 3-*C*), 101.4 (C$_q$,

---

[14] Only traces of the compound were identified as TFA salt.

4a-C_{ar}), 91.6 (+, C_{ar}H), 91.1 (+, C_{ar}H), 55.7 (+, OCH_3), 53.4 (–, NCH_2), 39.2 (+, NCH_3), 32.7 (–, CCH_2), 31.8 (–, CH_2), 26.8 (–, 2 × CH_2), 22.7 (–, CH_2), 14.1 (+, CH_3) ppm. – **^{19}F NMR** (471 MHz, CDCl$_3$) δ = –59.8 (s, CF_3), –75.8 (s, TFA)[15] ppm. – **IR** (ATR, ṽ) = 2953, 2922, 2853, 1710, 1605, 1561, 1517, 1487, 1456, 1400, 1367, 1312, 1258, 1156, 1113, 1060, 1037, 966, 802, 768, 722, 653 cm^{-1}. – **HRMS** (ESI, $C_{25}H_{28}O_3NF_3$) = calc.: 448.2094 [M+H]$^+$; found: 448.2080 [M+H]$^+$.

5.1.8 Synthesis and characterisation of 3-benzyl-7-(1,1-dimethyl)alkylcoumarins

3-Benzyl-7-(*tert*-butyl)-5-methoxy-2*H*-chromen-2-on (21ka)

According to **GP8**, 4-(*tert*-butyl)-2-hydroxy-6-methoxybenzaldehyde (**19k**, 40.0 mg, 0.19 mmol, 1.00 equiv.), cinnamaldehyde (0.06 ml, 63.0 mg, 0.48 mmol, 2.50 equiv.), K_2CO_3 (32.0 mg, 0.23 mmol, 1.20 equiv.) and 1,3-dimethylimidazolium dimethyl phosphate (0.05 ml, 64.0 mg, 0.29 mmol, 1.50 equiv.) in toluene (0.63 ml) were heated at 110 °C for 50 min *via* microwave irradiation. Purification by flash column chromatography (*c*Hex/EtOAc 30:1) resulted in the title compound as a white solid (19.6 mg, 60.1 μmol, 32%).

R_f (*c*Hex/EtOAc 10:1) = 0.41. – **^{1}H NMR** (500 MHz, CDCl$_3$) δ = 7.69 (s, 1H, C*H*), 7.29 – 7.23 (m, 4H, 4 × C*H*$_{ar}$), 7.22 – 7.16 (m, 1H, C*H*$_{ar}$), 6.88 (d, ^4J = 1.5 Hz, 1H, C*H*$_{ar}$), 6.65 (d, ^4J = 1.5 Hz, 1H, C*H*$_{ar}$), 3.84 (s, 3H, OC*H*$_3$), 3.83 (s, 2H, CC*H*$_2$), 1.28 (s, 9H, C(C*H*$_3$)$_3$) ppm. – **^{13}C NMR** (126 MHz, CDCl$_3$) δ = 162.2 (C_q, *C*OO), 156.1 (C_q, 7-C_{ar}C), 155.4 (C_q, 5-C_{ar}OCH$_3$), 154.2 (C_q, 8a-C_{ar}), 138.5 (C_q, CH$_2$$C_{ar}$), 134.6 (+, *C*H), 129.3 (+, 2 × C_{ar}H), 128.7 (+, 2 × C_{ar}H), 126.7 (+, C_{ar}H), 126.3 (C_q, 3-*C*), 107.9 (C_q, 4a-C_{ar}), 106.0 (+, C_{ar}H), 102.7 (+, C_{ar}H), 55.8 (+, O*C*H$_3$), 36.9 (–, C*C*H$_2$), 35.7 (C_q, *C*(CH$_3$)$_3$), 31.2 (+, C(*C*H$_3$)$_3$) ppm. – **IR** (ATR, ṽ) = 3065, 3029, 3001, 2963, 2925, 2865, 1712, 1613, 1494, 1465, 1416, 1313, 1248, 1163, 1114, 1051, 922, 836, 754, 725, 697, 665, 567, 529, 474, 388 cm^{-1}. – **HRMS** (ESI, $C_{21}H_{22}O_3$) = calc.: 323.1642 [M+H]$^+$; found: 323.1636 [M+H]$^+$.

7-(*tert*-Butyl)-5-methoxy-3-(2-methylbenzyl)-2*H*-chromen-2-one (21kb)

According to **GP8**, 4-(*tert*-butyl)-2-hydroxy-6-methoxybenzaldehyde (**19k**, 40.0 mg, 0.19 mmol, 1.00 equiv.), 2-methylcinnamaldehyde (70.0 mg, 0.48 mmol, 2.50 equiv.), K_2CO_3 (32.0 mg, 0.23 mmol, 1.20 equiv.) and 1,3-dimethylimidazolium dimethyl phosphate (0.05 ml, 64.0 mg, 0.29 mmol, 1.50

---

[15] Only traces of the compound were identified as TFA salt.

equiv.) in toluene (0.63 ml) were heated at 110 °C for 50 min *via* microwave irradiation. Purification by flash column chromatography (*c*Hex/EtOAc 30:1) resulted in the title compound as a white solid (44.3 mg, 0.13 mmol, 68%).

R_f (*c*Hex/EtOAc 10:1) = 0.44. – ^{1}H NMR (500 MHz, CDCl$_3$) δ = 7.43 (s, 1H, C*H*), 7.23 – 7.17 (m, 4H, 4 × C*H*$_{ar}$), 6.94 (s, 1H, C*H*$_{ar}$), 6.68 (s, 1H, C*H*$_{ar}$), 3.87 (s, 2H, CC*H*$_2$), 3.84 (s, 3H, OC*H*$_3$), 2.28 (s, 3H, C$_{ar}$C*H*$_3$), 1.33 (s, 9H, C(C*H*$_3$)$_3$) ppm. – ^{13}C NMR (126 MHz, CDCl$_3$) δ = 162.3 (C$_q$, COO), 156.0 (C$_q$, 7-*C*$_{ar}$C), 155.3 (C$_q$, 5-*C*$_{ar}$OCH$_3$), 154.0 (C$_q$, 8a-*C*$_{ar}$), 137.0 (C$_q$, *C*$_{ar}$), 136.3 (C$_q$, *C*$_{ar}$), 134.1 (+, *C*H), 130.6 (+, *C*$_{ar}$H), 130.3(+, *C*$_{ar}$H), 127.1 (+, *C*$_{ar}$H), 126.4 (+, *C*$_{ar}$H), 125.6 (C$_q$, 3-*C*), 107.9 (C$_q$, 4a-*C*$_{ar}$), 106.0 (+, *C*$_{ar}$H), 102.7 (+, *C*$_{ar}$H), 55.8 (+, O*C*H$_3$), 35.7 (C$_q$, *C*(CH$_3$)$_3$), 34.1 (–, *C*H$_2$), 31.2 (+, C(*C*H$_3$)$_3$), 19.7 (+, C$_{ar}$*C*H$_3$) ppm. – IR (ATR, ṽ) = 3068, 3008, 2951, 2868, 1703, 1614, 1493, 1460, 1415, 1362, 1295, 1251, 1227, 1204, 1157, 1109, 1053, 993, 906, 839, 759, 728, 667, 603, 577, 555, 460 cm^{-1}. – HRMS (ESI, C$_{22}$H$_{24}$O$_3$) = calc.: 337.1798 [M+H]$^+$; found: 337.1791 [M+H]$^+$.

7-(*tert*-Butyl)-5-methoxy-3-(2-methoxybenzyl)-2H-chromen-2-one (21kc)

According to **GP8**, 4-(*tert*-butyl)-2-hydroxy-6-methoxybenzaldehyde (**19k**, 40.0 mg, 0.19 mmol, 1.00 equiv.), 2-methoxycinnamaldehyde (78.0 mg, 0.48 mmol, 2.50 equiv.), K$_2$CO$_3$ (32.0 mg, 0.23 mmol, 1.20 equiv.) and 1,3-dimethylimidazolium dimethyl phosphate (0.04 ml, 51.0 mg, 0.23 mmol, 1.20 equiv.) in toluene (0.63 ml) were heated at 110 °C for 50 min *via* microwave irradiation. Purification by flash column chromatography (*c*Hex/EtOAc 10:1) resulted in the title compound as an off-white solid (42.3 mg, 0.12 mmol, 63%).

R_f (*c*Hex/EtOAc 10:1) = 0.25. – ^{1}H NMR (500 MHz, CDCl$_3$) δ = 7.64 (s, 1H, C*H*), 7.28 – 7.21 (m, 2H, 2 × C*H*$_{ar}$), 6.95 – 6.87 (m, 3H, 3 × C*H*$_{ar}$), 6.68 (s, 1H, C*H*$_{ar}$), 3.87 (s, 6H, 2 × OC*H*$_3$), 3.81 (s, 2H, CC*H*$_2$), 1.32 (s, 9H, C(C*H*$_3$)$_3$) ppm. – ^{13}C NMR (126 MHz, CDCl$_3$) δ = 162.8 (C$_q$, COO), 158.2 (C$_q$, *C*$_{ar}$OCH$_3$), 156.1 (C$_q$, 7-*C*$_{ar}$C), 155.8 (C$_q$, 5-*C*$_{ar}$OCH$_3$), 154.5 (C$_q$, 8a-*C*$_{ar}$), 134.7 (+, *C*H), 131.5 (+, *C*$_{ar}$H), 128.5 (+, *C*$_{ar}$H), 127.1 (C$_q$, *C*), 126.1 (C$_q$, *C*), 121.1 (+, *C*$_{ar}$H), 111.1 (+, *C*$_{ar}$H), 108.6 (C$_q$, 4a-*C*$_{ar}$), 106.4 (+, *C*$_{ar}$H), 103.1 (+, *C*$_{ar}$H), 56.3 (+, O*C*H$_3$), 55.9 (+, O*C*H$_3$), 36.1 (C$_q$, *C*(CH$_3$)$_3$), 31.7 (+, C(*C*H$_3$)$_3$), 31.5 (–, C*C*H$_2$) ppm. – IR (ATR, ṽ) = 3064, 3018, 2955, 2922, 2867, 1715, 1613, 1572, 1492, 1462, 1438, 1416, 1298, 1242, 1196, 1163, 1104, 1053, 1027, 928, 849, 810, 753, 701, 669, 584, 556, 444, 389 cm^{-1}. – HRMS (ESI, C$_{22}$H$_{24}$O$_4$) = calc.: 353.1747 [M+H]$^+$; found: 353.1742 [M+H]$^+$.

7-(*tert*-Butyl)-5-methoxy-3-(4-methoxybenzyl)-2*H*-chromen-2-one (21kd)

According to **GP8**, 4-(*tert*-butyl)-2-hydroxy-6-methoxybenzalde-hyde (**19k**, 40.0 mg, 0.19 mmol, 1.00 equiv.), 4-methox-ycinnamaldehyde (78.0 mg, 0.48 mmol, 2.50 equiv.), K_2CO_3 (32.0 mg, 0.23 mmol, 1.20 equiv.) and 1,3-dimethylimidazolium dimethyl phosphate (0.04 ml, 51.0 mg, 0.23 mmol, 1.20 equiv.) in toluene (0.63 ml) were heated at 110 °C for 50 min *via* microwave irradiation. Purification by flash column chromatography (*c*Hex/EtOAc 40:1 to 30:1) resulted in the title compound as an off-white solid (9.1 mg, 0.03 mmol, 13%).

R_f (*c*Hex/EtOAc 10:1) = 0.26. – **^{1}H NMR** (500 MHz, CDCl$_3$) δ = 7.70 (s, 1H, C*H*), 7.21 (d, ^3J = 8.6 Hz, 2H, 2 × *m*-C*H*$_{ar}$), 6.92 (d, ^4J = 1.5 Hz, 1H, C*H*$_{ar}$), 6.86 (d, ^3J = 8.6 Hz, 2H, 2 × *o*-C*H*$_{ar}$), 6.69 (d, ^4J = 1.5 Hz, 1H, C*H*$_{ar}$), 3.89 (s, 3H, OC*H*$_3$), 3.81 (s, 2H, CC*H*$_2$), 3.80 (s, 3H, OC*H*$_3$), 1.32 (s, 9H, C(C*H*$_3$)$_3$) ppm. – **^{13}C NMR** (126 MHz, CDCl$_3$) δ = 162.3 (C$_q$, *C*OO), 158.4 (C$_q$, *C*$_{ar}$OCH$_3$), 156.0 (C$_q$, 7-*C*$_{ar}$C), 155.4 (C$_q$, 5-*C*$_{ar}$OCH$_3$), 154.2 (C$_q$, 8a-*C*$_{ar}$), 134.3 (+,*C*H), 130.5 (C$_q$, CH$_2$*C*$_{ar}$), 130.3 (+, 2 × *m*-*C*$_{ar}$H), 126.8 (C$_q$, 3-*C*), 114.2 (+, 2 × *o*-*C*$_{ar}$H), 108.0 (C$_q$, 4a-*C*$_{ar}$), 106.0 (+, *C*$_{ar}$H), 102.7 (+, *C*$_{ar}$H), 55.9 (+, O*C*H$_3$), 55.4 (+, O*C*H$_3$), 36.1 (–, *C*CH$_2$), 35.7 (C$_q$, *C*(CH$_3$)$_3$), 31.2 (+, C(*C*H$_3$)$_3$) ppm. – **IR** (ATR, ṽ) = 2962, 2953, 2927, 2902, 2867, 2854, 2836, 1707, 1613, 1581, 1510, 1493, 1480, 1463, 1451, 1429, 1417, 1377, 1361, 1299, 1245, 1230, 1162, 1111, 1105, 1050, 1035, 1027, 997, 962, 953, 936, 926, 902, 843, 836, 822, 806, 788, 771, 762, 751, 724, 686, 666, 582, 561, 533, 510, 436, 387 cm^{-1}. – **HRMS** (ESI, C$_{22}$H$_{24}$O$_4$) = calc.: 353.1747 [M+H]$^+$; found: 353.1738 [M+H]$^+$.

7-(*tert*-Butyl)-3-(2-fluorobenzyl)-5-methoxy-2*H*-chromen-2-one (21ke)

According to **GP8**, 4-(*tert*-butyl)-2-hydroxy-6-methoxybenzaldehyde (**19k**, 40.0 mg, 0.19 mmol, 1.00 equiv.), 2-fluorocinnamaldehyde (72.0 mg, 0.48 mmol, 2.50 equiv.), K_2CO_3 (32.0 mg, 0.23 mmol, 1.20 equiv.) and 1,3-dimethylimidazolium dimethyl phosphate (0.05 ml, 64.0 mg, 0.29 mmol, 1.50 equiv.) in toluene (0.63 ml) were heated at 110 °C for 50 min *via* microwave irradiation. Purification by flash column chromatography (*c*Hex/EtOAc 25:1) resulted in the title compound as an off-white solid (19.5 mg, 56.9 μmol, 30%).

R_f (*c*Hex/EtOAc 10:1) = 0.25. – **^{1}H NMR** (500 MHz, CDCl$_3$) δ = 7.75 (s, 1H, C*H*), 7.39 – 7.31 (m, 1H, C*H*$_{ar}$), 7.25 – 7.19 (m, 1H, C*H*$_{ar}$), 7.10 (t, ^3J = 7.5 Hz, 1H, C*H*$_{ar}$), 7.05 (t, ^3J = 9.3 Hz, 1H, C*H*$_{ar}$), 6.92 (d, ^4J = 1.6 Hz, 1H, C*H*$_{ar}$), 6.69 (d, ^4J = 1.6 Hz, 1H, C*H*$_{ar}$), 3.90 (s, 2H, CC*H*$_2$), 3.88 (s, 3H, OC*H*$_3$), 1.32 (s, 9H, C(C*H*$_3$)$_3$) ppm. – **^{13}C NMR** (126 MHz, CDCl$_3$) δ = 162.1 (C$_q$,

COO), 161.4 (C$_q$, d, ^1J = 246.0 Hz, C$_{ar}$F), 156.2 (C$_q$, 7-C$_{ar}$C), 155.4 (C$_q$, 5-C$_{ar}$OCH$_3$), 154.2 (C$_q$, 8a-C$_{ar}$), 134.8 (+, CH), 131.7 (+, d, ^3J = 4.4 Hz, m-C$_{ar}$H), 128.6 (+, d, ^3J = 8.0 Hz, m-C$_{ar}$H), 125.4 (C$_q$, d, ^2J = 15.4 Hz, o-C$_{ar}$), 124.7 (C$_q$, 3-C), 124.3 (+, d, ^4J = 3.6 Hz, p-C$_{ar}$H), 115.5 (+, d, ^2J = 21.7 Hz, o-C$_{ar}$H), 107.9 (C$_q$, 4a-C$_{ar}$), 106.0 (+, C$_{ar}$H), 102.7 (+, C$_{ar}$H), 55.9 (+, OCH$_3$), 35.7 (C$_q$, C(CH$_3$)$_3$), 31.2 (+, C(CH$_3$)$_3$), 30.2 (–, d, ^3J = 3.1 Hz, CCH$_2$) ppm. – ^{19}F NMR (471 MHz, CDCl$_3$) δ = –117.6 (s, CF) ppm. – IR (ATR, ṽ) = 2963, 2924, 2868, 1704, 1616, 1492, 1450, 1416, 1362, 1293, 1248, 1228, 1161, 1112, 1052, 995, 939, 835, 756, 666, 602, 575, 555, 517, 475, 434, 385 cm^{-1}. – HRMS (ESI, C$_{21}$H$_{21}$O$_3$F) = calc.: 341.1547 [M+H]$^+$; found: 341.1541 [M+H]$^+$.

7-(*tert*-Butyl)-3-(4-fluorobenzyl)-5-methoxy-2*H*-chromen-2-one (21kf)

According to GP8, 4-(*tert*-butyl)-2-hydroxy-6-methoxybenzalde-hyde (19k, 40.0 mg, 0.19 mmol, 1.00 equiv.), 4-fluorocinnamalde-hyde (0.06 ml, 72.0 mg, 0.48 mmol, 2.50 equiv.), K$_2$CO$_3$ (32.0 mg, 0.23 mmol, 1.20 equiv.) and 1,3-dimethylimidazolium dimethyl phosphate (0.05 ml, 64.0 mg, 0.29 mmol, 1.50 equiv.) in toluene (0.63 ml) were heated at 110 °C for 50 min *via* microwave irradiation. Purification by flash column chromatography (*c*Hex/EtOAc 30:1) resulted in the title compound as a white solid (29.5 mg, 86.7 µmol, 45%).

R_f (*c*Hex/EtOAc 10:1) = 0.33. – ^{1}H NMR (500 MHz, CDCl$_3$) δ = 7.72 (s, 1H, CH), 7.30 – 7.29 (m, 2H, 2 × CH_{ar}), 6.99 (t, ^3J = 8.7 Hz, 2H, 2 × CH_{ar}), 6.92 (d, ^4J = 1.5 Hz, 1H, CH_{ar}), 6.71 (d, ^4J = 1.5 Hz, 1H, CH_{ar}), 3.90 (s, 3H, OCH_3), 3.83 (s, 2H, CCH_2), 1.33 (s, 9H, C(CH_3)$_3$) ppm. – ^{13}C NMR (126 MHz, CDCl$_3$) δ = 162.1 (C$_q$, COO), 161.8 (C$_q$, d, ^1J = 244.5 Hz, CF), 156.2 (C$_q$, 7-C$_{ar}$C), 155.4 (C$_q$, 5-C$_{ar}$OCH$_3$), 154.2 (C$_q$, 8a-C$_{ar}$), 134.6 (+, CH), 134.2 (C$_q$, d, ^4J = 3.2 Hz, p-C$_{ar}$), 130.7 (+, 2 × m-C$_{ar}$H), 126.1 (C$_q$, 3-C), 115.4 (+, d, ^2J = 21.3 Hz, 2 × o-C$_{ar}$H), 107.8 (C$_q$, 4a-C$_{ar}$), 106.0 (+, C$_{ar}$H), 102.8 (+, C$_{ar}$H), 55.9 (+, OCH$_3$), 36.2 (–, CCH_2), 35.7 (C$_q$, C(CH$_3$)$_3$), 31.2 (+, C(CH$_3$)$_3$) ppm. – IR (ATR, ṽ) = 3061, 3005, 2969, 2931, 2903, 2869, 1715, 1614, 1509, 1467, 1418, 1314, 1224, 1157, 1107, 1049, 996, 840, 778, 757, 724, 665, 578, 558, 531, 480, 452, 392 cm^{-1}. – HRMS (ESI, C$_{21}$H$_{22}$O$_3$F) = calc.: 341.1547 [M+H]$^+$; found: 341.1542 [M+H]$^+$.

7-(*tert*-Butyl)-5-methoxy-3-(2-(trifluoromethyl)benzyl)-2*H*-chromen-2-one (21kg)

According to GP8, 4-(*tert*-butyl)-2-hydroxy-6-methoxybenzaldehyde (19k, 40.0 mg, 0.19 mmol, 1.00 equiv.), 2-(trifluoromethyl)cinnamal-dehyde (96.0 mg, 0.48 mmol, 2.50 equiv.), K$_2$CO$_3$ (32.0 mg, 0.23

mmol, 1.20 equiv.) and 1,3-dimethylimidazolium dimethyl phosphate (0.05 ml, 64.0 mg, 0.29 mmol, 1.50 equiv.) in toluene (0.63 ml) were heated at 110 °C for 50 min *via* microwave irradiation. Purification by flash column chromatography (*c*Hex/EtOAc 30:1 to 10:1) resulted in the title compound as an off-white solid (19.0 mg, 58.2 μmol, 31%).

R_f (*c*Hex/EtOAc 10:1) = 0.35. – **^{1}H NMR** (500 MHz, CDCl$_3$) δ = 7.70 (d, ^3J = 7.8 Hz, 1H, CH_{ar}), 7.52 – 7.47 (m, 2H, CH and CH_{ar}), 7.38 (t, ^3J = 8.1 Hz, 2H, 2 × CH_{ar}), 6.95 (d, ^4J = 1.5 Hz, 1H, CH_{ar}), 6.69 (d, ^4J = 1.5 Hz, 1H, CH_{ar}), 4.08 (s, 2H, CH_2), 3.85 (s, 3H, OCH_3), 1.33 (s, 9H, C(CH_3)$_3$) ppm. – **^{13}C NMR** (126 MHz, CDCl$_3$) δ = 162.2 (C$_q$, COO), 156.4 (C$_q$, 7-C_{ar}C), 155.4 (C$_q$, 5-C_{ar}OCH$_3$), 154.1 (C$_q$, 8a-C_{ar}), 136.9 (C$_q$, d, ^3J = 1.9 Hz, C_{ar}), 135.2 (+, CH), 132.2 (+, C_{ar}H), 131.9 (+, C_{ar}H), 129.2 (C$_q$, q, ^2J = 29.9 Hz, C_{ar}CF$_3$), 127.0 (+, C_{ar}H), 126.4 (+, q, ^3J = 5.7 Hz, C_{ar}H), 125.3 (C$_q$, 3-C), 124.5 (C$_q$, q, ^1J = 273.9 Hz, CF$_3$), 107.8 (C$_q$, 4a-C_{ar}), 106.0 (+, C_{ar}H), 102.8 (+, C_{ar}H), 55.8 (+, OCH$_3$), 35.7 (C$_q$, C(CH$_3$)$_3$), 33.1 (–, d, ^4J = 2.1 Hz, CCH$_2$), 31.2 (+, C(CH$_3$)$_3$) ppm. – **IR** (ATR, ṽ) = 3077, 2962, 2871, 1706, 1613, 1496, 1454, 1418, 1310, 1252, 1229, 1157, 1103, 1062, 1036, 950, 906, 872, 842, 765, 737, 668, 652, 569 cm^{-1}. – **HRMS** (ESI, C$_{22}$H$_{21}$O$_3$F$_3$) = calc.: 391.1516 [M+H]$^+$; found: 391.1504 [M+H]$^+$.

3-Benzyl-5-methoxy-7-(2-methylhexan-2-yl)-2*H*-chromen-2-one (21la)

According to **GP8**, 2-hydroxy-6-methoxy-4-(2-methylhexan-2-yl)benzaldehyde (**19l**, 39.0 mg, 0.16 mmol, 1.00 equiv.), cinnamaldehyde (0.05 ml, 53.0 mg, 0.40 mmol, 2.50 equiv.), K$_2$CO$_3$ (27.0 mg, 0.19 mmol, 1.20 equiv.) and 1,3-dimethylimidazolium dimethyl phosphate (0.04 ml, 53.0 mg, 0.24 mmol, 1.50 equiv.) in toluene (0.53 ml) were heated at 110 °C for 50 min *via* microwave irradiation. Purification by flash column chromatography (*c*Hex/EtOAc 40:1 to 35:1) resulted in the title compound as a yellow oil (10.9 mg, 29.6 μmol, 19%).

R_f (*c*Hex/EtOAc 10:1) = 0.17. – **^{1}H NMR** (500 MHz, CDCl$_3$) δ = 7.73 (s, 1H, CH), 7.36 – 7.29 (m, 4H, 4 × CH_{ar}), 7.25 – 7.19 (m, 1H, CH_{ar}), 6.87 (d, ^4J = 1.5 Hz, 1H, CH_{ar}), 6.63 (d, ^4J = 1.5 Hz, 1H, CH_{ar}), 3.89 – 3.85 (m, 5H, 3-CCH_2 and OCH_3), 1.61 – 1.57 (m, 2H, 7-CCH_2), 1.29 (s, 6H, C(CH_3)$_2$), 1.26 – 1.15 (m, 2H, CH_2), 1.04 – 0.96 (m, 2H, CH_2), 0.81 (t, ^3J = 7.3 Hz, 3H, CH_3) ppm. – **^{13}C NMR** (126 MHz, CDCl$_3$) δ = 162.3 (C$_q$, COO), 155.4 (C$_q$, 5-C_{ar}OCH$_3$), 154.9 (C$_q$, 7-C_{ar}C), 154.2 (C$_q$, 8a-C_{ar}), 138.6 (C$_q$, CH$_2C_{ar}$), 134.7 (+, CH), 129.3 (+, 2 × C_{ar}H), 128.7 (+, 2 × C_{ar}H), 126.7 (+, C_{ar}H), 126.3 (C$_q$, 3-C), 107.9 (C$_q$, 4a-C_{ar}), 106.7 (+, C_{ar}H), 103.1 (+, C_{ar}H), 55.9 (+, OCH$_3$), 44.2 (–, 7-CCH$_2$), 38.7 (C$_q$, C(CH$_3$)$_2$), 36.9 (–, 3-CCH$_2$), 29.0 (+, C(CH$_3$)$_2$), 27.0 (–, CH$_2$), 23.4 (–, CH$_2$), 14.1 (+, CH$_3$) ppm. – **IR** (ATR, ṽ) = 3341, 2957, 2928, 2859, 1721, 1685, 1615, 1569, 1494, 1454, 1418, 1349, 1254, 1164, 1115, 1049, 844, 755, 735,

700, 673, 597 cm^{-1}. – **HRMS** (ESI, C$_{24}$H$_{29}$O$_3$) calc.: 365.2111 [M+H]$^+$; found: 365.2101 [M+H]$^+$.

5-Methoxy-3-(2-methylbenzyl)-7-(2-methylhexan-2-yl)-2*H*-chromen-2-on (21lb)

According to **GP8**, 2-hydroxy-6-methoxy-4-(2-methylhexan-2-yl)benzaldehyde (**19l**, 40.0 mg, 0.16 mmol, 1.00 equiv.), 2-methylcinnamaldehyde (58.0 mg, 0.40 mmol, 2.50 equiv.), K$_2$CO$_3$ (27.0 mg, 0.19 mmol, 1.20 equiv.) and 1,3-dimethylimidazolium dimethyl phosphate (0.04 ml, 53.0 mg, 0.24 mmol, 1.50 equiv.) in toluene (0.53 ml) were heated at 110 °C for 50 min *via* microwave irradiation. Purification by flash column chromatography (*c*Hex/EtOAc 40:1 to 20:1) resulted in the title compound as a white solid (29.0 mg, 77.3 µmol, 48%, purity ≥92%).

R$_f$ (*c*Hex/EtOAc 10:1) = 0.42. – **^{1}H NMR** (500 MHz, CDCl$_3$) δ = 7.44 (d, ^4J = 1.5 Hz, 1H, C*H*), 7.24 – 7.16 (m, 4H, 4 × C*H*$_{ar}$), 6.90 (d, ^4J = 1.5 Hz, 1H, C*H*$_{ar}$), 6.63 (d, ^4J = 1.5 Hz, 1H, C*H*$_{ar}$), 3.87 (d, ^4J = 1.5 Hz, 2H, 3-CC*H*$_2$), 3.83 (s, 3H, OC*H*$_3$), 2.29 (s, 3H, C$_{ar}$C*H*$_3$), 1.63 – 1.57 (m, 2H, 7-CC*H*$_2$), 1.30 (s, 6H, C(C*H*$_3$)$_2$), 1.21 (quin., ^3J = 7.4 Hz, 2H, C*H*$_2$), 1.06 – 0.98 (m, 2H, C*H*$_2$), 0.81 (t, ^3J = 7.4 Hz, 3H, C*H*$_3$) ppm. – **^{13}C NMR** (126 MHz, CDCl$_3$) δ = 162.3 (C$_q$, *C*OO), 155.3 (C$_q$, 5-C$_{ar}$OCH$_3$), 154.8 (C$_q$, 7-C$_{ar}$C), 154.0 (C$_q$, 8a-C$_{ar}$), 136.9 (C$_q$, C$_{ar}$), 136.2 (C$_q$, C$_{ar}$), 134.1 (+, *C*H), 130.6 (+, C$_{ar}$H), 130.3 (+, C$_{ar}$H), 127.1 (+, C$_{ar}$H), 126.3 (+, C$_{ar}$H), 125.5 (C$_q$, 3-*C*), 107.8 (C$_q$, 4a-C$_{ar}$), 106.6 (+, C$_{ar}$H), 103.1 (+, C$_{ar}$H), 55.8 (+, OCH$_3$), 44.2 (–, 7-C*C*H$_2$), 38.7 (C$_q$, *C*(CH$_3$)$_2$), 34.1 (–, 3-C*C*H$_2$), 29.0 (+, C(*C*H$_3$)$_2$), 27.0 (–, *C*H$_2$), 23.4 (–, *C*H$_2$), 19.6 (+, C$_{ar}$*C*H$_3$), 14.1 (+, CH$_2$*C*H$_3$) ppm. – **IR** (ATR, ṽ) = 2955, 2929, 2869, 1702, 1614, 1569, 1493, 1443, 1415, 1289, 1248, 1191, 1161, 1104, 1045, 947, 919, 895, 838, 760, 743, 724, 670, 597, 549, 461, 384 cm^{-1}. – **HRMS** (ESI, C$_{25}$H$_{30}$O$_3$) = calc.: 379.2268 [M+H]$^+$; found: 379.2258 [M+H]$^+$.

5-Methoxy-3-(2-methoxybenzyl)-7-(2-methylhexan-2-yl)-2*H*-chromen-2-one (21lc)

According to **GP8**, 2-hydroxy-6-methoxy-4-(2-methylhexan-2-yl)benzaldehyde (**19l**, 40.0 mg, 0.16 mmol, 1.00 equiv.), 2-methoxycinnamaldehyde (65.0 mg, 0.40 mmol, 2.50 equiv.), K$_2$CO$_3$ (27.0 mg, 0.19 mmol, 1.20 equiv.) and 1,3-dimethylimidazolium dimethyl phosphate (0.04 ml, 53.0 mg, 0.24 mmol, 1.50 equiv.) in toluene (0.53 ml) were heated at 110 °C for 50 min *via* microwave irradiation. Purification by flash column chromatography (*c*Hex/EtOAc

35:1 to 25:1) resulted in the title compound as a brown solid (23.0 mg, 58.4 µmol, 37%, purity ≥90%).

R_f (cHex/EtOAc 10:1) = 0.29. – **^{1}H NMR** (500 MHz, CDCl$_3$) δ = 7.65 (d, ^4J = 1.4 Hz, 1H, C*H*), 7.30 – 7.21 (m, 2H, 2 × C*H*$_{ar}$), 6.96 – 6.83 (m, 3H, 3 × C*H*$_{ar}$), 6.62 (d, ^4J = 1.6 Hz, 1H, C*H*$_{ar}$), 3.87 (s, 2H, 3-CC*H*$_2$), 3.86 (s, 3H, OC*H*$_3$), 3.81 (s, 3H, OC*H*$_3$), 1.62 – 1.55 (m, 2H, 7-CC*H*$_2$), 1.29 (s, 6H, C(C*H*$_3$)$_2$), 1.27 – 1.17 (m, 2H, C*H*$_2$), 1.06 – 0.96 (m, 2H, C*H*$_2$), 0.82 (t, ^3J = 7.3 Hz, 3H, C*H*$_3$) ppm. – **^{13}C NMR** (126 MHz, CDCl$_3$) δ = 162.4 (C$_q$, COO), 157.7 (C$_q$, C$_{ar}$OCH$_3$), 155.3 (C$_q$, 5-C$_{ar}$OCH$_3$), 154.5 (C$_q$, 7-C$_{ar}$C), 154.1 (C$_q$, 8a-C$_{ar}$), 134.3 (+, *C*H), 131.1 (+, C$_{ar}$H), 128.1 (+, C$_{ar}$H), 126.7 (C$_q$, *C*), 125.6 (C$_q$, *C*), 120.7 (+, C$_{ar}$H), 110.6 (+, C$_{ar}$H), 108.1 (C$_q$, 4a-C$_{ar}$), 106.6 (+, C$_{ar}$H), 103.0 (+, C$_{ar}$H), 55.8 (+, OCH$_3$), 55.5 (+, OCH$_3$), 44.2 (–, 7-C*C*H$_2$), 38.7 (C$_q$, *C*(CH$_3$)$_2$), 31.1 (–, 3-C*C*H$_2$), 29.0 (+, C(*C*H$_3$)$_2$), 27.0 (–, *C*H$_2$), 23.4 (–, *C*H$_2$), 14.1 (+, *C*H$_3$) ppm. – **IR** (ATR, ṽ) = 2955, 2928, 2859, 1718, 1614, 1569, 1493, 1461, 1416, 1347, 1289, 1244, 1194, 1159, 1114, 1031, 923, 842, 752, 672, 605, 561, 469 cm^{-1}. – **HRMS** (ESI, C$_{25}$H$_{30}$O$_4$) = calc.: 395.2217 [M+H]$^+$; found: 395.2205 [M+H]$^+$.

5-Methoxy-3-(4-methoxybenzyl)-7-(2-methylhexan-2-yl)-2*H*-chromen-2-one (21ld)

According to **GP8**, 2-hydroxy-6-methoxy-4-(2-methylhexan-2-yl)benzaldehyde (**19l**, 40.0 mg, 0.16 mmol, 1.00 equiv.), 4-methoxycinnamaldehyde (65.0 mg, 0.40 mmol, 2.50 equiv.), K$_2$CO$_3$ (27.0 mg, 0.19 mmol, 1.20 equiv.) and 1,3-dimethylimidazolium dimethyl phosphate (0.03 ml, 43.0 mg, 0.19 mmol, 1.20 equiv.) in toluene (0.53 ml) were heated at 110 °C for 50 min *via* microwave irradiation. Purification by flash column chromatography (cHex/EtOAc 40:1 to 30:1) resulted in the title compound as a cloudy oil (16.7 mg, 0.04 mmol, 26%).

R_f (cHex/EtOAc 10:1) = 0.31. – **^{1}H NMR** (500 MHz, CDCl$_3$) δ = 7.70 (s, 1H, C*H*), 7.24 – 7.20 (m, 2H, 2 × *m*-C*H*$_{ar}$), 6.87 (s, 2H, 2 × *o*-C*H*$_{ar}$), 6.86 – 6.85 (m, 1H, C*H*$_{ar}$), 6.63 (d, ^4J = 1.5 Hz, 1H, C*H*$_{ar}$), 3.88 (s, 3H, OC*H*$_3$), 3.81 (d, ^4J = 1.2 Hz, 2H, CC*H*$_2$), 3.80 (s, 3H, OC*H*$_3$), 1.62 – 1.55 (m, 2H, 7-CC*H*$_2$), 1.29 (s, 6H, C(C*H*$_3$)$_2$), 1.21 (q, ^3J = 7.4 Hz, 2H, C*H*$_2$), 1.05 – 0.97 (m, 2H, C*H*$_2$CH$_3$), 0.81 (t, ^3J = 7.3 Hz, 3H, CH$_2$C*H*$_3$) ppm. – **^{13}C NMR** (126 MHz, CDCl$_3$) δ = 162.3 (C$_q$, COO), 158.4 (C$_q$, C$_{ar}$OCH$_3$), 155.4 (C$_q$, 5-C$_{ar}$OCH$_3$), 154.8 (C$_q$, 7-C$_{ar}$C), 154.2 (C$_q$, 8a-C$_{ar}$), 134.4 (+, *C*H), 130.6 (C$_q$, CH$_2$*C*$_{ar}$), 130.3 (+, 2 × *m*-C$_{ar}$H), 126.7 (C$_q$, 3-*C*), 114.2 (+, 2 × *o*-C$_{ar}$H), 107.9 (C$_q$, 4a-C$_{ar}$), 106.7 (+, C$_{ar}$H), 103.1 (+, C$_{ar}$H), 55.9 (+, OCH$_3$), 55.4 (+, OCH$_3$), 44.2 (–, 7-C*C*H$_2$), 38.8 (C$_q$, *C*(CH$_3$)$_2$), 36.1 (–, 3-C*C*H$_2$), 29.0 (+, C(*C*H$_3$)$_2$), 27.0 (–, *C*H$_2$), 23.4 (–, *C*H$_2$), 14.1 (+, *C*H$_3$) ppm. – **IR** (ATR, ṽ) = 2955, 2928, 2868, 2859, 2837, 1717, 1613, 1585,

1568, 1510, 1493, 1462, 1417, 1388, 1377, 1366, 1343, 1313, 1300, 1289, 1245, 1208, 1176, 1162, 1113, 1040, 952, 935, 922, 898, 874, 843, 817, 790, 766, 752, 727, 687, 671, 636, 586, 562, 538, 533, 513 cm^{-1}. – **HRMS** (ESI, C$_{25}$H$_{30}$O$_4$) = calc.: 395.2217 [M+H]$^+$; found: 395.2208 [M+H]$^+$.

3-(2-Fluorobenzyl)-5-methoxy-7-(2-methylhexan-2-yl)-2H-chromen-2-one (21le)

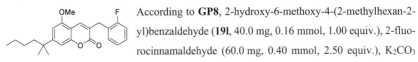

According to **GP8**, 2-hydroxy-6-methoxy-4-(2-methylhexan-2-yl)benzaldehyde (**19l**, 40.0 mg, 0.16 mmol, 1.00 equiv.), 2-fluorocinnamaldehyde (60.0 mg, 0.40 mmol, 2.50 equiv.), K$_2$CO$_3$ (27.0 mg, 0.19 mmol, 1.20 equiv.) and 1,3-dimethylimidazolium dimethyl phosphate (0.04 ml, 53.0 mg, 0.24 mmol, 1.50 equiv.) in toluene (0.53 ml) were heated at 110 °C for 50 min *via* microwave irradiation. Purification by flash column chromatography (*c*Hex/EtOAc 40:1 to 30:1) resulted in the title compound as a yellow oil (13.9 mg, 36.5 µmol, 23%).

R_f (*c*Hex/EtOAc 10:1) = 0.48. – **^{1}H NMR** (500 MHz, CDCl$_3$) δ = 7.75 (d, ^4J = 1.3 Hz, 1H, CH), 7.36 (td, $^{3,4}J$ = 7.6, 1.7 Hz, 1H, CH_{ar}), 7.23 (tdd, $^{3,3,4}J$ = 7.6, 5.3, 1.7 Hz, 1H, CH_{ar}), 7.10 (td, $^{3,4}J$ = 7.6, 1.2 Hz, 1H, CH_{ar}), 7.07 – 7.02 (m, 1H, CH_{ar}), 6.87 (d, ^4J = 1.5 Hz, 1H, CH_{ar}), 6.64 (d, ^4J = 1.5 Hz, 1H, CH_{ar}), 3.90 (s, 2H, 3-CCH_2), 3.88 (s, 3H, OCH_3), 1.65 – 1.56 (m, 2H, 7-CCH_2), 1.29 (s, 6H, C(CH_3)$_2$), 1.27 – 1.16 (m, 2H, CH_2), 1.06 – 0.96 (m, 2H, CH_2), 0.81 (t, ^3J = 7.3 Hz, 3H, CH_3) ppm. – **^{13}C NMR** (126 MHz, CDCl$_3$) δ = 162.1 (C$_q$, COO), 161.4 (C$_q$, d, ^1J = 246.1 Hz, C$_{ar}$F), 155.4 (C$_q$, 5-C$_{ar}$OCH$_3$), 155.0 (C$_q$, 7-C$_{ar}$C), 154.2 (C$_q$, 8a-C$_{ar}$), 134.8 (+, CH), 131.8 (+, d, ^3J = 4.4 Hz, m-C$_{ar}$H), 128.6 (+, d, ^3J = 8.1 Hz, m-C$_{ar}$H), 125.4 (C$_q$, d, ^2J = 15.4 Hz, o-C$_{ar}$), 124.6 (C$_q$, 3-C), 124.3 (+, d, ^4J = 3.6 Hz, p-C$_{ar}$H), 115.5 (+, d, ^2J = 21.8 Hz, o-C$_{ar}$H), 107.9 (C$_q$, 4a-C$_{ar}$), 106.7 (+, C$_{ar}$H), 103.1 (+, C$_{ar}$H), 55.9 (+, OCH$_3$), 44.2 (–, 7-CCH_2), 38.8 (C$_q$, C(CH$_3$)$_2$), 30.2 (–, d, ^3J = 3.1 Hz, 3-CCH_2), 29.0 (+, C(CH_3)$_2$), 27.0 (–, CH_2), 23.4 (–, CH_2), 14.1 (+, CH_3) ppm. – **IR** (ATR, ṽ) = 2956, 2929, 2859, 1720, 1614, 1569, 1491, 1455, 1416, 1290, 1247, 1231, 1162, 1114, 1049, 925, 843, 756, 672, 603, 576, 553, 436 cm^{-1}. – **HRMS** (ESI, C$_{24}$H$_{27}$O$_3$F) = calc.: 383.2017 [M+H]$^+$; found: 383.2004 [M+H]$^+$.

3-(4-Fluorobenzyl)-5-methoxy-7-(2-methylhexan-2-yl)-2H-chromen-2-one (21lf)

According to **GP8**, 2-hydroxy-6-methoxy-4-(2-methylhexan-2-yl)benzaldehyde (**19l**, 40.0 mg, 0.16 mmol, 1.00 equiv.), 4-fluorocinnamaldehyde (0.05 ml, 60.0 mg, 0.40 mmol, 2.50 equiv.), K$_2$CO$_3$ (27.0 mg, 0.19 mmol, 1.20 equiv.) and 1,3-dimethylimidazolium dimethyl phosphate (0.04 ml, 53.0 mg, 0.24 mmol, 1.50 equiv.) in toluene (0.53 ml) were heated at 110 °C for 50 min *via* microwave irradiation. Purification by flash column chromatography (*c*Hex/EtOAc 40:1 to 30:1) resulted in the title compound as an off-white solid (8.5 mg, 22.0 µmol, 13%).

R_f (*c*Hex/EtOAc 10:1) = 0.44. – **^{1}H NMR** (500 MHz, CDCl$_3$) δ = 7.72 (s, 1H, C*H*), 7.28 – 7.24 (m, 2H, 2 × C*H*$_{ar}$), 7.02 – 6.97 (m, 2H, 2 × C*H*$_{ar}$), 6.87 (d, ^4J = 1.5 Hz, 1H, C*H*$_{ar}$), 6.64 (d, ^4J = 1.5 Hz, 1H, C*H*$_{ar}$), 3.89 (s, 3H, OC*H*$_3$), 3.83 (s, 2H, 3-CC*H*$_2$), 1.62 – 1.56 (m, 2H, 7-CC*H*$_2$), 1.29 (s, 6H, C(C*H*$_3$)$_2$), 1.28 – 1.17 (m, 2H, C*H*$_2$), 1.05 – 0.97 (m, 2H, C*H*$_2$), 0.81 (t, ^3J = 7.3 Hz, 3H, C*H*$_3$) ppm. – **^{13}C NMR** (126 MHz, CDCl$_3$) δ = 162.1 (C$_q$, *C*OO), 161.8 (C$_q$, d, ^1J = 244.4 Hz, *C*F), 155.4 (C$_q$, 7-*C*$_{ar}$C), 155.1 (C$_q$, 5-*C*$_{ar}$OCH$_3$), 154.2 (C$_q$, 8a-*C*$_{ar}$), 134.7 (+, *C*H), 134.2 (C$_q$, d, ^4J = 3.2 Hz, *p*-*C*$_{ar}$), 130.7 (+, d, ^3J = 8.0 Hz, 2 × *m*-*C*$_{ar}$H), 126.1 (C$_q$, 3-*C*), 115.5 (+, d, ^2J = 21.3 Hz, 2 × *o*-*C*$_{ar}$), 107.8 (C$_q$, 4a-*C*$_{ar}$), 106.7 (+, *C*$_{ar}$H), 103.2 (+, *C*$_{ar}$H), 55.9 (+, O*C*H$_3$), 44.2 (–, 7-C*C*H$_2$), 38.8 (C$_q$, *C*(CH$_3$)$_2$), 36.2 (–, 3-C*C*H$_2$), 29.0 (+, C(*C*H$_3$)$_2$), 27.0 (–, *C*H$_2$), 23.4 (–, *C*H$_2$), 14.1 (+, *C*H$_3$) ppm. – **IR** (ATR, ṽ) = 2931, 1711, 1613, 1567, 1508, 1464, 1417, 1343, 1315, 1250, 1221, 1159, 1112, 1045, 1000, 922, 854, 839, 778, 760, 724, 667, 549, 483, 389 cm^{-1}. – **HRMS** (ESI, C$_{24}$H$_{27}$O$_3$F) = calc.: 383.2017 [M+H]$^+$; found: 383.2010 [M+H]$^+$.

5-Methoxy-7-(2-methylhexan-2-yl)-3-(2-(trifluoromethyl)benzyl)-2*H*-chromen-2-one (21lg)

According to **GP8**, 2-hydroxy-6-methoxy-4-(2-methylhexan-2-yl)benzaldehyde (**21l**, 40.0 mg, 0.16 mmol, 1.00 equiv.), 2-(tri-fluoromethyl)cinnamaldehyde (80.0 mg, 0.40 mmol, 2.50 equiv.), K$_2$CO$_3$ (27.0 mg, 0.19 mmol, 1.20 equiv.) and 1,3-dimethylimidazolium dimethyl phosphate (0.04 ml, 53.0 mg, 0.24 mmol, 1.50 equiv.) in toluene (0.53 ml) were heated at 110 °C for 50 min *via* microwave irradiation. Purification by flash column chromatography (*c*Hex/EtOAc 40:1 to 30:1) resulted in the title compound as a white solid (11.1 mg, 25.7 µmol, 16%).

R_f (*c*Hex/EtOAc 10:1) = 0.48. – **^{1}H NMR** (500 MHz, CDCl$_3$) δ = 7.70 (d, ^3J = 7.1 Hz, 1H, C*H*$_{ar}$), 7.54 – 7.47 (m, 2H, C*H* and C*H*$_{ar}$), 7.38 (t, ^3J = 7.1 Hz, 2H, 2 × C*H*$_{ar}$), 6.90 (d, ^4J = 1.5 Hz, 1H, C*H*$_{ar}$), 6.63 (d, ^4J = 1.5 Hz, 1H, C*H*$_{ar}$), 4.09 (s, 2H, CC*H*$_2$), 3.84 (s, 3H, OC*H*$_3$), 1.64 – 1.56 (m, 2H, 7-CC*H*$_2$), 1.29 (s, 6H, C(C*H*$_3$)$_2$), 1.27 – 1.16 (m, 2H, C*H*$_2$), 1.08 – 0.98 (m, 2H, C*H*$_2$), 0.82 (t, ^3J = 7.4 Hz, 3H, C*H*$_3$) ppm. – **^{13}C NMR** (126 MHz, CDCl$_3$) δ = 162.2 (C$_q$, *C*OO), 155.4 (C$_q$, 7-*C*$_{ar}$C), 155.2 (C$_q$, 5-*C*$_{ar}$OCH$_3$), 154.1 (C$_q$, 8a-*C*$_{ar}$), 136.9 (C$_q$, d, ^3J = 1.7 Hz, *C*$_{ar}$), 135.3 (+, *C*H), 132.2 (+, *C*$_{ar}$H), 131.9 (+, *C*$_{ar}$H), 129.3 (C$_q$, q, ^2J = 29.8 Hz, *C*$_{ar}$CF$_3$), 127.0 (+, *C*$_{ar}$H), 126.4 (+, q, ^3J = 5.7 Hz, *C*$_{ar}$H), 125.3 (C$_q$, 3-*C*), 124.5 (C$_q$, q, ^1J = 273.9 Hz, *C*$_{ar}$F$_3$), 107.7

(C$_q$, 4a-C_{ar}), 106.7 (+, C_{ar}H), 103.2 (+, C_{ar}H), 55.9 (+, OCH$_3$), 44.2 (–, 7-CCH$_2$), 38.8 (C$_q$, C(CH$_3$)$_2$), 33.1 (–, d, ^4J = 2.2 Hz, 3-CCH_2), 29.0 (+, C(CH$_3$)$_2$), 27.0 (–, CH$_2$), 23.4 (–, CH$_2$), 14.1 (+, CH$_3$) ppm. – **IR** (ATR, ṽ) = 2961, 2929, 2873, 1711, 1614, 1496, 1455, 1417, 1308, 1250, 1193, 1151, 1035, 946, 919, 866, 843, 765, 728, 673, 652, 573, 479, 405 cm^{-1}. – **HRMS** (ESI, C$_{25}$H$_{27}$O$_3$F$_3$) = calc.: 433.1985 [M+H]$^+$; found: 433.1972 [M+H]$^+$.

3-Benzyl-5-methoxy-7-(2-methyloctan-2-yl)-2H-chromen-2-one (21ma)

According to **GP8**, 2-hydroxy-6-methoxy-4-(2-methyloctan-2-yl)benzaldehyde (**19m**, 40.0 mg, 0.14 mmol, 1.00 equiv.), cinnamaldehyde (0.05 ml, 47.0 mg, 0.36 mmol, 2.50 equiv.), K$_2$CO$_3$ (24.0 mg, 0.17 mmol, 1.20 equiv.) and 1,3-dimethylimidazolium dimethyl phosphate (0.03 ml, 38.0 mg, 0.17 mmol, 1.20 equiv.) in toluene (0.47 ml) were heated at 110 °C for 50 min *via* microwave irradiation. Purification by flash column chromatography (*c*Hex/EtOAc 40:1 to 35:1) resulted in the title compound as a yellow oil (25.9 mg, 0.07 mmol, 46%).[158]

R_f (*c*Hex/EtOAc 10:1) = 0.41. – **^{1}H NMR** (500 MHz, CDCl$_3$) δ = 7.73 (s, 1H, CH), 7.35 – 7.29 (m, 4H, 4 × CH_{ar}), 7.26 – 7.22 (m, 1H, CH_{ar}), 6.87 (d, ^4J = 1.5 Hz, 1H, CH_{ar}), 6.63 (d, ^4J = 1.5 Hz, 1H, CH_{ar}), 3.88 – 3.87 (m, 5H, OCH_3 and CCH_2), 1.62 – 1.56 (m, 2H, 7-CCH_2), 1.29 (s, 6H, C(CH_3)$_2$), 1.24 – 1.15 (m, 6H, 3 × CH_2), 1.06 – 0.98 (m, 2H, CH_2CH$_3$), 0.84 (t, ^3J = 7.0 Hz, 3H, CH$_2$CH_3) ppm. – **^{13}C NMR** (126 MHz, CDCl$_3$) δ = 162.2 (C$_q$, COO), 155.4 (C$_q$, 5-C_{ar}OCH$_3$), 154.9 (C$_q$, 7-C_{ar}C), 154.2 (C$_q$, 8a-C_{ar}), 138.6 (C$_q$, CH$_2$$C_{ar}$), 134.7 (+, CH), 129.3 (+, 2 × C_{ar}H), 128.7 (+, 2 × C_{ar}H), 126.7 (+, C_{ar}H), 126.3 (C$_q$, 3-C), 107.9 (C$_q$, 4a-C_{ar}), 106.7 (+, C_{ar}H), 103.1 (+, C_{ar}H), 55.9 (+, OCH$_3$), 44.5 (–, 7-CCH$_2$), 38.8 (C$_q$, C(CH$_3$)$_2$), 36.9 (–, 3-CCH$_2$), 31.9 (–, CH$_2$), 30.0 (–, CH$_2$), 29.0 (+, C(CH$_3$)$_2$), 24.8 (–, CH$_2$), 22.8 (–, CH$_2$), 14.2 (+, CH$_2$CH$_3$) ppm. – **IR** (ATR, ṽ) = 2956, 2927, 2856, 1720, 1615, 1568, 1493, 1460, 1453, 1417, 1388, 1375, 1366, 1344, 1292, 1252, 1203, 1187, 1162, 1113, 1077, 1047, 1001, 843, 826, 755, 735, 700, 671, 598 cm^{-1}. – **HRMS** (ESI, C$_{26}$H$_{32}$O$_3$) = calc.: 393.2424 [M+H]$^+$; found: 393.2404 [M+H]$^+$.

5-Methoxy-3-(2-methylbenzyl)-7-(2-methyloctan-2-yl)-2H-chromen-2-one (21mb)

According to **GP8**, 2-hydroxy-6-methoxy-4-(2-methyloctan-2-yl)benzaldehyde (**19m**, 40.0 mg, 0.14 mmol, 1.00 equiv.), 2-methylcinnamaldehyde (53.0 mg, 0.36 mmol, 2.50 equiv.), K$_2$CO$_3$ (24.0 mg, 0.17 mmol, 1.20 equiv.) and 1,3-dimethylimidazolium dimethyl phosphate (0.04 ml, 48.0 mg, 0.22 mmol, 1.50 equiv.) in toluene (0.47 ml) were heated at 110 °C for 50 min *via* microwave irradiation. Purification by flash column chromatography (*c*Hex/EtOAc 40:1 to 30:1) resulted in the title compound as a yellow oil (18.7 mg, 0.05 mmol, 32%, purity ≤90%).

R_f (*c*Hex/EtOAc 10:1) = 0.39. – **¹H NMR** (500 MHz, CDCl₃) δ = 7.44 (d, ^4J = 1.5 Hz, 1H, C*H*), 7.23 – 7.17 (m, 4H, 4 × C*H*$_{ar}$), 6.89 (d, ^4J = 1.5 Hz, 1H, C*H*$_{ar}$), 6.62 (d, ^4J = 1.5 Hz, 1H, C*H*$_{ar}$), 3.87 (d, ^4J = 1.5 Hz, 2H, 3-CC*H₂*), 3.83 (s, 3H, OC*H₃*), 2.28 (s, 3H, C$_{ar}$C*H₃*), 1.64 – 1.56 (m, 2H, 7-CC*H₂*), 1.29 (s, 6H, C(C*H₃*)₂), 1.23 – 1.16 (m, 6H, 3 × C*H₂*), 1.09 – 0.99 (m, 2H, C*H₂*CH₃), 0.84 (t, ^3J = 7.1 Hz, 3H, CH₂C*H₃*) ppm. – **¹³C NMR** (126 MHz, CDCl₃) δ = 162.3 (C$_q$, *C*OO), 155.3 (C$_q$, 5-*C*$_{ar}$OCH₃), 154.9 (C$_q$, 7-*C*$_{ar}$C), 154.0 (C$_q$, 8a-*C*$_{ar}$), 137.0 (C$_q$, *C*$_{ar}$), 136.3 (C$_q$, *C*$_{ar}$), 134.1 (+, *C*H), 130.6 (+, *C*$_{ar}$H), 130.3 (+, *C*$_{ar}$H), 127.1 (+, *C*$_{ar}$H), 126.4 (+, *C*$_{ar}$H), 125.5 (C$_q$, 3-*C*), 107.9 (C$_q$, 4a-*C*$_{ar}$), 106.7 (+, *C*$_{ar}$H), 103.1 (+, *C*$_{ar}$H), 55.8 (+, O*C*H₃), 44.5 (–, 7-C*C*H₂), 38.8 (C$_q$, *C*(CH₃)₂), 34.1 (–, 3-C*C*H₂), 31.9 (–, *C*H₂), 30.1 (–, *C*H₂), 29.0 (+, C(*C*H₃)₂), 24.8 (–, *C*H₂), 22.8 (–, *C*H₂), 19.7 (+, C$_{ar}$*C*H₃), 14.2 (+, CH₂*C*H₃) ppm. – **IR** (ATR, ṽ) = 2955, 2927, 2856, 1723, 1684, 1615, 1568, 1493, 1460, 1415, 1387, 1377, 1366, 1347, 1315, 1293, 1248, 1230, 1190, 1159, 1115, 1052, 1041, 999, 953, 935, 919, 843, 799, 778, 761, 744, 727, 671 cm⁻¹. – **HRMS** (ESI, C₂₇H₃₄O₃) = calc.: 407.2581 [M+H]⁺; found: 407.2562 [M+H]⁺.

5-Methoxy-3-(2-methoxybenzyl)-7-(2-methyloctan-2-yl)-2*H*-chromen-2-one (21mc)

According to **GP8**, 2-hydroxy-6-methoxy-4-(2-methyloctan-2-yl)benzaldehyde (**19m**, 40.0 mg, 0.14 mmol, 1.00 equiv.), 2-methoxycinnamaldehyde (58.0 mg, 0.36 mmol, 2.50 equiv.), K₂CO₃ (24.0 mg, 0.17 mmol, 1.20 equiv.) and 1,3-dimethylimid-azolium dimethyl phosphate (0.03 ml, 38.0 mg, 0.17 mmol, 1.20 equiv.) in toluene (0.47 ml) were heated at 110 °C for 50 min *via* microwave irradiation. Purification by flash column chro-matography (*c*Hex/EtOAc 40:1 to 35:1) resulted in a mixture of the product and a methyl ester as impurity. Stirring in 2 M aq. NaOH (25 ml) for several hours, followed by extraction with EtOAc, drying over Na₂SO₄, filtration and removal of the volatiles, gave the title compound as a cloudy oil (27.9 mg, 0.07 mmol, 46%).[158]

R_f (*c*Hex/EtOAc 10:1) = 0.37. – **¹H NMR** (500 MHz, CDCl₃) δ = 7.44 (d, ^4J = 1.5 Hz, 1H, C*H*), 7.23 – 7.17 (m, 4H, 4 × C*H*$_{ar}$), 6.89 (d, ^4J = 1.5 Hz, 1H, C*H*$_{ar}$), 6.62 (d, ^4J = 1.5 Hz, 1H, C*H*$_{ar}$), 3.87 (d, ^4J = 1.5 Hz, 2H, 3-CC*H₂*), 3.86 (s, 3H, OC*H₃*), 3.81 (s, 3H, OC*H₃*), 1.64 – 1.56 (m, 2H, 7-CC*H₂*), 1.29 (s, 6H, C(C*H₃*)₂), 1.23 – 1.16 (m, 6H, 3 × C*H₂*), 1.09 – 0.99 (m, 2H, C*H₂*CH₃), 0.84 (t, ^3J = 7.1 Hz, 3H, CH₂C*H₃*) ppm. – **¹³C NMR** (126 MHz, CDCl₃) δ = 162.4

(C_q, COO), 157.8 (C_q, $C_{ar}OCH_3$), 155.3 (C_q, 5-$C_{ar}OCH_3$), 154.5 (C_q, 7-$C_{ar}C$), 154.1 (C_q, 8a-C_{ar}), 134.3 (+, CH), 131.1 (+, $C_{ar}H$), 128.1 (+, $C_{ar}H$), 126.7 (C_q, CH_2C_{ar}), 125.7 (C_q, 3-C), 120.7 (+, $C_{ar}H$), 110.7 (+, $C_{ar}H$), 108.1 (C_q, 4a-C_{ar}), 106.7 (+, $C_{ar}H$), 103.0 (+, $C_{ar}H$), 55.9 (+, OCH_3), 55.5 (+, OCH_3), 44.5 (–, 7-CCH_2), 38.7 (C_q, $C(CH_3)_2$), 31.9 (–, CH_2), 31.1 (–, 3-CCH_2), 30.1 (–, CH_2), 29.0 (+, $C(CH_3)_2$), 24.8 (–, CH_2), 22.8 (–, CH_2), 14.2 (+, CH_3) ppm. – **IR** (ATR, $\tilde{v}$) = 2955, 2927, 2856, 1718, 1613, 1588, 1570, 1493, 1460, 1439, 1417, 1289, 1244, 1191, 1160, 1113, 1050, 1031, 932, 841, 752, 728, 671 cm^{-1}. – **HRMS** (ESI, $C_{27}H_{34}O_4$) = calc.: 423.2530 [M+H]$^+$; found: 423.2508 [M+H]$^+$.

5-Methoxy-3-(4-methoxybenzyl)-7-(2-methyloctan-2-yl)-2*H*-chromen-2-one (21md)

According to **GP8**, 2-hydroxy-6-methoxy-4-(2-methyloctan-2-yl)benzaldehyde (**19m**, 40.0 mg, 0.14 mmol, 1.00 equiv.), 4-methoxycinnamaldehyde (58.0 mg, 0.36 mmol, 2.50 equiv.), K_2CO_3 (24.0 mg, 0.17 mmol, 1.20 equiv.) and 1,3-dimethylimidazolium dimethyl phosphate (0.03 ml, 38.0 mg, 0.17 mmol, 1.20 equiv.) in toluene (0.47 ml) were heated at 110 °C for 50 min *via* microwave irradiation. Purification by flash column chromatography (*c*Hex/EtOAc 40:1 to 30:1) resulted in the title compound as a cloudy oil (21.0 mg, 0.05 mmol, 35%).

R$_f$ (*c*Hex/EtOAc 10:1) = 0.37. – **^{1}H NMR** (500 MHz, CDCl$_3$) δ = 7.70 (s, 1H, C*H*), 7.24 – 7.20 (m, 2H, 2 × *m*-C*H*$_{ar}$), 6.89 – 6.84 (m, 3H, 2 × *o*-C*H*$_{ar}$ and C*H*$_{ar}$), 6.63 (d, ^4J = 1.5 Hz, 1H, C*H*$_{ar}$), 3.88 (s, 3H, OC*H*$_3$), 3.81 (d, ^4J = 1.2 Hz, 2H, CC*H*$_2$), 3.80 (s, 3H, OC*H*$_3$), 1.62 – 1.56 (m, 2H, 7-CC*H*$_2$), 1.28 (s, 6H, C(C*H*$_3$)$_2$), 1.24 – 1.15 (m, 6H, 3 × C*H*$_2$), 1.06 – 0.98 (m, 2H, C*H*$_2$CH$_3$), 0.83 (t, ^3J = 7.0 Hz, 3H, CH$_2$C*H*$_3$) ppm. – **^{13}C NMR** (126 MHz, CDCl$_3$) δ = 162.1 (C_q, COO), 158.4 (C_q, $C_{ar}OCH_3$), 155.4 (C_q, 5-$C_{ar}OCH_3$), 154.8 (C_q, 7-$C_{ar}C$), 154.2 (C_q, 8a-C_{ar}), 134.4 (+, C*H*), 130.6 (C_q, CH_2C_{ar}), 130.3 (+, 2 × *m*-$C_{ar}H$), 126.7 (C_q, 3-C), 114.2 (+, 2 × *o*-$C_{ar}H$), 107.9 (C_q, 4a-C_{ar}), 106.7 (+, $C_{ar}H$), 103.1 (+, $C_{ar}H$), 55.9 (+, OCH_3), 55.4 (+, OCH_3), 44.5 (–, 7-CCH_2), 38.8 (C_q, $C(CH_3)_2$), 36.1 (–, 3-CCH_2), 31.9 (–, CH_2), 30.1 (–, CH_2), 29.0 (+, $C(CH_3)_2$), 24.8 (–, CH_2), 22.8 (–, CH_2), 14.2 (+, CH_3) ppm. – **IR** (ATR, $\tilde{v}$) = 2955, 2927, 2856, 1717, 1613, 1585, 1570, 1510, 1493, 1462, 1417, 1388, 1375, 1364, 1343, 1300, 1245, 1207, 1176, 1160, 1113, 1040, 999, 949, 933, 919, 843, 817, 790, 766, 752, 727, 687, 671, 636, 586, 562, 534, 513 cm^{-1}. – **HRMS** (ESI, $C_{27}H_{34}O_4$) = calc.: 423.2530 [M+H]$^+$; found: 423.2518 [M+H]$^+$.

3-(2-Fluorobenzyl)-5-methoxy-7-(2-methyloctan-2-yl)-2H-chromen-2-one (21me)

According to **GP8**, 2-hydroxy-6-methoxy-4-(2-methyloctan-2-yl)benzaldehyde (**19m**, 40.0 mg, 0.14 mmol, 1.00 equiv.), 2-fluorocinnamaldehyde (54.0 mg, 0.36 mmol, 2.50 equiv.), K_2CO_3 (24.0 mg, 0.17 mmol, 1.20 equiv.) and 1,3-dimethylimidazolium dimethyl phosphate (0.03 ml, 38.0 mg, 0.17 mmol, 1.20 equiv.) in toluene (0.47 ml) were heated at 110 °C for 50 min *via* microwave irradiation. Purification by flash column chromatography (*c*Hex/EtOAc 40:1 to 30:1) resulted in the title compound as a yellow oil (27.9 mg, 0.07 mmol, 47%, purity 89%).

R_f (*c*Hex/EtOAc 10:1) = 0.37. – ^{1}H NMR (500 MHz, CDCl$_3$) δ = 7.78 – 7.74 (m, 1H, C*H*), 7.36 (td, $^{3,4}J$ = 7.6, 1.8 Hz, 1H, *m*-C*H*$_{ar}$), 7.26 – 7.19 (m, 1H, *m*-C*H*$_{ar}$), 7.10 (td, $^{3,4}J$ = 7.6, 1.2 Hz, 1H, *p*-C*H*$_{ar}$), 7.10 (ddd, $^{3,3,4}J$ = 9.6, 8.2, 1.2 Hz, 1H, *o*-C*H*$_{ar}$), 6.87 (d, ^4J = 1.5 Hz, 1H, C*H*$_{ar}$), 6.64 (d, ^4J = 1.5 Hz, 1H, C*H*$_{ar}$), 3.90 (d, ^4J = 1.4 Hz, 2H, 3-CC*H*$_2$), 3.88 (s, 3H, OC*H*$_3$), 1.63 – 1.55 (m, 2H, 7-CC*H*$_2$), 1.29 (s, 6H, C(C*H*$_3$)$_2$), 1.25 – 1.15 (m, 6H, 3 × C*H*$_2$), 1.06 – 0.99 (m, 2H, C*H*$_2$CH$_3$), 0.83 (t, ^3J = 7.0 Hz, 3H, CH$_2$C*H*$_3$) ppm. – ^{13}C NMR (126 MHz, CDCl$_3$) δ = 162.1 (C$_q$, COO), 161.4 (C$_q$, d, ^1J = 246.0 Hz, C$_{ar}$F), 155.4 (C$_q$, 5-C$_{ar}$OCH$_3$), 155.0 (C$_q$, 7-C$_{ar}$C), 154.2 (C$_q$, 8a-C$_{ar}$), 134.8 (+, d, ^5J = 1.6 Hz, C*H*), 131.8 (+, d, ^3J = 4.5 Hz, *m*-C$_{ar}$H), 128.6 (+, d, ^3J = 8.1 Hz, *m*-C$_{ar}$H), 125.4 (C$_q$, d, ^2J = 15.2 Hz, CH$_2$C$_{ar}$), 124.7 (C$_q$, 3-C), 124.3 (+, d, ^3J = 3.6 Hz, *p*-C$_{ar}$H), 115.5 (+, d, ^2J = 21.7 Hz, *o*-C$_{ar}$H), 107.9 (C$_q$, 4a-C$_{ar}$), 106.7 (+, C$_{ar}$H), 103.1 (+, C$_{ar}$H), 55.9 (+, OCH$_3$), 44.5 (–, 7-CCH$_2$), 38.8 (C$_q$, C(CH$_3$)$_2$), 31.8 (–, CH$_2$), 30.2 (–, d, ^3J = 3.6 Hz, 3-CCH$_2$), 30.0 (–, CH$_2$), 29.0 (+, C(CH$_3$)$_2$), 24.8 (–, CH$_2$), 22.8 (–, CH$_2$), 14.2 (+, CH$_3$) ppm. – **IR** (ATR, ṽ) = 2956, 2927, 2856, 1721, 1615, 1585, 1568, 1492, 1455, 1417, 1388, 1375, 1366, 1347, 1315, 1292, 1248, 1231, 1188, 1176, 1162, 1115, 1050, 999, 941, 924, 843, 795, 785, 755, 737, 727, 687, 671, 603, 575, 554, 436 cm^{-1}. – **HRMS** (ESI, C$_{26}$H$_{31}$O$_3$F) = calc.: 411.2330 [M+H]$^+$; found: 411.2310 [M+H]$^+$.

3-(4-Fluorobenzyl)-5-methoxy-7-(2-methyloctan-2-yl)-2H-chromen-2-one (2mf)

According to **GP8**, 2-hydroxy-6-methoxy-4-(2-methyloctan-2-yl)benzaldehyde (**19m**, 40.0 mg, 0.14 mmol, 1.00 equiv.), 4-fluorocinnamaldehyde (0.05 ml, 54.0 mg, 0.36 mmol, 2.50 equiv.), K_2CO_3 (24.0 mg, 0.17 mmol, 1.20 equiv.) and 1,3-dimethylimidazolium dimethyl phosphate (0.03 ml, 38.0 mg, 0.17 mmol, 1.20 equiv.) in toluene (0.47 ml) were heated at 110 °C for 50 min *via* microwave irradiation. Purification by flash column chromatography (*c*Hex/EtOAc 40:1 to 35:1) resulted in the title compound as a yellow oil (25.5 mg, 0.06 mmol, 43%).

R_f (*c*Hex/EtOAc 10:1) = 0.44. – **^{1}H NMR** (500 MHz, CDCl$_3$) δ = 7.72 (s, 1H, C*H*), 7.30 – 7.23 (m, 2H, 2 × *m*-C*H*$_{ar}$), 7.00 (t, ^3J = 8.7 Hz, 2H, 2 × *o*-C*H*$_{ar}$), 6.87 (d, ^4J = 1.5 Hz, 1H, C*H*$_{ar}$), 6.64 (d, ^4J = 1.5 Hz, 1H, C*H*$_{ar}$), 3.89 (s, 3H, OC*H*$_3$), 3.83 (s, 2H, CC*H*$_2$), 1.62 – 1.56 (m, 2H, 7-CC*H*$_2$), 1.29 (s, 6H, C(C*H*$_3$)$_2$), 1.24 – 1.15 (m, 6H, 3 × C*H*$_2$), 1.06 – 0.99 (m, 2H, C*H*$_2$CH$_3$), 0.83 (t, ^3J = 7.0 Hz, 3H, CH$_2$C*H*$_3$) ppm. – **^{13}C NMR** (126 MHz, CDCl$_3$) δ = 162.1 (C$_q$, *C*OO), 161.8 (C$_q$, d, ^1J = 244.4 Hz, *C*$_{ar}$F), 155.4 (C$_q$, 5-*C*$_{ar}$OCH$_3$), 155.1 (C$_q$, 7-*C*$_{ar}$C), 154.2 (C$_q$, 8a-*C*$_{ar}$), 134.7 (+, *C*H), 134.2 (C$_q$, d, ^4J = 3.2 Hz, CH$_2$*C*$_{ar}$), 130.7 (+, d, ^3J = 7.8 Hz, 2 × *m*-*C*$_{ar}$H), 126.1 (C$_q$, 3-*C*), 115.5 (+, d, ^2J = 21.3 Hz, *o*-*C*$_{ar}$H), 107.8 (C$_q$, 4a-*C*$_{ar}$), 106.7 (+, *C*$_{ar}$H), 103.2 (+, *C*$_{ar}$H), 55.9 (+, O*C*H$_3$), 44.5 (–, 7-C*C*H$_2$), 38.8 (C$_q$, *C*(CH$_3$)$_2$), 36.2 (–, 3-C*C*H$_2$), 31.9 (–, *C*H$_2$), 30.0 (–, *C*H$_2$), 29.0 (+, C(*C*H$_3$)$_2$), 24.8 (–, *C*H$_2$), 22.8 (–, *C*H$_2$), 14.2 (+, *C*H$_3$) ppm. – **IR** (ATR, ṽ) = 2956, 2927, 2856, 1718, 1613, 1568, 1507, 1494, 1462, 1417, 1293, 1252, 1221, 1181, 1156, 1113, 1050, 1016, 999, 935, 922, 841, 824, 795, 765, 754, 728, 671, 584, 517, 487 cm^{-1}. – **HRMS** (ESI, C$_{26}$H$_{31}$O$_3$F) = calc.: 411.2330 [M+H]$^+$; found: 411.2310 [M+H]$^+$.

5-Methoxy-7-(2-methyloctan-2-yl)-3-(2-(trifluoromethyl)benzyl)-2*H*-chromen-2-one (21mg)

According to **GP8**, 2-hydroxy-6-methoxy-4-(2-methyloctan-2-yl)benzaldehyde (**21m**, 40.0 mg, 0.14 mmol, 1.00 equiv.), 2-(trifluoromethyl)cinnamaldehyde (72.0 mg, 0.36 mmol, 2.50 equiv.), K$_2$CO$_3$ (24.0 mg, 0.17 mmol, 1.20 equiv.) and 1,3-dimethylimidazo-lium dimethyl phosphate (0.03 ml, 38.0 mg, 0.17 mmol, 1.20 equiv.) in toluene (0.47 ml) were heated at 110 °C for 50 min *via* microwave irradiation. Purification by flash column chroma-tography (*c*Hex/EtOAc 40:1 to 35:1) resulted in the title compound as a yellow solid (27.3 mg, 0.06 mmol, 41%).

R_f (*c*Hex/EtOAc 10:1) = 0.38. – **^{1}H NMR** (500 MHz, CDCl$_3$) δ = 7.71 (dd, $^{3,4}J$ = 8.1, 1.3 Hz, 1H, C*H*$_{ar}$), 7.53 – 7.47 (m, 2H, C*H* and C*H*$_{ar}$), 7.42 – 7.35 (m, 2H, 2 × C*H*$_{ar}$), 6.89 (d, ^4J = 1.5 Hz, 1H, C*H*$_{ar}$), 6.63 (d, ^4J = 1.5 Hz, 1H, C*H*$_{ar}$), 4.09 (s, 2H, CC*H*$_2$), 3.84 (s, 3H, OC*H*$_3$), 1.62 – 1.56 (m, 2H, 7-CC*H*$_2$), 1.29 (s, 6H, C(C*H*$_3$)$_2$), 1.24 – 1.16 (m, 6H, 3 × C*H*$_2$), 1.07 – 1.00 (m, 2H, C*H*$_2$CH$_3$), 0.84 (t, ^3J = 7.0 Hz, 3H, CH$_2$C*H*$_3$) ppm. – **^{13}C NMR** (126 MHz, CDCl$_3$) δ = 162.2 (C$_q$, *C*OO), 155.4 (C$_q$, 5-*C*$_{ar}$OCH$_3$), 155.2 (C$_q$, 7-*C*$_{ar}$C), 154.1 (C$_q$, 8a-*C*$_{ar}$), 136.9 (C$_q$, d, ^3J = 1.8 Hz, CH$_2$*C*$_{ar}$), 135.3 (+, *C*H), 132.2 (+, *C*$_{ar}$H), 131.9 (+, *C*$_{ar}$H), 129.3 (C$_q$, d, ^2J = 30.3 Hz, *C*$_{ar}$CF$_3$), 127.0 (+, *C*$_{ar}$H), 126.4 (+, d, ^3J = 5.8 Hz, *o*-*C*$_{ar}$H), 125.3 (C$_q$, 3-*C*), 124.6 (C$_q$, q, ^1J

= 274.0 Hz, CF_3), 107.8 (C_q, 4a-C_{ar}), 106.7 (+, $C_{ar}H$), 103.2 (+, $C_{ar}H$), 55.9 (+, OCH_3), 44.5 (–, 7-CCH_2), 38.8 (C_q, $C(CH_3)_2$), 33.1 (–, d, ^4J = 2.1 Hz, 3-CCH_2), 31.9 (–, CH_2), 30.1 (–, CH_2), 29.0 (+, $C(CH_3)_2$), 24.8 (–, CH_2), 22.8 (–, CH_2), 14.2 (+, CH_3) ppm. – **IR** (ATR, ṽ) = 2956, 2927, 2870, 2857, 1708, 1612, 1571, 1496, 1477, 1455, 1417, 1312, 1273, 1259, 1248, 1218, 1191, 1169, 1152, 1105, 1060, 1051, 1035, 996, 948, 844, 766, 724, 673, 653 cm^{-1}. – **HRMS** (ESI, $C_{27}H_{31}O_3F_3$) = calc.: 461.2298 [M+H]$^+$; found: 461.2274 [M+H]$^+$.

3-Benzyl-7-(*tert*-butyl)-5-hydroxy-2*H*-chromen-2-one (59ka)

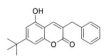

According to **GP9**, to a solution of the 5-methoxycoumarin **21ka** (7.30 mg, 21.7 μmol, 1.00 equiv.) in abs. DCM (0.43 ml), was added BBr$_3$ (1 M in DCM, 0.11 ml, 27.0 mg, 109 μmol, 5.00 equiv.) at –78 °C. After stirring for 30 min at this temperature, the reaction mixture was allowed to warm to rt and stirred overnight. Filtration over a small silica plug (*n*Pen/EtOAc 2:1) resulted in the title compound as a white solid (6.00 mg, 18.6 μmol, 86%).

^{1}H NMR (500 MHz, acetone-D$_6$) δ = 9.33 (s, 1H, O*H*), 7.81 (d, ^4J = 1.2 Hz, 1H, C*H*), 7.38 – 7.28 (m, 4H, 4 × C*H*$_{ar}$), 7.26 – 7.19 (m, 1H, C*H*$_{ar}$), 6.85 (d, ^4J = 1.7 Hz, 1H, C*H*$_{ar}$), 6.82 (d, ^4J = 1.5 Hz, 1H, C*H*$_{ar}$), 3.85 (d, ^4J = 1.2 Hz, 2H, CC*H*$_2$), 1.29 (s, 9H, C(C*H*$_3$)$_3$) ppm. – **^{13}C NMR** (126 MHz, acetone-D$_6$) δ = 162.2 (C_q, *C*OO), 156.7 (C_q, C_{ar}), 155.2 (C_q, C_{ar}), 154.5 (C_q, 8a-C_{ar}), 139.8 (C_q, CH$_2$$C_{ar}$), 135.1 (+, *C*H), 129.9 (+, 2 × $C_{ar}H$), 129.3 (+, 2 × $C_{ar}H$), 127.2 (+, $C_{ar}H$), 126.8 (C_q, 3-*C*), 108.0 (+, $C_{ar}H$), 107.6 (C_q, 4a-C_{ar}), 105.0 (+, $C_{ar}H$), 37.2 (–, 3-CCH_2), 35.7 (C_q, *C*(CH$_3$)$_3$), 31.2 (+, C(*C*H$_3$)$_3$) ppm. – **IR** (ATR, ṽ) = 3321, 2954, 1678, 1623, 1495, 1423, 1358, 1272, 1204, 1175, 1104, 1060, 949, 897, 871, 843, 759, 727, 698, 669, 554, 521, 474, 439, 409 cm^{-1}. – **HRMS** (ESI, $C_{20}H_{20}O_3$) = calc.: 309.1485 [M+H]$^+$; found: 309.1479 [M+H]$^+$.

7-(*tert*-Butyl)-5-hydroxy-3-(2-methylbenzyl)-2*H*-chromen-2-one (59kb)

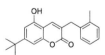

According to **GP9**, to a solution of the 5-methoxycoumarin **21kb** (5.00 mg, 14.3 μmol, 1.00 equiv.) in abs. DCM (0.29 ml), was added BBr$_3$ (1 M in DCM, 0.07 ml, 18.0 mg, 71.3 μmol, 5.00 equiv.) at –78 °C. After stirring for 30 min at this temperature, the reaction mixture was allowed to warm to rt and stirred overnight. Filtration over a small silica plug (*n*Pen/EtOAc 2:1) resulted in the title compound as a white solid (2.90 mg, 8.29 μmol, 58%).

^{1}H NMR (500 MHz, acetone-D$_6$) δ = 9.28 (s, 1H, O*H*), 7.50 (d, ^4J = 1.5 Hz, 1H, C*H*), 7.27 – 7.22 (m, 2H, 2 × C*H*$_{ar}$), 7.22 – 7.18 (m, 2H, 2 × C*H*$_{ar}$), 6.84 (s, 2H, 2 × C*H*$_{ar}$), 3.84 (s, 2H, C*H*$_2$),

2.30 (s, 3H, $C_{ar}CH_3$), 1.30 (s, 9H, $C(CH_3)_3$) ppm. – ^{13}C NMR (126 MHz, acetone-D_6) δ = 162.1 (C_q, COO), 156.7 (C_q, C_{ar}), 155.1 (C_q, C_{ar}), 154.4 (C_q, 8a-C_{ar}), 137.5 (C_q, C_{ar}), 137.4 (C_q, C_{ar}), 134.4 (+, CH), 131.2 (+, $C_{ar}H$), 131.0 (+, $C_{ar}H$), 127.8 (+, $C_{ar}H$), 127.1 (+, $C_{ar}H$), 126.1 (C_q, 3-C), 108.1 (+, $C_{ar}H$), 107.1 (C_q, 4a-C_{ar}), 105.2 (+, $C_{ar}H$), 35.7 (C_q, $C(CH_3)_3$), 34.5 (–, 3-CCH_2), 31.3 (+, $C(CH_3)_3$), 19.5 (+, $C_{ar}CH_3$) ppm. – IR (ATR, $\tilde{v}$) = 3325, 2923, 1679, 1622, 1454, 1422, 1353, 1261, 1172, 1103, 1061, 949, 904, 871, 844, 798, 760, 746, 726, 669, 555, 525, 503, 457, 407 cm^{-1}. – HRMS (ESI, $C_{21}H_{22}O_3$) = calc.: 323.1642 [M+H]$^+$; found: 323.1632 [M+H]$^+$.

7-(*tert*-Butyl)-5-hydroxy-3-(2-hydroxybenzyl)-2*H*-chromen-2-one (59kc)

According to **GP9**, to a solution of the dimethoxycoumarin **21kc** (17.7 mg, 50.2 µmol, 1.00 equiv.) in abs. DCM (1.13 ml), was added BBr$_3$ (1 M in DCM, 0.57 ml, 142 mg, 0.57 mmol, 11.4 equiv.) at –78 °C. After stirring for 30 min at this temperature, the reaction mixture was allowed to warm to rt and stirred overnight. Filtration over a small silica plug (*n*Pen/EtOAc 2:1) resulted in the title compound as a white solid (7.90 mg, 24.8 µmol, 49%).

^1H NMR (500 MHz, acetone-D_6) δ = 9.21 (s, 1H, OH), 8.44 (s, 1H, OH), 7.77 (s, 1H, CH), 7.25 (dd, $^{3,4}J$ = 7.5, 1.7 Hz, 1H, CH_{ar}), 7.11 (td, $^{3,4}J$ = 7.5, 1.7 Hz, 1H, CH_{ar}), 6.89 (dd, $^{3,4}J$ = 8.1, 1.2 Hz, 1H, CH_{ar}), 6.85 – 6.82 (m, 3H, 3 × CH_{ar}), 3.83 (d, ^4J = 1.2 Hz, 2H, CCH_2), 1.29 (s, 9H, C(CH_3)$_3$) ppm. – ^{13}C NMR (126 MHz, acetone-D_6) δ = 162.7 (C_q, COO), 156.6 (C_q, C_{ar}OH), 156.3 (C_q, C_{ar}OH), 155.1 (C_q, 7-$C_{ar}C$), 154.4 (C_q, 8a-C_{ar}), 134.8 (+, CH), 132.0 (+, $C_{ar}H$), 128.9 (+, $C_{ar}H$), 126.2 (C_q, C), 125.7 (C_q, C), 120.7 (+, $C_{ar}H$), 116.5 (+, $C_{ar}H$), 108.1 (+, $C_{ar}H$), 107.8 (C_q, 4a-C_{ar}), 105.2 (+, $C_{ar}H$), 35.8 (C_q, $C(CH_3)_3$), 31.7 (–, 3-CCH_2), 31.3 (+, $C(CH_3)_3$) ppm. – IR (ATR, $\tilde{v}$) = 3078, 2960, 1667, 1619, 1488, 1426, 1348, 1287, 1243, 1181, 1073, 1039, 950, 925, 856, 831, 806, 751, 667, 578, 557, 532, 501, 481, 432, 403 cm^{-1}. – HRMS (ESI, $C_{20}H_{20}O_4$) = calc.: 325.1434 [M+H]$^+$; found: 325.1429 [M+H]$^+$.

7-(*tert*-Butyl)-3-(2-fluorobenzyl)-5-hydroxy-2*H*-chromen-2-one (59ke)

According to **GP9**, to a solution of the 5-methoxycoumarin **21ke** (6.80 mg, 20.0 µmol, 1.00 equiv.) in abs. DCM (0.40 ml), was added BBr$_3$ (1 M in DCM, 0.10 ml, 25.0 mg, 0.10 mmol, 5.00 equiv.) at –78 °C. After stirring for 30 min at this temperature, the reaction mixture was allowed to warm to rt and stirred overnight. Filtration over a small silica plug (*n*Pen/EtOAc 2:1) resulted in the title compound as an off-white solid (5.00 mg, 14.3 µmol, 71%).

¹H NMR (500 MHz, acetone-D₆) δ = 9.29 (s, 1H, O*H*), 7.75 (s, 1H, C*H*), 7.39 – 7.31 (m, 1H, C*H*ar), 7.25 – 7.19 (m, 1H, C*H*ar), 7.10 (t, ³*J* = 7.5 Hz, 1H, C*H*ar), 7.05 (t, ³*J* = 9.3 Hz, 1H, C*H*ar), 6.92 (d, ⁴*J* = 1.6 Hz, 1H, C*H*ar), 6.69 (d, ⁴*J* = 1.6 Hz, 1H, C*H*ar), 3.90 (s, 2H, C*H*₂), 1.32 (s, 9H, C(C*H*₃)₃) ppm. – **¹³C NMR** (126 MHz, acetone-D₆) δ = 162.2 (Cq, d, ¹*J* = 244.4 Hz, *C*arF), 161.9 (Cq, *C*OO), 157.0 (Cq, *C*ar), 155.3 (Cq, *C*ar), 154.6 (Cq, 8a-*C*ar), 135.2 (+, *C*H), 132.6 (+, d, ³*J* = 4.4 Hz, *m*-*C*arH), 129.7 (+, d, ³*J* = 8.2 Hz, *m*-*C*arH), 126.4 (Cq, d, ²*J* = 15.5 Hz, *o*-*C*ar), 125.3 (+, d, ⁴*J* = 3.6 Hz, *p*-*C*arH), 125.2 (Cq, 3-*C*), 116.2 (+, d, ²*J* = 21.8 Hz, *o*-*C*arH), 108.2 (+, *C*arH), 107.6 (Cq, 4a-*C*ar), 105.3 (+, *C*arH), 35.9 (– and Cq, *C*H₂ and *C*(CH₃)₃), 31.3 (+, C(*C*H₃)₃) ppm. – **19F NMR** (471 MHz, CDCl₃) δ = –117.6 (s, C*F*) ppm. – **IR** (ATR, ṽ) = 3314, 2955, 1678, 1623, 1491, 1453, 1423, 1352, 1267, 1231, 1175, 1100, 1062, 949, 843, 780, 753, 726, 669, 517, 471, 435, 408 cm⁻¹. – **HRMS** (ESI, C₂₀H₁₉O₃F) = calc.: 327.1391 [M+H]⁺; found: 327.1386 [M+H]⁺.

7-(*tert*-Butyl)-3-(4-fluorobenzyl)-5-hydroxy-2*H*-chromen-2-one (59kf)

According to **GP9**, to a solution of the 5-methoxycoumarin **21kf** (10.0 mg, 29.4 µmol, 1.00 equiv.) in abs. DCM (0.59 ml), was added BBr₃ (1 M in DCM, 0.15 ml, 37.0 mg, 0.15 mmol, 5.00 equiv.) at – 78 °C. After stirring for 30 min at this temperature, the reaction mixture was allowed to warm to rt and stirred overnight. Filtration over a small silica plug (*n*Pen/EtOAc 2:1) resulted in the title compound as an off-white solid (9.80 mg, 28.8 µmol, 98%).

¹H NMR (500 MHz, acetone-D₆) δ = 9.27 (s, 1H, O*H*), 7.83 (s, 1H, C*H*), 7.42 – 7.36 (m, 2H, 2 × C*H*ar), 7.10 – 7.04 (m, 2H, 2 × C*H*ar), 6.85 (d, ⁴*J* = 1.7 Hz, 1H, C*H*ar), 6.81 (d, ⁴*J* = 1.7 Hz, 1H, C*H*ar), 3.84 (s, 2H, C*H*₂), 1.29 (s, 9H, C(C*H*₃)₃) ppm. – **¹³C NMR** (126 MHz, acetone-D₆) δ = 162.6 (d, ¹*J* = 242.5 Hz, Cq, C*F*), 162.0 (Cq, *C*OO), 156.9 (Cq, *C*ar), 155.4 (Cq, *C*ar), 154.6 (Cq, 8a-*C*ar), 136.0 (Cq, d, ⁴*J* = 3.2 Hz, *p*-*C*ar), 135.2 (+, *C*H), 131.8 (+, d, ³*J* = 8.2 Hz, 2 × *m*-*C*arH), 126.8 (Cq, 3-*C*), 116.0 (+, d, ²*J* = 21.3 Hz, 2 × *o*-*C*ar), 108.0 (+, *C*arH), 107.7 (Cq, 4a-*C*ar), 105.3 (+, *C*arH), 36.5 (–, *C*H₂), 35.8 (Cq, *C*(CH₃)₃), 31.3 (+, C(*C*H₃)₃) ppm. – **IR** (ATR, ṽ) = 3265, 2958, 1675, 1621, 1510, 1423, 1270, 1226, 1176, 1103, 1081, 1062, 948, 887, 848, 781, 729, 671, 578, 537, 522, 480, 454, 409 cm⁻¹. – **HRMS** (ESI, C₂₀H₁₉O₃F) calc.: 327.1391 [M+H]⁺; found. 327.1387 [M+H]⁺.

7-(*tert*-Butyl)-5-hydroxy-3-(2-(trifluoromethyl)benzyl)-2*H*-chromen-2-one (59kg)

According to **GP9**, to a solution of the 5-methoxycoumarin **21kg** (6.80 mg, 17.4 μmol, 1.00 equiv.) in abs. DCM (0.35 ml), was added BBr$_3$ (1 M in DCM, 0.09 ml, 22.0 mg, 0.09 mmol, 5.00 equiv.) at –78 °C. After stirring for 30 min at this temperature, the reaction mixture was allowed to warm to rt and stirred overnight. Filtration over a small silica plug (*n*Pen/EtOAc 2:1) resulted in the title compound as a white solid (6.00 mg, 14.9 μmol, 86%).

^{1}H NMR (500 MHz, acetone-D$_6$) δ = 9.23 (s, 1H. ,O*H*), 7.79 (d, ^3J = 7.7 Hz, 1H, C*H*$_{ar}$), 7.75 – 7.47 (m, 4H, C*H* and 3 × C*H*$_{ar}$), 6.88 – 6.84 (m, 2H, 2 × C*H*$_{ar}$), 4.07 (s, 2H, CC*H*$_2$), 1.30 (s, 9H, C(C*H*$_3$)$_3$) ppm. – **^{13}C NMR** (126 MHz, acetone-D$_6$) δ = 161.9 (C$_q$, *C*OO), 157.1 (C$_q$, *C*$_{ar}$), 155.3 (C$_q$, *C*$_{ar}$), 154.6 (C$_q$, 8a-*C*$_{ar}$), 138.0 (C$_q$, CH$_2$*C*$_{ar}$), 135.6 (+, *C*H), 133.5 (+, *C*$_{ar}$H), 130.0 (+, *C*$_{ar}$H), 129.4 (C$_q$, q, ^2J = 29.8 Hz, *C*$_{ar}$CF$_3$), 128.2 (+, *C*$_{ar}$H), 127.1 (+, q, ^3J = 5.8 Hz, *C*$_{ar}$H), 125.8 (C$_q$, 3-*C*), 125.7 (C$_q$, q, ^1J = 273.4 Hz, *C*$_{ar}$F$_3$), 108.2 (+, *C*$_{ar}$H), 107.8 (C$_q$, 4a-*C*$_{ar}$), 105.3 (+, *C*$_{ar}$H), 35.9 (C$_q$, *C*(CH$_3$)$_3$), 33.8 (–, CC*H*$_2$), 31.3 (+, C(*C*H$_3$)$_3$) ppm. – **IR** (ATR, ṽ) = 3320, 2966, 1677, 1622, 1423, 1348, 1313, 1268, 1158, 1112, 1081, 1064, 1037, 948, 873, 850, 770, 730, 672, 555, 522, 472, 436, 407 cm^{-1}. – **HRMS** (ESI, C$_{21}$H$_{19}$O$_3$F$_3$) = calc.: 377.1359 [M+H]$^+$; found: 377.1354 [M+H]$^+$.

3-Benzyl-5-hydroxy-7-(2-methylhexan-2-yl)-2*H*-chromen-2-one (59la)

According to **GP9**, to a solution of the 5-methoxycoumarin **21la** (5.00 mg, 14.3 μmol, 1.00 equiv.) in abs. DCM (0.29 ml), was added BBr$_3$ (1 M in DCM, 0.07 ml, 18.0 mg, 0.07 mmol, 5.00 equiv.) at –78 °C. After stirring for 30 min at this temperature, the reaction mixture was allowed to warm to rt and stirred overnight. Filtration over a small silica plug (*n*Pen/EtOAc 2:1) resulted in the title compound as an off-white solid (2.90 mg, 8.29 μmol, 58%).

^{1}H NMR (500 MHz, acetone-D$_6$) δ = 9.26 (s, 1H, O*H*), 7.82 (s, 1H, C*H*), 7.39 – 7.29 (m, 4H, 4 × C*H*$_{ar}$), 7.27 – 7.19 (m, 1H, C*H*$_{ar}$), 6.81 (d, ^4J = 1.6 Hz, 1H, C*H*$_{ar}$), 6.78 (d, ^4J = 1.6 Hz, 1H, C*H*$_{ar}$), 3.86 (s, 3H, 3-CC*H*$_2$), 1.63 – 1.58 (m, 2H, 7-CC*H*$_2$), 1.27 (s, 6H, C(C*H*$_3$)$_2$), 1.24 – 1.17 (m, 2H, C*H*$_2$), 1.07 – 0.99 (m, 2H, C*H*$_2$), 0.80 (t, ^3J = 7.4 Hz, 3H, C*H*$_3$) ppm. – **^{13}C NMR** (126 MHz, acetone-D$_6$) δ = 162.1 (C$_q$, *C*OO), 155.5 (C$_q$, *C*$_{ar}$), 154.3 (C$_q$, *C*$_{ar}$), 154.5 (C$_q$, 8a-*C*$_{ar}$), 139.9 (C$_q$, CH$_2$*C*$_{ar}$), 135.1 (+, *C*H), 130.1 (+, 2 × *C*$_{ar}$H), 129.4 (+, 2 × *C*$_{ar}$H), 127.3 (+, *C*$_{ar}$H), 127.0 (C$_q$, 3-*C*), 108.6 (+, *C*$_{ar}$H), 107.7 (C$_q$, 4a-*C*$_{ar}$), 105.9 (+, *C*$_{ar}$H), 44.7 (–, 7-CC*H*$_2$), 38.9 (C$_q$, *C*(CH$_3$)$_2$), 37.3 (–, 3-CC*H*$_2$), 29.2 (+, C(*C*H$_3$)$_2$), 27.8 (–, *C*H$_2$), 24.0 (–, *C*H$_2$), 14.3 (+, *C*H$_3$)

ppm. – **IR** (ATR, ṽ) = 3336, 2923, 1681, 1620, 1495, 1454, 1422, 1351, 1261, 1171, 1056, 957, 874, 844, 802, 754, 729, 699, 675, 557, 520, 500, 476, 435 cm^{-1}. – **HRMS** (ESI, C$_{23}$H$_{26}$O$_3$) = calc.: 351.1955 [M+H]$^+$; found: 351.1946 [M+H]$^+$.

5-Hydroxy-3-(2-methylbenzyl)-7-(2-methylhexan-2-yl)-2*H*-chromen-2-one (59lb)

According to **GP9**, to a solution of the 5-methoxycoumarin **21lb** (5.60 mg, 14.8 µmol, 1.00 equiv.) in abs. DCM (0.30 ml), was added BBr$_3$ (1 M in DCM, 0.07 ml, 19.0 mg, 0.07 mmol, 5.00 equiv.) at –78 °C. After stirring for 30 min at this temperature, the reaction mixture was allowed to warm to rt and stirred overnight. Filtration over a small silica plug (*n*Pen/EtOAc 2:1) resulted in the title compound as a white solid (4.80 mg, 13.2 µmol, 89%).

¹H NMR (500 MHz, acetone-D$_6$) δ = 9.19 (s, 1H, O*H*), 7.51 – 7.49 (m, 1H, C*H*), 7.27 – 7.21 (m, 2H, 2 × C*H*$_{ar}$), 7.21 – 7.18 (m, 2H, 2 × C*H*$_{ar}$), 6.80 (s, 2H, 2 × C*H*$_{ar}$), 3.84 (d, ^4J = 1.5 Hz, 2H, CC*H*$_2$), 2.30 (s, 3H, C$_{ar}$C*H*$_3$), 1.65 – 1.58 (m, 2H, 7-CC*H*$_2$), 1.28 (s, 6H, C(C*H*$_3$)$_2$), 1.21 (quin., ^3J = 7.2 Hz, 2H, C*H*$_2$), 1.09 – 0.99 (m, 2H, C*H*$_2$), 0.81 (t, ^3J = 7.4 Hz, 3H, C*H*$_3$) ppm. – **¹³C NMR** (126 MHz, acetone-D$_6$) δ = 162.1 (C$_q$, *C*OO), 155.4 (C$_q$, *C*$_{ar}$), 155.2 (C$_q$, *C*$_{ar}$), 154.4 (C$_q$, 8a-*C*$_{ar}$), 137.6 (C$_q$, *C*$_{ar}$), 137.4 (C$_q$, *C*$_{ar}$), 134.4 (+, *C*H), 131.3 (+, *C*$_{ar}$H), 131.0 (+, *C*$_{ar}$H), 127.8 (+, *C*$_{ar}$H), 127.1 (+, *C*$_{ar}$H), 126.1 (C$_q$, 3-*C*), 108.6 (+, *C*$_{ar}$H), 107.6 (C$_q$, 4a-*C*$_{ar}$), 105.9 (+, *C*$_{ar}$H), 44.7 (–, 7-C*C*H$_2$), 38.9 (C$_q$, *C*(CH$_3$)$_2$), 34.5 (–, 3-C*C*H$_2$), 29.2 (+, C(*C*H$_3$)$_2$), 27.7 (–, *C*H$_2$), 24.0 (–, *C*H$_2$), 19.6 (+, C$_{ar}$*C*H$_3$), 14.3 (+, *C*H$_3$) ppm. – **IR** (ATR, ṽ) = 3289, 2957, 2927, 2859, 1675, 1618, 1492, 1460, 1422, 1340, 1269, 1171, 1106, 1058, 937, 843, 760, 743, 727, 674, 528, 460, 383 cm^{-1}. – **HRMS** (ESI, C$_{24}$H$_{28}$O$_3$) calc.: 365.2111 [M+H]$^+$; found. 365.2107 [M+H]$^+$.

5-Hydroxy-3-(2-hydroxybenzyl)-7-(2-methylhexan-2-yl)-2*H*-chromen-2-one (59lc)

According to **GP9**, to a solution of the dimethoxycoumarin **21lc** (6.40 mg, 16.2 µmol, 1.00 equiv.) in abs. DCM (0.32 ml), was added BBr$_3$ (1 M in DCM, 0.16 ml, 41.0 mg, 0.16 mmol, 10.0 equiv.) at –78 °C. After stirring for 30 min at this temperature, the reaction mixture was allowed to warm to rt and stirred overnight. Filtration over a small silica plug (*n*Pen/EtOAc 2:1) resulted in the title compound as a white solid (2.00 mg, 5.45 µmol, 34%).

¹H NMR (500 MHz, acetone-D$_6$) δ = 9.21 (bs, 1H, O*H*), 8.45 (bs, 1H, O*H*), 7.77 (s, 1H, C*H*), 7.25 (dd, $^{3,4}J$ = 7.5, 1.7 Hz, 1H, C*H*$_{ar}$), 7.10 (td, $^{3,4}J$ = 7.5, 1.7 Hz, 1H, C*H*$_{ar}$), 6.89 (dd, $^{3,4}J$ = 8.1, 1.2 Hz, 1H, C*H*$_{ar}$), 6.83 (td, $^{3,4}J$ = 7.5, 1.2 Hz, 1H, C*H*$_{ar}$), 6.83 (d, ^4J = 1.5 Hz, 1H, C*H*$_{ar}$),

6.78 (d, ^4J = 1.5 Hz, 1H, CH_{ar}), 3.83 (d, ^4J = 1.2 Hz, 2H, CCH_2), 1.63 – 1.58 (m, 2H, 7-CCH_2), 1.27 (s, 6H, C(CH_3)$_2$), 1.23 – 1.17 (m, 2H, CH_2), 1.06 – 0.99 (m, 2H, CH_2), 0.79 (t, ^3J = 7.4 Hz, 3H, CH_3) ppm. – ^{13}C NMR (126 MHz, acetone-D$_6$) δ = 162.8 (C$_q$, COO), 156.3 (C$_q$, C_{ar}OH), 155.3 (C$_q$, C_{ar}), 155.2 (C$_q$, C_{ar}), 154.4 (C$_q$, 8a-C_{ar}), 134.8 (+, CH), 132.1 (+, C_{ar}H), 128.9 (+, C_{ar}H), 126.2 (C$_q$, C), 125.8 (C$_q$, C), 120.7 (+, C_{ar}H), 116.5 (+, C_{ar}H), 108.6 (+, C_{ar}H), 107.8 (C$_q$, 4a-C_{ar}), 105.9 (+, C_{ar}H), 44.7 (–, 7-CCH$_2$), 38.9 (C$_q$, C(CH$_3$)$_2$), 31.7 (–, 3-CCH$_2$), 29.2 (+, C(CH_3)$_2$), 27.8 (–, CH$_2$), 24.0 (–, CH$_2$), 14.3 (+, CH$_3$) ppm. – IR (ATR, ṽ) = 3296, 2956, 2925, 2855, 1677, 1616, 1489, 1456, 1422, 1343, 1236, 1170, 1099, 1060, 936, 845, 803, 751, 731, 672, 526 cm^{-1}. – HRMS (ESI, C$_{23}$H$_{26}$O$_4$) = calc.: 367.1899 [M+H]$^+$; found: 367.1899 [M+H]$^+$.

5-Hydroxy-3-(4-hydroxybenzyl)-7-(2-methylhexan-2-yl)-2H-chromen-2-one (59ld)

According to **GP9**, to a solution of the dimethoxycoumarin **21ld** (10.2 mg, 25.9 µmol, 1.00 equiv.) in abs. DCM (0.52 ml), was added BBr$_3$ (1 M in DCM, 0.26 ml, 65.0 mg, 260 µmol, 10.0 equiv.) at –78 °C. After stirring for 30 min at this temperature, the reaction mixture was allowed to warm to rt and stirred overnight. Purification by flash column chromatography (nPen/EtOAc 5:1) resulted in the title compound as a cloudy oil (9.2 mg, 25.1 µmol, 97%).

R_f (nPen/EtOAc 5:1) = 0.19. – ^{1}H NMR (500 MHz, acetone-D$_6$) δ = 9.05 (s, 1H, OH), 8.28 (s, 1H, OH), 7.75 (s, 1H, CH), 7.20 – 7.14 (m, 2H, 2 × CH_{ar}), 6.82 – 6.78 (m, 3H, 3 × CH_{ar}), 6.77 (d, ^4J = 1.5 Hz, 1H, CH_{ar}), 3.75 (s, 2H, CCH_2), 1.64 – 1.58 (m, 2H, 7-CCH_2), 1.27 (s, 6H, C(CH_3)$_2$), 1.25 – 1.17 (m, 2H, CH_2), 1.08 – 1.00 (m, 2H, CH_2), 0.80 (t, ^3J = 7.3 Hz, 3H, CH$_2$CH_3) ppm. – ^{13}C NMR (126 MHz, acetone-D$_6$) δ = 162.2 (C$_q$, COO), 157.0 (C$_q$, C_{ar}OH), 155.4 (C$_q$, C_{ar}), 155.3 (C$_q$, C_{ar}), 154.5 (C$_q$, 8a-C_{ar}), 134.7 (+, CH), 131.2 (+, 2 × C_{ar}H), 130.3 (C$_q$, CH$_2$$C_{ar}$), 127.7 (C$_q$, 3-$C$), 116.3 (+, 2 × C_{ar}H), 108.6 (+, C_{ar}H), 107.8 (C$_q$, 4a-C_{ar}), 105.9 (+, C_{ar}H), 44.8 (–, 7-CCH$_2$), 38.9 (C$_q$, C(CH$_3$)$_2$), 36.5 (–, 3-CCH$_2$), 29.2 (+, C(CH_3)$_2$), 27.8 (–, CH$_2$), 24.0 (–, CH$_2$), 14.4 (+, CH$_3$) ppm. – IR (ATR, ṽ) = 3296, 2956, 2931, 2859, 1677, 1618, 1575, 1511, 1422, 1346, 1241, 1171, 1103, 1061, 1018, 938, 840, 819, 798, 730, 670, 640, 530 cm^{-1}. – HRMS (ESI, C$_{23}$H$_{26}$O$_4$) = calc.: 367.1904 [M+H]$^+$; found: 367.1892 [M+H]$^+$.

3-(2-Fluorobenzyl)-5-hydroxy-7-(2-methylhexan-2-yl)-2H-chromen-2-one (59le)

According to **GP9**, to a solution of the 5-methoxycoumarin **21le** (4.50 mg, 11.8 µmol, 1.00 equiv.) in abs. DCM (0.24 ml), was added BBr$_3$ (1 M in DCM, 0.06 ml, 15.0 mg, 0.06 mmol, 5.00

equiv.) at –78 °C. After stirring for 30 min at this temperature, the reaction mixture was allowed to warm to rt and stirred overnight. Filtration over a small silica plug (*n*Pen/EtOAc 2:1) resulted in the title compound as an off-white solid (4.00 mg, 10.9 µmol, 92%).

¹H NMR (500 MHz, acetone-D₆) δ = 9.27 (s, 1H, O*H*), 7.77 (s, 1H, C*H*), 7.42 (td, $^{3,4}J$ = 7.5, 1.8 Hz, 1H, C*H*$_{ar}$), 7.32 (tdd, $^{3,3,4}J$ = 7.5, 5.3, 1.8 Hz, 1H, C*H*$_{ar}$), 7.19 – 7.11 (m, 2H, 2 × C*H*$_{ar}$), 6.81 (d, ^4J = 1.7 Hz, 1H, C*H*$_{ar}$), 6.79 (d, ^4J = 1.7 Hz, 1H, C*H*$_{ar}$), 3.89 (s, 2H, C*H*₂), 1.63 – 1.59 (m, 2H, 7-CC*H*₂), 1.27 (s, 6H, C(C*H*₃)₂), 1.21 (quin., ^3J = 7.2 Hz, 2H, C*H*₂), 1.07 – 1.00 (m, 2H, C*H*₂), 0.80 (t, ^3J = 7.4 Hz, 3H, C*H*₃) ppm. – **¹³C NMR** (126 MHz, acetone-D₆) δ = 162.2 (C_q, d, ^1J = 244.4 Hz, C$_{ar}$F), 161.9 (C_q, COO), 155.7 (C_q, C$_{ar}$), 155.3 (C_q, C$_{ar}$), 154.5 (C_q, 8a-C$_{ar}$), 135.2 (+, C*H*), 132.6 (+, d, ^3J = 4.4 Hz, *m*-C$_{ar}$H), 129.7 (+, d, ^3J = 8.2 Hz, *m*-C$_{ar}$H), 126.4 (C_q, d, ^2J = 15.4 Hz, *o*-C$_{ar}$), 125.3 (+, d, ^4J = 3.6 Hz, *p*-C$_{ar}$H), 125.2 (C_q, 3-C), 116.2 (+, d, ^2J = 22.0 Hz, *o*-C$_{ar}$H), 108.7 (+, C$_{ar}$H), 107.6 (C_q, 4a-C$_{ar}$), 105.9 (+, C$_{ar}$H), 44.7 (–, 7-CC*H*₂), 39.0 (C_q, *C*(CH₃)₂), 29.2 (+, C(*C*H₃)₂), 27.8 (–, *C*H₂), 24.0 (–, *C*H₂), 14.3 (+, *C*H₃) ppm. – **IR** (ATR, ṽ) = 3275, 2957, 2927, 2858, 1679, 1617, 1491, 1455, 1422, 1341, 1283, 1232, 1174, 1099, 1059, 938, 845, 754, 673, 558, 522, 432, 394 cm⁻¹. – **HRMS** (ESI, C₂₃H₂₅O₃F) = calc.: 369.1860 [M+H]⁺; found: 369.1858 [M+H]⁺.

3-(4-Fluorobenzyl)-5-hydroxy-7-(2-methylhexan-2-yl)-2*H*-chromen-2-one (59lf)

According to **GP9**, to a solution of the 5-methoxycoumarin **21lf** (8.00 mg, 20.7 µmol, 1.00 equiv.) in abs. DCM (0.41 ml), was added BBr₃ (1 M in DCM, 0.10 ml, 26.0 mg, 0.10 mmol, 5.00 equiv.) at –78 °C. After stirring for 30 min at this temperature, the reaction mixture was allowed to warm to rt and stirred overnight. Filtration over a small silica plug (*n*Pen/EtOAc 2:1) resulted in the title compound as an off-white solid (5.80 mg, 14.9 µmol, 72%).

¹H NMR (500 MHz, acetone-D₆) δ = 9.31 (s, 1H, O*H*), 7.84 (s, 1H, C*H*), 7.42 – 7.36 (m, 2H, 2 × C*H*$_{ar}$), 7.08 (t, ^3J = 8.9 Hz, 2H, 2 × C*H*$_{ar}$), 6.82 (d, ^4J = 1.7 Hz, 1H, C*H*$_{ar}$), 6.77 (d, ^4J = 1.7 Hz, 1H, C*H*$_{ar}$), 3.85 (s, 2H, CC*H*₂), 1.65 – 1.57 (m, 2H, 7-CC*H*₂), 1.27 (s, 6H, C(C*H*₃)₂), 1.24 – 1.16 (m, 2H, C*H*₂), 1.07 – 0.99 (m, 2H, C*H*₂), 0.80 (t, ^3J = 7.4 Hz, 3H, C*H*₃) ppm. – **¹³C NMR** (126 MHz, acetone-D₆) δ = 162.5 (C_q, d, ^1J = 242.5 Hz, C$_{ar}$F), 162.0 (C_q, COO), 155.6 (C_q, C$_{ar}$), 155.3 (C_q, C$_{ar}$), 154.6 (C_q, 8a-C$_{ar}$), 136.0 (C_q, d, ^4J = 3.2 Hz, *p*-C$_{ar}$), 135.2 (+, C*H*), 131.8 (+, d, ^3J = 8.1 Hz, 2 × *m*-C$_{ar}$H), 126.7 (C_q, 3-C), 116.0 (+, d, ^2J = 21.4 Hz, 2 × *o*-C$_{ar}$), 108.6 (+, C$_{ar}$H), 107.6 (C_q, 4a-C$_{ar}$), 105.9 (+, C$_{ar}$H), 44.7 (–, 7-CC*H*₂), 38.9 (C_q, *C*(CH₃)₂), 36.5 (–, 3-CC*H*₂), 29.1 (+, C(*C*H₃)₂), 27.7 (–, *C*H₂), 24.0 (–, *C*H₂), 14.3 (+, *C*H₃) ppm. – **IR** (ATR, ṽ) = 3351, 2962, 2926, 2872, 2856, 1682, 1615, 1572, 1507, 1423, 1338, 1272, 1221, 1157, 1100,

1060, 838, 792, 727, 669, 583, 526, 487, 451, 406 cm^{-1}. – **HRMS** (ESI, C$_{23}$H$_{25}$O$_3$F) = calc.: 369.1860 [M+H]$^+$; found: 369.1849 [M+H]$^+$.

5-Hydroxy-7-(2-methylhexan-2-yl)-3-(2-(trifluoromethyl)benzyl)-2*H*-chromen-2-one (59lg)

According to **GP9**, to a solution of the 5-methoxycoumarin **21lg** (4.20 mg, 9.71 μmol, 1.00 equiv.) in abs. DCM (0.19 ml), was added BBr$_3$ (1 M in DCM, 0.05 ml, 12.0 mg, 0.05 mmol, 5.00 equiv.) at –78 °C. After stirring for 30 min at this temperature, the reaction mixture was allowed to warm to rt and stirred overnight. Filtration over a small silica plug (*n*Pen/EtOAc 2:1) resulted in the title compound as an off-white solid (4.00 mg, 9.47 μmol, 98%, purity 75%).

^{1}H NMR (500 MHz, acetone-D$_6$) δ = 9.22 (s, 1H. ,O*H*), 7.78 (d, ^3J = 7.3 Hz, 1H, C*H*$_{ar}$), 7.65 (t, ^3J = 7.6 Hz, 1H, C*H*$_{ar}$), 7.57 – 7.46 (m, 2H, 2 × C*H*$_{ar}$), 6.80 (s, 2H, 2 × C*H*$_{ar}$), 4.07 (s, 2H, CC*H*$_2$), 1.65 – 1.56 (m, 2H, 7-CC*H*$_2$), 1.27 (s, 6H, C(C*H*$_3$)$_2$), 1.24 – 1.17 (m, 2H, C*H*$_2$), 1.08 – 0.98 (m, 2H, C*H*$_2$), 0.80 (t, ^3J = 7.3 Hz, 3H, C*H*$_3$) ppm. – **^{13}C NMR** (126 MHz, acetone-D$_6$) δ = 161.9 (C$_q$, *C*OO), 155.8 (C$_q$, *C*$_{ar}$), 155.3 (C$_q$, *C*$_{ar}$), 154.5 (C$_q$, 8a-*C*$_{ar}$), 137.9 (C$_q$, CH$_2$*C*$_{ar}$), 135.6 (+, *C*H), 133.5 (+, *C*$_{ar}$H), 133.0 (+, *C*$_{ar}$H), 129.4 (C$_q$, d, ^2J = 30.1 Hz, *C*$_{ar}$CF$_3$), 128.2 (+, *C*$_{ar}$H), 127.1 (+, q, ^3J = 6.2 Hz, *o*-*C*$_{ar}$H), 125.8 (C$_q$, 3-*C*), 125.7 (C$_q$, q, ^1J = 273.4 Hz, *C*$_{ar}$F$_3$), 108.7 (+, *C*$_{ar}$H), 107.5 (C$_q$, 4a-*C*$_{ar}$), 105.9 (+, *C*$_{ar}$H), 44.7 (–, 7-C*C*H$_2$), 39.0 (C$_q$, *C*(CH$_3$)$_2$), 33.8 (–, ^4J = 6.2 Hz, 3-C*C*H$_2$), 29.2 (+, C(*C*H$_3$)$_2$), 27.8 (–, *C*H$_2$), 24.0 (–, *C*H$_2$), 14.4 (+, *C*H$_3$) ppm. – **IR** (ATR, ṽ) = 3342, 2924, 2854, 1678, 1622, 1455, 1424, 1350, 1313, 1269, 1160, 1109, 1058, 1036, 955, 870, 845, 768, 729, 675, 649, 558, 524, 474, 431 cm^{-1}. – **HRMS** (ESI, C$_{24}$H$_{25}$O$_3$F$_3$) = calc.: 419.1829 [M+H]$^+$; found: 419.1823 [M+H]$^+$.

3-Benzyl-5-hydroxy-7-(2-methyloctan-2-yl)-2*H*-chromen-2-one (59ma)

According to **GP9**, to a solution of the 5-methoxycoumarin **21ma** (15.5 mg, 39.5 μmol, 1.00 equiv.) in abs. DCM (0.79 ml), was added BBr$_3$ (1 M in DCM, 0.20 ml, 49.0 mg, 200 μmol, 5.00 equiv.) at –78 °C. After stirring for 30 min at this temperature, the reaction mixture was allowed to warm to rt and stirred overnight. Purification by flash column chromatography (*n*Pen/EtOAc 5:1) resulted in the title compound as an off-white solid (10.6 mg, 28.1 μmol, 71%).[158]

R$_f$ (*n*Pen/EtOAc 7:1) = 0.23. – **^{1}H NMR** (500 MHz, acetone-D$_6$) δ = 9.19 (s, 1H, O*H*), 7.81 (s, 1H, C*H*), 7.39 – 7.29 (m, 4H, 4 × C*H*$_{ar}$), 7.26 – 7.20 (m, 1H, C*H*$_{ar}$), 6.81 (d, ^4J = 1.6 Hz, 1H,

CH_{ar}), 6.78 (d, ^4J = 1.6 Hz, 1H, CH_{ar}), 3.86 (s, 2H, CCH_2), 1.65 – 1.58 (m, 2H, 7-CCH_2), 1.27 (s, 6H, C(CH_3)$_2$), 1.25 – 1.15 (m, 6H, 3 × CH_2), 1.10 – 1.02 (m, 2H, CH_2), 0.82 (t, ^3J = 6.8 Hz, 3H, CH_3) ppm. – ^{13}C NMR (126 MHz, acetone-D$_6$) δ = 162.1 (C$_q$, COO), 155.6 (C$_q$, C_{ar}$), 155.4 (C$_q$, C_{ar}$), 154.5 (C$_q$, 8a-C_{ar}$), 140.0 (C$_q$, CH$_2$$C$_{ar}$), 135.2 (+, CH), 130.1 (+, 2 × C_{ar}$H), 129.5 (+, 2 × C_{ar}$H), 127.4 (+, C_{ar}$H), 127.0 (C$_q$, 3-$C$), 108.7 (+, C_{ar}$H), 107.7 (C$_q$, 4a-C_{ar}$), 106.0 (+, C_{ar}$H), 45.0 (–, 7-CCH_2), 39.0 (C$_q$, C(CH$_3$)$_2$), 37.3 (–, 3-CCH$_2$), 32.6 (–, CH$_2$), 30.8 (–, CH$_2$), 29.2 (+, C(CH$_3$)$_2$), 25.5 (–, CH$_2$), 23.4 (–, CH$_2$), 14.4 (+, CH$_3$) ppm. – IR (ATR, ṽ) = 3308, 2956, 2924, 2870, 2853, 1674, 1622, 1578, 1523, 1494, 1465, 1453, 1422, 1387, 1354, 1313, 1271, 1252, 1221, 1204, 1174, 1101, 1078, 1057, 1031, 953, 938, 922, 897, 873, 841, 756, 728, 698, 673, 605, 596, 557, 528, 510, 479, 439 cm^{-1}. – HRMS (ESI, C$_{25}$H$_{30}$O$_3$) = calc.: 379.2268 [M+H]$^+$; found: 379.2257 [M+H]$^+$.

5-Hydroxy-3-(2-methylbenzyl)-7-(2-methyloctan-2-yl)-2H-chromen-2-one (59mb)

According to GP9, to a solution of the 5-methoxycoumarin 21mb (13.1 mg, 32.2 μmol, 1.00 equiv.) in abs. DCM (0.64 ml), was added BBr$_3$ (1 M in DCM, 0.16 ml, 40.0 mg, 160 μmol, 5.00 equiv.) at –78 °C. After stirring for 30 min at this temperature, the reaction mixture was allowed to warm to rt and stirred overnight. Purification by flash column chromatography (nPen/EtOAc 7:1) resulted in the title compound as an off-white solid (8.3 mg, 21.3 μmol, 66%).

R_f (nPen/EtOAc 7:1) = 0.25. – ^{1}H NMR (500 MHz, acetone-D$_6$) δ = 9.12 (s, 1H, OH), 7.52 – 7.47 (m, 1H, CH), 7.24 (ddd, $^{3,3,4}J$ = 8.9, 5.4, 3.9 Hz, 2H, 2 × CH_{ar}), 7.20 – 7.17 (m, 2H, 2 × CH_{ar}), 6.79 (s, 2H, 2 × CH_{ar}), 3.84 (d, ^4J = 1.4 Hz, 2H, CCH_2), 2.30 (s, 3H, C$_{ar}$CH_3), 1.65 – 1.57 (m, 2H, 7-CCH_2), 1.27 (s, 6H, C(CH_3)$_2$), 1.23 – 1.17 (m, 6H, 3 × CH_2), 1.11 – 1.03 (m, 2H, CH_2), 0.86 – 0.80 (m, 3H, CH$_2$CH_3) ppm. – ^{13}C NMR (126 MHz, acetone-D$_6$) δ = 162.2 (C$_q$, COO), 155.5 (C$_q$, C_{ar}$), 155.2 (C$_q$, C_{ar}$), 154.4 (C$_q$, 8a-C_{ar}$), 137.6 (C$_q$, C_{ar}$), 137.5 (C$_q$, C_{ar}$), 134.5 (+, CH), 131.3 (+, C_{ar}$H), 131.1 (+, C_{ar}$H), 127.9 (+, C_{ar}$H), 127.2 (+, C_{ar}$H), 126.2 (C$_q$, 3-C), 108.7 (+, C_{ar}$H), 107.7 (C$_q$, 4a-$C$_{ar}$), 106.0 (+, C_{ar}$H), 45.0 (–, 7-C$CH_2$), 39.0 (C$_q$, C(CH$_3$)$_2$), 34.6 (–, 3-CCH$_2$), 32.6 (–, CH$_2$), 30.8 (–, CH$_2$), 29.2 (+, C(CH$_3$)$_2$), 25.5 (–, CH$_2$), 23.4 (–, CH$_2$), 19.6 (+, C$_{ar}$$CH_3$), 14.4 (+, CH$_3$) ppm. – IR (ATR, ṽ) = 3286, 2956, 2927, 2871, 2857, 1674, 1621, 1575, 1422, 1387, 1269, 1173, 1102, 1079, 1060, 844, 727, 674, 458 cm^{-1}. – HRMS (ESI, C$_{26}$H$_{32}$O$_3$) = calc.: 393.2424 [M+H]$^+$; found: 393.2409 [M+H]$^+$.

5-Hydroxy-3-(2-hydroxybenzyl)-7-(2-methyloctan-2-yl)-2*H*-chromen-2-one (59mc)

According to **GP9**, to a solution of the dimethoxycoumarin **21mc** (13.0 mg, 31.2 μmol, 1.00 equiv.) in abs. DCM (0.62 ml), was added BBr$_3$ (1 M in DCM, 0.31 ml, 78.0 mg, 310 μmol, 10.0 equiv.) at −78 °C. After stirring for 30 min at this temperature, the reaction mixture was allowed to warm to rt and stirred overnight. Purification by flash column chromatography (*n*Pen/EtOAc 5:1) resulted in the title compound as a cloudy oil (8.4 mg, 21.5 μmol, 69%).[158]

R$_f$ (*n*Pen/EtOAc 5:1) = 0.10. – **^{1}H NMR** (500 MHz, acetone-D$_6$) δ = 9.13 (s, 1H, O*H*), 8.43 (s, 1H, O*H*), 7.78 (d, ^4J = 1.4 Hz, 1H, C*H*), 7.26 (dd, $^{3,4}J$ = 7.5, 1.7 Hz, 1H, *o*-C*H*$_{ar}$), 7.11 (td, $^{3,4}J$ = 7.7, 1.7 Hz, 1H, *p*-C*H*$_{ar}$), 6.89 (dd, $^{3,4}J$ = 8.1, 1.2 Hz, 1H, *m*-C*H*$_{ar}$), 6.83 (td, $^{3,4}J$ = 7.4, 1.2 Hz, 1H, *m*-C*H*$_{ar}$), 6.80 (d, ^4J = 1.6 Hz, 1H, C*H*$_{ar}$), 6.79 (d, ^4J = 1.6 Hz, 1H, C*H*$_{ar}$), 3.83 (d, ^4J = 1.4 Hz, 2H, CC*H*$_2$), 1.65 – 1.56 (m, 2H, 7-CC*H*$_2$), 1.27 (s, 6H, C(C*H*$_3$)$_2$), 1.23 – 1.16 (m, 6H, 3 × C*H*$_2$), 1.09 – 1.01 (m, 2H, C*H*$_2$), 0.84 – 0.80 (m, 3H, CH$_2$C*H*$_3$) ppm. – **^{13}C NMR** (126 MHz, acetone-D$_6$) δ = 162.8 (C$_q$, *C*OO), 156.3 (C$_q$, *C*$_{ar}$OH), 155.4 (C$_q$, *C*$_{ar}$), 155.2 (C$_q$, *C*$_{ar}$), 154.4 (C$_q$, 8a-*C*$_{ar}$), 134.9 (+, *C*H), 132.1 (+, *C*$_{ar}$H), 129.0 (+, *C*$_{ar}$H), 126.3 (C$_q$, 3-*C*), 125.8 (C$_q$, CH$_2$*C*$_{ar}$), 120.8 (+, *C*$_{ar}$H), 116.5 (+, *C*$_{ar}$H), 108.7 (+, *C*$_{ar}$H), 107.9 (C$_q$, 4a-*C*$_{ar}$), 105.9 (+, *C*$_{ar}$H), 45.1 (−, 7-C*C*H$_2$), 39.0 (C$_q$, *C*(CH$_3$)$_2$), 32.6 (−, *C*H$_2$), 31.8 (−, 3-C*C*H$_2$), 30.8 (−, *C*H$_2$), 29.2 (+, C(*C*H$_3$)$_2$), 25.5 (−, *C*H$_2$), 23.4 (−, *C*H$_2$), 14.4 (+, *C*H$_3$) ppm. – **IR** (ATR, ṽ) = 3292, 2956, 2925, 2854, 1676, 1616, 1572, 1506, 1489, 1455, 1422, 1388, 1377, 1341, 1258, 1235, 1174, 1099, 1061, 1041, 955, 924, 846, 805, 751, 732, 673, 605, 577, 560, 527, 497, 476, 443, 407 cm^{-1}. – **HRMS** (ESI, C$_{25}$H$_{30}$O$_4$) = calc.: 395.2217 [M+H]$^+$; found: 395.2207 [M+H]$^+$.

5-Hydroxy-3-(4-hydroxybenzyl)-7-(2-methyloctan-2-yl)-2*H*-chromen-2-one (59md)

According to **GP9**, to a solution of the dimethoxycoumarin **21md** (10.3 mg, 24.4 μmol, 1.00 equiv.) in abs. DCM (0.49 ml), was added BBr$_3$ (1 M in DCM, 0.24 ml, 61.0 mg, 240 μmol, 10.0 equiv.) at −78 °C. After stirring for 30 min at this temperature, the reaction mixture was allowed to warm to rt and stirred overnight. Purification by flash column chromatography (*n*Pen/EtOAc 5:1) resulted in the title compound as a cloudy oil (9.6 mg, 24.4 μmol, quant.).

R$_f$ (*n*Pen/EtOAc 5:1) = 0.17. – **^{1}H NMR** (500 MHz, acetone-D$_6$) δ = 9.15 (s, 1H, O*H*), 8.19 (s, 1H, O*H*), 7.75 (d, ^4J = 1.5 Hz, 1H, C*H*), 7.21 – 7.14 (m, 2H, 2 × C*H*$_{ar}$), 6.82 – 6.78 (m, 3H, 3 × C*H*$_{ar}$), 6.77 (d, ^4J = 1.6 Hz, 1H, C*H*$_{ar}$), 3.75 (s, 2H, CC*H*$_2$), 1.65 – 1.57 (m, 2H, 7-CC*H*$_2$), 1.27

(s, 6H, C(CH$_3$)$_2$), 1.23 – 1.16 (m, 6H, 3 × CH$_2$), 1.10 – 1.02 (m, 2H, CH$_2$), 0.84 – 0.80 (m, 3H, CH$_2$CH$_3$) ppm. – ^{13}C NMR (126 MHz, acetone-D$_6$) δ = 162.2 (C$_q$, COO), 157.0 (C$_q$, C$_{ar}$OH), 155.4 (C$_q$, C$_{ar}$), 155.3 (C$_q$, C$_{ar}$), 154.5 (C$_q$, 8a-C$_{ar}$), 134.7 (+, CH), 131.2 (+, 2 × C$_{ar}$H), 130.3 (C$_q$, CH$_2$C$_{ar}$), 127.7 (C$_q$, 3-C), 116.3 (+, 2 × C$_{ar}$H), 108.6 (+, C$_{ar}$H), 107.8 (C$_q$, 4a-C$_{ar}$), 105.9 (+, C$_{ar}$H), 45.0 (–, 7-CCH$_2$), 39.0 (C$_q$, C(CH$_3$)$_2$), 36.5 (–, 3-CCH$_2$), 32.6 (–, CH$_2$), 30.8 (–, CH$_2$), 29.2 (+, C(CH$_3$)$_2$), 25.5 (–, CH$_2$), 23.4 (–, CH$_2$), 14.4 (+, CH$_3$) ppm. – IR (ATR, ṽ) = 3318, 2955, 2922, 2851, 1679, 1625, 1579, 1516, 1456, 1424, 1375, 1358, 1316, 1269, 1248, 1205, 1176, 1103, 1078, 1057, 955, 840, 819, 730, 674, 530 cm^{-1}. – HRMS (ESI, C$_{25}$H$_{30}$O$_4$) = calc.: 395.2217 [M+H]$^+$; found: 395.2204 [M+H]$^+$.

3-(2-Fluorobenzyl)-5-hydroxy-7-(2-methyloctan-2-yl)-2H-chromen-2-one (59me)

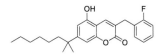

According to GP9, to a solution of the 5-methoxycoumarin 21me (13.0 mg, 32.4 μmol, 1.00 equiv.) in abs. DCM (0.65 ml), was added BBr$_3$ (1 M in DCM, 0.16 ml, 41.0 mg, 160 μmol, 5.00 equiv.) at –78 °C. After stirring for 30 min at this temperature, the reaction mixture was allowed to warm to rt and stirred overnight. Purification by flash column chromatography (nPen/EtOAc 7:1) resulted in the title compound as a brown solid (9.5 mg, 24.0 μmol, 74%).

R$_f$ (nPen/EtOAc 7:1) = 0.16. – ^{1}H NMR (500 MHz, acetone-D$_6$) δ = 9.21 (s, 1H, OH), 7.81 – 7.74 (m, 1H, CH), 7.42 (td, $^{3,4}J$ = 7.7, 1.8 Hz, 1H, m-CH$_{ar}$), 7.32 (tdd, $^{3,4}J$ = 7.7, 5.2, 1.8 Hz, 1H, m-CH$_{ar}$), 7.20 – 7.11 (m, 2H, p- and o-CH$_{ar}$), 6.82 (d, ^4J = 1.6 Hz, 1H, CH$_{ar}$), 6.79 (d, ^4J = 1.6 Hz, 1H, CH$_{ar}$), 3.89 (d, ^4J = 1.3 Hz, 2H, 3-CCH$_2$), 1.67 – 1.55 (m, 2H, 7-CCH$_2$), 1.27 (s, 6H, C(CH$_3$)$_2$), 1.24 – 1.17 (m, 6H, 3 × CH$_2$), 1.10 – 1.02 (m, 2H, CH$_2$CH$_3$), 0.84 – 0.79 (m, 3H, CH$_2$CH$_3$) ppm. – ^{13}C NMR (126 MHz, acetone-D$_6$) δ = 162.2 (C$_q$, d, ^1J = 244.4 Hz, C$_{ar}$F), 161.9 (C$_q$, COO), 155.8 (C$_q$, C$_{ar}$), 155.4 (C$_q$, C$_{ar}$), 154.6 (C$_q$, 8a-C$_{ar}$), 135.2 (+, CH), 132.7 (+, d, ^3J = 4.2 Hz, m-C$_{ar}$H), 129.7 (+, d, ^3J = 8.2 Hz, m-C$_{ar}$H), 126.5 (C$_q$, d, ^2J = 15.4 Hz, CH$_2$C$_{ar}$), 125.4 (+, d, ^4J = 3.6 Hz, p-C$_{ar}$H), 125.3 (C$_q$, 3-C), 116.2 (+, d, ^2J = 22.0 Hz, o-C$_{ar}$H), 108.7 (+, C$_{ar}$H), 107.6 (C$_q$, 4a-C$_{ar}$), 106.0 (+, C$_{ar}$H), 45.0 (–, 7-CCH$_2$), 39.1 (C$_q$, C(CH$_3$)$_2$), 32.6 (–, CH$_2$), 30.8 (–, CH$_2$), 29.2 (+, C(CH$_3$)$_2$), 25.5 (–, CH$_2$), 23.4 (–, CH$_2$), 14.4 (+, CH$_3$) ppm. – ^{19}F NMR (471 MHz, acetone-D$_6$) δ = –119.1 (s, CF) ppm. – IR (ATR, ṽ) = 3272, 2956, 2927, 2870, 2856, 1673, 1621, 1578, 1523, 1492, 1465, 1455, 1424, 1387, 1350, 1322, 1309, 1285, 1268, 1252, 1231, 1194, 1176, 1101, 1081, 1060, 1034, 956, 941, 843, 752, 737, 725, 674, 521, 470, 433 cm^{-1}. – HRMS (ESI, C$_{25}$H$_{29}$O$_3$F) = calc.: 397.2173 [M+H]$^+$; found: 397.2162 [M+H]$^+$.

3-(4-Fluorobenzyl)-5-hydroxy-7-(2-methyloctan-2-yl)-2*H*-chromen-2-one (59mf)

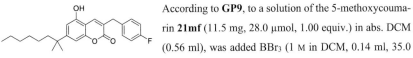

According to **GP9**, to a solution of the 5-methoxycoumarin **21mf** (11.5 mg, 28.0 μmol, 1.00 equiv.) in abs. DCM (0.56 ml), was added BBr$_3$ (1 M in DCM, 0.14 ml, 35.0 mg, 140 μmol, 5.00 equiv.) at –78 °C. After stirring for 30 min at this temperature, the reaction mixture was allowed to warm to rt and stirred overnight. Purification by flash column chromatography (*n*Pen/EtOAc 7:1) resulted in the title compound as an off-white solid (10.6 mg, 26.7 μmol, 95%).

R$_f$ (*n*Pen/EtOAc 7:1) = 0.23. – **^{1}H NMR** (500 MHz, acetone-D$_6$) δ = 9.21 (s, 1H, O*H*), 7.84 (s, 1H, C*H*), 7.44 – 7.37 (m, 2H, 2 × *m*-C*H*$_{ar}$), 7.08 (t, ^3J = 8.9 Hz, 2H, 2 × *o*-C*H*$_{ar}$), 6.81 (d, ^4J = 1.6 Hz, 1H, C*H*$_{ar}$), 6.78 (d, ^4J = 1.6 Hz, 1H, C*H*$_{ar}$), 3.85 (s, 2H, CC*H*$_2$), 1.65 – 1.58 (m, 2H, 7-CC*H*$_2$), 1.27 (s, 6H, C(C*H*$_3$)$_2$), 1.22 – 1.17 (m, 6H, 3 × C*H*$_2$), 1.10 – 1.02 (m, 2H, C*H*$_2$CH$_3$), 0.84 – 0.80 (m, 3H, CH$_2$C*H*$_3$) ppm. – **^{13}C NMR** (126 MHz, acetone-D$_6$) δ = 162.6 (C$_q$, d, ^1J = 242.3 Hz, C$_{ar}$F), 162.1 (C$_q$, COO), 155.7 (C$_q$, C$_{ar}$), 155.4 (C$_q$, C$_{ar}$), 154.6 (C$_q$, 8a-C$_{ar}$), 136.0 (C$_q$, d, ^4J = 3.2 Hz, CH$_2$C$_{ar}$), 135.3 (+, CH), 131.9 (+, d, ^3J = 7.7 Hz, 2 × *m*-C$_{ar}$H), 126.9 (C$_q$, 3-C), 116.0 (+, d, ^2J = 21.3 Hz, 2 × *o*-C$_{ar}$H), 108.7 (+, C$_{ar}$H), 107.7 (C$_q$, 4a-C$_{ar}$), 106.0 (+, C$_{ar}$H), 45.1 (–, 7-CC*H*$_2$), 39.1 (C$_q$, *C*(CH$_3$)$_2$), 36.6 (–, 3-CC*H*$_2$), 32.6 (–, *C*H$_2$), 30.8 (–, *C*H$_2$), 29.2 (+, C(*C*H$_3$)$_2$), 25.5 (–, *C*H$_2$), 23.4 (–, *C*H$_2$), 14.4 (+, *C*H$_3$) ppm. – **IR** (ATR, ṽ) = 3305, 2956, 2921, 2851, 1677, 1625, 1579, 1511, 1463, 1424, 1388, 1378, 1357, 1316, 1290, 1271, 1251, 1231, 1204, 1176, 1160, 1101, 1079, 1058, 1018, 953, 901, 885, 874, 844, 826, 782, 764, 728, 673, 659, 606, 567, 531, 503, 482, 453, 421 cm^{-1}. – **HRMS** (ESI, C$_{25}$H$_{29}$O$_3$F) = calc.: 397.2173 [M+H]$^+$; found: 397.2163 [M+H]$^+$.

5-Hydroxy-7-(2-methyloctan-2-yl)-3-(2-(trifluoromethyl)benzyl)-2*H*-chromen-2-one (59mg)

According to **GP9**, to a solution of the 5-methoxycoumarin 21mg (13.0 mg, 32.4 μmol, 1.00 equiv.) in abs. DCM (0.65 ml), was added BBr$_3$ (1 M in DCM, 0.16 ml, 41.0 mg, 160 μmol, 5.00 equiv.) at –78 °C. After stirring for 30 min at this temperature, the reaction mixture was allowed to warm to rt and stirred overnight. Purification by flash column chromatography (*n*Pen/EtOAc 5:1) resulted in a mixture of the product and an unidentified impurity. Another purification by prepHPLC (5-95% in 41 min, 10 ml/min, 218 nm) gave the title compound as a yellow oil (6.2 mg, 13.7 μmol, 58%).

R_t (5%–95% in 41 min) = 48.6 min. – **¹H NMR** (500 MHz, CDCl₃) δ = 7.70 (d, ^3J = 7.6 Hz, 1H, CH_{ar}), 7.54 – 7.47 (m, 2H, CH and CH_{ar}), 7.38 (t, ^3J = 6.8 Hz, 2H, 2 × CH_{ar}), 6.87 (d, ^4J = 1.5 Hz, 1H, CH_{ar}), 6.63 (d, ^4J = 1.5 Hz, 1H, CH_{ar}), 4.09 (s, 2H, 3-CCH_2), 1.59 – 1.52 (m, 2H, 7-CCH_2), 1.25 (s, 6H, C(CH_3)₂), 1.23 – 1.14 (m, 6H, 3 × CH_2), 1.05 – 0.97 (m, 2H, CH_2CH₃), 0.83 (t, ^3J = 7.0 Hz, 3H, CH₂CH_3) ppm. – **¹³C NMR** (126 MHz, CDCl₃) δ = 162.6 (C_q, COO), 155.4 (C_q, C_{ar}), 154.2 (C_q, C_{ar}), 151.9 (C_q, 8a-C_{ar}), 136.7 (C_q, CH₂C_{ar}), 135.3 (+, CH), 132.2 (+, C_{ar}H), 132.0 (+, C_{ar}H), 129.3 (C_q, q, ^2J = 29.9 Hz, C_{ar}CF₃), 127.1 (+, C_{ar}H), 126.4 (+, q, ^3J = 5.6 Hz, o-C_{ar}H), 125.4 (C_q, 3-C), 123.4 (C_q, q, ^1J = 273.9 Hz, CF₃), 108.2 (+, C_{ar}H), 106.9 (C_q, 4a-C_{ar}), 106.7 (+, C_{ar}H), 44.5 (–, 7-CCH₂), 38.5 (C_q, C(CH₃)₂), 33.1 (–, d, ^4J = 2.2 Hz, 3-CCH₂), 31.1 (–, CH₂), 30.1 (–, CH₂), 28.9 (+, C(CH₃)₂), 24.8 (–, CH₂), 22.8 (–, CH₂), 14.2 (+, CH₃) ppm. – **IR** (ATR, ṽ) = 3354, 2959, 2927, 2871, 2857, 1684, 1621, 1577, 1453, 1424, 1344, 1313, 1269, 1254, 1215, 1207, 1188, 1154, 1120, 1075, 1060, 1037, 958, 854, 768, 731, 677, 662, 650, 603, 560, 524, 514, 470 cm⁻¹. – **HRMS** (ESI, C₂₆H₂₉O₃F₃) = calc.: 447.2142 [M+H]⁺; found: 447.2129 [M+H]⁺.

5.1.9   Synthesis and characterisations of 3-benzylcoumarins

3-Benzyl-5,8-2H-chromen-2-one (21na)

According to **GP8**, 2-hydroxy-3,6-dimethylbenzaldehyde (**19n**, 50.0 mg, 0.33 mmol, 1.00 equiv.), cinnamaldehyde (0.10 ml, 110 mg, 0.83 mmol, 2.50 equiv.), K₂CO₃ (55.0 mg, 0.40 mmol, 1.20 equiv.) and 1,3-dimethylimidazolium dimethyl phosphate (0.07 ml, 89.0 mg, 0.40 mmol, 1.20 equiv.) in toluene (1.10 ml) were heated at 110 °C for 50 min *via* microwave irradiation. Purification by flash column chromatography (cHex/EtOAc 10:1) resulted in the title compound as a brown solid (53.0 mg, 0.20 mmol, 60%).

R_f (cHex/EtOAc 10:1) = 0.20. – **¹H NMR** (400 MHz, CDCl₃) δ = 7.53 (s, 1H, CH), 7.39 – 7.23 (m, 5H, 5 × CH_{ar}), 7.18 (d, ^3J = 7.5 Hz, 1H, CH_{ar}), 6.95 (d, ^3J = 7.5 Hz, 1H, CH_{ar}), 3.92 (s, 2H, CCH_2), 2.40 (s, 3H, CH_3), 2.37 (s, 3 H, CH_3) ppm. – **¹³C NMR** (101 MHz, CDCl₃) δ = 161.9 (C_q, COO), 152.0 (C_q, 8a-C_{ar}), 138.2 (C_q, CH₂C_{ar}), 136.9 (+, C_{ar}H), 132.9 (C_q, C_{ar}), 131.9 (+, C_{ar}H), 129.3 (+, 2 × C_{ar}H), 128.8 (+, 2 × C_{ar}H), 128.2 (C_q, C_{ar}), 126.8 (+, C_{ar}H), 125.2 (+, C_{ar}H), 123.7 (C_q, C_{ar}), 118.0 (C_q, C_{ar}), 36.9 (–, CCH₂), 18.2 (+, CH₃), 15.5 (+, CH₃) ppm. – **IR** (ATR, ṽ) = 2923, 1687, 1598, 1489, 1453, 1377, 1273, 1189, 1160, 1067, 1035, 990, 814, 773, 755, 742, 700, 646, 586, 534, 511, 466, 415, 386 cm⁻¹. – **MS** (70 eV, EI, 60 °C) *m/z* (%) = 265 (15) [M+H]⁺, 264 (72) [M]⁺, 249 (5) [M-CH₃]⁺, 236 (13), 235 (17), 231 (14), 222 (12), 199 (31), 181 (54), 173 (18) [M-C₇H₇]⁺, 132 (13), 131 (81), 119 (13), 115 (11), 105 (16), 104 (43), 103

(11), 100 (13), 91 (37), 78 (11), 77 (16), 69 (100). – **HRMS** (EI, $C_{18}H_{16}O_2$) = calc.: 264.1145; found: 264.1146.

3-(2-Methoxybenzyl)-5,8-dimethyl-2*H*-chromen-2-one (21nc)

According to **GP8**, 2-hydroxy-3,6-dimethylbenzaldehyde (**19n**, 100 mg, 0.67 mmol, 1.00 equiv.), 2-methoxycinnamaldehyde (270 mg, 1.66 mmol, 2.50 equiv.), K_2CO_3 (110 mg, 0.80 mmol, 1.20 equiv.) and 1,3-dimethylimidazolium dimethyl phosphate (0.14 ml, 178 mg, 0.80 mmol, 1.20 equiv.) in toluene (2.20 ml) were heated at 110 °C for 50 min *via* microwave irradiation. Purification by flash column chromatography (*c*Hex/EtOAc 10:1) resulted in the title compound as a brown solid (86.0 mg, 0.29 mmol, 42%).

R_f (*c*Hex/EtOAc 10:1) = 0.25. – **^{1}H NMR** (300 MHz, CDCl$_3$) δ = 7.52 (s, 1H, C*H*), 7.33 – 7.13 (m, 3H, 3 × C*H*$_{ar}$), 7.00 – 6.86 (m, 3H, 3 × C*H*$_{ar}$), 3.91 (s, 2H, CC*H*$_2$), 3.82 (s, 3H, OC*H*$_3$), 2.40 (s, 3H, C*H*$_3$), 2.34 (s, 3 H, C*H*$_3$) ppm. – The analytical data are consistent with the literature.[141]

3-(4-Methoxybenzyl)-5,8-dimethyl-2*H*-chromen-2-one (21nd)

According to **GP8**, 2-hydroxy-3,6-dimethylbenzaldehyde (**19n**, 100 mg, 0.67 mmol, 1.00 equiv.), 4-methoxycinnamaldehyde (270 mg, 1.66 mmol, 2.50 equiv.), K_2CO_3 (110 mg, 0.80 mmol, 1.20 equiv.) and 1,3-dimethylimidazolium dimethyl phosphate (0.14 ml, 178 mg, 0.40 mmol, 1.20 equiv.) in toluene (2.20 ml) were heated at 110 °C for 50 min *via* microwave irradiation. Purification by flash column chromatography (*c*Hex/EtOAc 10:1) resulted in the title compound as a brown solid (101 mg, 0.34 mmol, 52%).

R_f (*c*Hex/EtOAc 10:1) = 0.22. – **^{1}H NMR** (400 MHz, CDCl$_3$) δ = 7.51 (s, 1H, C*H*), 7.22 (d, ^3J = 8.6 Hz, 2H, 2 × C*H*$_{ar}$), 7.18 (d, ^3J = 7.5 Hz, 1H, C*H*$_{ar}$), 6.94 (d, ^3J = 7.5 Hz, 1H, C*H*$_{ar}$), 6.88 (d, ^3J = 8.6 Hz, 2H, 2 × C*H*$_{ar}$), 3.85 (s, 2H, CC*H*$_2$), 3.80 (s, 3H, OC*H*$_3$), 2.40 (s, 3H, C*H*$_3$), 2.37 (s, 3H, C*H*$_3$) ppm. – **^{13}C-NMR** (101 MHz, CDCl$_3$) δ = 161.9 (C$_q$, COO), 158.5 (C$_q$, C$_{ar}$OCH$_3$), 152.0 (C$_q$, 8a-C$_{ar}$), 136.6 (+, CH), 132.9 (C$_q$, C$_{ar}$), 131.8 (+, C$_{ar}$H), 130.4 (+, 2 × C$_{ar}$H), 130.1 (C$_q$, C$_{ar}$), 128.6 (C$_q$, C$_{ar}$), 125.1 (+, C$_{ar}$H), 123.6 (C$_q$, C$_{ar}$), 118.0 (C$_q$, C$_{ar}$), 114.2 (+, 2 × C$_{ar}$H), 55.4 (+, OCH$_3$), 36.1 (−, CCH$_2$), 18.2 (+, CH$_3$), 15.5 (+, CH$_3$) ppm. – **IR** (ATR, ṽ) = 2917, 1704, 1600, 1504, 1460, 1376, 1296, 1234, 1193, 1173, 1108, 1074, 1031, 925, 831, 813, 787, 754, 727, 647, 605, 579, 566, 518, 493, 478, 415 cm^{-1}. – **MS** (70 eV, EI, 110 °C) *m/z* (%) = 295 (24) [M+H]$^+$, 294 (100) [M]$^+$, 279 (13) [M–CH$_3$]$^+$, 265 (12), 263 (6) [M–OCH$_3$]$^+$, 187 (4)

[M–C$_7$H$_7$O]$^+$, 173 (2) [M–C$_8$H$_9$O]$^+$, 121 (11). – **HRMS** (EI, C$_{19}$H$_{18}$O$_3$) = calc.: 294.1250; found: 294.1251.

3-(4-Fluorobenzyl)-5,8-dimethyl-2*H*-chromen-2-one (21nf)

 According to **GP8**, 2-hydroxy-3,6-dimethylbenzaldehyde (**19n**, 50.0 mg, 0.33 mmol, 1.00 equiv.), 4-fluorocinnamaldehyde (0.11 ml, 125 mg, 0.83 mmol, 2.50 equiv.), K$_2$CO$_3$ (55.0 mg, 0.40 mmol, 1.20 equiv.) and 1,3-dimethylimidazolium dimethyl phosphate (0.07 ml, 89.0 mg, 0.40 mmol, 1.20 equiv.) in toluene (1.10 ml) were heated at 110 °C for 50 min *via* microwave irradiation. Purification by flash column chromatography (*c*Hex/EtOAc 10:1) resulted in the title compound as a brown solid (76.0 mg, 0.27 mmol, 81%).

R$_f$ (*c*Hex/EtOAc 10:1) = 0.25. – **^{1}H NMR** (400 MHz, CDCl$_3$) δ = 7.78 (s, 1H, C*H*), 7.55 – 7.48 (m, 2H, 2 × C*H*$_{ar}$), 7.44 (d, ^3J = 7.6 Hz, 1H, C*H*$_{ar}$), 7.31 – 7.21 (m, 2H, 2 × C*H*$_{ar}$), 7.20 (d, ^3J = 7.6 Hz, 1H, C*H*$_{ar}$), 4.13 (s, 2H, CC*H*$_2$), 2.64 (s, 3H, C*H*$_3$), 2.63 (s, 3H, C*H*$_3$) ppm. – **^{13}C NMR** (101 MHz, CDCl$_3$) δ = 161.9 (C$_q$, d, ^1J = 244.8 Hz, C$_{ar}$F), 161.8 (C$_q$, COO), 152.1 (C$_q$, 8a-C$_{ar}$), 136.9 (+, CH), 133.9 (C$_q$, d, ^4J = 3.2 Hz, CH$_2$C$_{ar}$), 133.0 (C$_q$, C$_{ar}$), 132.1 (+, C$_{ar}$H), 130.8 (+, d, ^3J = 3.2 Hz, 2 × *m*-C$_{ar}$H), 128.0 (C$_q$, C$_{ar}$), 125.7 (+, C$_{ar}$H), 123.3 (C$_q$, C$_{ar}$), 117.9 (C$_q$, C$_{ar}$), 115.5 (+, d, ^3J = 21.3 Hz, 2 × *o*-C$_{ar}$H), 36.3 (–, CCH$_2$), 18.2 (+, CH$_3$), 15.5 (+, CH$_3$) ppm. – **^{19}F NMR** (376 MHz, CDCl$_3$) δ = –120.64 (C*F*) ppm. – **IR** (ATR, ṽ) = 2921, 1697, 1598, 1504, 1430, 1376, 1215, 1190, 1560, 1100, 1074, 1033, 992, 926, 854, 814, 793, 755, 731, 716, 645, 605, 580, 559, 534, 488, 447, 417 cm^{-1}. – **MS** (70 eV, EI, 100 °C) *m/z* (%) = 283 (21) [M+H]$^+$, 282 (100) [M]$^+$, 281 (10), 267 (8) [M–CH$_3$]$^+$, 254 (12), 253 (26), 239 (12), 181 (15), 159 (11), 131 (14), 109 (25), 69 (20). – **HRMS** (EI, C$_{18}$H$_{15}$O$_2$F) = calc.: 282.1051; found: 282.1053.

3-(4-Chlorobenzyl)-5,8-dimethyl-2*H*-chromen-2-one (21nh)

 According to **GP8**, 2-hydroxy-3,6-dimethylbenzaldehyde (**19n**, 50.0 mg, 0.33 mmol, 1.00 equiv.), 4-chlorocinnamaldehyde (138 mg, 0.83 mmol, 2.50 equiv.), K$_2$CO$_3$ (55.0 mg, 0.40 mmol, 1.20 equiv.) and 1,3-dimethylimidazolium dimethyl phosphate (0.07 ml, 89.0 mg, 0.40 mmol, 1.20 equiv.) in toluene (1.10 ml) were heated at 110 °C for 50 min *via* microwave irradiation. Purification by flash column chromatography (*c*Hex/EtOAc 10:1) resulted in the title compound as a brown solid (33.0 mg, 0.11 mmol, 33%).

R$_f$ (*c*Hex/EtOAc 10:1) = 0.40. – **^{1}H NMR** (400 MHz, CDCl$_3$) δ = 7.56 (s, 1H, C*H*), 7.30 (d, ^3J = 8.5 Hz, 2H, 2 × C*H*$_{ar}$), 7.26 (d, ^3J = 8.5 Hz, 2H, 2 × C*H*$_{ar}$), 7.20 (d, ^3J = 7.5, 1H, C*H*$_{ar}$), 6.96

(d, ^3J = 7.5 Hz, 1H, CH_{ar}), 3.88 (s, 2H, CCH_2), 2.40 (s, 6H, 2 × CH_3) ppm. – ^{13}C NMR (101 MHz, CDCl$_3$) δ = 161.7 (C$_q$, COO), 152.1 (C$_q$, 8a-C$_{ar}$), 137.0 (+, CH), 136.8 (C$_q$, C$_{ar}$Cl), 133.0 (C$_q$, C$_{ar}$), 132.7 (C$_q$, C$_{ar}$), 132.1 (+, C$_{ar}$H), 130.6 (+, 2 × C$_{ar}$H), 128.9 (+, 2 × C$_{ar}$H), 127.6 (C$_q$, C$_{ar}$), 125.3 (+, C$_{ar}$H), 123.8 (C$_q$, C$_{ar}$), 117.9 (C$_q$, C$_{ar}$), 36.5 (–, CCH$_2$), 18.4 (+, CH$_3$), 15.5 (+, CH$_3$) ppm. – **IR** (ATR, ṽ) = 2922, 1705, 1599, 1486, 1429, 1403, 1376, 1260, 1190, 1175, 1159, 1073, 1032, 1011, 916, 849, 829, 800, 775, 749, 716, 642, 604, 578, 546, 530, 479, 403 cm^{-1}. – **MS** (70 eV, EI, 100 °C) m/z (%) = 299 (10) [M+H]$^+$, 300/298 (16/50) [M]$^+$, 283 (3) [M–CH$_3$]$^+$, 274/272 (11/33), 263 (22) [M–Cl]$^+$, 148 (11), 149/147 (3/100), 131 (10), 125 (17), 119 (31), 71/69 (3/12). – **HRMS** (EI, C$_{18}$H$_{15}$O$_2$Cl) = calc.: 298.0755; found: 298.0756.

3-(4-(Dimethylamino)benzyl)-5,8-dimethyl-2*H*-chromen-2-one (21ni)

According to **GP8**, 2-hydroxy-3,6-dimethylbenzaldehyde (**19n**, 50.0 mg, 0.33 mmol, 1.00 equiv.), 4-(dimethylamino)cinnamaldehyde (146 mg, 0.83 mmol, 2.50 equiv.), K$_2$CO$_3$ (55.0 mg, 0.40 mmol, 1.20 equiv.) and 1,3-dimethylimidazolium dimethyl phosphate (0.07 ml, 89.0 mg, 0.40 mmol, 1.20 equiv.) in toluene (1.10 ml) were heated at 110 °C for 50 min *via* microwave irradiation. Purification by flash column chromatography (*c*Hex/EtOAc 7:1) resulted in the title compound as a brown solid (34.0 mg, 0.11 mmol, 33%).

R_f (CH/EE 7:1) = 0.28. – ^{1}H NMR (400 MHz, CDCl$_3$) δ = 7.53 (s, 1H, CH), 7.21 – 7.13 (m, 3H, 3 × CH_{ar}), 6.94 (d, ^3J = 7.6Hz, 1H, CH_{ar}), 6.73 (d, ^3J = 8.3 Hz, 2H, 2 × CH_{ar}), 3.82 (s, 2H, CCH_2), 2.94 (s, 6H, 2 × NCH_3), 2.40 (s, 3H, CH_3), 2.37 (s, 3H, CH_3) ppm. – ^{13}C NMR (101 MHz, CDCl$_3$) δ = 162.1 (C$_q$, COO), 152.0 (C$_q$, 8a-C$_{ar}$), 149.7 (C$_q$, C$_{ar}$N(CH$_3$)$_2$), 136.9 (+, C$_{ar}$H), 132.9 (C$_q$, C$_{ar}$), 131.7 (+, C$_{ar}$H), 130.1 (+, 2 × C$_{ar}$H), 129.1 (C$_q$, C$_{ar}$), 125.9 (C$_q$, C$_{ar}$), 125.1 (+, C$_{ar}$H), 123.6 (C$_q$, C$_{ar}$), 118.2 (C$_q$, C$_{ar}$), 113.2 (+, 2 × C$_{ar}$H), 40.9 (+, N(CH$_3$)$_2$), 36.0 (–, CCH$_2$), 18.3 (+, CH$_3$), 15.5 (+, CH$_3$) ppm. – **IR** (ATR, ṽ) = 2920, 1698, 1602, 1524, 1491, 1421, 1353, 1260, 1230, 1198, 1165, 1066, 1028, 948, 894, 857, 827, 807, 762, 750, 731, 664, 607, 574, 549, 511, 476 cm^{-1}. – **MS** (70 eV, EI, 100 °C) m/z (%) = 308 (21) [M+H]$^+$, 307 (100) [M]$^+$, 306 (17), 263 (7) [M-C$_2$H$_6$N]$^+$, 262 (11), 231 (12), 134 (19). – **HRMS** (EI, C$_{20}$H$_{21}$O$_2$N) = calc.: 307.1567; found: 307.1568.

3-(2-Hydroxybenzyl)-5,8-dimethyl-2*H*-chromen-2-one (60a)

According to **GP9**, to a solution of the methoxycoumarin **21nc** (21.0 mg, 71.3 μmol, 1.00 equiv.) in abs. DCM (1.43 ml), was added BBr$_3$ (1 M in DCM, 0.36 ml, 89.0 mg, 357 μmol, 5.00 equiv.) at –78 °C. After stirring for 30 min at this temperature, the reaction mixture was allowed to warm to rt and stirred over-night. Purification by flash column chromatography (cHex/EtOAc 3:1) resulted in the title com-pound as an off-white solid (14.8 mg, 52.8 µmol, 74%).

R_f (cHex/EtOAc 5:1) = 0.36. – ^{1}H NMR (400 MHz, CDCl$_3$) δ = 8.13 (bs, 1H, OH), 7.93 (s, 1H, CH), 7.24 – 7.20 (m, 2H, 2 × CH_{ar}), 7.15 (td, $^{3,4}J$ = 7.7, 1.7 Hz, 1H, CH_{ar}), 7.01 (d, ^3J = 7.5 Hz, 1H, CH_{ar}), 6.96 (dd, $^{3,4}J$ = 8.1, 1.3 Hz, 1H, CH_{ar}), 6.88 (td, $^{3,4}J$ = 7.5, 1.3 Hz, 1H, CH_{ar}), 3.89 (s, 2H, CCH_2), 2.51 (s, 3H, CH_3), 2.38 (s, 3H, CH_3) ppm. – ^{13}C-NMR (101 MHz, CDCl$_3$) δ = 164.7 (C$_q$, COO), 154.8 (C$_q$, C_{ar}OH), 151.9 (C$_q$, 8a-C_{ar}), 138.1 (+, CH), 133.2 (C$_q$, C_{ar}), 132.5 (+, C_{ar}H), 130.6 (+, C_{ar}H), 128.9 (+, C_{ar}H), 127.3 (C$_q$, C_{ar}), 125.8 (+, C_{ar}H), 125.1 (C$_q$, C_{ar}), 123.9 (C$_q$, C_{ar}), 121.0 (+, C_{ar}H), 118.4 (+, C_{ar}H), 118.2 (C$_q$, C_{ar}), 32.6 (–, CCH$_2$), 18.3 (+, CH$_3$), 15.4 (+, CH$_3$) ppm. – IR (ATR, ṽ) = 3266, 3075, 3041, 2950, 2924, 2854, 2726, 1681, 1593, 1487, 1456, 1368, 1237, 1193, 1168, 1106, 1082, 1040, 916, 859, 824, 798, 744, 689, 651, 610, 579, 503, 480, 451 cm^{-1}. – MS (70 eV, EI, 100 °C) m/z (%) = 281 (20) [M+H]$^+$, 280 (74) [M]$^+$, 251 (11), 174 (27), 159 (14), 146 (13), 121 (33), 119 (17), 115 (14), 100 (11), 91 (13), 77 (12), 57 (15). – HRMS (EI, C$_{18}$H$_{16}$O$_3$) = calc.: 280.1099; found: 280.1098.

3-(4-Hydroxybenzyl)-5,8-dimethyl-2H-chromen-2-one (60b)

According to **GP9**, to a solution of the methoxycoumarin **21nd** (15.0 mg, 51.0 µmol, 1.00 equiv.) in abs. DCM (1.02 ml), was added BBr$_3$ (1 M in DCM, 0.25 ml, 64.0 mg, 255 µmol, 5.00 equiv.) at –78 °C. After stirring for 30 min at this temperature, the reaction mixture was allowed to warm to rt and stirred overnight. Purification by flash column chromatography (cHex/EtOAc 5:1) resulted in the title compound as an off-white solid (9.0 mg, 32.6 µmol, 64%).

R_f (cHex/EtOAc 5:1) = 0.25. – ^{1}H NMR (400 MHz, CDCl$_3$) δ = 7.53 (d, ^4J = 1.3 Hz, 1H, CH), 7.17 (t, ^3J = 7.9 Hz, 3H, 3 × CH_{ar}), 6.95 (d, ^3J = 7.5 Hz, 1H, CH_{ar}), 6.82 (d, ^4J = 2.1 Hz, 1H, CH_{ar}), 6.80 (d, ^4J = 2.1 Hz, 1H, CH_{ar}), 5.12 (bs, 1H, OH), 3.84 (d, ^4J = 1.3 Hz, 2H, CCH_2), 2.39 (s, 3H, CH_3), 2.38 (s, 3H, CH_3) ppm. – ^{13}C-NMR (101 MHz, CDCl$_3$) δ = 162.1 (C$_q$, COO), 154.6 (C$_q$, C_{ar}OH), 152.0 (C$_q$, 8a-C_{ar}), 136.8 (+, CH), 132.9 (C$_q$, C_{ar}), 131.9 (+, C_{ar}H), 130.5 (+, 2 × C_{ar}H), 130.1 (C$_q$, C_{ar}), 128.5 (C$_q$, C_{ar}), 125.2 (+, C_{ar}H), 123.7 (C$_q$, C_{ar}), 118.0 (C$_q$, C_{ar}), 115.7 (+, 2 × C_{ar}H), 36.1 (–, CCH$_2$), 18.2 (+, CH$_3$), 15.5 (+, CH$_3$) ppm. – IR (ATR, ṽ) = 3371, 3018, 2918, 2858, 1690, 1595, 1515, 1488, 1440, 1416, 1380, 1263, 1225, 1180, 1080, 1033, 944, 891, 866, 822, 790, 765, 733, 668, 552, 496, 476, 429 cm^{-1}. – MS (70 eV, EI, 120 °C) m/z (%) = 281 (24) [M+H]$^+$, 280 (100) [M]$^+$, 279 (11) [M–CH$_3$]$^+$, 252 (11), 251 (31), 237 (13), 231 (13),

159 (12), 119 (15), 115 (12), 107 (17), 100 (12). – **HRMS** (EI, $C_{18}H_{16}O_3$) = calc.: 280.1099; found: 280.1098.

6-Methoxy-3-(4-methoxybenzyl)-5,7,8-trimethyl-2*H*-chromen-2-one (21oc)

According to **GP8**, 2-hydroxy-5-methoxy-3,4,6-trimethylbenzaldehyde (**19o**, 100 mg, 0.52 mmol, 1.00 equiv.), 4-methoxycinnamaldehyde (209 mg, 1.29 mmol, 2.50 equiv.), K_2CO_3 (85.0 mg, 0.62 mmol, 1.20 equiv.) and 1,3-dimethylimidazolium dimethyl phosphate (0.11 ml, 137 mg, 0.62 mmol, 1.20 equiv.) in toluene (2.20 ml) were heated at 110 °C for 50 min *via* microwave irradiation. Purification by flash column chromatography (*c*Hex/EtOAc 5:1) resulted in the title compound as a brown oil (107 mg, 0.32 mmol, 61%).

R_f (*c*Hex/EtOAc 5:1) = 0.24. – **^{1}H NMR** (400 MHz, CDCl$_3$) δ = 7.49 (s, 1H, C*H*), 7.22 (d, ^3J = 8.6 Hz, 2H, 2 × C*H*$_{ar}$), 6.87 (d, ^3J = 8.6 Hz, 2H, 2 × C*H*$_{ar}$), 3.84 (s, 2H, CC*H*$_2$), 3.80 (s, 3H, OC*H*$_3$), 3.65 (s, 3H, OC*H*$_3$), 2.33 (s, 3H, C*H*$_3$), 2.30 (s, 3H, C*H*$_3$), 2.29 (s, 3H, C*H*$_3$) ppm. – **^{13}C NMR** (101 MHz, CDCl$_3$) δ = 162.1 (C$_q$, *C*OO), 158.5 (C$_q$, *C*$_{ar}$OCH$_3$), 152.9 (C$_q$, 8a-*C*$_{ar}$), 148.4 (C$_q$, *C*$_{ar}$OCH$_3$), 136.9 (+, *C*H), 134.1 (C$_q$, CH$_2$*C*$_{ar}$), 130.4 (+, 2 × *C*$_{ar}$H), 130.3 (C$_q$, *C*$_{ar}$), 127.8 (C$_q$, *C*$_{ar}$), 124.2 (C$_q$, *C*$_{ar}$), 123.3 (C$_q$, *C*$_{ar}$), 116.8 (C$_q$, *C*$_{ar}$), 114.2 (+, 2 × *C*$_{ar}$H), 60.7 (+, O*C*H$_3$), 55.4 (+, O*C*H$_3$), 36.0 (–, *C*CH$_2$), 13.4 (+, *C*H$_3$), 11.9 (+, *C*H$_3$), 11.2 (+, *C*H$_3$) ppm. – **IR** (ATR, ṽ) = 2925, 2834, 1708, 1597, 1509, 1453, 1373, 1333, 1284, 1243, 1174, 1135, 1092, 1032, 999, 819, 790, 765, 651, 620, 597, 576, 420 cm^{-1}. – **MS** (70 eV, EI, 100 °C) *m/z* (%) = 339 (19) [M+H]$^+$, 338 (85) [M]$^+$, 331 (15), 323 (20) [M–CH$_3$]$^+$, 281 (11), 262 (12), 243 (13), 231 (21) [M–C$_7$H$_7$O]$^+$, 192 (13), 181 (77), 167 (11), 166 (100), 162 (13), 161 (12), 152 (11), 151 (97), 135 (10), 131 (67), 123 (10), 121 (21), 119 (15), 100 (12), 83 (23), 82 (46), 77 (11), 69 (86), 67 (20), 55 (19). – **HRMS** (EI, $C_{21}H_{25}O_4$) = calc.: 338.1518; found: 338.1514.

3-(4-Chlorobenzyl)-6-methoxy-5,7,8-trimethyl-2*H*-chromen-2-one (21oh)

According to **GP8**, 2-hydroxy-5-methoxy-3,4,6-trimethylbenzaldehyde (**19o**, 100 mg, 0.52 mmol, 1.00 equiv.), 4-chlorocinnamaldehyde (0.16 ml, 170 mg, 1.29 mmol, 2.50 equiv.), K_2CO_3 (85.0 mg, 0.62 mmol, 1.20 equiv.) and 1,3-dimethylimidazolium dimethyl phosphate (0.11 ml, 137 mg, 0.62 mmol, 1.20 equiv.) in toluene (2.20 ml) were heated at 110 °C for 50 min *via* microwave irradiation. Purification by flash column chromatography (*c*Hex/EtOAc 10:1) resulted in the title compound as a white solid (115 mg, 0.34 mmol, 65%).

R_f (cHex/EtOAc 10:1) = 0.24. – ^{1}H NMR (400 MHz, CDCl$_3$) δ = 7.54 (s, 1H, CH), 7.29 (d, ^3J = 8.5 Hz, 2H, 2 × CH_{ar}), 7.24 (d, ^3J = 8.5 Hz, 2H, 2 × CH_{ar}), 3.86 (s, 2H, CH_2), 3.66 (s, 3H, OCH_3), 2.33 (s, 6H, 2 × CH_3), 2.30 (s, 3H, CH_3) ppm. – ^{13}C NMR (101 MHz, CDCl$_3$) δ = 161.9 (C$_q$, COO), 153.0 (C$_q$, 8a-C_{ar}), 148.5 (C$_q$, C_{ar}OCH$_3$), 137.3 (+, CH), 136.9 (C$_q$, C_{ar}), 134.4 (C$_q$, C_{ar}), 132.6 (C$_q$, C_{ar}), 130.6 (+, 2 × C_{ar}H), 128.9 (+, 2 × C_{ar}H), 126.8 (C$_q$, C_{ar}), 124.3 (C$_q$, C_{ar}), 123.4 (C$_q$, C_{ar}), 116.6 (C$_q$, C_{ar}), 60.7 (+, OCH$_3$), 36.4 (–, CH$_2$), 13.4 (+, CH$_3$), 11.9 (+, CH$_3$), 11.3 (+, CH$_3$) ppm. – IR (ATR, ṽ) = 2918, 1702, 1598, 1488, 1457, 1381, 1334, 1283, 1203, 1166, 1135, 1082, 1038, 1001, 927, 879, 841, 803, 781, 707, 644, 592, 575, 514, 486, 448, 399 cm^{-1}. – MS (70 eV, EI, 120 °C) m/z (%) = 345/343 (8/24) [M+H]$^+$, 344/342 (36/100) [M]$^+$, 329/327 (12/36) [M–CH$_3$]$^+$, 307 (8) [M–Cl]$^+$. – HRMS (EI, C$_{20}$H$_{19}$ClO$_3$) = calc.: 342.1023; found: 342.1018.

3-(4-(Dimethylamino)benzyl)-6-methoxy-5,7,8-trimethyl-2H-chromen-2-one (21oi)

According to **GP8**, 2-hydroxy-5-methoxy-3,4,6-trimethylbenzaldehyde (**19o**, 100 mg, 0.52 mmol, 1.00 equiv.), 4-(dimethylamino)cinnamaldehyde (226 mg, 1.29 mmol, 2.50 equiv.), K$_2$CO$_3$ (85.0 mg, 0.62 mmol, 1.20 equiv.) and 1,3-dimethylimidazolium dimethyl phosphate (0.11 ml, 137 mg, 0.62 mmol, 1.20 equiv.) in toluene (2.20 ml) were heated at 110 °C for 50 min *via* microwave irradiation. Purification by flash column chromatography (cHex/EtOAc 5:1) resulted in the title compound as an orange oil (74.0 mg, 0.21 mmol, 41%).

R_f (cHex/EtOAc 5:1) = 0.27. – ^{1}H NMR (400 MHz, CDCl$_3$) δ = 7.50 (s, 1H, CH), 7.17 (d, ^3J = 8.7 Hz, 2H, 2 × CH_{ar}), 6.73 (d, ^3J = 8.7 Hz, 2H, 2 × CH_{ar}), 3.81 (s, 2H, CCH_2), 3.65 (s, 3H, OCH_3), 2.93 (s, 6H, 2 × NCH_3), 2.34 (s, 3H, CH_3), 2.30 (s, 3H, CH_3), 2.29 (s, 3H, CH_3) ppm. – ^{13}C NMR (101 MHz, CDCl$_3$) δ = 162.2 (C$_q$, COO), 152.9 (C$_q$, 8a-C_{ar}), 149.6 (C$_q$, C_{ar}N(CH$_3$)$_2$), 148.4 (C$_q$, C_{ar}OCH$_3$), 136.7 (+, CH), 133.8 (C$_q$, CH$_2$$C_{ar}$), 130.0 (+, 2 × C_{ar}H), 128.2 (C$_q$, C_{ar}), 126.2 (C$_q$, C_{ar}), 124.1 (C$_q$, C_{ar}), 123.2 (C$_q$, C_{ar}), 116.9 (C$_q$, C_{ar}), 113.2 (+, 2 × C_{ar}H), 60.7 (+, OCH$_3$), 41.0 (+, N(CH$_3$)$_2$), 35.9 (–, CCH$_2$), 13.4 (+, CH$_3$), 12.0 (+, CH$_3$), 11.3 (+, CH$_3$) ppm. – IR (ATR, ṽ) = 2923, 1709, 1597, 1519, 1450, 1334, 1283, 1167, 1134, 1092, 1034, 1000, 946, 807, 784, 651, 619, 595, 575 cm^{-1}. – MS (70 eV, EI, 110 °C) m/z (%) = 352 (25) [M+H]$^+$, 351 (100) [M]$^+$, 336 (3) [M–CH$_3$]$^+$, 307 (5) [M–C$_2$H$_6$N]$^+$, 231 (9) [M–C$_8$H$_{10}$N]$^+$, 181 (40), 167 (11), 166 (15), 149 (15), 148 (17), 134 (30), 131 (35), 83 (18), 82 (38), 69 (56), 67 (12), 55 (12). – HRMS (EI, C$_{22}$H$_{25}$O$_3$N) = calc.: 351.1834; found: 351.1829.

6-Hydroxy-3-(4-hydroxybenzyl)-5,7,8-trimethyl-2H-chromen-2-one (61a)

According to **GP9**, to a solution of the dimethoxycoumarin **21oc** (14.0 mg, 41.4 μmol, 1.00 equiv.) in abs. DCM (0.83 ml), was added BBr$_3$ (1 M in DCM, 0.21 ml, 52.0 mg, 207 μmol, 5.00 equiv.) at –78 °C. After stirring for 30 min at this temperature, the reaction mixture was allowed to warm to rt and stirred overnight. Purification by flash column chromatography (cHex/EtOAc 2:1) resulted in the title compound as an off-white solid (2.6 mg, 8.28 μmol, 20%, purity ≤90%).

R_f (cHex/EtOAc 5:1) = 0.29. – **^{1}H NMR** (400 MHz, CDCl$_3$) δ = 8.02 (s, 1H, OH), 7.49 (d, ^4J = 1.3 Hz, 1H, CH), 7.17 (d, ^3J = 8.5 Hz, 2H, 2 × CH_{ar}), 6.81 (d, ^3J = 8.5 Hz, 2H, 2 × CH_{ar}), 3.83 (s, 2H, CCH_2), 2.96 (s, 6H, 2 × CH_3), 2.89 (s, 3H, CH_3) ppm. – **^{13}C NMR** insufficient quantities available. – **IR** (ATR, ṽ) = 3313, 3150, 3019, 2920, 2851, 1659, 1591, 1512, 1448, 1364, 1234, 1206, 1170, 1098, 917, 834, 795, 726, 656, 629, 602, 538, 496, 432 cm^{-1}. – **MS** (70 eV, EI, 150 °C) m/z (%) = 311 (16) [M+H]$^+$, 310 (71) [M]$^+$, 295 (11) [M–CH$_3$]$^+$, 231 (10), 151 (10), 125 (13), 123 (10), 119 (15), 111 (23), 109 (16), 107 (15), 97 (31), 95 (22), 85 (22), 83 (31), 81 (18), 70 (10), 67 (11), 57 (48), 56 (11), 55 (26). – **HRMS** (EI, C$_{19}$H$_{18}$O$_4$) = calc.: 310.1205; found: 310.1205.

3-(4-Chlorobenzyl)-6-hydroxy-5,7,8-trimethyl-2H-chromen-2-one (61b)

According to **GP9**, to a solution of the 6-methoxycoumarin **21oh** (21.0 mg, 61.3 μmol, 1.00 equiv.) in abs. DCM (1.23 ml), was added BBr$_3$ (1 M in DCM, 0.31 ml, 77.0 mg, 306 μmol, 5.00 equiv.) at –78 °C. After stirring for 30 min at this temperature, the reaction mixture was allowed to warm to rt and stirred overnight. Purification by flash column chromatography (cHex/EtOAc 3:1) resulted in the title compound as an off-white solid (14.0 mg, 42.8 μmol, 70%).

R_f (cHex/EtOAc 5:1) = 0.24. – **^{1}H NMR** (400 MHz, CDCl$_3$) δ = 7.54 (d, ^3J = 1.2 Hz, 1H, CH), 7.35 – 7.21 (m, 4H, 4 × CH_{ar}), 4.67 (bs, 1H, OH), 3.87 (s, 2H, CCH_2), 2.36 (s, 3H, CH_3), 2.28 (s, 6H, 2 × CH_3) ppm. – **^{13}C NMR** (101 MHz, CDCl$_3$) δ = 162.0 (C$_q$, COO), 148.2 (C$_q$, C_{ar}), 148.2 (C$_q$, C_{ar}), 137.1 (+, CH), 136.9 (C$_q$, C_{ar}), 132.6 (C$_q$, C_{ar}), 130.7 (+, 2 × C_{ar}H), 128.9 (+, 2 × C_{ar}H), 127.3 (C$_q$, C_{ar}), 127.0 (C$_q$, C_{ar}), 122.8 (C$_q$, C_{ar}), 116.2 (C$_q$, C_{ar}), 115.9 (C$_q$, C_{ar}), 36.4 (–, CCH$_2$), 13.0 (+, CH$_3$), 12.0 (+, CH$_3$), 10.9 (+, CH$_3$) ppm. – **IR** (ATR, ṽ) = 3452, 2924, 2850, 1675, 1603, 1567, 1494, 1426, 1386, 1324, 1290, 1206, 1184, 1096, 1020, 950, 896, 853, 812, 782, 758, 732, 689, 648, 578, 480, 399 cm^{-1}. – **MS** (70 eV, EI, 130 °C) m/z (%) = 331/329

(7/24) [M+H]$^+$, 330/328 (37/100) [M]$^+$, 293 (29) [M–Cl]$^+$, 189 (13), 125 (12). – **HRMS** (EI, $C_{19}H_{17}O_3{}^{35}Cl$) = calc.: 328.0866; found: 328.0867.

3-(4-(Dimethylamino)benzyl)-6-hydroxy-5,7,8-trimethyl-2*H*-chromen-2-one (61c)

According to **GP9**, to a solution of the 6-methoxycoumarin **21oi** (17.0 mg, 48.4 μmol, 1.00 equiv.) in abs. DCM (0.97 ml), was added BBr$_3$ (1 M in DCM, 0.24 ml, 61.0 mg, 242 μmol, 5.00 equiv.) at –78 °C. After stirring for 30 min at this temperature, the reaction mixture was allowed to warm to rt and stirred overnight. Purification by flash column chromatography (*c*Hex/EtOAc 3:1) resulted in the title compound as a red solid (11.1 mg, 33.4 μmol, 69%).

R_f (*c*Hex/EtOAc 5:1) = 0.20. – **^{1}H NMR** (400 MHz, CDCl$_3$) δ = 7.50 (d, ^4J = 1.3 Hz, 1H, C*H*), 7.20 – 7.14 (m, 2H, 2 × C*H*$_{ar}$), 6.77 – 6.70 (m, 2H, 2 × C*H*$_{ar}$), 4.77 (bs, 1H, O*H*), 3.81 (d, ^4J = 1.3 Hz, 2H, CC*H*$_2$), 2.93 (s, 6H, 2 × NC*H*$_3$), 2.35 (s, 3H, C*H*$_3$), 2.27 (s, 3H, C*H*$_3$), 2.24 (s, 3H, C*H*$_3$) ppm. – **^{13}C NMR** (101 MHz, CDCl$_3$) δ = 162.4 (C$_q$, *C*OO), 148.1 (C$_q$, *C*$_{ar}$), 146.4 (C$_q$, *C*$_{ar}$), 136.6 (+, *C*H), 130.1 (+, 2 × *C*$_{ar}$H), 128.3 (C$_q$, *C*$_{ar}$), 126.9 (C$_q$, *C*$_{ar}$), 126.2 (C$_q$, *C*$_{ar}$), 122.5 (C$_q$, *C*$_{ar}$), 116.5 (C$_q$, *C*$_{ar}$), 115.8 (C$_q$, *C*$_{ar}$), 113.2 (+, 2 × *C*$_{ar}$H), 41.0 (+, N(*C*H$_3$)$_2$), 35.9 (–, *C*CH$_2$), 13.0 (+, *C*H$_3$), 12.0 (+, *C*H$_3$), 10.9 (+, *C*H$_3$) ppm. – **IR** (ATR, ṽ) = 3431, 3072, 2991, 2922, 2856, 2793, 1674, 1617, 1567, 1522, 1444, 1386, 1322, 1291, 1183, 1097, 1066, 1026, 947, 897, 844, 812, 784, 758, 733, 705, 668, 579, 484 cm^{-1}. – **MS** (70 eV, EI, 150 °C) *m/z* (%) = 338 (24) [M+H]$^+$, 337 (100) [M]$^+$, 336 (3) [M–H]$^+$, 308 (11), 134 (33), 69 (26). – **HRMS** (EI, $C_{21}H_{23}O_3N$) = calc.: 337.1678; found: 337.1678.

3-(2-Fluorobenzyl)-5-methoxy-7-pentyl-2*H*-chromen-2-one (65a)

According to **GP8**, 2-hydroxy-5-methoxy-7-pentylbenzalde-hyde (50.0 mg, 0.225 mmol, 1.00 equiv.), 2-fluorocinnamaldehyde (84.4 mg, 0.562 mmol, 2.50 equiv.), K$_2$CO$_3$ (37.3 mg, 0.270 mmol, 1.20 equiv.) and 1,3-dimethylimidazolium dimethyl phosphate (47.0 μl, 60.0 mg, 0.270 mmol, 1.20 equiv.) in toluene (0.74 ml) were heated at 110 °C for 50 min *via* microwave irra-diation. Purification by flash column chromatography (*c*Hex/EtOAc 20:1) resulted in the title compound as a yellow solid (26.8 mg, 0.08 mmol, 34%).

R_f (*c*Hex/EtOAc 10:1) = 0.37. – **^{1}H NMR** (500 MHz, CDCl$_3$) δ = 7.75 (s, 1H, C*H*), 7.35 (td, $^{3,4}J$ = 7.6, 1.5 Hz, 1H, C*H*$_{ar}$), 7.26 – 7.20 (d, ^3J = 8.5 Hz, 2H, 2 × C*H*$_{ar}$), 7.10 (td, $^{3,4}J$ = 7.6, 1.5 Hz, 1H, C*H*$_{ar}$), 7.07 – 7.01 (m, 1H, C*H*$_{ar}$), 6.72 (d, ^4J = 1.1 Hz, 1H, C*H*$_{ar}$), 6.49 (d, ^4J = 1.1 Hz, 1H, C*H*$_{ar}$), 3.89 (s, 2H, CC*H*$_2$), 3.86 (s, 3H, OC*H*$_3$), 2.63 (t, ^3J = 7.6 Hz, 2H, 7-C$_{ar}$C*H*$_2$), 1.62

(p, ^3J = 7.6 Hz, 2H, CH_2), 1.36 – 1.28 (m, 4H, 2 × CH_2), 0.89 (t, ^3J = 7.0 Hz, 3H, CH_3) ppm. – ^{13}C NMR (126 MHz, CDCl$_3$) δ = 162.0 (C$_q$, COO), 161.4 (C$_q$, d, ^1J = 246.0 Hz, C$_{ar}$F), 155.6 (C$_q$, C$_{ar}$OCH$_3$), 154.3 (C$_q$, 8a-C$_{ar}$), 147.9 (C$_q$, 7-C$_{ar}$), 134.9 (+, d, ^5J = 1.4 Hz, CH), 131.8 (+, d, ^3J = 4.3 Hz, m-C$_{ar}$H), 128.6 (+, d, ^3J = 8.0 Hz, m-C$_{ar}$H), 125.4 (C$_q$, d, ^2J = 15.4 Hz, o-C$_{ar}$), 124.3 (+, d, ^4J = 2.4 Hz, p-C$_{ar}$), 124.3 (C$_q$, C3), 115.5 (+, d, ^2J = 21.7 Hz, o-C$_{ar}$H), 108.5 (+, C$_{ar}$H), 108.0 (C$_q$, 4a-C$_{ar}$), 105.7 (+, C$_{ar}$H), 55.9 (+, OCH$_3$), 36.7 (–, 7-C$_{ar}$CH$_2$), 31.5 (–, CH$_2$), 30.8 (–, CH$_2$), 30.1 (–, d, ^3J = 3.0 Hz, CCH$_2$), 22.6 (–, CH$_2$), 14.1 (+, CH$_3$) ppm. – ^{19}F NMR (471 MHz, CDCl$_3$) δ = –117.5 (q, ^1J = 7.8 Hz, C$_{ar}$F) ppm. – IR (ATR, ṽ) = 2925, 2857, 1706, 1616, 1490, 1452, 1421, 1353, 1298, 1247, 1225, 1173, 1142, 1111, 1095, 1053, 943, 922, 835, 796, 750, 601, 570, 552, 518, 435, 394 cm^{-1}. – HRMS (ESI, C$_{22}$H$_{23}$O$_3$F) = calc.: 355.1704 [M+H]$^+$, found: 355.1696 [M+H]$^+$.

5-Methoxy-7-pentyl-3-(2-(trifluoromethyl)benzyl)-2H-chromen-2-one (65b)

According to GP8, 2-hydroxy-5-methoxy-7-pentylbenzalde-hyde (50.0 mg, 0.225 mmol, 1.00 equiv.), 2-(trifluorome-thyl)cinnamaldehyde (113 mg, 0.562 mmol, 2.50 equiv.), K$_2$CO$_3$ (37.3 mg, 0.270 mmol, 1.20 equiv.) and 1,3-dimethylimidazolium dimethyl phosphate (47.0 μl, 60.0 mg, 0.270 mmol, 1.20 equiv.) in toluene (0.74 ml) were heated at 110 °C for 50 min via microwave irradiation. Purification by flash column chromatography (cHex/EtOAc 40:1) resulted in the title compound as a yellow solid (27.0 mg, 0.07 mmol, 30%).

R_f (cHex/EtOAc 20:1) = 0.26. – ^{1}H NMR (500 MHz, CDCl$_3$) δ = 7.70 (d, ^3J = 7.4 Hz, 1H, CH_{ar}), 7.54 – 7.47 (m, 2H, CH and CH_{ar}), 7.42 – 7.34 (m, 2H, 2 × CH_{ar}), 6.75 (d, ^4J = 1.3 Hz, 1H, CH_{ar}), 6.49 (d, ^4J = 1.3 Hz, 1H, CH_{ar}), 4.08 (s, 2H, CH_2), 3.82 (s, 3H, OCH_3), 2.63 (t, ^3J = 7.7 Hz, 2H, 7-C$_{ar}$CH_2), 1.69 – 1.55 (m, 2H, CH_2), 1.39 – 1.28 (m, 4H, 2 × CH_2), 0.89 (t, ^3J = 6.9 Hz, 3H, CH_3) ppm. – ^{13}C NMR (126 MHz, CDCl$_3$) δ = 162.1 (C$_q$, COO), 155.6 (C$_q$, C$_{ar}$OCH$_3$), 154.2 (C$_q$, 8a-C$_{ar}$), 148.1 (C$_q$, 7-C$_{ar}$), 136.9 (C$_q$, d, ^3J = 1.8 Hz, m-C$_{ar}$H), 135.4 (+, CH), 132.2 (+, C$_{ar}$H), 131.9 (+, C$_{ar}$H), 129.2 (C$_q$, q, ^2J = 29.9 Hz, C$_{ar}$CF$_3$), 127.0 (+, C$_{ar}$H), 126.3 (+, q, ^3J = 5.7 Hz, o-C$_{ar}$H), 125.0 (C$_q$, C3), 124.5 (C$_q$, q, ^1J = 273.9 Hz, CF$_3$), 108.5 (+, C$_{ar}$H), 107.9 (C$_q$, 4a-C$_{ar}$), 105.7 (+, C$_{ar}$H), 55.9 (+, OCH$_3$), 36.7 (–, 7-C$_{ar}$CH$_2$), 33.1 (–, q, ^4J = 2.1 Hz, CCH$_2$), 31.5 (–, CH$_2$), 30.9 (–, CH$_2$), 22.6 (–, CH$_2$), 14.1 (+, CH$_3$) ppm. – ^{19}F NMR (471 MHz, CDCl$_3$) δ = –59.8 (CF$_3$) ppm. – IR (ATR, ṽ) = 2926, 2856, 1706, 1615, 1498, 1455, 1429, 1310, 1249, 1144, 1103, 1062, 1036, 953, 912, 876, 838, 796, 763, 727, 653, 596, 569, 529, 398 cm^{-1}. – HRMS (ESI, C$_{23}$H$_{23}$O$_3$F$_3$) = calc.: 405.1672 [M+H]$^+$, found: 405.1665 [M+H]$^+$.

5.1.10 Synthesis and characterisation of 3-alkylcoumarins

5,7-Dimethoxy-2*H*-chromen-2-one (16aa)

According to **GP11**, 2-hydroxy-4,6-dimethoxybenzaldehyde (**19a**, 100 mg, 0.55 mmol, 1.00 equiv.), K_2CO_3 (4.00 mg, 0.03 mmol, 0.05 equiv.) and acetic anhydride (0.18 ml, 196 mg, 1.92 mmol, 3.50 equiv.) were heated at 180 °C for 65 min *via* microwave irradiation. Purification by flash column chromatography (cHex/EtOAc 3:1) resulted in the title compound as an off-white solid (100 mg, 0.49 mmol, 88%).

R_f (cHex/EtOAc 3:1) = 0.28. – **¹H NMR** (400 MHz, CDCl₃) δ = 7.96 (d, ^3J = 9.6 Hz, 1H, C*H*), 6.41 (d, ^4J = 2.2 Hz, 1H, C*H*ar), 6.28 (d, ^4J = 2.2 Hz, 1H, C*H*ar), 6.15 (d, ^3J = 9.6 Hz, 1H, C*H*), 3.88 (s, 3H, OC*H*₃), 3.85 (s, 3H, OC*H*₃) ppm. – **¹³C NMR** (101 MHz, CDCl₃) δ = 163.8 (C_q, *C*OO), 161.7 (C_q, 7-*C*arOCH₃), 157.1 (C_q, 5-*C*arOCH₃), 157.0 (C_q, 8a-*C*ar), 138.9 (+, *C*H), 111.1 (+, *C*H), 104.2 (C_q, 4a-*C*ar), 95.0 (+, *C*arH), 92.9 (+, *C*arH), 56.1 (+, O*C*H₃), 55.9 (+, O*C*H₃) ppm. – **IR** (ATR, ṽ) = 3085, 3048, 2979, 2948, 2927, 2918, 2843, 1708, 1694, 1606, 1562, 1496, 1469, 1453, 1421, 1397, 1361, 1312, 1239, 1221, 1204, 1184, 1150, 1115, 1096, 1047, 1000, 989, 943, 899, 857, 812, 762, 748, 701, 681, 650, 640, 623, 565, 537, 510, 466, 416, 398 cm⁻¹. – **MS** (70 eV, EI, 40 °C) *m/z* (%) = 207 (23) [M+H]⁺, 206 (100) [M]⁺, 181 (18), 178 (75), 163 (36), 135 (17), 131 (14), 69 (33). – **HRMS** (EI, C₁₁H₁₀O₄) = calc.: 206.0579; found: 206.0578.

5,7-Dimethoxy-3-methyl-2*H*-chromen-2-one (16ab)

According to **GP11**, 2-hydroxy-4,6-dimethoxybenzaldehyde (**19a**, 100 mg, 0.55 mmol, 1.00 equiv.), K_2CO_3 (4.00 mg, 0.03 mmol, 0.05 equiv.) and propionic anhydride (0.25 ml, 250 mg, 1.92 mmol, 3.50 equiv.) were heated at 180 °C for 65 min *via* microwave irradiation. Purification by flash column chromatography (cHex/EtOAc 2:1) resulted in the title compound as an off-white solid (77.0 mg, 0.35 mmol, 64%).

R_f (cHex/EtOAc 1:1) = 0.50. – **¹H NMR** (400 MHz, CDCl₃) δ = 7.80 – 7.78 (m, 1H, C*H*), 6.41 (d, ^4J = 2.1 Hz, 1H, C*H*ar), 6.28 (d, ^4J = 2.1 Hz, 1H, C*H*ar), 3.88 (s, 3H, OC*H*₃), 3.84 (s, 3H, OC*H*₃), 2.16 (d, ^4J = 1.3 Hz, 3H, C*H*₃) ppm. – **¹³C NMR** (101 MHz, CDCl₃) δ = 163.0 (C_q, *C*OO), 162.6 (C_q, 7-*C*arOCH₃), 156.4 (C_q, 5-*C*arOCH₃), 155.9 (C_q, 8a-*C*ar), 134.8 (+, *C*H), 120.3 (C_q, 3-*C*), 104.6 (C_q, 4a-*C*ar), 94.9 (+, *C*arH), 92.6 (+, *C*arH), 56.0 (+, O*C*H₃), 55.9 (+, O*C*H₃), 17.1 (+, *C*H₃) ppm. – **IR** (ATR, ṽ) = 3002, 2952, 2917, 2849, 1720, 1704, 1611, 1579, 1504,

1476, 1456, 1434, 1366, 1349, 1320, 1264, 1232, 1204, 1190, 1150, 1109, 1078, 1041, 1003, 992, 946, 926, 813, 790, 761, 724, 711, 647, 633, 557, 528, 501, 377 cm^{-1}. – **MS** (70 eV, EI, 60 °C) *m/z* (%) = 221 (13) [M+H]$^{+}$, 220 (100) [M]$^{+}$, 192 (48), 191 (11), 181 (17), 177 (65), 149 (10), 131 (21), 69 (46), 58 (11). – **HRMS** (EI, C$_{12}$H$_{12}$O$_4$) = calc.: 220.0736; found: 220.0736.

3-Ethyl-5,7-dimethoxy-2*H*-chromen-2-one (16ac)

According to **GP11**, 2-hydroxy-4,6-dimethoxybenzaldehyde (**19a**, 100 mg, 0.55 mmol, 1.00 equiv.), K$_2$CO$_3$ (4.00 mg, 0.03 mmol, 0.05 equiv.) and butyric anhydride (0.31 ml, 304 mg, 1.92 mmol, 3.50 equiv.) were heated at 180 °C for 65 min *via* microwave irradiation. Purification by flash column chromatography (*c*Hex/EtOAc 5:1) resulted in the title compound as a white solid (110 mg, 0.47 mmol, 85%).

R$_f$ (*c*Hex/EtOAc 5:1) = 0.32. – **^{1}H NMR** (400 MHz, CDCl$_3$) δ = 7.76 (s, 1H, C*H*), 6.42 (d, ^4J = 2.2 Hz, 1H, C*H*$_{ar}$), 6.28 (d, ^4J = 2.2 Hz, 1H, C*H*$_{ar}$), 3.89 (s, 3H, OC*H*$_3$), 3.84 (s, 3H, OC*H*$_3$), 2.55 (qd, $^{3,4}J$ = 7.4, 1.2 Hz, 2H, C*H*$_2$), 1.22 (t, ^3J = 7.4 Hz, 3H, C*H*$_3$) ppm. – **^{13}C NMR** (101 MHz, CDCl$_3$) δ = 163.4 (C$_q$, *C*OO), 163.3 (C$_q$, 7-*C*$_{ar}$OCH$_3$), 157.3 (C$_q$, 5-*C*$_{ar}$OCH$_3$), 156.4 (C$_q$, 8a-*C*$_{ar}$), 133.8 (+, *C*H), 126.6 (C$_q$, 3-*C*), 105.3 (C$_q$, 4a-*C*$_{ar}$), 95.6 (+, *C*$_{ar}$H), 93.3 (+, *C*$_{ar}$H), 56.8 (+, O*C*H$_3$), 56.6 (+, O*C*H$_3$), 24.7 (–, *C*H$_2$CH$_3$), 13.5 (+, *C*H$_3$) ppm. – **IR** (ATR, ṽ) = 2964, 2843, 1703, 1616, 1580, 1494, 1451, 1417, 1377, 1318, 1301, 1247, 1203, 1185, 1150, 1112, 1080, 1046, 999, 960, 922, 812, 751, 717, 685, 661, 640, 590, 557, 517, 474, 413 cm^{-1}. – **MS** (70 eV, EI, 60 °C) *m/z* (%) = 234 (65) [M]$^{+}$, 220 (27), 219 (100) [M–CH$_3$]$^{+}$, 191 (20). – **HRMS** (EI, C$_{13}$H$_{14}$O$_4$) = calc.: 234.0892; found: 234.0893.

3-Propyl-5,7-dimethoxy-2*H*-chromen-2-one (16ad)

According to **GP11**, 2-hydroxy-4,6-dimethoxybenzaldehyde (**19a**, 100 mg, 0.55 mmol, 1.00 equiv.), K$_2$CO$_3$ (4.00 mg, 0.03 mmol, 0.05 equiv.) and valeric anhydride (0.38 ml, 358 mg, 1.92 mmol, 3.50 equiv.) were heated at 180 °C for 65 min *via* microwave irradiation. Purification by flash column chromatography (*c*Hex/EtOAc 5:1) resulted in the title compound as a white solid (100 mg, 0.40 mmol, 74%).

R$_f$ (*c*Hex/EtOAc 5:1) = 0.39. – **^{1}H NMR** (400 MHz, CDCl$_3$) δ = 7.75 (d, ^4J = 0.8 Hz, 1H, C*H*), 6.41 (d, ^4J = 2.2 Hz, 1H, C*H*$_{ar}$), 6.28 (d, ^4J = 2.1 Hz, 1H, C*H*$_{ar}$), 3.89 (s, 3H, OC*H*$_3$), 3.84 (s, 3H, OC*H*$_3$), 2.49 (td, $^{3,4}J$ = 7.5, 1.3 Hz, 2H, CC*H*$_2$), 1.65 (sex., ^3J = 7.5 Hz, 2H, C*H*$_2$CH$_3$), 0.98 (t, ^3J = 7.5 Hz, 3H, C*H*$_3$) ppm. – **^{13}C NMR** (101 MHz, CDCl$_3$) δ = 163.4 (C$_q$, *C*OO), 163.4 (C$_q$,

7-$C_{ar}OCH_3$), 157.3 (C_q, 5-$C_{ar}OCH_3$), 156.5 (C_q, 8a-C_{ar}), 134.8 (+, CH), 125.1 (C_q, 3-C), 105.3 (C_q, 4a-C_{ar}), 95.6 (+, $C_{ar}H$), 93.3 (+, $C_{ar}H$), 56.8 (+, OCH_3), 55.6 (+, OCH_3), 33.7 (–, CCH_2), 22.4 (–, CH_2CH_3), 14.7 (+, CH_3) ppm. – **IR** (ATR, $\tilde{v}$) = 2958, 2925, 2876, 2844, 1714, 1629, 1608, 1581, 1497, 1485, 1472, 1448, 1425, 1383, 1366, 1340, 1305, 1247, 1227, 1201, 1187, 1166, 1142, 1116, 1102, 1043, 1023, 999, 928, 904, 871, 841, 799, 772, 744, 715, 659, 643, 611, 558, 528, 483, 424 cm^{-1}. – **MS** (70 eV, EI, 60 °C) m/z (%) = 248 (39) [M]$^+$, 220 (16), 219 (100) [M–C_2H_5]$^+$, 181 (16), 69 (13). – **HRMS** (EI, $C_{14}H_{16}O_4$) = calc.: 248.1049; found: 248.1049.

3-Butyl-5,7-dimethoxy-2*H*-chromen-2-one (16ae)

According to **GP11**, 2-hydroxy-4,6-dimethoxybenzaldehyde (**19a**, 100 mg, 0.55 mmol, 1.00 equiv.), K_2CO_3 (4.00 mg, 0.03 mmol, 0.05 equiv.) and hexanoic anhydride (0.44 ml, 412 mg, 1.92 mmol, 3.50 equiv.) were heated at 180 °C for 65 min *via* microwave irradiation. Purification by flash column chromatography (*c*Hex/EtOAc 5:1) resulted in the title compound as a white solid (112 mg, 0.43 mmol, 78%, purity 94%).

R_f (*c*Hex/EtOAc 5:1) = 0.39. – **^{1}H NMR** (400 MHz, CDCl$_3$) δ = 7.74 (s, 1H, C*H*), 6.41 (d, ^4J = 2.2 Hz, 1H, C*H*$_{ar}$), 6.28 (d, ^4J = 2.2 Hz, 1H, C*H*$_{ar}$), 3.89 (s, 3H, OC*H*$_3$), 3.84 (s, 3H, OC*H*$_3$), 2.51 (td, $^{3,4}J$ = 7.9, 1.1 Hz, 2H, CC*H*$_2$), 1.65 – 1.54 (m, 2H, CH$_2$C*H*$_2$), 1.40 (sex., ^3J = 7.4 Hz, 3H, C*H*$_2$CH$_3$), 0.94 (t, ^3J = 7.4 Hz, 3H, C*H*$_3$) ppm. – **^{13}C NMR** (101 MHz, CDCl$_3$) δ = 162.6 (C_q, COO), 162.6 (C_q, 7-$C_{ar}OCH_3$), 156.5 (C_q, 5-$C_{ar}OCH_3$), 155.7 (C_q, 8a-C_{ar}), 133.9 (+, CH), 124.6 (C_q, 3-C), 104.6 (C_q, 4a-C_{ar}), 94.8 (+, $C_{ar}H$), 92.5 (+, $C_{ar}H$), 56.0 (+, OCH_3), 55.9 (+, OCH_3), 30.6 (–, CH_2), 30.5 (–, CH_2), 22.6 (–, CH_2CH_3), 14.1 (+, CH_3) ppm. – **IR** (ATR, $\tilde{v}$) = 3021, 2987, 2951, 2927, 2870, 2856, 1711, 1677, 1618, 1579, 1494, 1466, 1451, 1434, 1417, 1387, 1370, 1329, 1305, 1259, 1244, 1220, 1198, 1181, 1163, 1147, 1116, 1101, 1071, 1044, 992, 933, 912, 870, 843, 816, 812, 795, 756, 732, 713, 700, 662, 639, 609, 558, 528, 480, 433, 421, 387 cm^{-1}. – **MS** (70 eV, EI, 70 °C) m/z (%) = 262 (47) [M]$^+$, 233 (12) [M–C_2H_5]$^+$, 220 (37), 219 (100) [M–C_3H_7]$^+$. – **HRMS** (EI, $C_{15}H_{18}O_4$) = calc.: 262.1205; found: 262.1206.

3-Pentyl-5,7-dimethoxy-2*H*-chromen-2-one (16af)

According to **GP11**, 2-hydroxy-4,6-dimethoxybenzaldehyde (**19a**, 100 mg, 0.55 mmol, 1.00 equiv.), K_2CO_3 (4.00 mg, 0.03 mmol, 0.05 equiv.) and heptanoic anhydride (0.51 ml, 466 mg, 1.92 mmol, 3.50

equiv.) were heated at 180 °C for 65 min *via* microwave irradiation. Purification by flash column chromatography (*c*Hex/EtOAc 5:1) resulted in a mixture of the product and heptanoic acid as impurity. After washing the product fractions with aq. NaHCO₃, drying over Na₂SO₄ and removal of the volatiles resulted in the title compound as colourless crystals (124 mg, 0.45 mmol, 82%).

R_f (*c*Hex/EtOAc 5:1) = 0.39. – **¹H NMR** (400 MHz, CDCl₃) δ = 7.74 (s, 1H, C*H*), 6.41 (d, ^4J = 2.1 Hz, 1H, C*H*$_{ar}$), 6.28 (d, ^4J = 2.1 Hz, 1H, C*H*$_{ar}$), 3.89 (s, 3H, OC*H₃*), 3.83 (s, 3H, OC*H₃*), 2.56 – 2.45 (m, 2H, CC*H₂*), 1.67 – 1.57 (m, 2H, C*H₂*), 1.39 – 1.31 (m, 4H, 2 × C*H₂*), 0.94 – 0.86 (m, 3H, C*H₃*) ppm. – **¹³C NMR** (101 MHz, CDCl₃) δ = 162.6 (C$_q$, *C*OO), 162.6 (C$_q$, 7-*C*$_{ar}$OCH₃), 156.5 (C$_q$, 5-*C*$_{ar}$OCH₃), 155.7 (C$_q$, 8a-*C*$_{ar}$), 133.9 (+, *C*H), 124.6 (C$_q$, 3-*C*), 104.6 (C$_q$, 4a-*C*$_{ar}$), 94.8 (+, *C*$_{ar}$H), 92.5 (+, *C*$_{ar}$H), 56.0 (+, O*C*H₃), 55.9 (+, O*C*H₃), 31.6 (–, *C*H₂), 30.9 (–, *C*H₂), 28.1 (–, *C*H₂), 22.6 (–, *C*H₂CH₃), 14.2 (+, *C*H₃) ppm. – **IR** (ATR, ṽ) = 3014, 2980, 2956, 2925, 2870, 2854, 1728, 1710, 1632, 1609, 1578, 1497, 1485, 1470, 1458, 1425, 1384, 1367, 1340, 1309, 1296, 1251, 1225, 1201, 1190, 1163, 1140, 1111, 1067, 1044, 1017, 996, 972, 932, 922, 904, 870, 843, 800, 775, 735, 714, 662, 643, 611, 560, 528, 490, 446, 421, 384 cm⁻¹. – **MS** (70 eV, EI, 80 °C) *m/z* (%) = 276 (47) [M]⁺, 233 (11) [M–C₃H₇]⁺, 220 (39), 219 (100) [M–C₄H₉]⁺. – **HRMS** (EI, C₁₆H₂₀O₄) = calc.: 276.1362; found: 276.1362.

3-Hexyl-5,7-dimethoxy-2*H*-chromen-2-one (16ag)

According to **GP11**, 2-hydroxy-4,6-dimethoxybenzaldehyde (**19a**, 100 mg, 0.55 mmol, 1.00 equiv.), K₂CO₃ (4.00 mg, 0.03 mmol, 0.05 equiv.) and caprylic anhydride (0.57 ml, 520 mg, 1.92 mmol, 3.50 equiv.) were heated at 180 °C for 65 min *via* microwave irradiation. Purification by flash column chromatography (*c*Hex/EtOAc 5:1) resulted in a mixture of the product and caprylic acid as impurity. After washing the product fractions with aq. NaHCO₃, drying over Na₂SO₄ and removal of the volatiles resulted in the title compound as colourless crystals (93.0 mg, 0.32 mmol, 58%).

R_f (*c*Hex/EtOAc 5:1) = 0.39. – **¹H NMR** (400 MHz, CDCl₃) δ = 7.74 (d, ^4J = 0.8 Hz, 1H, C*H*), 6.41 (d, ^4J = 2.2 Hz, 1H, C*H*$_{ar}$), 6.28 (d, ^4J = 2.2 Hz, 1H, C*H*$_{ar}$), 3.89 (s, 3H, OC*H₃*), 3.83 (s, 3H, OC*H₃*), 2.50 (td, $^{3,4}J$ = 7.8, 1.1 Hz, 2H, CC*H₂*), 1.66 – 1.56 (m, 2H, C*H₂*), 1.39 – 1.27 (m, 6H, 3 × C*H₂*), 0.91 – 0.86 (m, 3H, C*H₃*) ppm. – **¹³C NMR** (101 MHz, CDCl₃) δ = 162.6 (C$_q$, *C*OO), 162.6 (C$_q$, 7-*C*$_{ar}$OCH₃), 156.5 (C$_q$, 5-*C*$_{ar}$OCH₃), 155.7 (C$_q$, 8a-*C*$_{ar}$), 133.9 (+, *C*H), 124.6 (C$_q$, 3-*C*), 104.6 (C$_q$, 4a-*C*$_{ar}$), 94.8 (+, *C*$_{ar}$H), 92.5 (+, *C*$_{ar}$H), 56.0 (+, O*C*H₃), 55.9 (+, O*C*H₃), 31.8 (–, *C*H₂), 30.9 (–, *C*H₂), 29.1 (–, *C*H₂), 28.3 (–, *C*H₂), 22.7 (–, *C*H₂CH₃), 14.2 (+, *C*H₃)

Experimental part

ppm. – **IR** (ATR, ṽ) = 3014, 2980, 2956, 2925, 2870, 2854, 1710, 1632, 1609, 1578, 1497, 1470, 1458, 1425, 1384, 1367, 1340, 1309, 1296, 1251, 1225, 1201, 1190, 1163, 1140, 1111, 1067, 1044, 1017, 996, 932, 922, 904, 843, 800, 775, 735, 714, 662, 643, 611, 560, 528, 490, 446, 421, 384 cm⁻¹. – **MS** (70 eV, EI, 90 °C) *m/z* (%) = 290 (50) [M]⁺, 233 (11) [M–C₄H₉], 220 (44), 219 (100) [M–C₅H₁₁], 69 (11). – **HRMS** (EI, C₁₇H₂₂O₄) = calc.: 290.1518; found: 290.1517.

7-Ethoxy-5-methoxy-2*H*-chromen-2-one (16ba)

 According to **GP11**, 4-ethoxy-2-hydroxy-6-methoxybenzaldehyde (**19b**, 50.0 mg, 0.25 mmol, 1.00 equiv.), K₂CO₃ (2.00 mg, 0.01 mmol, 0.05 equiv.) and acetic anhydride (0.08 ml, 91.0 mg, 0.89 mmol, 3.50 equiv.) were heated at 180 °C for 65 min *via* microwave irradiation. Purification by flash column chromatography (cHex/EtOAc 3:1) resulted in the title compound as a white solid (45.0 mg, 0.20 mmol, 80%).

*R*f (cHex/EtOAc 2:1) = 0.50. – **¹H NMR** (400 MHz, CDCl₃) δ = 7.96 (d, ³*J* = 9.6 Hz, 1H, C*H*), 6.40 (d, ⁴*J* = 2.1 Hz, 1H, C*H*ₐᵣ), 6.28 (d, ⁴*J* = 2.1 Hz, 1H, C*H*ₐᵣ), 6.15 (d, ³*J* = 9.6 Hz, 1H, C*H*), 4.07 (q, ³*J* = 7.0 Hz, 2H, OC*H₂*), 3.88 (s, 3H, OC*H₃*), 1.45 (t, ³*J* = 7.0 Hz, 3H, C*H₃*) ppm. – **¹³C NMR** (101 MHz, CDCl₃) δ = 163.2 (Cq, *C*OO), 161.8 (Cq, 7-*C*ₐᵣOCH₂), 157.1 (Cq, 5-*C*ₐᵣOCH₃), 157.0 (Cq, 8a-*C*ₐᵣ), 138.9 (+, *C*H), 111.1 (+, *C*H), 104.2 (Cq, 4a-*C*ₐᵣ), 95.3 (+, *C*ₐᵣH), 93.4 (+, *C*ₐᵣH), 64.3 (–, O*C*H₂), 56.1 (+, O*C*H₃), 14.7 (+, *C*H₃) ppm. – **IR** (ATR, ṽ) = 3089, 3077, 3017, 2980, 2939, 2902, 2830, 1718, 1609, 1562, 1497, 1455, 1421, 1391, 1377, 1360, 1309, 1241, 1218, 1197, 1184, 1156, 1116, 1106, 1095, 1045, 987, 972, 898, 880, 834, 815, 762, 747, 686, 666, 640, 622, 555, 509, 465, 443, 438, 419, 395 cm⁻¹. – **MS** (70 eV, EI, 70 °C) *m/z* (%) = 221 (22) [M+H]⁺, 220 (100) [M]⁺, 192 (38), 181 (12), 164 (63), 149 (18), 119 (12), 69 (39). – **HRMS** (EI, C₁₂H₁₈O₄) = calc.: 220.0736; found: 220.0735.

7-Ethoxy-5-methoxy-3-methyl-2*H*-chromen-2-one (16bb)

According to **GP11**, 4-ethoxy-2-hydroxy-6-methoxybenzaldehyde (**19b**, 50.0 mg, 0.25 mmol, 1.00 equiv.), K₂CO₃ (2.00 mg, 0.01 mmol, 0.05 equiv.) and propionic anhydride (0.11 ml, 116 mg, 0.89 mmol, 3.50 equiv.) were heated at 180 °C for 65 min *via* microwave irradiation. Purification by flash column chromatography (cHex/EtOAc 5:1) resulted in the title compound as white crystals (47.0 mg, 0.20 mmol, 78%).

R_f (cHex/EtOAc 5:1) = 0.31. – **¹H NMR** (400 MHz, CDCl₃) δ = 7.78 (d, ^4J = 1.3 Hz, 1H, C*H*), 6.40 (d, ^4J = 2.1 Hz, 1H, C*H*ₐᵣ), 6.28 (d, ^4J = 2.1 Hz, 1H, C*H*ₐᵣ), 4.05 (q, ^3J = 7.0 Hz, 2H, OC*H₂*), 3.88 (s, 3H, OC*H₃*), 2.16 (d, ^4J = 1.3 Hz, 3H, CC*H₃*), 1.44 (t, ^3J = 7.0 Hz, 3H, C*H₃*) ppm. – **¹³C NMR** (101 MHz, CDCl₃) δ = 163.0 (C_q, COO), 162.0 (C_q, 7-*C*ₐᵣOCH₂), 156.4 (C_q, 5-*C*ₐᵣOCH₃), 155.9 (C_q, 8a-*C*ₐᵣ), 134.8 (+, *C*H), 120.1 (C_q, 3-*C*), 104.5 (C_q, 4a-*C*ₐᵣ), 95.2 (+, *C*ₐᵣH), 93.1 (+, *C*ₐᵣH), 64.2 (–, O*C*H₂), 56.0 (+, O*C*H₃), 17.1 (+, C*C*H₃), 14.7 (+, OCH₂*C*H₃) ppm. – **IR** (ATR, ṽ) = 2989, 2969, 2948, 2919, 2876, 2853, 1700, 1633, 1612, 1581, 1497, 1466, 1453, 1448, 1425, 1378, 1370, 1356, 1299, 1256, 1225, 1193, 1153, 1118, 1105, 1062, 1045, 999, 973, 922, 887, 857, 817, 792, 764, 722, 686, 662, 636, 609, 554, 524, 439, 399 cm⁻¹. – **MS** (70 eV, EI, 50 °C) *m/z* (%) = 235 (18) [M+H]⁺, 234 (100) [M]⁺, 220 (19), 208 (14), 206 (42), 205 (12) [M–C₂H₅]⁺, 191 (18), 184 (39), 181 (20), 178 (58), 177 (20), 169 (14), 162 (33), 155 (11), 148 (17), 134 (14), 127 (11), 125 (13), 118 (12), 111 (24), 109 (14), 99 (11), 97 (31), 96 (11), 95 (18), 85 (32), 84 (11), 83 (25), 82 (12), 81 (15), 71 (44), 70 (14), 69 (54), 58 (15), 57 (74), 56 (12), 55 (27). – **HRMS** (EI, C₁₃H₁₄O₄) = calc.: 234.0887; found: 234.0888.

7-Ethoxy-3-ethyl-5-methoxy-2*H*-chromen-2-one (16bc)

According to **GP11**, 4-ethoxy-2-hydroxy-6-methoxybenzaldehyde (**19b**, 50.0 mg, 0.25 mmol, 1.00 equiv.), K₂CO₃ (2.00 mg, 0.01 mmol, 0.05 equiv.) and butyric anhydride (0.15 ml, 141 mg, 0.89 mmol, 3.50 equiv.) were heated at 180 °C for 65 min *via* microwave irradiation. Purification by flash column chromatography (cHex/EtOAc 5:1) resulted in the title compound as white crystals (48.0 mg, 0.19 mmol, 76%).

R_f (cHex/EtOAc 3:1) = 0.65. – **¹H NMR** (400 MHz, CDCl₃) δ = 7.76 (s, 1H, C*H*), 6.40 (d, ^4J = 2.2 Hz, 1H, C*H*ₐᵣ), 6.28 (d, ^4J = 2.2 Hz, 1H, C*H*ₐᵣ), 4.06 (q, ^3J = 7.0 Hz, 2H, OC*H₂*), 3.89 (s, 3H, OC*H₃*), 2.55 (qd, $^{3,4}J$ = 7.5, 1.3 Hz, 2H, CC*H₂*), 1.44 (t, ^3J = 7.0 Hz, 3H, OCH₂C*H₃*), 1.23 (t, ^3J = 7.5 Hz, 3H, C*H₃*) ppm. – **¹³C NMR** (101 MHz, CDCl₃) δ = 162.6 (C_q, COO), 162.0 (C_q, 7-*C*ₐᵣOCH₂), 156.5 (C_q, 5-*C*ₐᵣOCH₃), 155.6 (C_q, 8a-*C*ₐᵣ), 133.1 (+, *C*H), 125.7 (C_q, 3-*C*), 104.5 (C_q, 4a-*C*ₐᵣ), 95.2 (+, *C*ₐᵣH), 93.0 (+, *C*ₐᵣH), 64.2 (–, O*C*H₂), 56.0 (+, O*C*H₃), 23.9 (–, C*C*H₂), 14.7 (+, OCH₂*C*H₃), 12.7 (+, *C*H₃) ppm. – **IR** (ATR, ṽ) = 2975, 2936, 2924, 2902, 1700, 1612, 1578, 1496, 1466, 1456, 1425, 1377, 1363, 1316, 1296, 1247, 1228, 1203, 1154, 1115, 1079, 1044, 975, 955, 918, 887, 843, 817, 800, 754, 715, 687, 666, 637, 619, 560, 524, 470, 381 cm⁻¹. – **MS** (70 eV, EI, 60 °C) *m/z* (%) = 249 (16) [M+H]⁺, 248 (100) [M]⁺, 234 (16), 233 (77) [M–CH₃]⁺, 220 (12), 219 (15) [M–C₂H₅]⁺, 208 (15), 206 (14), 205 (73), 191 (15), 184 (19), 181

(17), 177 (15), 168 (11), 69 (34), 58 (27), 57 (12). – **HRMS** (EI, $C_{14}H_{16}O_4$) = calc.: 248.1043; found: 248.1043.

7-Ethoxy-5-methoxy-3-propyl-2*H*-chromen-2-one (19bd)

According to **GP11**, 4-ethoxy-2-hydroxy-6-methoxybenzaldehyde (**19b**, 50.0 mg, 0.25 mmol, 1.00 equiv.), K_2CO_3 (2.00 mg, 0.01 mmol, 0.05 equiv.) and valeric anhydride (0.17 ml, 166 mg, 0.89 mmol, 3.50 equiv.) were heated at 180 °C for 65 min *via* microwave irradiation. Purification by flash column chromatography (*c*Hex/EtOAc 5:1) resulted in the title compound as an off-white solid (46.0 mg, 0.17 mmol, 69%).

R_f (*c*Hex/EtOAc 5:1) = 0.42. – **¹H NMR** (400 MHz, CDCl₃) δ = 7.74 (d, ^4J = 0.9 Hz, 1H, C*H*), 6.39 (d, ^4J = 2.2 Hz, 1H, C*H*ₐᵣ), 6.27 (d, ^4J = 2.2 Hz, 1H, C*H*ₐᵣ), 4.05 (q, ^3J = 7.0 Hz, 2H, OC*H*₂), 3.88 (s, 3H, OC*H*₃), 2.49 (td, $^{3,4}J$ = 7.6, 0.9 Hz, 2H, CC*H*₂), 1.70 – 1.59 (m, 2H, CCH₂C*H*₂), 1.44 (t, ^3J = 7.0 Hz, 3H, OCH₂C*H*₃), 0.98 (t, ^3J = 7.4 Hz, 3H, C*H*₃) ppm. – **¹³C NMR** (101 MHz, CDCl₃) δ = 162.7 (C_q, *C*OO), 162.0 (C_q, 7-*C*ₐᵣOCH₂), 156.5 (C_q, 5-*C*ₐᵣOCH₃), 155.7 (C_q, 8a-*C*ₐᵣ), 134.1 (+, *C*H), 124.2 (C_q, 3-*C*), 104.4 (C_q, 4a-*C*ₐᵣ), 95.2 (+, *C*ₐᵣH), 93.0 (+, *C*ₐᵣH), 64.2 (–, O*C*H₂), 56.0 (+, O*C*H₃), 32.9 (–, *C*H₂), 21.6 (–, *C*H₂), 14.7 (+, OCH₂*C*H₃), 13.9 (+, *C*H₃) ppm. – **IR** (ATR, ṽ) = 3019, 2987, 2968, 2952, 2924, 2874, 1720, 1710, 1615, 1574, 1497, 1463, 1455, 1422, 1388, 1375, 1363, 1340, 1305, 1282, 1245, 1224, 1197, 1170, 1154, 1116, 1098, 1047, 1031, 1007, 972, 931, 905, 892, 870, 823, 785, 771, 738, 710, 663, 633, 609, 579, 558, 540, 517, 483, 449, 433, 414, 405, 377 cm⁻¹. – **MS** (70 eV, EI, 40 °C) *m/z* (%) = 263 (10) [M+H]⁺, 262 (59) [M]⁺, 234 (20), 233 (100) [M–C₂H₅]⁺, 205 (51), 184 (20), 68 (12). – **HRMS** (EI, $C_{15}H_{18}O_4$) = calc.: 262.1205; found: 262.1206.

3-Butyl-7-ethoxy-5-methoxy-2*H*-chromen-2-one (16be)

According to **GP11**, 4-ethoxy-2-hydroxy-6-methoxybenzaldehyde (50.0 mg, 0.25 mmol, 1.00 equiv.), K_2CO_3 (2.00 mg, 0.01 mmol, 0.05 equiv.) and hexanoic anhydride (0.21 ml, 191 mg, 0.89 mmol, 3.50 equiv.) were heated at 180 °C for 65 min *via* microwave irradiation. Purification by flash column chromatography (*c*Hex/EtOAc 5:1) resulted in the title compound as a tan solid (60.0 mg, 0.22 mmol, 86%).

R_f (*c*Hex/EtOAc 5:1) = 0.46. – **¹H NMR** (400 MHz, CDCl₃) δ = 7.74 (s, 1H, C*H*), 6.39 (d, ^4J = 2.2 Hz, 1H, C*H*ₐᵣ), 6.28 (d, ^4J = 2.2 Hz, 1H, C*H*ₐᵣ), 4.05 (q, ^3J = 7.0 Hz, 2H, OC*H*₂), 3.88 (s, 3H, OC*H*₃), 2.51 (td, $^{3,4}J$ = 7.5, 1.1 Hz, 2H, CC*H*₂), 1.68 – 1.54 (m, 2H, C*H*₂), 1.49 – 1.34 [m,

5H cont.: 1.44 (t, ^3J = 7.0 Hz, 3H, OCH$_2$CH_3), 1.45 – 1.34 (m, 2H, CH_2)], 0.94 (t, ^3J = 7.3 Hz, 3H, CH_3) ppm. – ^{13}**C NMR** (101 MHz, CDCl$_3$) δ = 162.7 (C$_q$, COO), 162.0 (C$_q$, 7-C$_{ar}$OCH$_2$), 156.5 (C$_q$, 5-C$_{ar}$OCH$_3$), 155.7 (C$_q$, 8a-C$_{ar}$), 133.9 (+, CH), 124.5 (C$_q$, 3-C), 104.5 (C$_q$, 4a-C$_{ar}$), 95.2 (+, C$_{ar}$H), 93.0 (+, C$_{ar}$H), 64.2 (–, OCH$_2$), 56.0 (+, OCH$_3$), 30.6 (–, CH$_2$), 30.5 (–, CH$_2$), 22.6 (–, CH$_2$), 14.7 (+, OCH$_2$CH$_3$), 14.1 (+, CH$_3$) ppm. – **IR** (ATR, ṽ) = 3027, 2985, 2962, 2952, 2918, 2871, 2856, 1710, 1616, 1574, 1497, 1456, 1435, 1422, 1387, 1377, 1363, 1324, 1303, 1256, 1242, 1224, 1197, 1170, 1154, 1116, 1102, 1069, 1045, 1020, 990, 973, 943, 912, 890, 864, 837, 822, 798, 786, 758, 732, 710, 666, 639, 630, 611, 582, 558, 540, 480, 429, 377 cm^{-1}. – **MS** (70 eV, EI, 60 °C) m/z (%) = 277 (12) [M+H]$^+$, 276 (66) [M]$^+$, 247 (12) [M–C$_2$H$_5$]$^+$, 234 (39), 233 (100) [M–C$_3$H$_7$]$^+$, 206 (12), 205 (48), 168 (11), 69 (13). – **HRMS** (EI, C$_{16}$H$_{20}$O$_4$) = calc.: 276.1356; found: 276.1354.

7-Ethoxy-5-methoxy-3-pentyl-2*H*-chromen-2-one (16bf)

According to **GP11**, 4-ethoxy-2-hydroxy-6-methoxybenzaldehyde (**19b**, 50.0 mg, 0.25 mmol, 1.00 equiv.), K$_2$CO$_3$ (2.00 mg, 0.01 mmol, 0.05 equiv.) and heptanoic anhydride (0.23 ml, 216 mg, 0.89 mmol, 3.50 equiv.) were heated at 180 °C for 65 min *via* microwave irradiation. Purification by flash column chromatography (*c*Hex/EtOAc 6:1) resulted in a mixture of the product and heptanoic acid as impurity. After washing the product fractions with 2 M aq. NaOH, drying over Na$_2$SO$_4$ and removal of the volatiles, gave the title compound as a white solid (53.0 mg, 0.18 mmol, 72%).

R_f (*c*Hex/EtOAc 6:1) = 0.44. – ^{1}**H NMR** (400 MHz, CDCl$_3$) δ = 7.74 (s, 1H, CH), 6.39 (d, ^4J = 2.1 Hz, 1H, CH_{ar}), 6.28 (d, ^4J = 2.1 Hz, 1H, CH_{ar}), 4.05 (q, ^3J = 7.0 Hz, 2H, OCH_2), 3.88 (s, 3H, OCH_3), 2.50 (td, $^{3,4}J$ = 7.6, 1.2 Hz, 2H, CCH_2), 1.67 – 1.56 (m, 2H, CH_2), 1.44 (t, ^3J = 7.0 Hz, 3H, OCH$_2$CH_3), 1.40 – 1.31 (m, 4H, 2 × CH_2), 0.90 (t, ^3J = 7.0 Hz, 3H, CH_3) ppm. – ^{13}**C NMR** (101 MHz, CDCl$_3$) δ = 162.7 (C$_q$, COO), 162.0 (C$_q$, 7-C$_{ar}$OCH$_2$), 156.5 (C$_q$, 5-C$_{ar}$OCH$_3$), 155.7 (C$_q$, 8a-C$_{ar}$), 133.9 (+, CH), 124.5 (C$_q$, 3-C), 104.5 (C$_q$, 4a-C$_{ar}$), 95.2 (+, C$_{ar}$H), 93.0 (+, C$_{ar}$H), 64.2 (–, OCH$_2$), 56.0 (+, OCH$_3$), 31.7 (–, CH$_2$), 30.9 (–, CH$_2$), 28.1 (–, CH$_2$), 22.6 (–, CH$_2$), 14.7 (+, OCH$_2$CH$_3$), 14.1 (+, CH$_3$) ppm. – **IR** (ATR, ṽ) = 3024, 2986, 2965, 2951, 2918, 2871, 2856, 1718, 1711, 1616, 1574, 1499, 1463, 1456, 1436, 1422, 1387, 1377, 1364, 1343, 1307, 1296, 1249, 1234, 1222, 1197, 1170, 1154, 1118, 1106, 1069, 1047, 1028, 1013, 994, 972, 922, 905, 890, 864, 837, 823, 785, 775, 710, 666, 639, 633, 560, 540, 492, 441 cm^{-1}. – **MS** (70 eV, EI, 80 °C) m/z (%) = 291 (14) [M+H]$^+$, 290 (70) [M]$^+$, 247 (11) [M–C$_3$H$_7$]$^+$, 234 (39),

233 (100) [M–C$_4$H$_9$]$^+$, 206 (12), 205 (46). – **HRMS** (EI, C$_{17}$H$_{22}$O$_4$) = calc.: 290.1518; found: 290.1519.

7-Ethoxy-3-hexyl-5-methoxy-2*H*-chromen-2-one (16bg)

According to **GP11**, 4-ethoxy-2-hydroxy-6-methoxybenzalde-hyde (**19b**, 50.0 mg, 0.25 mmol, 1.00 equiv.), K$_2$CO$_3$ (2.00 mg, 0.01 mmol, 0.05 equiv.) and caprylic anhydride (0.27 ml, 241 mg, 0.89 mmol, 3.50 equiv.) were heated at 180 °C for 65 min *via* microwave irradiation. Purification by flash column chromatography (cHex/EtOAc 6:1) resulted in the title compound as a tan solid (26.0 mg, 0.08 mmol, 33%).

R_f (cHex/EtOAc 6:1) = 0.28. – **^{1}H NMR** (500 MHz, CDCl$_3$) δ = 7.73 (s, 1H, C*H*), 6.38 (d, ^4J = 2.1 Hz, 1H, C*H*$_{ar}$), 6.27 (d, ^4J = 2.1 Hz, 1H, C*H*$_{ar}$), 4.05 (q, ^3J = 7.0 Hz, 2H, OC*H*$_2$), 3.88 (s, 3H, OC*H*$_3$), 2.50 (t, ^3J = 7.8 Hz, 2H, CC*H*$_2$), 1.60 (quin., ^3J = 7.4 Hz, 2H, C*H*$_2$), 1.43 (t, ^3J = 7.0 Hz, 3H, OCH$_2$C*H*$_3$), 1.40 – 1.28 (m, 6H, 3 × C*H*$_2$), 0.88 (t, ^3J = 7.0 Hz, 3H, C*H*$_3$) ppm. – **^{13}C NMR** (126 MHz, CDCl$_3$) δ = 162.7 (C$_q$, *C*OO), 161.9 (C$_q$, 7-*C*$_{ar}$OCH$_2$), 156.5 (C$_q$, 5-*C*$_{ar}$OCH$_3$), 155.7 (C$_q$, 8a-*C*$_{ar}$), 133.9 (+, *C*H), 124.5 (C$_q$, 3-*C*), 104.5 (C$_q$, 4a-*C*$_{ar}$), 95.1 (+, *C*$_{ar}$H), 93.0 (+, *C*$_{ar}$H), 64.2 (–, O*C*H$_2$), 56.0 (+, O*C*H$_3$), 31.8 (–, *C*H$_2$), 30.9 (–, *C*H$_2$), 29.1 (–, *C*H$_2$), 28.3 (–, *C*H$_2$), 22.7 (–, *C*H$_2$), 14.7 (+, OCH$_2$*C*H$_3$), 14.2 (+, *C*H$_3$) ppm. – **IR** (ATR, ṽ) = 3024, 2986, 2962, 2951, 2918, 2880, 2871, 2853, 1717, 1710, 1619, 1574, 1499, 1456, 1422, 1387, 1377, 1363, 1329, 1305, 1281, 1248, 1220, 1198, 1170, 1154, 1118, 1109, 1071, 1047, 1009, 992, 973, 928, 891, 881, 867, 837, 822, 796, 786, 762, 738, 710, 667, 639, 633, 609, 560, 543, 492, 441, 421, 375 cm^{-1}. – **MS** (70 eV, EI, 90 °C) *m/z* (%) = 305 (20) [M+H]$^+$, 304 (100) [M]$^+$, 247 (10) [M–C$_4$H$_9$]$^+$, 234 (37), 233 (86) [M–C$_5$H$_{11}$]$^+$, 205 (41), 181 (19), 131 (15), 69 (18). – **HRMS** (EI, C$_{18}$H$_{24}$O$_4$) = calc.: 304.1675; found: 304.1676.

5-Methoxy-7-propoxy-2*H*-chromen-2-one (16ca)

According to **GP11**, 4-propoxy-2-hydroxy-6-methoxybenzaldehyde (**19c**, 50.0 mg, 0.24 mmol, 1.00 equiv.), K$_2$CO$_3$ (2.00 mg, 0.01 mmol, 0.05 equiv.) and acetic anhydride (0.08 ml, 85.0 mg, 0.83 mmol, 3.50 equiv.) were heated at 180 °C for 65 min *via* microwave irradiation. Purification by flash column chromatography (cHex/EtOAc 5:1) resulted in the title compound as a tan solid (43.0 mg, 0.18 mmol, 77%).

R_f (cHex/EtOAc 5:1) = 0.38. – **^{1}H NMR** (400 MHz, CDCl$_3$) δ = 7.95 (d, ^3J = 9.6 Hz, 1H, C*H*), 6.38 (d, ^4J = 2.1 Hz, 1H, C*H*$_{ar}$), 6.27 (d, ^4J = 2.1 Hz, 1H, C*H*$_{ar}$), 6.13 (d, ^3J = 9.6 Hz, 1H, C*H*),

3.94 (t, ^3J = 6.5 Hz, 2H, OCH_2), 3.87 (s, 3H, OCH_3), 1.82 (sex., ^3J = 7.4 Hz, 2H, CH_2CH$_3$), 1.04 (t, ^3J = 7.4 Hz, 3H, CH_3) ppm. – ^{13}C NMR (101 MHz, CDCl$_3$) δ = 164.1 (C$_q$, COO), 162.5 (C$_q$, 7-C$_{ar}$OCH$_2$), 157.8 (C$_q$, 5-C$_{ar}$OCH$_3$), 157.7 (C$_q$, 8a-C$_{ar}$), 139.7 (+, CH), 111.6 (+, CH), 104.7 (C$_q$, 4a-C$_{ar}$), 96.0 (+, C$_{ar}$H), 94.2 (+, C$_{ar}$H), 71.0 (–, OCH$_2$), 56.8 (+, OCH$_3$), 23.2 (–, CH$_2$CH$_3$), 11.4 (+, CH$_3$) ppm. – IR (ATR, ṽ) = 2968, 2944, 2927, 2874, 2854, 1714, 1609, 1497, 1453, 1422, 1401, 1388, 1363, 1310, 1282, 1238, 1218, 1191, 1160, 1147, 1115, 1095, 1064, 1044, 1030, 1001, 984, 950, 938, 924, 894, 817, 802, 772, 742, 684, 666, 636, 620, 557, 511, 476, 463, 446, 404 cm^{-1}. – MS (70 eV, EI, 70 °C) m/z (%) = 234 (59) [M]$^+$, 192 (75), 165 (10), 164 (100), 163 (13), 149 (26), 69 (16). – HRMS (EI, C$_{13}$H$_{14}$O$_4$) = calc.: 234.0892; found: 234.0891.

3-Butyl-5-methoxy-7-propoxy-2*H*-chromen-2-one (16ce)

According to GP11, 4-propoxy-2-hydroxy-6-methoxybenzalde-hyde (19c, 50.0 mg, 0.24 mmol, 1.00 equiv.), K$_2$CO$_3$ (2.00 mg, 0.01 mmol, 0.05 equiv.) and hexanoic anhydride (0.19 ml, 178 mg, 0.83 mmol, 3.50 equiv.) were heated at 180 °C for 65 min *via* microwave irradiation. Purification by flash column chromatography (cHex/EtOAc 30:1 to 25:1) resulted in a mixture of the product and hexanoic acid as impurities. Another purification by prepHPLC (50–95_alkyl) gave the title compound as an off-white solid (10.6 mg, 0.04 mmol, 15%).

R_f (cHex/EtOAc 10:1) = 0.50. – ^{1}H NMR (500 MHz, CDCl$_3$) δ = 7.67 (s, 1H, CH), 6.32 (d, ^4J = 2.1 Hz, 1H, CH_{ar}), 6.21 (d, ^4J = 2.1 Hz, 1H, CH_{ar}), 3.87 (t, ^3J = 6.6 Hz, 2H, OCH_2), 3.82 (s, 3H, OCH_3), 2.44 (t, ^3J = 7.7 Hz, 2H, CCH_2), 1.76 (sex., ^3J = 7.1 Hz, 2H, CH_2), 1.61 – 1.48 (m, 2H, CH_2), 1.33 (sex., ^3J = 7.3 Hz, 2H, CH_2), 0.98 (t, ^3J = 7.4 Hz, 3H, CH_3), 0.88 (t, ^3J = 7.4 Hz, 3H, CH_3) ppm. – ^{13}C NMR (126 MHz, CDCl$_3$) δ = 162.7 (C$_q$, COO), 162.2 (C$_q$, 7-C$_{ar}$OCH$_2$), 156.4 (C$_q$, 5-C$_{ar}$OCH$_3$), 155.7 (C$_q$, 8a-C$_{ar}$), 133.9 (+, CH), 124.4 (C$_q$, 3-C), 104.4 (C$_q$, 4a-C$_{ar}$), 95.2 (+, C$_{ar}$H), 93.0 (+, C$_{ar}$H), 70.1 (–, OCH$_2$), 56.0 (+, OCH$_3$), 30.6 (–, CH$_2$), 30.5 (–, CH$_2$), 22.6 (–, CH$_2$), 22.5 (–, CH$_2$), 14.1 (+, CH$_3$), 10.6 (+, CH$_3$) ppm. – IR (ATR, ṽ) = 2961, 2922, 2873, 2860, 1706, 1632, 1609, 1579, 1497, 1465, 1452, 1422, 1402, 1388, 1364, 1326, 1298, 1258, 1242, 1227, 1194, 1154, 1122, 1103, 1069, 1044, 990, 946, 909, 871, 820, 815, 798, 768, 738, 711, 669, 640, 626, 557, 544, 487, 469 cm^{-1}. – HRMS (ESI, C$_{17}$H$_{22}$O$_4$) = calc.: 291.1591 [M+H]$^+$; found: 291.1578 [M+H]$^+$.

7-Butoxy-5-methoxy-3-propyl-2*H*-chromen-2-one (16dd)

According to **GP11**, 4-butoxy-2-hydroxy-6-methoxybenzaldehyde (**19d**, 50.0 mg, 0.22 mmol, 1.00 equiv.), K_2CO_3 (2.00 mg, 0.01 mmol, 0.05 equiv.) and valeric anhydride (0.15 ml, 145 mg, 0.78 mmol, 3.50 equiv.) were heated at 180 °C for 65 min *via* microwave irradiation. Purification by flash column chromatography (cHex/EtOAc 30:1 to 25:1) resulted in a mixture of the product and valeric acid as impurities. Another purification by prepHPLC (50–95_alkyl) gave the title compound as a yellow solid (21.0 mg, 0.07 mmol, 32%).

R_f (cHex/EtOAc 30:1) = 0.19. – **¹H NMR** (500 MHz, CDCl₃) δ = 7.73 (s, 1H, C*H*), 6.38 (d, ^4J = 2.2 Hz, 1H, C*H*ₐᵣ), 6.27 (d, ^4J = 2.2 Hz, 1H, C*H*ₐᵣ), 3.97 (t, ^3J = 6.6 Hz, 2H, OC*H₂*), 3.88 (s, 3H, OC*H₃*), 2.48 (t, ^3J = 7.4 Hz, 2H, CC*H₂*), 1.77 (quin., ^3J = 7.4, 6.6 Hz, 2H, OCH₂C*H₂*), 1.64 (sex., ^3J = 7.4 Hz, 2H, C*H₂*), 1.49 (sex., ^3J = 7.4 Hz, 2H, C*H₂*), 0.97 (t, ^3J = 7.4 Hz, 6H, 2 × C*H₃*) ppm. – **¹³C NMR** (126 MHz, CDCl₃) δ = 162.5 (C_q, *C*OO), 162.1 (C_q, 7-*C*ₐᵣOCH₂), 156.3 (C_q, 5-*C*ₐᵣOCH₃), 155.6 (C_q, 8a-*C*ₐᵣ), 134.0 (+, *C*H), 124.0 (C_q, 3-*C*), 104.2 (C_q, 4a-*C*ₐᵣ), 95.0 (+, *C*ₐᵣH), 92.9 (+, *C*ₐᵣH), 68.2 (–, O*C*H₂), 55.9 (+, O*C*H₃), 32.8 (–, C*C*H₂), 31.0 (–, O*C*H₂CH₂), 21.5 (–, *C*H₂), 19.2 (–, *C*H₂), 13.8 (+, 2 × *C*H₃) ppm. – **IR** (ATR, ṽ) = 2962, 2948, 2925, 2908, 2870, 1715, 1710, 1611, 1578, 1496, 1466, 1456, 1421, 1381, 1367, 1305, 1282, 1247, 1224, 1196, 1167, 1153, 1113, 1098, 1044, 1018, 997, 949, 931, 901, 871, 829, 806, 795, 778, 744, 713, 666, 640, 630, 613, 594, 557, 540, 500, 467, 418, 382 cm⁻¹. – **HRMS** (ESI, C₁₇H₂₂O₄) = calc.: 291.1591 [M+H]⁺; found: 291.1582 [M+H]⁺.

3-Ethyl-5-methoxy-7-(pentyloxy)-2*H*-chromen-2-one (16ec)

According to **GP11**, 4-butoxy-2-hydroxy-6-methoxybenzaldehyde (**19e**, 50.0 mg, 0.22 mmol, 1.00 equiv.), K_2CO_3 (2.00 mg, 0.01 mmol, 0.05 equiv.) and butanoic anhydride (0.15 ml, 145 mg, 0.78 mmol, 3.50 equiv.) were heated at 180 °C for 65 min *via* microwave irradiation. After extraction, the product was used without further purification as an orange solid (32.0 mg, 0.11 mmol, 86%).

¹H NMR (500 MHz, CDCl₃) δ = 7.75 (d, ^4J = 1.4 Hz, 1H, C*H*), 6.39 (d, ^4J = 2.2 Hz, 1H, C*H*ₐᵣ), 6.27 (d, ^4J = 2.2 Hz, 1H, C*H*ₐᵣ), 3.97 (t, ^3J = 6.6 Hz, 2H, OC*H₂*), 3.88 (s, 3H, OC*H₃*), 2.54 (qd, $^{3,4}J$ = 7.4, 1.4 Hz, 2H, CC*H₂*), 1.80 (quin., ^3J = 6.7 Hz, 2H, OCH₂C*H₂*), 1.51 – 1.33 (m, 4H, 2 × C*H₂*), 1.22 (t, ^3J = 7.4 Hz, 3H, CCH₂C*H₃*), 0.94 (t, ^3J = 7.1 Hz, 3H, C*H₃*) ppm. – **¹³C NMR** (126 MHz, CDCl₃) δ = 162.5 (C_q, *C*OO), 162.0 (C_q, 7-*C*ₐᵣOCH₂), 156.3 (C_q, 5-*C*ₐᵣOCH₃), 155.5

(C_q, 8a-C_{ar}), 133.0 (+, CH), 125.5 (C_q, 3-C), 104.3 (C_q, 4a-C_{ar}), 95.0 (+, $C_{ar}H$), 92.9 (+, $C_{ar}H$), 68.5 (–, OCH_2), 55.9 (+, OCH_3), 28.7 (–, CH_2), 28.1 (–, CH_2), 23.8 (–, CCH_2), 22.5 (–, CH_2), 14.0 (+, CH_3), 12.6 (+, CCH_2CH_3) ppm. – **IR** (ATR, ṽ) = 2949, 2922, 2870, 2856, 1707, 1615, 1575, 1500, 1470, 1455, 1429, 1411, 1400, 1383, 1370, 1343, 1333, 1299, 1249, 1227, 1191, 1173, 1150, 1112, 1081, 1035, 993, 966, 950, 928, 892, 850, 822, 802, 775, 752, 731, 715, 698, 673, 637, 596, 557, 535, 503, 467, 415, 390 cm^{-1}. – **HRMS** (ESI, $C_{17}H_{22}O_4$) = calc.: 291.1591 [M+H]$^+$; found: 291.1581 [M+H]$^+$.

7-(Hexyloxy)-5-methoxy-3-methyl-2*H*-chromen-2-one (16fb)

According to **GP11**, 4-(hexyloxy)-2-hydroxy-6-methoxybenzaldehyde (**19f**, 30.0 mg, 0.12 mmol, 1.00 equiv.), K_2CO_3 (1.00 mg, 0.01 mmol, 0.05 equiv.) and propionic anhydride (0.05 ml, 54.0 mg, 0.42 mmol, 3.50 equiv.) were heated at 180 °C for 65 min *via* microwave irradiation. Purification by flash column chromatography (*c*Hex/EtOAc 15:1) resulted in the title compound as an off-white solid (16.0 mg, 0.06 mmol, 46%).

R_f (*c*Hex/EtOAc 15:1) = 0.17. – **^{1}H NMR** (500 MHz, CDCl$_3$) δ = 7.77 (d, ^4J = 1.5 Hz, 1H, C*H*), 6.38 (d, ^4J = 2.2 Hz, 1H, C*H*$_{ar}$), 6.27 (d, ^4J = 2.2 Hz, 1H, C*H*$_{ar}$), 3.96 (t, ^3J = 6.5 Hz, 2H, OC*H*$_2$), 3.87 (s, 3H, OC*H*$_3$), 2.15 (s, 3H, CC*H*$_3$), 1.85 – 1.73 (m, 2H, OCH$_2$C*H*$_2$), 1.50 – 1.41 (m, 2H, C*H*$_2$), 1.39 – 1.31 (m, 4H, 2 × C*H*$_2$), 0.91 (t, ^3J = 6.5 Hz, 3H, C*H*$_3$) ppm. – **^{13}C NMR** (126 MHz, CDCl$_3$) δ = 163.0 (C_q, C*O*O), 162.1 (C_q, 7-C_{ar}OCH$_2$), 156.3 (C_q, 5-C_{ar}OCH$_3$), 155.8 (C_q, 8a-C_{ar}), 134.8 (+, CH), 120.0 (C_q, 3-C), 104.4 (C_q, 4a-C_{ar}), 95.2 (+, $C_{ar}H$), 93.1 (+, $C_{ar}H$), 68.7 (–, OC*H*$_2$), 56.0 (+, OC*H*$_3$), 31.7 (–, CH_2), 29.1 (–, CH_2), 25.8 (–, CH_2), 22.7 (–, CH_2), 17.1 (+, CH_3), 14.2 (+, 3-C*C*H$_3$) ppm. – **IR** (ATR, ṽ) = 2970, 2949, 2935, 2919, 2904, 2870, 2853, 1717, 1626, 1579, 1500, 1456, 1424, 1404, 1370, 1305, 1255, 1231, 1200, 1163, 1118, 1068, 1033, 1014, 994, 941, 921, 894, 823, 802, 761, 724, 663, 637, 609, 554, 534, 499, 462, 428, 404, 385 cm^{-1}. – **HRMS** (ESI, $C_{17}H_{22}O_4$) = calc.: 291.1591 [M+H]$^+$; found: 291.1582 [M+H]$^+$.

5,8-Dimethyl-2*H*-chromen-2-one (16na)

According to **GP11**, 2-hydroxy-3,6-dimethylbenzaldehyde (**19n**, 100 mg, 0.67 mmol, 1.00 equiv.), K_2CO_3 (3.00 mg, 0.03 mmol, 0.05 equiv.) and acetic anhydride (0.22 ml, 238 mg, 2.33 mmol, 3.50 equiv.) were heated at 180 °C for 65 min *via* microwave irradiation. Purification by flash column chromatography (*c*Hex/EtOAc 5:1) resulted in the title compound as a white solid (80.0 mg, 0.43 mmol, 64%).

R_f (cHex/EtOAc 5:1) = 0.28. – ^{1}H NMR (400 MHz, CDCl$_3$) δ = 7.91 (d, ^3J = 9.7 Hz, 1H, CH_{ar}), 7.26 (d, ^3J = 7.5 Hz, 1H, CH_{ar}), 6.99 (d, ^3J = 7.5 Hz, 1H, CH_{ar}), 6.41 (d, ^3J = 9.7 Hz, 1H, CH), 2.48 (s, 3H, CH_3), 2.40 (s, 3H, CH_3) ppm. – ^{13}C-NMR (101 MHz, CDCl$_3$) δ = 161.1 (C$_q$, COO) 153.0 (C$_q$, 8a-C_{ar}), 140.9 (+, CH), 134.0 (C$_q$, C_{ar}), 133.0 (+, C_{ar}H), 125.3 (+, C_{ar}H), 124.1 (C$_q$, C_{ar}), 117.5 (C$_q$, C_{ar}), 115.7 (+, CH), 18.2 (+, CH$_3$), 15.4 (+, CH$_3$) ppm. – IR (ATR, ṽ) = 2922, 1952, 1712, 1593, 1486, 1443, 1378, 1243, 1182, 1124, 1061, 957, 886, 843, 819, 764, 728, 677, 608, 581, 566, 535, 463, 416 cm^{-1}. – MS (70 eV, EI, 40 °C) m/z (%) = 175 (13) [M+H]$^+$, 174 (97) [M]$^+$, 159 (5) [M–CH$_3$]$^+$, 150 (55), 146 (62), 145 (32), 131 (100), 117 (14), 115 (21), 91 (18), 69 (62). – HRMS (EI, C$_{11}$H$_{10}$O$_2$) = calc.: 174.0675; found: 174.0676.

3,5,8-Trimethyl-2H-chromen-2-one (16nb)

According to GP11, 2-hydroxy-3,6-dimethylbenzaldehyde (19n, 75.0 mg, 0.50 mmol, 1.00 equiv.), K$_2$CO$_3$ (3.00 mg, 0.02 mmol, 0.05 equiv.) and propionic anhydride (0.22 ml, 227 mg, 1.75 mmol, 3.50 equiv.) were heated at 180 °C for 65 min via microwave irradiation. Purification by flash column chromatography (cHex/EtOAc 10:1) resulted in the title compound as a white solid (74.0 mg, 0.39 mmol, 79%).

R_f (cHex/EtOAc 10:1) = 0.30. – ^{1}H-NMR (400 MHz, CDCl$_3$) δ = 7.69 (s, 1H, CH), 7.17 (d, ^3J = 7.5 Hz, 1H, CH_{ar}), 6.96 (d, ^3J = 7.5 Hz, 1H, CH_{ar}), 2.46 (s, 3H, CH_3), 2.40 (s, 3H, CH_3), 2.23 (s, 3H, 3-CCH_3) ppm. – ^{13}C NMR (101 MHz, CDCl$_3$) δ = 162.4 (C$_q$, COO), 152.1 (C$_q$, 8a-C_{ar}), 136.7 (+, CH), 132.5 (C$_q$, C_{ar}), 131.6 (+, C_{ar}H), 125.1 (+, C_{ar}H), 124.7 (C$_q$, C_{ar}), 123.7 (C$_q$, C_{ar}), 118.1 (C$_q$, C_{ar}), 18.3 (+, CH$_3$), 17.5 (+, 3-CH$_3$), 15.5 (+, CH$_3$) ppm. – IR (ATR, ṽ) = 2920, 1711, 1601, 1488, 1444, 1375, 1261, 1186, 1085, 1032, 998, 902, 816, 765, 736, 700, 632, 576, 475, 399 cm^{-1}. – MS (70 eV, EI, 40 °C) m/z (%) = 189 (14) [M+H]$^+$, 188 (100) [M]$^+$, 181 (16), 173 (3) [M–CH$_3$]$^+$, 160 (40), 159 (37), 145 (32), 131 (18), 115 (16), 91 (11), 69 (29). – HRMS (EI, C$_{12}$H$_{12}$O$_2$) = calc.: 188.0832; found: 188.0833.

3-Ethyl-5,8-dimethyl-2H-chromen-2-one (16nc)

According to GP11, 2-hydroxy-3,6-dimethylbenzaldehyde (19n, 75.0 mg, 0.50 mmol, 1.00 equiv.), K$_2$CO$_3$ (3.00 mg, 0.02 mmol, 0.05 equiv.) and butanoic anhydride (0.22 ml, 227 mg, 1.75 mmol, 3.50 equiv.) were heated at 180 °C for 65 min via microwave irradiation. Purification by flash column chromatography (cHex/EtOAc 40:1) resulted in the title compound as a white solid (94.0 mg, 0.47 mmol, 93%).

R_f (cHex/EtOAc 40:1) = 0.21. – ^{1}H NMR (400 MHz, CDCl$_3$) δ = 7.65 (s, 1H, CH), 7.18 (d, ^3J = 7.5 Hz, 1H, CH_{ar}), 6.97 (d, ^3J = 7.5 Hz, 1H, CH_{ar}), 2.62 (q, ^3J = 7.4 Hz, 2H, CCH_2), 2.48 (s,

3H, C*H₃*), 2.40 (s, 3H, C*H₃*), 1.26 (t, ^3J = 7.4 Hz, 3H, CH₂C*H₃*) ppm. – **¹³C NMR** (101 MHz, CDCl₃) δ = 162.0 (C$_q$, *C*OO), 152.0 (C$_q$, 8a-*C*$_{ar}$), 135.0 (+, *C*H), 132.7 (C$_q$, *C*$_{ar}$), 131.6 (+, *C*$_{ar}$H), 130.3 (C$_q$, *C*$_{ar}$), 125.1 (+, *C*$_{ar}$H), 123.6 (C$_q$, *C*$_{ar}$), 118.1 (C$_q$, *C*$_{ar}$), 24.4 (–, *C*H₂), 18.3 (+, *C*H₃), 15.5 (+, *C*H₃), 12.8 (+, CH₂*C*H₃) ppm. – **IR** (ATR, ṽ) = 2919, 1714, 1596, 1487, 1432, 1377, 1271, 1254, 1175, 1087, 1047, 1035, 989, 959, 907, 825, 759, 740, 722, 688, 647, 610, 578, 541, 474, 449, 433 cm⁻¹. – **MS** (70 eV, EI, 50 °C) *m/z* (%) = 203 (14) [M+H]⁺, 202 (100) [M]⁺, 187 (17) [M–CH₃]⁺, 159 (12). – **HRMS** (EI, C₁₃H₁₄O₂) = calc.: 202.0988; found: 202.0988.

5,8-Dimethyl-3-propyl-2*H*-chromen-2-one (16nd)

According to **GP11**, 2-hydroxy-3,6-dimethylbenzaldehyde (**19n**, 75.0 mg, 0.50 mmol, 1.00 equiv.), K₂CO₃ (3.00 mg, 0.02 mmol, 0.05 equiv.) and va-leric anhydride (0.35 ml, 325 mg, 1.75 mmol, 3.50 equiv.) were heated at 180 °C for 65 min *via* microwave irradiation. Purification by flash column chromatography (*c*Hex/EtOAc 40:1) resulted in the title compound as a white solid (108 mg, 0.50 mmol, quant.).

R$_f$ (CH/EE 40:1) = 0.25. – **¹H NMR** (400 MHz, CDCl₃) δ = 7.65 (s, 1H, C*H*), 7.18 (d, ^3J = 7.5 Hz, 1H, C*H$_{ar}$*), 6.96 (d, ^3J = 7.5 Hz, 1H, C*H$_{ar}$*), 2.61 – 2.52 (m, 2H, CC*H₂*), 2.48 (s, 3H, C*H₃*), 2.40 (s, 3H, C*H₃*), 1.69 (sex., ^3J = 7.4 Hz, 2H, C*H₂*CH₃), 1.00 (t, ^3J = 7.4 Hz, 3H, CH₂C*H₃*) ppm. – **¹³C NMR** (101 MHz, CDCl₃) δ = 162.0 (C$_q$, *C*OO) 152.0 (C$_q$, 8a-*C*$_{ar}$), 136.0 (+, *C*H), 132.7 (C$_q$, *C*$_{ar}$), 131.6 (+, *C*$_{ar}$H), 128.8 (C$_q$, *C*$_{ar}$), 125.1 (+, *C*$_{ar}$H), 123.6 (C$_q$, *C*$_{ar}$), 118.1 (C$_q$, *C*$_{ar}$), 33.3 (–, *C*H₂), 21.6 (–, *C*H₂), 18.3 (+, *C*H₃), 15.5 (+, *C*H₃), 13.9 (+, CH₂*C*H₃) ppm. – **IR** (ATR, ṽ) = 2959, 2926, 1714, 1598, 1489, 1427, 1376, 1272, 1256, 1172, 1110, 1064, 1025, 992, 920, 902, 868, 830, 774, 749, 718, 672, 580, 479, 436 cm⁻¹. – **MS** (70 eV, EI, 50 °C) *m/z* (%) = 217 (17) [M+H]⁺, 216 (100) [M]⁺, 201 (22) [M–CH₃]⁺, 188 (26), 187 [M–C₂H₅]⁺. – **HRMS** (EI, C₁₄H₁₆O₂) = calc.: 216.1145; found: 216.1143.

3-Butyl-5,8-dimethyl-2*H*-chromen-2-one (16ne)

According to **GP11**, 2-hydroxy-3,6-dimethylbenzaldehyde (**19n**, 75.0 mg, 0.50 mmol, 1.00 equiv.), K₂CO₃ (3.00 mg, 0.02 mmol, 0.05 equiv.) and hex-anoic anhydride (0.40 ml, 375 mg, 1.75 mmol, 3.50 equiv.) were heated at 180 °C for 65 min *via* microwave irradiation. Purification by flash column chromatography (*c*Hex/EtOAc 20:1) resulted in the title compound as a white solid (71.0 mg, 0.31 mmol, 62%).

R$_f$ (*c*Hex/EtOAc 20:1) = 0.36. – **¹H NMR** (400 MHz, CDCl₃) δ = 7.65 (s, 1H, C*H*), 7.18 (d, ^3J = 7.5 Hz, 1H, C*H$_{ar}$*), 6.96 (d, ^3J = 7.5 Hz, 1H, C*H$_{ar}$*), 2.63 – 2.54 (m, 2H, 3-CC*H₂*), 2.48 (s, 3H, C*H₃*), 2.40 (s, 3H, C*H₃*), 1.70 – 1.57 (m, 2H, C*H₂*), 1.42 (sex., ^3J = 7.4 Hz, 2H, C*H₂*), 0.96 (t,

^3J = 7.4 Hz, 3H, CH$_2$CH_3) ppm. – **^{13}C NMR** (101 MHz, CDCl$_3$) δ = 162.0 (C$_q$, COO), 152.0 (C$_q$, 8a-$C$$_{ar}$), 135.8 (+, CH), 132.6 (C$_q$, $C$$_{ar}$), 131.6 (+, $C$$_{ar}$H), 129.0 (C$_q$, $C$$_{ar}$), 125.1 (+, $C$$_{ar}$H), 123.6 (C$_q$, $C$$_{ar}$), 118.1 (C$_q$, $C$$_{ar}$), 31.0 (–, CH$_2$), 30.5 (–, CH$_2$), 22.6 (–, CH$_2$), 18.3 (+, CH$_3$), 15.5 (+, CH$_3$), 14.1 (+, CH$_2CH_3$) ppm. – **IR** (ATR, ṽ) = 2951, 2926, 2867, 1712, 1599, 1487, 1462, 1418, 1373, 1271, 1247, 1170, 1111, 1075, 1034, 996, 943, 899, 822, 800, 764, 742, 714, 667, 616, 578, 539, 478, 424 cm^{-1}. – **MS** (70 eV, EI, 50 °C) *m/z* (%) = 231 (8) [M+H]$^+$, 230 (43) [M]$^+$, 215 (7) [M–CH$_3$]$^+$, 201 (29) [M–C$_2$H$_5$]$^+$, 199 (12) [M–C$_3$H$_7$]$^+$, 160 (12), 159 (17), 115 (10), 69 (15). – **HRMS** (EI, C$_{15}$H$_{18}$O$_2$) = calc.: 230.1301; found: 230.1303.

5,8-Dimethyl-3-pentyl-2*H*-chromen-2-one (16nf)

According to **GP11**, 2-hydroxy-3,6-dimethylbenzaldehyde (**19n**, 75.0 mg, 0.50 mmol, 1.00 equiv.), K$_2$CO$_3$ (3.00 mg, 0.02 mmol, 0.05 equiv.) and heptanoic anhydride (0.46 ml, 423 mg, 1.75 mmol, 3.50 equiv.) were heated at 180 °C for 65 min *via* microwave irradiation. Purification by flash column chromatography (*c*Hex/EtOAc 25:1) resulted in the title compound as a white solid (120 mg, 0.49 mmol, 98%).

R$_f$ (*c*Hex/EtOAc 25:1) = 0.29. – **^{1}H NMR** (500 MHz, CDCl$_3$) δ = 7.64 (s, 1H, CH), 7.17 (d, ^3J = 7.5 Hz, 1H, CH_{ar}), 6.96 (d, ^3J = 7.5 Hz, 1H, CH_{ar}), 2.63 – 2.54 (m, 2H, 3-CCH_2), 2.48 (s, 3H, CH_3), 2.40 (s, 3H, CH_3), 1.71 – 1.59 (m, 2H, CH_2), 1.44 – 1.31 (m, 4H, 2 × CH_2), 0.96 – 0.86 (m, 3H, CH$_2$CH_3) ppm. – **^{13}C NMR** (101 MHz, CDCl$_3$) δ = 161.9 (C$_q$, COO), 152.0 (C$_q$, 8a-$C$$_{ar}$), 135.7 (+, CH), 132.6 (C$_q$, $C$$_{ar}$), 131.5 (+, $C$$_{ar}$H), 129.0 (C$_q$, $C$$_{ar}$), 125.1 (+, $C$$_{ar}$H), 123.6 (C$_q$, $C$$_{ar}$), 118.1 (C$_q$, $C$$_{ar}$), 31.6 (–, CH$_2$), 31.2 (–, CH$_2$), 28.1 (–, CH$_2$), 22.6 (–, CH$_2$), 18.2 (+, CH$_3$), 15.4 (+, CH$_3$), 14.1 (+, CH$_2CH_3$) ppm. – **IR** (ATR, ṽ) = 2922, 2855, 1710, 1599, 1489, 1465, 1428, 1379, 1272, 1258, 1237, 1171, 1115, 1077, 1020, 996, 904, 829, 776, 755, 740, 718, 669, 615, 579, 542, 478, 443, 419 cm^{-1}. – **MS** (70 eV, EI, 70 °C) *m/z* (%) = 245 (17) [M+H]$^+$, 244 (87) [M]$^+$, 229 (5) [M–CH$_3$]$^+$, 215 (9) [M–C$_2$H$_5$]$^+$, 201 (15) [M–C$_3$H$_7$]$^+$, 189 (17), 188 (100), 187 (12) [M–C$_4$H$_9$]$^+$. – **HRMS** (EI, C$_{16}$H$_{20}$O$_2$) = calc.: 244.1458; found; 244.1457.

3-Hexyl-5,8-dimethyl-2*H*-chromen-2-one (16ng)

According to **GP11**, 2-hydroxy-3,6-dimethylbenzaldehyde (**19n**, 75.0 mg, 0.50 mmol, 1.00 equiv.), K$_2$CO$_3$ (3.00 mg, 0.02 mmol, 0.05 equiv.) and caprylic anhydride (0.52 ml, 472 mg, 1.75 mmol, 3.50 equiv.) were heated at 180 °C for 65 min *via* microwave irradiation. Purification by flash column chromatography (*c*Hex/EtOAc 40:1) resulted in the title compound as a white solid (104 mg, 0.40 mmol, 80%).

R_f (*c*Hex/EtOAc 40:1) = 0.31. – **¹H NMR** (400 MHz, CDCl₃) δ = 7.65 (s, 1H, C*H*), 7.18 (d, ^3J = 7.5 Hz, 1H, C*H*ₐᵣ), 6.97 (d, ^3J = 7.5 Hz, 1H, C*H*ₐᵣ), 2.62 – 2.52 (m, 2H, 3-CC*H*₂), 2.48 (s, 3H, C*H*₃), 2.41 (s, 3H, C*H*₃), 1.72 – 1.55 (m, 2H, C*H*₂), 1.46 – 1.19 (m, 6H, 3 × C*H*₂), 0.96 – 0.81 (m, 3H, CH₂C*H*₃) ppm. – **¹³C NMR** (101 MHz, CDCl₃) δ = 162.0 (C_q, *C*OO) 152.0 (C_q, 8a-*C*ₐᵣ), 135.8 (+, *C*H), 132.6 (C_q, *C*ₐᵣ), 131.6 (+, *C*ₐᵣH), 129.1 (C_q, *C*ₐᵣ), 125.1 (+, *C*ₐᵣH), 123.7 (C_q, *C*ₐᵣ), 118.1 (C_q, *C*ₐᵣ), 31.6 (–,*C*H₂), 31.2 (–, *C*H₂), 29.2 (–, *C*H₂), 28.4 (–, *C*H₂), 22.8 (–, *C*H₂), 18.3 (+, *C*H₃), 15.5 (+, *C*H₃), 14.2 (+, CH₂*C*H₃) ppm. – **IR** (ATR, ṽ) = 2952, 2921, 2847, 1717, 1598, 1488, 1453, 1376, 1271, 1255, 1224, 1171, 1118, 1080, 1014, 989, 918, 878, 826, 765, 730, 716, 672, 617, 579, 540, 479, 444, 401 cm⁻¹. – **MS** (70 eV, EI, 60 °C) *m/z* (%) = 259 (18) [M+H]⁺, 258 (100) [M]⁺, 243 (3) [M–CH₃]⁺, 229 (3) [M–C₂H₅]⁺, 215 (5) [M–C₃H₇]⁺, 201 (10) [M–C₄H₉]⁺, 189 (12), 188 (77), 187 (7) [M–C₅H₁₁]⁺. – **HRMS** (EI, C₁₇H₂₂O₂) = calc.; 258.1614; found: 258.1616.

5,8-Dimethyl-3-isopropyl-2*H*-chromen-2-one (16nh)

According to **GP11**, 2-hydroxy-3,6-dimethylbenzaldehyde (**19n**, 75.0 mg, 0.50 mmol, 1.00 equiv.), K₂CO₃ (3.00 mg, 0.02 mmol, 0.05 equiv.) and isovaleric anhydride (0.35 ml, 325 mg, 1.75 mmol, 3.50 equiv.) were heated at 180 °C for 3 h *via* microwave irradiation. Purification by flash column chromatography (*c*Hex/EtOAc 20:1) resulted in the title compound as a white solid (105 mg, 0.48 mmol, 97%).

R_f (*c*Hex/EtOAc 20:1) = 0.27. – **¹H NMR** (400 MHz, CDCl₃) δ = 7.64 (s, 1H, C*H*), 7.17 (d, ^3J = 7.5 Hz, 1H, C*H*ₐᵣ), 6.97 (d, ^3J = 7.5 Hz, 1H, C*H*ₐᵣ), 3.14 (hept., ^3J = 6.8 Hz, 1H, CC*H*(CH₃)₂), 2.49 (s, 3H, C*H*₃), 2.40 (s, 3H, C*H*₃), 1.27 (d, ^3J = 6.8 Hz, 6H, CH(C*H*₃)₂) ppm. – **¹³C NMR** (101 MHz, CDCl₃) δ = 161.6 (C_q, *C*OO), 151.7 (C_q, 8a-*C*ₐᵣ), 134.6 (C_q, *C*ₐᵣ), 133.2 (+, *C*ₐᵣH), 132.8 (C_q, *C*ₐᵣ), 131.6 (+, *C*ₐᵣH), 125.1 (+, *C*ₐᵣH), 123.5 (C_q, *C*ₐᵣ), 118.1 (C_q, *C*ₐᵣ), 29.0 (+, *C*H(CH₃)₂), 21.7 (+, CH(*C*H₃)₂), 18.3 (+, *C*H₃), 15.5 (+, *C*H₃) ppm. – **IR** (ATR, ṽ) = 2950, 1698, 1599, 1487, 1460, 1383, 1295, 1272, 1191, 1156, 1068, 1056, 1010, 989, 921, 806, 782, 741, 685, 610, 581, 510, 473, 441, 410 cm⁻¹. – **MS** (70 eV, EI, 50 °C) *m/z* (%) = 217 (17) [M+H]⁺, 216 (100) [M]⁺, 201 (54) [M–CH₃]⁺, 188 (15), 173 (5) [M–C₃H₇]⁺. – **HRMS** (EI, C₁₄H₁₆O₂) = calc.: 216.1145; found: 216.1146.

3-Ethyl-6-methoxy-5,7,8-trimethyl-2*H*-chromen-2-one (16oc)

According to **GP11**, 2-hydroxy-5-methoxy-3,4,6-trimethylbenzaldehyde (**19o**, 100 mg, 0.52 mmol, 1.00 equiv.), K_2CO_3 (4.00 mg, 0.03 mmol, 0.05 equiv.) and butanoic anhydride (0.29 ml, 285 mg, 1.80 mmol, 3.50 equiv.) were heated at 180 °C for 65 min *via* microwave irradiation. Purification by flash column chromatography (*c*Hex/EtOAc 10:1) resulted in the title compound as a white solid (104 mg, 0.42 mmol, 82%).

R_f (*c*Hex/EtOAc 10:1) = 0.30 – **¹H NMR** (400 MHz, CDCl₃) δ = 7.64 (s, 1H, C*H*) 3.68 (s, 3H, OC*H₃*), 2.61 (q, ^3J = 7.3 Hz, 2H, 3-C*H₂*), 2.42 (s, 3H, C*H₃*), 2.34 (s, 3H, C*H₃*), 2.30 (s, 3H, C*H₃*), 1.25 (t, ^3J = 7.3 Hz, 3H, CH₂C*H₃*) ppm. – **¹³C NMR** (101 MHz, CDCl₃) δ = 162.2 (C_q, *C*OO), 152.9 (C_q, 8a-*C*_ar), 148.3 (C_q, *C*_arOCH₃), 135.3 (+, *C*H), 133.7 (C_q, *C*_ar), 129.4 (C_q, *C*_ar), 123.9 (C_q, *C*_ar), 123.2 (C_q, *C*_ar), 116.8 (C_q, *C*_ar), 60.7 (+, O*C*H₃), 24.3 (–, *C*H₂CH₃), 13.3 (+, *C*H₃), 12.8 (+, *C*H₃), 11.9 (+, *C*H₃), 11.3 (+, *C*H₃) ppm. – **IR** (ATR, ṽ) = 2926, 1698, 1596, 1446, 1380, 1335, 1285, 1199, 1138, 1088, 1033, 1002, 942, 895, 799, 780, 710, 654, 619, 588, 565 cm⁻¹. – **MS** (70 eV, EI, 60 °C) *m/z* (%) = 247 (13) [M+H]⁺, 246 (100) [M]⁺, 232 (13), 231 (100) [M–CH₃]⁺, 203 (23), 194 (28), 181 (29), 179 (20), 131 (33), 119 (12), 69 (27). – **HRMS** (EI, $C_{15}H_{18}O_3$) = calc.: 246.1250; found: 246.1252.

6-Methoxy-5,7,8-trimethyl-3-propyl-2*H*-chromen-2-one (16od)

According to **GP11**, 2-hydroxy-5-methoxy-3,4,6-trimethylbenzaldehyde (**19o**, 100 mg, 0.52 mmol, 1.00 equiv.), K_2CO_3 (4.00 mg, 0.03 mmol, 0.05 equiv.) and valeric anhydride (0.36 ml, 336 mg, 1.80 mmol, 3.50 equiv.) were heated at 180 °C for 65 min *via* microwave irradiation. Purification by flash column chromatography (*c*Hex/EtOAc 20:1) resulted in the title compound as a white solid (126 mg, 0.48 mmol, 94%).

R_f (*c*Hex/EtOAc 20:1) = 0.20. – **¹H NMR** (400 MHz, CDCl₃) δ = 7.63 (s, 1H, C*H*), 3.68 (s, 3H, OC*H₃*), 2.59 – 2.50 (m, 2H, 3-CC*H₂*), 2.41 (s, 3H, C*H₃*), 2.34 (s, 3H, C*H₃*), 2.30 (s, 3H, C*H₃*), 1.68 (sex., ^3J = 7.4 Hz, 2H, C*H₂*), 0.99 (t, ^3J = 7.4 Hz, 3H, CH₂C*H₃*) ppm. – **¹³C NMR** (101 MHz, CDCl₃) δ = 162.2 (C_q, *C*OO), 152.9 (C_q, 8a-*C*_ar), 148.4 (C_q, *C*_arOCH₃), 136.3 (+, *C*H), 133.7 (C_q, *C*_ar), 127.9 (C_q, *C*_ar), 123.9 (C_q, *C*_ar), 123.2 (C_q, *C*_ar), 116.8 (C_q, *C*_ar), 60.7 (+, O*C*H₃), 33.2 (–, *C*H₂), 21.7 (–, *C*H₂), 13.9 (+, *C*H₃), 13.3 (+, *C*H₃), 11.9 (+, *C*H₃), 11.3 (+, *C*H₃) ppm. – **IR** (ATR, ṽ) = 2921, 1704, 1597, 1455, 1379, 1333, 1291, 1276, 1197, 1137, 1089, 1040, 1029, 1000, 937, 881, 782, 754, 655, 619, 596, 578, 456, 407 cm⁻¹. – **MS** (70 eV, EI, 50 °C)

m/z (%) = 261 (18) [M+H]$^+$, 260 (100) [M]$^+$, 245 (54) [M–CH$_3$]$^+$, 232 (20), 231 (59) [M–C$_2$H$_5$]$^+$, 217 (10) [M–C$_3$H$_7$]$^+$, 181 (19), 131 (17), 69 (23). – **HRMS** (EI, C$_{16}$H$_{20}$O$_3$) = calc.: 260.1407; found 260.1408.

3-Butyl-6-methoxy-5,7,8-trimethyl-2*H*-chromen-2-one (16oe)

According to **GP11**, 2-hydroxy-5-methoxy-3,4,6-trimethylbenzalde-hyde (**19o**, 100 mg, 0.52 mmol, 1.00 equiv.), K$_2$CO$_3$ (4.00 mg, 0.03 mmol, 0.05 equiv.) and hexanoic anhydride (0.42 ml, 386 mg, 1.80 mmol, 3.50 equiv.) were heated at 180 °C for 65 min *via* microwave irradiation. Purification by flash column chromatography (*c*Hex/EtOAc 20:1) resulted in the title compound as an off-white solid (97.0 mg, 0.35 mmol, 69%).

R_f(*c*Hex/EtOAc 20:1) = 0.21. – **^{1}H NMR** (400 MHz, CDCl$_3$) δ = 7.62 (s, 1H, C*H*), 3.68 (s, 3H, OC*H$_3$*), 2.61 – 2.52 (m, 2H, 3-CC*H$_2$*), 2.41 (s, 3H, C*H$_3$*), 2.34 (s, 3H, C*H$_3$*), 2.30 (s, 3H, C*H$_3$*), 1.68 – 1.56 (m, 2H, C*H$_2$*), 1.41 (sex., ^3J = 7.4 Hz, 2H, C*H$_2$*), 0.95 (t, ^3J = 7.4 Hz, 3H, CH$_2$C*H$_3$*) ppm. – **^{13}C NMR** (101 MHz, CDCl$_3$) δ = 162.2 (C$_q$, *C*OO), 152.9 (C$_q$, 8a-*C*$_{ar}$), 148.4 (C$_q$, *C*$_{ar}$OCH$_3$), 136.1 (+, *C*H), 133.7 (C$_q$, *C*$_{ar}$), 128.2 (C$_q$, *C*$_{ar}$), 123.9 (C$_q$, *C*$_{ar}$), 123.2 (C$_q$, *C*$_{ar}$), 116.8 (C$_q$, *C*$_{ar}$), 60.7 (+, O*C*H$_3$), 30.9 (–, *C*H$_2$), 30.5 (–, *C*H$_2$), 22.6 (–, *C*H$_2$), 14.1 (+, *C*H$_3$), 13.3 (+, *C*H$_3$), 11.9 (+, *C*H$_3$), 11.3 (+, *C*H$_3$) ppm. – **IR** (ATR, ṽ) = 2952, 2867, 1703, 1598, 1454, 1380, 1327, 1286, 1197, 1139, 1092, 1037, 1006, 946, 928, 885, 781, 731, 654, 619, 593, 577, 432 cm^{-1}. – **MS** (70 eV, EI) m/z (%) = 275 (20) [M+H]$^+$, 274 (100) [M]$^+$, 259 (26) [M–CH$_3$]$^+$, 245 (24) [M–C$_2$H$_5$]$^+$, 233 (10), 232 (69), 231 (55) [M–C$_3$H$_7$]$^+$, 217 (24) [M–C$_4$H$_9$]$^+$, 196 (11), 194 (16), 188 (12), 181 (14), 131 (12), 69 (13). – **HRMS** (EI, C$_{17}$H$_{22}$O$_3$) calc.: 274.1563; found: 274.1564.

6-Methoxy-5,7,8-trimethyl-3-pentyl-2*H*-chromen-2-one (16of)

According to **GP11**, 2-hydroxy-5-methoxy-3,4,6-trimethylbenzal-dehyde (**19o**, 100 mg, 0.52 mmol, 1.00 equiv.), K$_2$CO$_3$ (4.00 mg, 0.03 mmol, 0.05 equiv.) and heptanoic anhydride (0.48 ml, 437 mg, 1.80 mmol, 3.50 equiv.) were heated at 180 °C for 65 min *via* microwave irradiation. Purification by flash column chromatography (*c*Hex/EtOAc 10:1) resulted in the title compound as a white solid (109 mg, 0.37 mmol, 74%).

R_f(*c*Hex/EtOAc 10:1) = 0.33. – **^{1}H NMR** (400 MHz, CDCl$_3$) δ = 7.62 (s, 1H, C*H*), 3.68 (s, 3H, OC*H$_3$*), 2.61 – 2.52 (m, 2H, 3-CC*H$_2$*), 2.42 (s, 3H, C*H$_3$*), 2.34 (s, 3H, C*H$_3$*), 2.30 (s, 3H, C*H$_3$*), 1.70 – 1.59 (m, 2H, C*H$_2$*), 1.43 – 1.30 (m, 4H, 2 × C*H$_2$*), 0.95 – 0.86 (m, 3H, CH$_2$C*H$_3$*) ppm. –

^{13}C NMR (101 MHz, CDCl$_3$) δ = 162.2 (C$_q$, COO), 152.9 (C$_q$, 8a-C$_{ar}$), 148.4 (C$_q$, C$_{ar}$OCH$_3$), 136.1 (+, CH), 133.7 (C$_q$, C$_{ar}$), 128.2 (C$_q$, C$_{ar}$), 123.9 (C$_q$, C$_{ar}$), 123.2 (C$_q$, C$_{ar}$), 116.8 (C$_q$, C$_{ar}$), 60.7 (+, OCH$_3$), 31.6 (–, CH$_2$), 31.2 (–, CH$_2$), 28.1 (–, CH$_2$), 22.6 (–, CH$_2$), 14.2 (+, CH$_3$), 13.3 (+, CH$_3$), 11.9 (+, CH$_3$), 11.3 (+, CH$_3$) ppm. – **IR** (ATR, $\tilde{v}$) = 2927, 2861, 1704, 1598, 1457, 1380, 1335, 1304, 1286, 1195, 1139, 1092, 1038, 1004, 935, 892, 847, 783, 728, 691, 655, 619, 596, 578, 422 cm^{-1}. – **MS** (70 eV, EI, 70 °C) m/z (%) = 289 (19) [M+H]$^+$, 288 (100) [M]$^+$, 273 (17) [M–CH$_3$]$^+$, 271 (12), 259 (10) [M–C$_2$H$_5$]$^+$, 245 (25) [M–C$_3$H$_7$]$^+$, 233 (11), 232 (72), 231 (50) [M–C$_4$H$_9$]$^+$, 217 (24) [M–C$_5$H$_{11}$]$^+$, 203 (10), 196 (74), 194 (18), 188 (13), 181 (41), 166 (52), 151 (10), 131 (20), 69 (26). – **HRMS** (EI, C$_{18}$H$_{24}$O$_3$) = calc.: 288.1720; found 288.1722.

3-Hexyl-6-methoxy-5,7,8-trimethyl-2H-chromen-2-one (16og)

According to **GP11**, 2-hydroxy-5-methoxy-3,4,6-trimethylbenzaldehyde (**19o**, 100 mg, 0.52 mmol, 1.00 equiv.), K$_2$CO$_3$ (4.00 mg, 0.03 mmol, 0.05 equiv.) and heptanoic anhydride (0.54 ml, 487 mg, 1.80 mmol, 3.50 equiv.) were heated at 180 °C for 65 min *via* microwave irradiation. Purification by flash column chromatography (cHex/EtOAc 10:1) resulted in the title compound as an off-white solid (72.0 mg, 0.24 mmol, 46%).

R_f (cHex/EtOAc 10:1) = 0.28. – ^{1}**H NMR** (400 MHz, CDCl$_3$) δ = 7.62 (s, 1H, CH), 3.68 (s, 3H, OCH$_3$), 2.56 (t, ^3J = 7.5 Hz, 2H, 3-CCH$_2$), 2.41 (s, 3H, CH$_3$), 2.34 (s, 3H, CH$_3$), 2.30 (s, 3H, CH$_3$), 1.63 (quin., ^3J = 7.5 Hz, 2H, CH$_2$), 1.44 – 1.27 (m, 6H, 3 × CH$_2$), 0.93 – 0.84 (m, 3H, CH$_2$CH$_3$) ppm. – ^{13}**C NMR** (101 MHz, CDCl$_3$) δ = 162.2 (C$_q$, COO) 152.9 (C$_q$, 8a-C$_{ar}$), 148.4 (C$_q$, C$_{ar}$OCH$_3$), 136.1 (+, CH), 133.7 (C$_q$, C$_{ar}$), 128.2 (C$_q$, C$_{ar}$), 123.9 (C$_q$, C$_{ar}$), 123.2 (C$_q$, C$_{ar}$), 116.8 (C$_q$, C$_{ar}$), 60.7 (+, OCH$_3$), 31.8 (–, CH$_2$), 31.2 (–, CH$_2$), 29.1 (–, CH$_2$), 28.4 (–, CH$_2$), 22.8 (–, CH$_2$), 14.2 (+, CH$_3$), 13.3 (+, CH$_3$), 11.9 (+, CH$_3$), 11.3 (+, CH$_3$) ppm. – **IR** (ATR, $\tilde{v}$) = 2927, 2857, 1703, 1598, 1455, 1380, 1336, 1288, 1201, 1138, 1111, 1090, 1036, 1004, 938, 903, 783, 725, 655, 620, 596, 578, 395 cm^{-1}. – **MS** (70 eV, EI, 80 °C) m/z (%) = 303 (20) [M+H]$^+$, 302 (100) [M]$^+$, 287 (9) [M–CH$_3$]$^+$, 273 (2) [M–C$_2$H$_5$]$^+$, 259 (10) [M–C$_3$H$_7$]$^+$, 245 (17) [M–C$_4$H$_9$]$^+$, 233 (11), 232 (58), 231 (41) [M–C$_5$H$_{11}$]$^+$, 217 (19) [M–C$_6$H$_{13}$]$^+$, 196 (15), 181 (37), 166 (10), 131 (29), 69 (37). – **HRMS** (EI, C$_{19}$H$_{26}$O$_3$) = calc.: 302.1876; found: 302.1875.

3-Isopropyl-6-methoxy-5,7,8-trimethyl-2H-chromen-2-one (16oh)

According to **GP11**, 2-hydroxy-5-methoxy-3,4,6-trimethylbenzaldehyde (**19o**, 100 mg, 0.52 mmol, 1.00 equiv.), K$_2$CO$_3$ (4.00 mg, 0.03 mmol, 0.05 equiv.) and isovaleric anhydride (0.36 ml, 336 mg, 1.80 mmol, 3.50 equiv.)

were heated at 180 °C for 3 h *via* microwave irradiation. Purification by flash column chromatography (*c*Hex/EtOAc 20:1) resulted in the title compound as a white solid (37.0 mg, 0.14 mmol, 27%).

R_f (*c*Hex/EtOAc 20:1) = 0.22. – **^{1}H NMR** (400 MHz, CDCl$_3$) δ = 7.63 (s, 1H, C*H*), 3.68 (s, 3H, OC*H$_3$*), 3.13 (sex., ^3J = 6.9 Hz, 1H, C$_{ar}$C*H*(CH$_3$)$_2$), 2.43 (s, 3H, C*H$_3$*), 2.35 (s, 3H, C*H$_3$*), 2.31 (s, 3H, C*H$_3$*), 1.27 (d, ^3J = 6.9 Hz, 6H, CH(C*H$_3$*)$_2$) ppm. – **^{13}C NMR** (101 MHz, CDCl$_3$) δ = 161.8 (C$_q$, *C*OO), 152.9 (C$_q$, 8a-*C*$_{ar}$), 148.1 (C$_q$, *C*$_{ar}$OCH$_3$), 133.8 (C$_q$, *C*$_{ar}$), 133.7 (C$_q$, *C*$_{ar}$), 133.5 (+, *C*H), 124.1 (C$_q$, *C*$_{ar}$), 123.2 (C$_q$, *C*$_{ar}$), 116.7 (C$_q$, *C*$_{ar}$), 60.7 (+, O*C*H$_3$), 28.9 (+, *C*H(CH$_3$)$_2$), 21.7 (+, CH(*C*H$_3$)$_2$), 13.3 (+, *C*H$_3$), 11.9 (+, *C*H$_3$), 11.3 (+, *C*H$_3$) ppm. – **IR** (ATR, ṽ) = 2950, 1760, 1687, 1594, 1452, 1373, 1335, 1282, 1195, 1158, 1090, 1005, 931, 909, 892, 852, 785, 764, 696, 681, 621, 557, 473, 411 cm^{-1}. – **MS** (70 eV, EI, 50 °C) *m/z* (%) = 261 (16) [M+H]$^+$, 260 (100) [M]$^+$, 246 (14), 245 (90) [M–CH$_3$]$^+$, 217 (13) [M–C$_3$H$_7$]$^+$, 69 (11). – **HRMS** (EI, C$_{16}$H$_{20}$O$_3$) = calc.: 260.1407; found: 260.1406.

3-Ethyl-6-hydroxy-5,7,8-trimethyl-2*H*-chromen-2-one (63a)

According to **GP9**, to a solution of the 6-methoxycoumarin **16oc** (23.0 mg, 87.0 μmol, 1.00 equiv.) in abs. DCM (1.74 ml), was added BBr$_3$ (1 M in DCM, 0.44 ml, 109 mg, 435 μmol, 5.00 equiv.) at –78 °C. After stirring for 30 min at this temperature, the reaction mixture was allowed to warm to rt and stirred overnight. Purification by flash column chromatography (*c*Hex/EtOAc 3:1) resulted in the title compound as a white solid (16.5 mg, 72.2 μmol, 83%).

R_f (*c*Hex/EtOAc 5:1) = 0.28. – **^{1}H NMR** (400 MHz, CDCl$_3$) δ = 7.65 (d, ^4J = 1.2 Hz, 1H, C*H*), 4.91 (s, 1H, O*H*), 2.61 (qd, $^{3,4}J$ = 7.4, 1.2 Hz, 2H, 3-CC*H$_2$*), 2.37 (s, 3H, C*H$_3$*), 2.35 (s, 3H, C*H$_3$*), 2.28 (s, 3H, C*H$_3$*), 1.25 (t, ^3J = 7.4 Hz, 3H, CH$_2$C*H$_3$*) ppm. – **^{13}C NMR** (101 MHz, CDCl$_3$) δ = 162.4 (C$_q$, *C*OO) 148.1 (C$_q$, *C*$_{ar}$), 146.4 (C$_q$, *C*$_{ar}$), 135.2 (+, *C*H), 129.5 (C$_q$, *C*$_{ar}$), 126.9 (C$_q$, *C*$_{ar}$), 122.5 (C$_q$, *C*$_{ar}$), 116.5 (C$_q$, *C*$_{ar}$), 115.8 (C$_q$, *C*$_{ar}$), 24.3 (–, *C*H$_2$), 13.0 (+, *C*H$_3$), 12.8 (+, *C*H$_3$), 11.9 (+, *C*H$_3$), 10.9 (+, *C*H$_3$) ppm. – **IR** (ATR, ṽ) = 3417, 2979, 2914, 1663, 1602, 1570, 1457, 1383, 1347, 1289, 1190, 1095, 1005, 881, 802, 754, 727, 680, 631, 555, 438, 398 cm^{-1}. – **MS** (70 eV, EI, 100 °C) *m/z* (%) = 233 (15) [M+H]$^+$, 232 (100) [M]$^+$, 217 (40) [M–CH$_3$]$^+$, 189 (34). – **HRMS** (EI, C$_{14}$H$_{16}$O$_3$) = calc.: 232.1099; found: 232.1101.

6-Hydroxy-5,7,8-trimethyl-3-propyl-2H-chromen-2-one (63b)

According to **GP9**, to a solution of the 6-methoxycoumarin **16od** (23.0 mg, 88.3 μmol, 1.00 equiv.) in abs. DCM (1.77 ml), was added BBr$_3$ (1 M in DCM, 0.44 ml, 111 mg, 442 μmol, 5.00 equiv.) at –78 °C. After stirring for 30 min at this temperature, the reaction mixture was allowed to warm to rt and stirred overnight. Purification by flash column chromatography (cHex/EtOAc 3:1) resulted in the title compound as a white solid (15.8 mg, 69.8 μmol, 79%).

R_f (cHex/EtOAc 5:1) = 0.28. – **^{1}H NMR** (400 MHz, CDCl$_3$) δ = 7.64 (d, ^4J = 1.1 Hz, 1H, CH), 4.73 (s, 1H, OH), 2.55 (td, $^{3,4}J$ = 7.4, 1.1 Hz, 2H, 3-CCH_2), 2.37 (s, 6H, 2 × CH_3), 2.28 (s, 3H, CH_3), 1.67 (sex., ^3J = 7.4 Hz, 3H, CH_2), 1.00 (t, ^3J = 7.4, 3H, CH$_2$CH_3) ppm. – **^{13}C NMR** (101 MHz, CDCl$_3$) δ = 162.3 (C$_q$, COO) 148.1 (C$_q$, C_{ar}), 146.5 (C$_q$, C_{ar}), 136.1 (+, CH), 128.1 (C$_q$, C_{ar}), 126.7 (C$_q$, C_{ar}), 122.6 (C$_q$, C_{ar}), 116.5 (C$_q$, C_{ar}), 115.6 (C$_q$, C_{ar}), 33.3 (–, CH$_2$), 21.7 (–, CH$_2$), 14.0 (+, CH$_3$), 13.0 (+, CH$_3$), 12.0 (+, CH$_3$), 10.9 (+, CH$_3$) ppm. – **IR** (ATR, ṽ) = 3452, 2958, 2923, 2876, 1678, 1603, 1566, 1458, 1382, 1356, 1304, 1274, 1171, 1119, 1096, 1024, 920, 894, 867, 776, 748, 721, 678, 630, 579, 532, 493, 415 cm^{-1}. – **MS** (70 eV, EI, 100 °C) m/z (%) = 247 (19) [M+H]$^+$, 246 (100) [M]$^+$, 231 (34) [M–CH$_3$]$^+$, 218 (30) [M–C$_2$H$_4$]$^+$, 217 (73) [M–C$_2$H$_5$]$^+$, 189 (29), 181 (11), 69 (20). – **HRMS** (EI, C$_{15}$H$_{18}$O$_3$) = calc.: 246.1256; found: 246.1258.

3-Butyl-6-hydroxy-5,7,8-trimethyl-2H-chromen-2-one (63c)

According to **GP9**, to a solution of the 6-methoxycoumarin **16oe** (23.0 mg, 83.8 μmol, 1.00 equiv.) in abs. DCM (1.68 ml), was added BBr$_3$ (1 M in DCM, 0.42 ml, 105 mg, 419 μmol, 5.00 equiv.) at –78 °C. After stirring for 30 min at this temperature, the reaction mixture was allowed to warm to rt and stirred overnight. Purification by flash column chromatography (cHex/EtOAc 10:1) resulted in the title compound as a white solid (13.7 mg, 57.5 μmol, 69%).

R_f (cHex/EtOAc 10:1) = 0.20. – **^{1}H NMR** (400 MHz, CDCl$_3$) δ = 7.64 (d, ^4J = 1.1, 1H, CH), 4.92 (s, 1H, OH), 2.62 – 2.52 (m, 2H, 3-CCH_2), 2.37 (s, 3H, CH_3), 2.35 (s, 3H, CH_3), 2.28 (s, 3H, CH_3), 1.66 – 1.58 (m, 2H, CH_2), 1.45 – 1.38 (m, 2H, CH_2), 0.95 (t, ^3J = 7.3 Hz, 3H, CH$_2$CH_3) ppm. – **^{13}C NMR** (101 MHz, CDCl$_3$) δ = 162.4 (C$_q$, COO) 148.1 (C$_q$, C_{ar}), 146.4 (C$_q$, C_{ar}), 136.0 (+, CH), 128.2 (C$_q$, C_{ar}), 126.9 (C$_q$, C_{ar}), 122.5 (C$_q$, C_{ar}), 116.5 (C$_q$, C_{ar}), 115.7 (C$_q$, C_{ar}), 31.0 (–, CH$_2$), 30.5 (–, CH$_2$), 22.6 (–, CH$_2$), 14.1 (+, CH$_3$), 13.0 (+, CH$_3$), 11.9 (+, CH$_3$), 10.9 (+, CH$_3$) ppm. – **IR** (ATR, ṽ) = 3370, 2951, 2926, 2867, 1678, 1603, 1573, 1457, 1400,

1375, 1345, 1286, 1192, 1119, 1095, 1000, 897, 795, 758, 715, 680, 631, 570, 490, 445, 394 cm^{-1}. – **MS** (70 eV, EI, 110 °C) m/z (%) = 261 (20) [M+H]$^+$, 260 (100) [M]$^+$, 231 (31) [M–C$_2$H$_5$]$^+$, 219 (15), 218 (93) [M–C$_3$H$_6$]$^+$, 217 (51) [M–C$_3$H$_7$]$^+$, 189 (28). – **HRMS** (EI, C$_{16}$H$_{20}$O$_3$) = calc.: 260.1412; found: 260.1412.

6-Hydroxy-5,7,8-trimethyl-3-pentyl-2*H*-chromen-2-one (63d)

According to **GP9**, to a solution of the 6-methoxycoumarin **16of** (21.0 mg, 72.8 µmol, 1.00 equiv.) in abs. DCM (1.46 ml), was added BBr$_3$ (1 M in DCM, 0.36 ml, 91.0 mg, 364 µmol, 5.00 equiv.) at –78 °C. After stirring for 30 min at this temperature, the reaction mixture was allowed to warm to rt and stirred overnight. Purification by flash column chromatography (*c*Hex/EtOAc 3:1) resulted in the title compound as a white solid (15.6 mg, 56.8 µmol, 78%).

R_f (*c*Hex/EtOAc 5:1) = 0.32. – **^{1}H NMR** (400 MHz, CDCl$_3$) δ = 7.64 (d, ^4J = 1.2, 1H, C*H*), 4.97 (s, 1H, O*H*), 2.59 – 2.53 (m, 2H, 3-CC*H*$_2$), 2.37 (s, 3H, C*H*$_3$), 2.34 (s, 3H, C*H*$_3$), 2.28 (s, 3H, C*H*$_3$), 1.69 – 1.59 (m, 2H, C*H*$_2$), 1.40 – 1.32 (m, 4H, 2 × C*H*$_2$), 0.94 – 0.87 (m, 3H, C*H*$_2$C*H*$_3$) ppm. – **^{13}C NMR** (101 MHz, CDCl$_3$) δ = 162.4 (C$_q$, *C*OO) 148.2 (C$_q$, *C*$_{ar}$), 146.4 (C$_q$, *C*$_{ar}$), 136.1 (+, *C*H), 128.3 (C$_q$, *C*$_{ar}$), 126.9 (C$_q$, *C*$_{ar}$), 122.5 (C$_q$, *C*$_{ar}$), 116.5 (C$_q$, *C*$_{ar}$), 115.8 (C$_q$, *C*$_{ar}$), 31.7 (–, *C*H$_2$), 31.2 (–, *C*H$_2$), 28.1 (–, *C*H$_2$), 22.6 (–, *C*H$_2$), 14.2 (+, *C*H$_3$), 13.0 (+, *C*H$_3$), 11.9 (+, *C*H$_3$), 11.0 (+, *C*H$_3$) ppm. – **IR** (ATR, ṽ) = 3401, 2955, 2926, 2854, 1678, 1601, 1571, 1454, 1398, 1343, 1287, 1189, 1096, 1000, 905, 773, 715, 680, 630, 571, 493, 468, 427, 400 cm^{-1}. – **MS** (70 eV, EI, 100 °C) m/z (%) = 275 (19) [M+H]$^+$, 274 (100) [M]$^+$, 257 (11), 245 (11), 231 (25), 219 (15), 218 (82) [M–C$_4$H$_8$]$^+$, 217 (39) [M–C$_4$H$_9$]$^+$, 189 (25), 181 (11), 69 (21). – **HRMS** (EI, C$_{17}$H$_{22}$O$_3$) = calc.: 274.1569; found: 274.1570.

3-Hexyl-6-hydroxy-5,7,8-trimethyl-2*H*-chromen-2-one (63e)

According to **GP9**, to a solution of the 6-methoxycoumarin **16og** (21.0 mg, 69.4 µmol, 1.00 equiv.) in abs. DCM (1.39 ml), was added BBr$_3$ (1 M in DCM, 0.35 ml, 87.0 mg, 347 µmol, 5.00 equiv.) at –78 °C. After stirring for 30 min at this temperature, the reaction mixture was allowed to warm to rt and stirred overnight. Purification by flash column chromatography (*c*Hex/EtOAc 3:1) resulted in the title compound as a white solid (16.1 mg, 56.2 µmol, 81%).

R_f (*c*Hex/EtOAc 5:1) = 0.36. – **^{1}H NMR** (400 MHz, CDCl$_3$) δ = 7.64 (d, ^4J = 1.2 Hz, 1H, C*H*), 4.91 (s, 1H, O*H*), 2.61 – 2.51 (m, 2H, 3-CC*H*$_2$), 2.37 (s, 3H, C*H*$_3$), 2.35 (s, 3H, C*H*$_3$), 2.28 (s, 3H, C*H*$_3$), 1.69 – 1.58 (m, 2H, C*H*$_2$), 1.43 – 1.27 (m, 6H, 3 × C*H*$_2$), 0.92 – 0.85 (m, 3H, C*H*$_2$C*H*$_3$)

ppm. – ^{13}C NMR (101 MHz, CDCl$_3$) δ = 162.4 (C$_q$, COO) 148.1 (C$_q$, C$_{ar}$), 146.4 (C$_q$, C$_{ar}$), 136.0 (+, CH), 128.3 (C$_q$, C$_{ar}$), 126.9 (C$_q$, C$_{ar}$), 122.5 (C$_q$, C$_{ar}$), 116.5 (C$_q$, C$_{ar}$), 115.7 (C$_q$, C$_{ar}$), 31.8 (–, CH$_2$), 31.2 (–, CH$_2$), 29.2 (–, CH$_2$), 28.4 (–, CH$_2$), 22.7 (–, CH$_2$), 14.2 (+, CH$_3$), 13.0 (+, CH$_3$), 11.9 (+, CH$_3$), 10.9 (+, CH$_3$) ppm. – IR (ATR, ṽ) = 3399, 2927, 2855, 1677, 1601, 1570, 1458, 1399, 1344, 1284, 1188, 1123, 1099, 1074, 1000, 915, 900, 764, 717, 679, 630, 572, 495, 400 cm^{-1}. – MS (70 eV, EI, 100 °C) m/z (%) = 289 (18) [M+H]$^+$, 288 (100) [M]$^+$, 271 (10), 231 (23), 219 (15), 218 (83) [M–C$_5$H$_{10}$]$^+$, 217 (40) [M–C$_5$H$_{11}$]$^+$, 189 (24), 181 (10), 69 (19). – HRMS (EI, C$_{18}$H$_{24}$O$_3$) = calc.: 288.1725; found: 288.1727.

3-Isopropyl-6-hydroxy-5,7,8-trimethyl-2*H*-chromen-2-one (63f)

According to GP9, to a solution of the 6-methoxycoumarin 16oh (11.0 mg, 42.3 μmol, 1.00 equiv.) in abs. DCM (0.85 ml), was added BBr$_3$ (1 M in DCM, 0.21 ml, 53.0 mg, 211 μmol, 5.00 equiv.) at –78 °C. After stirring for 30 min at this temperature, the reaction mixture was allowed to warm to rt and stirred overnight. Purification by flash column chromatography (*c*Hex/EtOAc 3:1) resulted in the title compound as a white solid (6.5 mg, 27.5 μmol, 65%).

R_f (*c*Hex/EtOAc 5:1) = 0.28. – ^{1}H NMR (400 MHz, CDCl$_3$) δ = 7.64 (d, ^4J = 1.0 Hz, 1H, C*H*), 4.70 (s, 1H, O*H*), 3.22 – 3.07 (m, 1H, C$_{ar}$C*H*(CH$_3$)$_2$), 2.38 (s, 3H, C*H*$_3$), 2.37 (s, 3H, C*H*$_3$), 2.28 (s, 3H, C*H*$_3$), 1.28 (s, 3H, CHC*H*$_3$), 1.27 (s, 3H, CHC*H*$_3$) ppm. – ^{13}C NMR (101 MHz, CDCl$_3$) δ = 161.9 (C$_q$, COO), 148.0 (C$_q$, C$_{ar}$), 146.2 (C$_q$, C$_{ar}$), 133.9 (C$_q$, C$_{ar}$), 133.3 (+, CH), 126.7 (C$_q$, C$_{ar}$), 122.5 (C$_q$, C$_{ar}$), 116.4 (C$_q$, C$_{ar}$), 115.8 (C$_q$, C$_{ar}$), 28.9 (+, CH(CH$_3$)$_2$), 21.7 (+, CH(CH$_3$)$_2$), 13.0 (+, CH$_3$), 12.0 (+, CH$_3$), 10.9 (+, CH$_3$) ppm. – IR (ATR, ṽ) = 3381, 3078, 2995, 2958, 2921, 2870, 1679, 1603, 1571, 1452, 1348, 1289, 1196, 1170, 1090, 1076, 911, 854, 781, 681, 632, 560, 474, 439 cm^{-1}. – MS (70 eV, EI, 90 °C) m/z (%) = 247 (18) [M+H]$^+$, 246 (100) [M]$^+$, 232 (17), 231 (99) [M–CH$_3$]$^+$, 217 (16) [M–C$_3$H$_7$]$^+$, 203 (16), 69 (13). – HRMS (EI, C$_{15}$H$_{18}$O$_3$) = calc.: 246.1256; found: 246.1258.

5.2   Pharmacological evaluation of synthetic cannabinoids

5.2.1   General remarks biology

The PathHunter® CHOK1hCB1_bgal and CHOK1hCB2_bgal (catalogue number 93-0959C2 and 93-0706C2) β-Arrestin cell lines cells were purchased from EUROFINS DISCOVERX (Fremont, CA). Cell culture plates were purchased from SARSTEDT (Nürnbrecht, Germany). Bicinchoninic acid (BCA) and the BSA protein assay reagents were purchased from PIERCE CHEMICAL COMPANY (Rochford, IL). [^{3}H]CP55,940 (specific activity 149 Ci/mmol), [^{35}S]GTPγS (specific activity 1250 Ci/mmol) and GF-B/GF-C plates were purchased from PERKIN ELMER (Waltham, MA). Cannabinoid receptor reference standards Rimonabant and AM630 were purchased from CAYMAN CHEMICAL COMPANY, CP55,940 were purchased from SIGMA ALDRICH (St. Louis, MO). S9 human liver fraction was purchased from CORNING (Woburn, MA). All solutions and buffers were prepared using Millipore water (deionisation by MilliQ A10 Biocel™, with a 0.22 μm filter). Buffers were prepared at room temperature and, if not stated otherwise, stored at 4 °C. All solvents and reagents were used as analytical grade. Different concentrations of compounds were added using a HP D300 DIGITAL DISPENSER (TECAN, Männedorf. Switzerland) and the DMSO stock solutions. In all assays, the final concentration of DMSO/assay point was limited to ≤1%. All values are expressed as means of at least three individual experiments in duplicates, if not stated otherwise. Errors are expressed as standard error of the mean (SEM).

5.2.2   Methods

Cell culture

CHOK1hCB1_bgal and CHOK1hCB2_bgal were cultured in modified Ham′s F12 Nutrient Mixture supplemented with GlutaMAX™ as glutamine source. Additional supplements were 10% foetal calf serum (FCS), 50 μg/ml penicillin, 50 μg/ml streptomycin, 300 mg/ml hygromycin and 800 μg/ml geneticin in a humidified atmosphere at 37°C and 5% CO_2. Cells were sub-cultured twice a week at a confluence of ~90% and at a ratio of 1:10 on 10-cm diameter plates by trypsinisation. Two days before membrane preparation the cells were sub-cultured 1:20 on 15-cm diameter plates.

Membrane preparation

Per batch of membranes, from thirty 15-cm diameter plates cells were detached from the bottom by scraping them into 5 ml phosphate-saline buffer (PBS), collected in 50 ml Falcon tubes and centrifuged for 5 min at 200 g (3,000 rpm). The pellets were resuspended in ice-cold 50 mM Tris-HCl buffer (pH 7.4) and 5 mM $MgCl_2$. For homogenizing the cell suspension, an UltraThurrax homogenizer (HEIDOLPH INSTRUMENTS, Schwabach, Germany) was used. The membrane fraction was separated from the cytosolic fraction by centrifugation at 100,00 g (31,000 rpm) in a BECKMAN OPTIMA LE-80K ultracentrifuge (BECKMAN COULTER INC., Fullerton, CA) at 4°C for 20 min. The pellet was resuspended in 10 ml ice-cold 50 mM Tris-HCl buffer (pH 7.4) and 5 mM $MgCl_2$ and the ultracentrifugation step was repeated. The final membrane pellet was resuspended in 10 ml ice-cold 50 mM Tris-HCl buffer (pH 7.4) and 5 mM $MgCl_2$ and aliquots of 200 μl (CHOK1hCB1_bgal) or 50 μl (CHOK1hCB2_bgal), respectively, were stored at –80°C until further use. The membrane concentrations were measured using the BCA method[229].

[^{3}H]CP55,940 equilibrium radioligand displacement assay

Membrane aliquots containing 5 μg (CHOK1hCB1_bgal) or 1.5 μg (CHOK1hCB2_bgal) protein were incubated under shaking (~400 rpm) in a total volume of 100 μl assay buffer (50 mM Tris-HCl buffer (pH 7.4), 5 mM $MgCl_2$ and 0.1% BSA) in the presence of ~1.5 nM [^{3}H]CP55,940 (CHOK1hCB1_bgal and CHOK1hCB2_bgal) at 25°C for 2 h. Nonspecific binding (NSB) was determined in the presence of 10 μM Rimonabant (CHOK1hCB1_bgal) or AM630 (CHOK1hCB2_bgal). Single point assays were performed at 1 μM of the competing ligand, full-curve assays were performed with ten concentrations of the competing ligand to determine the pKi values. Incubation were terminated by rapid filtration on 96-well GF/C plates as described below.

96-wells harvest procedure

Samples were harvested by rapid filtration on 96-well GF/C-filter plates (PERKIN ELMER, Groningen, the Netherlands), pre-coated with PEI (Polyethyleneimine), using a PERKIN ELMER 96-well harvester (PERKIN ELMER, Groningen, the Netherlands). To remove free radioligand the filters were washed ten times with ice-cold assay buffer (50 mM Tris-HCl buffer (pH 7.4), 5 mM $MgCl_2$ and 0.1% BSA) twice, followed by drying the filters at 55°C for 30 min. After 3 h pre-incubation in scintillation fluid, the filter-bound radioactivity was determined by scintillation spectrometry, using a MICROBETA2® 2450 microplate counter (PERKIN ELMER, Boston, MA).

[³H]CP55,940 association assay

Association kinetic assay of [³H]CP55,940 on CB₁ was performed by incubate membrane aliquots containing 5 µg of CHOK1hCB1_bgal membranes in a total volume of 100 µl assay buffer (50 mM Tris-HCl buffer (pH 7.4), 5 mM $MgCl_2$ and 0.1% BSA) with ~1.5 nM [³H]CP55,940 at 25°C for a range of time points (120, 90, 60, 30, 20, 15, 10, 8, 6, 4, 2, and 1 min). NSB was determined in the presence of 10 µM Rimonabant. Incubation were terminated by rapid filtration on 96-well GF/C plates as described above.

[³H]CP55,940 dissociation assay

Dissociation kinetic assay of [³H]CP55,940 on CB₁ was performed by pre-incubate membrane aliquots containing 5 µg of CHOK1hCB1_bgal membranes in a total volume of 100 µl assay buffer (50 mM Tris-HCl buffer (pH 7.4), 5 mM $MgCl_2$ and 0.1% BSA) with ~1.5 nM [³H]CP55,940 at 25°C for 2 h. Dissociation was initiated by addition of 10 µM (final concentration) Rimonabant at a range of time points (240, 180, 120, 90, 60, 30, 20, 10, 5, 2, and 1 min). NSB was determined in the presence of 10 µM Rimonabant. Incubation were terminated by rapid filtration on 96-well GF/C plates as described above.

[³⁵S]GTPγS assay

G protein activation measurements as following of receptor activity[66] were performed by pre-incubation of 5 µg CHOK1hCB1_bgal or CHOK1hCB2_bgal membranes in a total volume of 100 µl assay buffer (50 mM Tris-HCl buffer (pH 7.4), 5 mM $MgCl_2$, 150 mM NaCl, 1 mM EDTA, 0.05% BSA and 1 mM DTT, freshly prepared every day) supplemented with 1 µM GDP and 5 µg saponin (final concentration), with different concentrations of the ligands of interest for 30 min at room temperature. The basal level of [³⁵S]GTPγS binding was measured in untreated membrane samples, and the maximal level of [³⁵S]GTPγS binding was measured with 10 µM CP55,940 as reference. Subsequently after pre-incubation, [³⁵S]GTPγS (0.3 nM, final concentration) was added and incubation continued at 25°C and ~400 rpm for 90 min. Incubation were terminated by rapid filtration on 96-well GF/B plates (as described above), except instead using GF/B filter plates and washing buffer containing 50 mM Tris-HCl buffer (pH 7.4), 5 mM $MgCl_2$.

Activity based protein profiling

See Chapter 5.3.7

S9 metabolic stability assay

To measure the metabolic stability in the presence of liver enzymes a S9-fraction metabolic stability assay was performed in duplicates. Prior executing the assay the enzyme regeneration system containing one part 2 mM NADP, two parts 10 mM G6P and one part G6PD 10 U/ml in potassium hydrogensulphate buffer (pH 7.4) was prepared and incubated under mild shaking for 30 min at rt. Followed by preparation of the assay buffer by mixing 100 µl S9-fraction (100 µg/assay point), 200 µl regeneration system, 200 µl 50 mM $MgCl_2$ solution (10 mM final concentration) and supplemented by 200 µl 0.5 mM UDPGA (0.1 mM final concentration) and 200 µl 0.5 mM PAPS (0.1 mM final concentration). The assay was started by mixing 90 µl of master mix with 10 µl of the ligand solution (10 µM in 20% ACN/H_2O) and incubation for several time points (60, 30, 15, 5 and 0 min) at 37°C. The assay was terminated by adding 200 µl ice-cold stop solution (100% ACN) containing 0.75 mM Amitriptyline (0.5 mM final concentration) as internal standard. Negative control to check degradation without enzyme activity was performed by aliquoting 100 µl S9-fraction (10 mg/ml) in 800 µl potassium hydrogensulphate buffer (pH 7.4) into two 90 µl aliquots per tested compound and denature at 95°C for 15 min. Followed by adding of the ligand solution (10 µM in 20% ACN/H_2O) and incubation for 60 min at 37°C. To stop the incubation 200 µl stop solution were added (see above). Subsequently all samples were centrifuged at 17,000 g for 5 min and 200 µl of the supernatant were transferred into LC-MS vials with a glass inlet and stored at 4°C until further measurement. Analysis was performed by LC-MS/MS with an individual optimization for each compound. Quantification was based on a standard curve with known concentrations from 10 nM to 500 nM in comparison to the internal standard. The results are expressed as percentage of remaining ligand, normalized to t = 0 as 100%.[180]

Data analysis

CLogP and tPSA values were calculated using ChemDraw® Professional 16.0 (PERKIN ELMER). All experimental data from the assays were analysed with GraphPad Prism (GRAPHPAD SOFTWARE INC., San Diego, CA, version 7 and 8). For [³H]CP55,940 displacement assays, non-linear regression analysis for "one site – Fit Ki" was used to obtain logK_i values, which were calculated by direct application of the Cheng-Prusoff equation[169]: $K_i = IC_{50}/(1+([L]/K_D))$, where [L] described the exact concentration of [³H]CP55,940 (determined each experiment, ~1.5 nM). The kinetic K_D was calculated by using the equation $K_D = k_{off}/k_{on}$ and was determined for CB_1 (0.41 ± 0.08 nM) using an association ($K_{on} = 4.49 ± 0.21 × 10^7$ M^{-1} s^{-1}) and dissociation assay ($K_{off} = 1.85 ± 0.41 × 10^{-2}$ s^{-1}), respectively (three individual experiments in duplicates) and for

CB_2 (1.24 ± 0.10 nM) as previously reported.[170] The observed rate constant (k_{obs}) values from the kinetic experiments were converted by fitting them to an "one-phase exponential association analysis" for k_{on}, using the equation $k_{on} = (k_{obs} - k_{off})/[L]$, where [L] is the exact concentration of [^{3}H]CP55,940 for each experiment and an "one-phase exponential decay" for k_{off}. Results of the GTPγS assay were analysed with a nonlinear regression analysis "log (agonist) vs. response – variable slope" to calculate the potency (EC_{50}) and the efficacy ($E_{max.}$) of the ligands. The efficacy of agonistic ligands was normalized to the effect of 10 μM [^{3}H]CP55,940 as 100% and the basal activity as 0%. Half-life times for metabolic stability were analysed using the nonlinear regression "dissociation - one phase exponential decay" For statistical analysis of a correlation between two independent variables, a one-way ANOVA correlation analysis was applied, with a P-value of 0.05 as statistically significant.

Computational studies

Preparation steps and docking were performed using Schrödinger (SCHRÖDINGER, LC, New York, NY, 2018; version 2018-2)[230]. Crystal structures of CB1 (PDB: 5XRA)[172] and CB2 (PDB: 5ZTY)[173] were prepared using protein preparation by which disulphide bridges were created, and explicit hydrogens and missing side chains were added. Compounds were prepared for docking using Ligprep, generating states at pH 7. A maximum of ten docked poses were generated per compound. Docking was performed without constraints.

5.3 Synthesis and evaluation of MAGL inhibitors

5.3.1 General remarks chemistry

Air or moisture sensitive reactions were carried out in oven dried glass devices, sealed with rubber seals, under an Argon atmosphere according to SCHLENK-techniques. Liquids were added via plastic syringes and V2A-needles, volumes at small µl scale were added with pipets from EPPENDORF or GILSON or if not easy applicable exactly weighed. Solids were added in pulverized form. Reactions at 0 °C were cooled with a mixture of ice/water. Reactions at deeper temperatures were tempered with brine/ice mixture (–10 °C) or isopropanol/dry ice mixture (–78 °C). All reactions were monitored either by thin layer chromatography (TLC), LC-MS or GC-MS. Solvents were removed at 40 °C with a rotary evaporator. Used solvent mixtures were measured volumetrically. If not stated otherwise, solutions of inorganic salts are saturated aqueous solutions. If not otherwise specified, the crude products, were purified by flash column chromatography following the concepts of STILL et al.[225] using silica gel (SCREENING DEVICES B.V., pore size 60 Å, mesh 40 – 63 µm) and sand (calcined and purified with hydrochloric acid) as stationary phase. Used solvent grades were at least ≥95% for intermediates and p.a. for final compounds. Solvent mixtures were prepared individually, in terms of volume ratios are given as volumetric. The use of a gradient is indicated in the experimental procedures. Celite® for filtrations was purchased from ALFA AESAR (Celite® 545, treated with Na_2CO_3). If not stated otherwise, purities of the synthesized compounds were determined as ≥95%.

5.3.2 Analytics and equipment

Nuclear Magnetic Resonance (NMR)

NMR spectra have been recorded on following spectrometer:

	^{1}H MHz	^{13}C MHz	^{19}F MHz
BRUKER Avance 400	400	101	
BRUKER Avance 500 DRX 500	500	126	471

If not stated otherwise, all spectra were obtained at room temperature. As solvents, products obtained from SIGMA ALDRICH were used: chloroform-d_1 and dimethylsulfoxide-d_6. The chemical shift δ is expressed in parts per million (ppm) where the residual signal of the solvent has been used as secondary reference: chloroform (δ = 7.26 ppm) and dimethyl sulfoxide (δ = 2.50 ppm) in ^{1}H spectra and chloroform (δ = 77.0 ppm) and dimethyl sulfoxide (δ = 39.4 ppm) in

^{13}C spectra. The spectra were analysed according to first order. For central symmetrical signals the midpoint is given, for multiplets the range of the signal region is given. The multiplicities of the signals were abbreviated as follows: s = singlet, d = doublet, t = triplet, q = quartet, quin. = quintet, sex. = sextet, hept = heptet, bs = broad singlet, m = multiplet and combinations thereof. All coupling constants J are stated as modulus in Hertz [Hz] and the number of bonds is given as superscripted index. The assignment ensued for the ^{1}H NMR directly via the chemical shifts and for the ^{13}C NMR through the APT-technique (APT = Attached Proton Test) and is stated as follows: "+" = primary or tertiary carbon atoms (positive DEPT-signal), "–" = secondary or quaternary carbon atoms (negative DEPT-signal). For ^{19}F NMR an internal reference by the spectrometer was used only. Analysis of the NMR raw data was done with either BRUKER TOPSPIN (Version 1.3) or MESTRENOVA (11.0).

High Resolution Mass Spectrometry (ESI-HRMS)

ESI-HRMS spectra were recorded on a WATERS SYNAPT G2-SI QTOF in positive mode and by direct injection into the electron spray ionisation source with Leu-enkephalin as lock mass and calibration prior measurement to the spectrum of Glu-1-fibrinopeptide B. Notation of molecular fragments is given as mass to charge ratio (m/z); the intensities of the signals are noted in percent relative to the base signal (100%). For HRMS (High Resolution Mass Spectrometry) following abbreviations were used: calc. = expected value (calculated); found = value found in analysis.

Preparative High-Performance Liquid Chromatography (prepHPLC/MS)

A WATERS preparative LC-MS system, equipped with a WATERS 515 HPLC pump M, a WATERS 515 HPLC pump L, a WATERS 2767 sample manager, a WATERS System Fluidics Organizer, a WATERS Acquity Ultra Performance LC, Single Quad Detector and WATERS Binary Gradient Module. As stationary phase a PHENOMENEX Gemini C18 5µm 110 Å 21.2 × 150 mm preparative column with acetonitrile and H_2O + 0.1% TFA as mobile phase, was used. The desired fraction was detected by their m/z ratio.

Liquid Chromatography Mass Spectroscopy (LC-MS)

For purity determination and reaction control, LC-MS measurements were recorded on a FINNIGAN Surveyor HPLC system with a PHENOMENEX Gemini C18 5µm 110 Å 4.6 × 150 mm analytical column, measuring UV absorbance (at 200 – 600 nm), coupled to a FINNIGAN LCQ Advantage Max mass spectrometer with ESI source. The applied buffer was ACN in H_2O + 0.1%, with a linear gradient from 5 – 95%.

Thin Layer Chromatography (TLC)

Analytical thin layer chromatography (TLC, prepTLC) was done on TLC plates purchased from MERK (silica gel 60 on aluminium plate, fluorescence indicator F254, layer thickness analytical: 0.25 mm, preparative: 1.00 mm). Detection was carried out under UV-light at $\lambda = 254$ nm and $\lambda = 366$ nm. Alternatively, the TLC plates were stained by spraying solutions of SEEBACH (2.5% phosphor molybdic acid, 1.0% cer(IV)sulphate-tetrahydrate, 6.0% conc. H_2SO_4, 90.5% H_2O) or potassium permanganate (1.50 g $KMnO_4$, 10 g K_2CO_3 and 1.25 mL 10% NaOH in 200 mL H_2O) and dried in a hot air stream.

5.3.3   General procedures

General Procedure: Protection of benzoic acids (**GP12**)

A solution of the respective benzoic acid (1.00 equiv.), $Boc_2(O)$ (2.50 equiv.) and DMAP (0.30 equiv.) in ^{t}BuOH (0.33 M) were stirred for 3 h at 60 °C. After cooling to rt the volatiles were removed in vacuo and the crude product was purified by filtration over short silica pad.

General Procedure: Synthesis of aromatic thioethers (**GP13**)

To a solution of the respective tert-butyl benzoate (1.00 equiv.) in ACN (0.25 M), K_2CO_3 (2.00 equiv.) and ethyl 2-mercaptoacetate (2.00 equiv.) were added. After stirring overnight at 60 °C, the reaction mixture was diluted with EtOAc, washed with H_2O and brine, dried over Na_2SO_4, filtrated and the volatiles were removed under reduce pressure. The crude product was purified by flash column chromatography.

General Procedure: Oxidation of thioethers to sulfones (**GP14**)

To a solution of respective thiol ether (1.00 equiv.) in MeOH (0.12 M final concentration) was added a solution of Oxone in H_2O (1.00 equiv.) at 0 °C. Then the mixture was allowed to warm up to rt and stirred for 1 h. The reaction mixture was diluted with H_2O, extracted with EtOAc, the combined organic layers were dried over Na_2SO_4, filtrated and the volatiles were removed in vacuo. The crude product was purified by flash column chromatography.

General Procedure: Peptide coupling (**GP15**)

To a solution of the respective benzoic acid (1.00 eq.) and DIPEA (3.00 eq.) in DCM (0.05 M), HATU (29.3 mg, 77.0 μmol, 1.00 eq.) was added and stirred for 1h at rt. Followed by addition of 1-(3-chlorophenyl)-2,3-dimethylpiperazine (1.00 eq.) and stirring overnight. The reaction mixture was concentrated under reduced pressure and the crude product was either purified by prep-HPLC or flash column chromatography.

5.3.4 Synthesis and characterisation of carboxylic part

tert-Butyl 4-fluoro-3-(trifluoromethyl)benzoate (78a)

According to **GP12**, a solution of 4-fluoro-3-(trifluoromethyl)benzoic acid (**77a**, 300 mg, 1.44 mmol, 1.00 equiv.), Boc$_2$(O) (0.77 ml, 787 mg, 3.60 mmol, 2.50 equiv.) and DMAP (52.8 mg, 0.43 mmol, 0.30 equiv.) in ^{t}BuOH (4.4 ml), was stirred at 60°C for 4 h. Filtration over a small silica plug (*n*Pen/DCM 1:1) resulted in the product as a colourless oil (345 mg, 1.31 mmol, 91%).

^{1}H NMR (500 MHz, CDCl$_3$) δ = 8.25 (dd, $^{3,4}J$ = 7.0, 2.2 Hz, 1H, CH_{ar}), 8.18 (ddd, J = 8.5, 4.8, 2.2 Hz, 1H, CH_{ar}), 7.27 – 7.23 (m, 1H, CH_{ar}), 1.60 (s, 9H, C(CH_3)$_3$) ppm. – **^{13}C NMR** (126 MHz, CDCl$_3$) δ = 163.7 (–, C$_q$, COO), 162.4 (–, dd, $^{1,2}J$ = 262.9, 1.8 Hz, C$_q$, C$_{ar}$F), 135.5 (+, d, ^3J = 9.7 Hz, *m*-C$_{ar}$H), 129.1 (+, dd, ^3J = 4.6, 2.6 Hz, *m*-C$_{ar}$H), 128.6 (–, d, ^4J = 3.5 Hz, C$_q$, *p*-C$_{ar}$), 122.3 (–, d, ^1J = 272.7 Hz, C$_q$, CF$_3$), 118.9 – 118.4 (–, m, C$_q$, *o*-C$_{ar}$CF$_3$), 117.1 (+, d, ^2J = 21.2 Hz, *o*-C$_{ar}$H), 82.4 (–, C$_q$, C(CH$_3$)$_3$), 28.2 (+, C(CH$_3$)$_3$) ppm. – **^{19}F NMR** (471 MHz, CDCl$_3$) δ = –61.7 (d, ^4J = 12.5 Hz, CF$_3$), –108.2 – –108.9 (m, CF) ppm.

tert-Butyl 4-fluoro-3-chlorobenzoate (78b)

According to **GP12**, a solution of 4-fluoro-3-chlorobenzoic acid (**77b**, 1.00 g, 5.73 mmol, 1.00 equiv.), Boc$_2$(O) (3.06 ml, 3.13 g, 14.3 mmol, 2.50 equiv.) and DMAP (210 mg, 1.72 mmol, 0.30 equiv.) in ^{t}BuOH (17 ml), was stirred at 60°C for 4 h. Filtration over a small silica plug (*n*Pen/DCM 1:1) resulted in the product as a colourless oil (1.16 g, 5.04 mmol, 88%).

^{1}H NMR (500 MHz, CDCl$_3$) δ = 8.03 (dd, $^{3,4}J$ = 7.2, 2.2 Hz, 1H, CH_{ar}), 7.88 (ddd, J = 8.6, 4.7, 2.2 Hz, 1H, CH_{ar}), 7.17 (t, ^3J = 8.6 Hz, 1H, CH_{ar}), 1.59 (s, 9H, C(CH_3)$_3$) ppm. – **^{13}C NMR** (126 MHz, CDCl$_3$) δ = 163.9 (–, C$_q$, COO), 160.9 (–, d, ^1J = 255.2 Hz, C$_q$, C$_{ar}$F), 132.3 (+, *m*-C$_{ar}$H), 129.9 (+, d, ^3J = 8.4 Hz, *m*-C$_{ar}$H), 129.3 (–, d, ^4J = 3.6 Hz, C$_q$, *p*-C$_{ar}$), 121.3 (–, d, ^2J = 18.2 Hz, C$_q$, *o*-C$_{ar}$Cl), 116.5 (+, d, ^2J = 21.6 Hz, *o*-C$_{ar}$H), 82.1 (–, C$_q$, C(CH$_3$)$_3$), 28.3 (+, C(CH$_3$)$_3$) ppm. – **^{19}F NMR** (471 MHz, CDCl$_3$) δ = –109.2 (td, J = 7.8, 4.7 Hz, CF) ppm.

tert-Butyl 4-((2-ethoxy-2-oxoethyl)thio)-3-(trifluoromethyl)benzoate (79a)

According to **GP13**, to a solution of *tert*-butyl 4-fluoro-3-(trifluoromethyl)benzoate (**78a**, 100 mg, 0.38 mmol, 1.00 equiv.) in ACN (1.5 ml), K_2CO_3 (105 mg, 0.76 mmol, 2.00 equiv.) and ethyl 2-mercaptoacetate (0.08 ml, 91.0 mg, 0.76 mmol, 2.00 equiv.) were added and stirred at 60°C overnight. Purification by flash column chromatography (*n*Pen/DCM 2:1) resulted in the title compound as a colourless liquid (124 mg, 0.34 mmol, 90%).

R_f (*n*Pen/DCM 2:1) = 0.14. – **^{1}H NMR** (400 MHz, CDCl$_3$) δ = 8.24 (d, ^4J = 1.9 Hz, 1H, C*H*$_{ar}$), 8.06 (dd, $^{3,4}J$ = 8.3, 1.9 Hz, 1H, C*H*$_{ar}$), 7.55 (d, ^3J = 8.3 Hz, 1H, C*H*$_{ar}$), 4.19 (q, ^3J = 7.1 Hz, 2H, OC*H*$_2$), 3.75 (s, 2H, SC*H*$_2$), 1.59 (s, 9H, C(C*H*$_3$)$_3$), 1.24 (t, ^3J = 7.1 Hz, 3H, C*H*$_3$) ppm. – **^{13}C NMR** (101 MHz, CDCl$_3$) δ = 168.7 (–, C$_q$, *C*OO), 164.2 (–, C$_q$, *C*OO), 140.7 (–, C$_q$, *C*$_{ar}$), 132.8 (+, *C*$_{ar}$H), 129.9 (–, C$_q$, *C*$_{ar}$), 129.3 (+, *C*$_{ar}$H), 129.0 (–, d, ^2J = 31.2 Hz, C$_q$, *C*$_{ar}$CF$_3$), 128.0 (+, q, ^3J = 5.8 Hz, *C*$_{ar}$H), 123.4 (–, q, ^1J = 273.9 Hz, C$_q$, *C*F$_3$), 82.1 (–, C$_q$, *C*(CH$_3$)$_3$), 62.1 (–, O*C*H$_2$), 35.9 (–, S*C*H$_2$), 28.2 (+, C(*C*H$_3$)$_3$), 14.2 (+, *C*H$_3$) ppm.

tert-Butyl 3-chloro-4-((2-ethoxy-2-oxoethyl)thio)benzoate (79b)

According to **GP13**, to a solution of *tert*-butyl 3-chloro-4-fluorobenzoate (**78b**, 1.16 g, 5.04 mmol, 1.00 equiv.) in ACN (20 ml), K_2CO_3 (1.39 g, 10.1 mmol, 2.00 equiv.) and ethyl 2-mercaptoacetate (1.11 ml, 1.21 g, 10.1 mmol, 2.00 equiv.) were added and stirred at 60°C overnight. Purification by flash column chromatography (*n*Pen/DCM 1:1) resulted in the title compound as a colourless oil (1.62 g, 4.60 mmol, 91%, purity 94%).

R_f (*n*Pen/DCM 1:1) = 0.33. – **^{1}H NMR** (400 MHz, CDCl$_3$) δ = 7.94 (d, ^4J = 1.8 Hz, 1H, C*H*$_{ar}$), 7.83 (dd, $^{3,4}J$ = 8.3, 1.8 Hz, 1H, C*H*$_{ar}$), 7.32 (d, ^3J = 8.3 Hz, 1H, C*H*$_{ar}$), 4.20 (q, ^3J = 7.1 Hz, 2H, OC*H*$_2$), 3.74 (s, 2H, SC*H*$_2$), 1.58 (s, 9H, C(C*H*$_3$)$_3$), 1.26 (t, ^3J = 7.1 Hz, 3H, C*H*$_3$) ppm. – **^{13}C NMR** (101 MHz, CDCl$_3$) δ = 168.8 (–, C$_q$, *C*OO), 164.4 (–, C$_q$, *C*OO), 140.6 (–, C$_q$, *C*$_{ar}$), 132.2 (–, C$_q$, *C*$_{ar}$), 130.6 (–, C$_q$, *C*$_{ar}$), 130.5 (+, *C*$_{ar}$H), 128.2 (+, *C*$_{ar}$H), 126.6 (+, *C*$_{ar}$H), 81.8 (–, C$_q$, *C*(CH$_3$)$_3$), 62.2 (–, O*C*H$_2$), 34.5 (–, S*C*H$_2$), 28.3 (+, C(*C*H$_3$)$_3$), 14.2 (+, *C*H$_3$) ppm.

Ethyl 2-((2-chloro-4-formylphenyl)thio)acetate (83)

Similar to **GP13**, to a solution of 3-chloro-4-fluorobenzaldehyde (**82**, 300 mg, 1.89 mmol, 1.00 equiv.) in DMF (5.0 ml), K_2CO_3 (523 mg, 3.78 mmol, 2.00 equiv.) and ethyl 2-mercaptoacetate (0.21 ml, 227 mg,

1.89 mmol, 2.00 equiv.) were added and stirred at rt overnight. Purification by flash column chromatography (nPen/Et$_2$O 9:1 to 2:1) resulted in the title compound as a colourless oil (1.62 g, 4.60 mmol, 91%, purity >92%).

R_f (nPen/Et$_2$O 4:1) = 0.16. – **^{1}H NMR** (400 MHz, CDCl$_3$) δ = 9.89 (s, 1H, CHO), 7.84 (d, ^4J = 1.8 Hz, 1H, CH_{ar}), 7.72 (dd, $^{3,4}J$ = 8.3, 1.8 Hz, 1H, CH_{ar}), 7.42 (d, ^3J = 8.3 Hz, 1H, CH_{ar}), 4.22 (q, ^3J = 7.1 Hz, 2H, OCH_2), 3.78 (s, 2H, SCH_2), 1.27 (t, ^3J = 7.1 Hz, 3H, CH_3) ppm. – **^{13}C NMR** (126 MHz, CDCl$_3$) δ = 190.1 (–, C$_q$, CHO), 168.5 (–, C$_q$, COO), 143.9 (–, C$_q$, C_{ar}), 134.6 (–, C$_q$, C_{ar}), 132.6 (–, C$_q$, C_{ar}), 130.1 (+, C_{ar}H), 128.4 (+, C_{ar}H), 126.3 (+, C_{ar}H), 62.3 (–, OCH$_2$), 34.2 (–, SCH$_2$), 14.2 (+, CH$_3$) ppm.

tert-Butyl 4-((2-ethoxy-2-oxoethyl)sulfinyl)-3-(trifluoromethyl)benzoate (80a)

According to **GP14**, to a solution of the thioether **79a** (104 mg, 0.29 mmol, 1.00 equiv.) in MeOH (1.75 ml), was added a solution of Oxone (175 mg, 0.29 mmol, 1.00 equiv.) in H$_2$O (0.70 ml) at 0 °C and the mixture was stirred at rt for 1 h. Filtration over a small silica pad (nPen/Et$_2$O 1:1) resulted in the title compound as a yellow oil (109 mg, 0.29 mmol, quant.), which was used without further purification.

^{1}H NMR (400 MHz, CDCl$_3$) δ = 8.46 – 8.26 (m, 3H, 3 × CH_{ar}), 4.23 (qq, J = 7.1, 3.7 Hz, 2H, OCH_2), 3.83 (d, J = 13.9 Hz, 1H, SCHH), 3.60 (d, J = 13.9 Hz, 1H, SCHH), 1.62 (s, 9H, C(CH_3)$_3$), 1.28 (t, ^3J = 7.1 Hz, 3H, CH_3) ppm. – **^{13}C NMR** (126 MHz, CDCl$_3$) δ = 164.3 (–, C$_q$, COO), 163.4 (–, C$_q$, COO), 147.2 (–, C$_q$, C_{ar}), 135.5 (–, C$_q$, C_{ar}), 133.9 (+, C_{ar}H), 127.8 (+, d, ^3J = 5.5 Hz, C_{ar}H), 126.1 (+, C_{ar}H), 83.1 (–, C$_q$, C(CH$_3$)$_3$), 62.6 (–, CH$_2$), 61.3 (–, CH$_2$), 28.2 (+, C(CH$_3$)$_3$), 14.2 (+, CH$_3$) ppm.

tert-Butyl 3-chloro-4-((2-ethoxy-2-oxoethyl)sulfinyl)benzoate (80b)

According to **GP14**, to a solution of the thioether **79b** (140 mg, 0.42 mmol, 1.00 equiv.) in MeOH (2.5 ml), was added a solution of Oxone (260 mg, 0.42 mmol, 1.00 equiv.) in H$_2$O (1.0 ml) at 0 °C and the mixture was stirred at rt for 1 h. Filtration over a small silica pad (nPen/Et$_2$O 1:1) resulted in the title compound as a yellow oil (147 mg, 0.42 mmol, quant.), which was used without further purification.

^{1}H NMR (400 MHz, CDCl$_3$) δ = 8.12 (d, ^3J = 8.1 Hz, 1H, CH_{ar}), 8.06 – 7.92 (m, 2H, CH_{ar}), 4.29 – 4.18 (m, 2H, OCH_2), 4.03 (d, J = 13.1 Hz, 1H, SCHH), 3.69 (d, J = 13.8 Hz, 1H, SCHH),

1.61 (s, 9H, C(CH_3)$_3$), 1.26 (t, ^3J = 7.0 Hz, 3H, CH_3) ppm. – ^{13}C NMR (126 MHz, CDCl$_3$) δ = 164.5 (–, C$_q$, COO), 163.6 (–, C$_q$, COO), 145.2 (–, C$_q$, C$_{ar}$), 136.5 (–, C$_q$, C$_{ar}$), 130.8 (+, C$_{ar}$H), 130.1 (–, C$_q$, C$_{ar}$), 128.8 (+, C$_{ar}$H), 126.5 (+, C$_{ar}$H), 82.7 (–, C$_q$, C(CH$_3$)$_3$), 62.4 (–, CH$_2$), 58.1 (–, CH$_2$), 28.2 (+, C(CH_3)$_3$), 14.2 (+, CH_3) ppm.

4-((2-Ethoxy-2-oxoethyl)sulfinyl)-3-(trifluoromethyl)benzoic acid (81a)

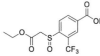 To a solution of the sulfoxide **80a** (109 g, 0.29 mmol, 1.00 equiv.) in DCM (1.8 ml) was added an excess of TFA (0.20 ml, 296 mg, 2.60 mmol, 9,06 equiv.) and stirred at rt overnight. Then the reaction mixture was concentrated under reduced pressure and the obtained residue was purified by flash column chromatography (DCM/MeOH 20:1) to get the product as a white solid (55.0 mg, 0.17 mmol, 59%).

R$_f$ (DCM/MeOH 20:1) = 0.01. – ^{1}H NMR (500 MHz, CDCl$_3$) δ = 8.50 – 8.44 (m, 1H, CH_{ar}), 8.42 – 8.33 (m, 2H, 2 × CH_{ar}), 4.28 – 4.16 (m, 2H, OCH_2), 3.84 (d, J = 13.9 Hz, 1H, SCHH), 3.63 (d, J = 13.9 Hz, 1H, SCHH), 1.29 – 1.24 (m, 3H, CH_3) ppm. – ^{13}C NMR (126 MHz, CDCl$_3$) δ = 168.5 (–, C$_q$, COO), 164.2 (–, C$_q$, COO), 148.7 (–, C$_q$, C$_{ar}$), 134.6 (+, C$_{ar}$H), 132.7 (–, C$_q$, C$_{ar}$), 128.5 (–, C$_q$, C$_{ar}$), 126.6 (+, 2 × C$_{ar}$H), 62.7 (–, CH$_2$), 61.1 (–, CH$_2$), 14.2 (+, CH_3) ppm.

3-Chloro-4-((2-ethoxy-2-oxoethyl)sulfinyl)benzoic acid (81b)

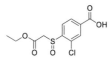 To a solution of the sulfoxide **80b** (1.39 g, 4.00 mmol, 1.00 equiv.) in DCM (18 ml) was added an excess of TFA (2.00 ml, 2.96 g, 26.0 mmol, 6,49 equiv.) and stirred at rt overnight. Then the reaction mixture was concentrated under reduced pressure and the obtained residue was purified by flash column chromatography (DCM/MeOH 20:1 to 15:1) to get the product as a white solid (1.04 g, 3.59 mmol, 90 %).

R$_f$ (DCM/MeOH 20:1) = 0.07. – ^{1}H NMR (500 MHz, CDCl$_3$) δ = 8.17 (dd, $^{3,4}J$ = 8.1, 1.5 Hz, 1H, CH_{ar}), 8.07 (d, ^4J = 1.5, 1H, CH_{ar}), 7.99 (d, ^3J = 8.1, 1H, CH_{ar}), 4.26 – 4.14 (m, 2H, OCH_2), 4.04 (d, J = 14.0 Hz, 1H, SCHH), 3.72 (d, J = 14.0 Hz, 1H, SCHH), 1.23 (t, ^3J = 7.1 Hz, 3H, CH_3) ppm. – ^{13}C NMR (126 MHz, CDCl$_3$) δ = 166.5 (–, C$_q$, COO), 164.5 (–, C$_q$, COO), 145.4 (–, C$_q$, C$_{ar}$), 135.0 (–, C$_q$, C$_{ar}$), 131.2 (+, C$_{ar}$H), 130.2 (–, C$_q$, C$_{ar}$), 129.3 (+, C$_{ar}$H), 126.7 (+, C$_{ar}$H), 62.5 (–, CH$_2$), 58.0 (–, CH$_2$), 14.1 (+, CH_3) ppm.

3-Chloro-4-formylbenzoic acid (85)

Under an inert atmosphere, to a solution of 4-bromo-3-chlorobenzoic acid (300 mg, 1.27 mmol, 1.00 equiv.) in abs. THF (5.0 mL) at –78 C, Turbo-GRIGNARD (1.3 M in THF, 2.94 ml, 3.82 mmol, 3.00 equiv.) was added. The temperature was raised to 0 C after 10 min and the reaction mixture was stirred at 0 C° for 1 h. Subsequently DMF (6.37 ml, 0.49 mmol, 5.00 equiv.) was added, the reaction mixture warmed up to room temperature and stirred further for 1.5 h. The reaction was quenched by addition of sat. aq. NH$_4$CI, followed by washing with EtOAc. Then the pH of the aqueous layer was adjusted with aq. 1 N HCl and extracted with EtOAc. The combined organic layers were neutralized with sat. aq. NaHCO$_3$, washed with brine, dried over Na$_2$SO$_4$ and concentrated under reduced pressure. The title compound was obtained without any further purification as colourless oil (165 mg, 0.89 mmol, 70%).

^{1}H NMR (500 MHz, DMSO-d_6) δ = 10.38 (s, 1H, C*H*O), 8.07 – 8.00 (m, 2H, 2 × C*H*$_{ar}$), 8.01 – 7.95 (m, 1H, C*H*$_{ar}$) ppm. – **^{13}C NMR** (126 MHz, DMSO-d_6) δ = 189.6 (–, C$_q$, CHO), 165.4 (–, C$_q$, COO), 136.8 (–, C$_q$, C$_{ar}$), 136.1 (–, C$_q$, C$_{ar}$), 134.8 (–, C$_q$, C$_{ar}$), 131.1 (+, C$_{ar}$H), 130.1 (+, C$_{ar}$H), 128.3 (+, C$_{ar}$H) ppm.

5.3.5   Synthesis and characterization of trans-(2,3)-dimethylpiperazine part

5,6-Dimethyl-2,3-dihydropyrazine (87)

A solution of ethylenediamine (6.68 ml, 6.01 g, 100 mmol, 1.00 equiv.) in Et$_2$O (250 ml) was cooled to 0 °C, 2,3-butanedione (**86**, 8.70 ml, 8.61 g, 100 mmol, 1.00 equiv.), dissolved in Et$_2$O (100 ml) was added drop wisely and the resulting suspension was stirred at rt overnight. The resulting clear mixture was dried with potassium hydroxide (13.5 g 240 mmol, 2.40 equiv.) for further 30 min. After filtration, the solution was concentrated and the product was distilled (max. 72°C) under vacuum into a –80 °C precooled flask to get the product as a clear pale-yellow oil (9.41 g, 85.0 mmol, 85%).[231]

^{1}H NMR (500 MHz, CDCl$_3$) δ = 3.37 (s, 4H, 2 × C*H*$_2$), 2.16 (s, 6H, 2 × C*H*$_3$) ppm. – Analytical data are consistent with literature.[205, 232]

(2*R*,3*R*)-2,3-Dimethylpiperazine (±75)

To a solution of 5,6-dimethyl-2,3-dihydropyrazine (**87**, 9.40 g, 85.0 mmol, 1.00 equiv.) in abs. EtOH (250 ml) was added sodium (23.0 g, 1.00 mol, 11.72 equiv.)

portion wise. After complete addition of the sodium, the mixture was refluxed for further 16 h and then neutralized by addition of acetic acid (50 ml, 875 mmol) at 0 °C. The resulting suspension was diluted with DCM and the precipitated sodium acetate was filtered off. The product was purified by flash column chromatography (DCM/MeOH 9:1 + 0.1% NH₄OH to Et₂O/MeOH/NH₄OH 10:4:1) resulted in the analytical pure *trans* configured title compound as yellow crystals (4.07 g, 35.7 mmol, 42%).[233]

R_f (MeOH/NH₄OH 94:6) = 0.34. – **¹H NMR** (400 MHz, CDCl₃) δ = 3.37 (s, 4H, 2 × CH_2), 2.16 (s, 6H, 2 × CH_3) ppm. – Analytical data are consistent with literature.[205, 234]

tert-Butyl (2*R*,3*R*)-2,3-dimethylpiperazine-1-carboxylate (±91a)

To a solution of *trans*-2,3-dimethylpiperazine (±**75**, 523 mg, 4.58 mmol, 2.50 equiv.) and Et₃N (0.95 ml, 695 mg, 6.87 mmol, 3.75 equiv.) in MeOH (50 ml) was dropwise added a solution of Boc₂(O) (0.43 ml, 400 mg, 1.83 mmol, 1.00 equiv.) in MeOH (50 ml) and the mixture was stirred at rt for 6 h. Then the mixture was concentrated under reduced pressure, followed by addition of EtOAc. The organic layer was washed three times with water, dried over Na₂SO₄ and the solvent removed under reduced pressure. Purification by flash column chromatography (DCM/MeOH 98:2 to 9:1) resulted in the analytical pure *trans* configured title compound as a yellow oil (178 mg, 2.06 mmol, 45%)

¹H NMR (400 MHz, CDCl₃) δ = 3.99 – 3.87 (m, 1H, CH_2), 3.86 – 3.75 (m, 1H, CH_2), 3.27 (bs, 1H, NH), 3.06 – 2.97 (m, 2H, CH_2), 2.92 – 2.85 (m, 1H, CH), 2.78 – 2.67 (m, 1H, CH), 1.44 (s, 9H, C(CH_3)₃), 1.24 (d, ^3J = 6.8 Hz, 6H, 2 × CH_3) ppm. – **¹³C NMR** (101 MHz, CDCl₃) δ = 155.4 (–, C_q, COO), 79.6 (–, C_q, C(CH₃)₃), 51.4 (+, CH), 51.1 (+, CH), 38.9 (–, CH₂), 38.1 (–, CH_2), 28.5 (+, C(CH_3)₃), 17.9 (+, CH_3), 16.2 (+, CH_3) ppm.

tert-Butyl -4-(3-chlorophenyl)-(2*R*,3*R*)-2,3-dimethylpiperazine-1-carboxylate (±92)

A mixture of *tert*-butyl 2,3-dimethylpiperazine-1-carboxylate (±**91a**, 50.0 mg, 0.23 mmol, 1.00 equiv.) and 1-bromo-3-chlorobenzene (33.0 μl, 53.6 mg, 0.28 mmol, 1.20 equiv.), dissolved in degassed 1,4-dioxane (1.0 ml, c=0.23 M) and in the presence of Pd₂(dba)₃ (8.55 mg, 9.33 μmol, 0.04 equiv.), BINAP (8.72 mg, 14.0 μmol, 0.06 equiv.) and Cs₂CO₃ (114 mg, 0.35 mmol, 1.50 equiv.) was stirred under an inert atmosphere at 85 °C for 3 d. After cooling to rt, the reaction mixture was filtered through a pad of Celite® to remove the inorganic impurities. Purification by flash column chromatography (*n*Pen/Et2O 19:1 to 9:1) resulted in the title compound as a yellow oil (53.0 mg, 0.12 mmol, 53%).

R_f (nPen/Et2O 19:1) = 0.01. – **¹H NMR** (400 MHz, CDCl₃) δ = 7.14 (t, ^3J = 8.1 Hz, 1H, CH_{ar}), 6.80 – 6.73 (m, 2H, 2 × CH_{ar}), 6.69 (dd, $^{3,4}J$ = 8.4, 2.4 Hz, 1H, CH_{ar}), 4.29 – 3.80 (m, 2H, CH_2), 3.77 – 3.62 (m, 1H, CH), 3.31 – 3.14 (m, 2H, CH_2), 3.13 – 3.03 (m, 1H, CH), 1.48 (s, 9H, C(CH_3)₃), 1.32 (d, ^3J = 6.8 Hz, 3H, CH_3), 1.06 (d, ^3J = 6.6 Hz, 3H, CH_3) ppm. – **¹³C NMR** (101 MHz, CDCl₃) δ = 151.8 (–, C_q, C_ar), 135.2 (–, C_q, C_ar), 130.2 (+, C_arH), 118.6 (+, C_arH), 115.6 (+, C_arH), 113.6 (+, C_arH), 79.9 (–, C_q, C(CH_3)₃), 55.6 (+, CH), 52.1 (+, CH), 40.9 (–, CH_2), 37.6 (–, CH_2), 28.6 (+, C(CH_3)₃), 17.1 (+, CH_3), 13.3 (+, CH_3) ppm.

(2R,3R)-4-(3-Chlorophenyl)-2,3-dimethylpiperazine (76a)

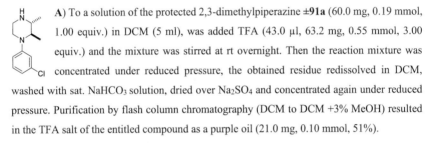

A) To a solution of the protected 2,3-dimethylpiperazine ±**91a** (60.0 mg, 0.19 mmol, 1.00 equiv.) in DCM (5 ml), was added TFA (43.0 µl, 63.2 mg, 0.55 mmol, 3.00 equiv.) and the mixture was stirred at rt overnight. Then the reaction mixture was concentrated under reduced pressure, the obtained residue redissolved in DCM, washed with sat. NaHCO₃ solution, dried over Na₂SO₄ and concentrated again under reduced pressure. Purification by flash column chromatography (DCM to DCM +3% MeOH) resulted in the TFA salt of the entitled compound as a purple oil (21.0 mg, 0.10 mmol, 51%).

B) To a mixture of *trans*-2,3-dimethylpiperazine (±**75**, 1.00 g, 8.76 mmol, 1.00 equiv.) and 1-bromo-chlorobenzene (1.03 ml, 1.68 g, 8.75 mmol, 1.00 equiv.), in abs. dioxane (50 ml, final ratio dioxane:THF 3:1, c=0.13 M), was added KHMDS (1.0 M in THF, 17.5 ml, 17.5 mmol, 2.00 equiv.) at 10°C, then the reaction mixture was stirred at rt overnight. The reaction was quenched by addition of H₂O, followed by extraction of the organic phase with DCM. The combined organic layers were dried over Na₂SO₄ and concentrated under reduced pressure. Purification by flash column chromatography (DCM/MeOH 99:1 to 9:1) resulted in the title compound as a yellow oil (486 mg, 2.16 mmol, 25 %, purity >90%).

¹H NMR (500 MHz, CDCl₃) δ = 7.23 (t, ^3J = 8.0 Hz, 1H, CH_{ar}), 7.06 (dd, $^{3,4}J$ = 8.0, 2.0 Hz, 1H, CH_{ar}), 7.02 (t, ^4J = 2.0 Hz, 1H, CH_{ar}), 6.93 (dd, $^{3,4}J$ = 8.0, 2.0 Hz, 1H, CH_{ar}), 3.88 (bs, 2H, NH_2)[16], 3.41 – 3.16 (m, 6H, 2 × CH_2 and 2 × CH), 1.49 (d, ^3J = 6.5 Hz, 3H, CH_3), 1.05 (d, ^3J = 6.4 Hz, 3H, CH_3) ppm. – **¹³C NMR** (126 MHz, CDCl₃) δ = 151.3 (–, C_q, C_ar), 135.1 (–, C_q, C_ar), 130.5 (+, C_arH), 124.3 (+, C_arH), 123.0 (+, C_arH), 120.7 (+, C_arH), 57.4 (+, CH), 55.5 (+, CH), 47.0 (–, CH_2), 41.7 (–, CH_2), 15.8 (+, CH_3), 14.9 (+, CH_3) ppm.

---

[16] Only for TFA salt.

5.3.6   Synthesis and characterization of final MAGL inhibitors

Ethyl 2-((4-((2R,3R)-4-(3-chlorophenyl)-2,3-dimethylpiperazine-1-carbonyl)-2-(trifluoro-methyl)phenyl)sulfinyl)acetate (33b)

According to **GP15**, the benzoic acid **81a** (25.0 mg, 77.0 μmol, 1.00 equiv.), DIPEA (40.0 μl, 29.9 mg, 231 μmol, 3.00 equiv.) and HATU (29.3 mg, 77.0 μmol, 1.00 equiv.) in DCM (2.0 ml), was stirred at rt for 1h. Then 1-(3-chloro-phenyl)-2,3-dimethylpiperazine (±**75**, 17.3 mg, 77.0 μmol, 1.00 equiv.) was added and the mixture was stirred overnight. Purification by prep-HPLC-MS resulted in the title compound as a yellow oil (12.0 mg, 22.3 μmol, 29 %).

^{1}**H NMR** (500 MHz, CDCl$_3$) δ = 8.37 (d, ^{3}J = 8.1 Hz, 1H, CH_{ar}), 7.88 – 7.73 (m, 2H, CH_{ar}), 7.17 (t, ^{3}J = 8.1 Hz, 1H, CH_{ar}), 6.84 – 6.78 (m, 2H, CH_{ar}), 6.75 – 6.68 (m, 1H, CH_{ar}), 4.83 (quin., ^{3}J = 7.3 Hz, 1H, CH), 4.65 (t, J = 10.3 Hz, 1H, CHH), 4.31 – 4.19 (m, 2H, SCH_2), 3.91 – 3.86 (m, 1H, CHH), 3.83 (dd, J = 14.0, 2.4 Hz, 1H, CHH), 3.73 – 3.45 (m, 3H, 3 × CH and 2 × CHH), 3.39 – 3.16 (m, 2H, 2 × CH_2 and CHH), 3.15 – 3.07 (m, 1H, CHH), 1.55 – 1.45 (m, 5H, CH_3), 1.41 (d, ^{3}J = 6.7 Hz, 1H, CH_3), 1.30 (t, ^{3}J = 7.2 Hz, 3H, CH_3), 1.17 – 098 (m, 3H, CH_3) ppm. – ^{13}**C NMR** (126 MHz, CDCl$_3$) δ = 164.4 (–, C$_q$, C$_{ar}$), 151.3 (–, C$_q$, C$_{ar}$), 144.6 (–, C$_q$, C$_{ar}$), 139.8 (–, C$_q$, C$_{ar}$), 135.3 (–, C$_q$, C$_{ar}$), 131.4 (+, C$_{ar}$H), 130.9 (+, C$_{ar}$H), 130.4 (+, C$_{ar}$H), 126.5 (+, d, J = 8.4 Hz, C$_{ar}$H), 125.3 (+, C$_{ar}$H), 124.8 (+, C$_{ar}$H), 119.6 (+, C$_{ar}$H), 116.3 (+, d, J = 2.5 Hz, C$_{ar}$H), 114.3 (+, C$_{ar}$H), 62.6 (–, CH$_2$), 61.5 (–, CH$_2$), 56.2 (+, CH), 55.7 (+, CH), 53.7 (+, CH), 49.8 (+, CH), 42.5 (–, CH$_2$), 42.1 (–, CH$_2$), 41.4 (–, CH$_2$), 40.5 (–, CH$_2$), 36.7 (–, CH$_2$), 18.7 (+, CH$_3$), 17.9 (+, CH$_3$), 17.6 (+, CH$_3$), 16.8 (+, CH$_3$), 14.2 (+, CH$_3$), 12.9 (+, CH$_3$), 12.6 (+, CH$_3$), 12.0 (+, CH$_3$) ppm. – Both NMR spectra showed a mixture of rotamers and were not separately assigned. – **HRMS** (ESI, C$_{24}$H$_{26}$ClF$_3$N$_2$O$_4$S) = calc.: 531.1327 [M+H]$^{+}$; found: 531.1705 [M+H]$^{+}$.

Ethyl   2-((2-chloro-4-((2R,3R)-4-(3-chlorophenyl)-2,3-dimethylpiperazine-1-car-bonyl)phenyl)sulfinyl)acetate (33c)

According to **GP15**, the benzoic acid **81b** (30.0 mg, 103 μmol, 1.00 equiv.), DIPEA (54.0 μl, 31.0 mg, 310 μmol, 3.00 equiv.) and HATU (39.2 mg, 103 μmol, 1.00 equiv.) in DCM (2.0 ml), was stirred at rt for 1h. Then 1-(3-chlorophenyl)-2,3-dimethylpiperazine (±**75**, 23.2 mg, 103 μmol, 1.00 equiv.) was added and the mixture was stirred overnight. Purification by prep-HPLC-MS resulted in the title compound as a yellow oil (38.3 mg, 77.3 μmol, 75 %).

^{1}H NMR (500 MHz, CDCl$_3$) δ = 8.01 (d, ^3J = 8.0 Hz, 1H, C*H*$_{ar}$), 7.54 (dd, ^3J = 8.0, 1.6 Hz, 1H, C*H*$_{ar}$), 7.51 – 7.42 (m, 1H, C*H*$_{ar}$), 7.16 (t, ^3J = 8.0 Hz, 1H, C*H*$_{ar}$), 6.83 – 6.78 (m, 2H, 2 × C*H*$_{ar}$), 6.70 (d, ^3J = 8.4 Hz, 1H, C*H*$_{ar}$), 4.80 (t, ^3J = 6.7 Hz, 1H, C*H*), 4.62 (s, 1H, C*H*H), 4.29 – 4.17 (m, 2H, SC*H*$_2$), 4.04 (dd, $^{3,4}J$ = 14.1, 1.7 Hz, 1H, CH*H*), 3.87 (d, J = 7.0 Hz, 1H, C*H*H), 3.69 (dd, $^{3,4}J$ = 14.0, 1.2 Hz, 1H, C*H*H), 3.67 – 3.60 (m, 1H, 2 × CH*H*), 3.57 – 3.49 (m, 1H, C*H*H), 3.37 – 3.06 (m, 3H, 2 × C*H*$_2$ and C*H*H), 1.52 – 1.44 (m, 3H, C*H*$_3$), 1.27 (t, ^3J = 7.1 Hz, 3H, C*H*$_3$), 1.16 – 097 (m, 3H, C*H*$_3$) ppm. – **^{13}C NMR** (126 MHz, CDCl$_3$) δ = 168.8 (–, C$_q$, C$_{ar}$), 168.3 (–, C$_q$, C$_{ar}$), 164.5 (–, C$_q$, C$_{ar}$), 151.3 (–, C$_q$, C$_{ar}$), 142.4 (–, C$_q$, C$_{ar}$), 140.5 (–, C$_q$, C$_{ar}$), 135.3 (–, C$_q$, C$_{ar}$), 130.8 (–, C$_q$, C$_{ar}$), 130.4 (+, C$_{ar}$H), 128.4 (+, C$_{ar}$H), 127.7 (+, C$_{ar}$H), 127.1 (+, C$_{ar}$H), 126.3 (+, C$_{ar}$H), 125.8 (+, C$_{ar}$H), 119.5 (+, C$_{ar}$H), 116.3 (+, C$_{ar}$H), 114.3 (+, C$_{ar}$H), 62.5 (–, CH$_2$), 58.4 (–, CH$_2$), 56.2 (+, CH), 55.6 (+, CH), 49.8 (+, CH), 42.4 (–, CH$_2$), 41.3 (–, CH$_2$), 40.5 (–, CH$_2$), 36.6 (–, CH$_2$), 17.8 (+, CH$_3$), 16.8 (+, CH$_3$), 14.2 (+, CH$_3$), 12.8 (+, CH$_3$), 12.6 (+, CH$_3$) ppm. – Both NMR spectra showed a mixture of rotamers and were not separately assigned. – **HRMS** (ESI, C$_{23}$H$_{27}$Cl$_2$N$_2$O$_4$S) = calc.: 497.1063/499.1034 [M+H]$^+$, 519.0883/521.0853 [M+Na]$^+$; found: 497.1065/499.1042 [M+H]$^+$, 519.0886/521.0862 [M+Na]$^+$.

Ethyl   2-((2-chloro-4-((2R,3R)-4-(3-chlorophenyl)-2,3-dimethylpiperazin-1-yl)methyl)phenyl)thio)acetate (93)

To a solution of ethyl-2-((2-chloro-4-formylphe-nyl)thio)acetate (**83**, 31.8 mg, 0.12 mmol, 1.20 equiv.) in 1,2-dichloroethane (1.0 ml) was added trans-1-(3-chlorophenyl)-2,3-dimethylpiperazine (±**75**, 23.0 mg, 0.10 mmol, 1.00 equiv.) and the reaction mixture was stirred at rt for 30min. Then sodium triacetoxyborohydride (65.1 mg, 0.31 mmol, 3.00 equiv.) was added and the reaction mixture was stirred further at rt overnight. The reaction was quenched by addition of H$_2$O, followed by extraction with DCM. The combined organic layers were washed with brine, dried over Na$_2$SO$_4$ and the volatiles were removed under reduced pressure. Purification by flash column chromatography (nPen/EtOAc 9:1) resulted in the title compound as a colourless oil (18.8 mg, 49.2 μmol, 40 %).

R$_f$ (nPen/EtOAc 9:1) = 0.52. – **^{1}H NMR** (400 MHz, CDCl$_3$) δ = 7.46 (s, 1H, C*H*$_{ar}$), 7.39 (d, ^3J = 8.0 Hz, 1H, C*H*$_{ar}$), 7.30 – 7.27 (m, 1H, C*H*$_{ar}$), 7.15 (t, ^3J = 8.0 Hz, 1H, C*H*$_{ar}$), 6.84 – 6.79 (m,

1H, CH_{ar}), 6.73 (t, ^3J = 8.0 Hz, 1H, CH_{ar}), 4.20 (q, ^3J = 7.2 Hz, 2H, CH_2), 3.75 – 3.67 (m, 3H, NCH_2 and CH), 3.57 (q, J = 13.9 Hz, 2H, CH_2), 3.27 – 3.09 (m, 2H, CH_2), 2.91 – 2.81 (m, 1H, CH), 2.75 (td, J = 11.4 Hz, 1H, CH), 2.52 (d, J = 11.5 Hz, 1H, CH), 1.27 (d, ^3J = 7.2 Hz, 3H, CH_3), 1.24 – 1.19 (m, 3H, CH_3), 1.15 (d, ^3J = 6.3 Hz, 3H, CH_3) ppm. – ^{13}C NMR (101 MHz, CDCl$_3$) δ = 169.4 (–, C$_q$, C$_{ar}$), 152.1 (–, C$_q$, C$_{ar}$), 140.2 (–, C$_q$, C$_{ar}$), 135.1 (–, C$_q$, C$_{ar}$), 134.6 (–, C$_q$, C$_{ar}$), 132.3 (–, C$_q$, C$_{ar}$), 130.5 (+, C$_{ar}$H), 130.1 (+, C$_{ar}$H), 129.9 (+, C$_{ar}$H), 127.6 (+, C$_{ar}$H), 118.0 (+, C$_{ar}$H), 115.5 (+, C$_{ar}$H), 113.6 (+, C$_{ar}$H), 61.8 (–, CH_2), 58.0 (–, CH_2), 56.7 (+, CH), 56.6 (+, CH), 44.9 (–, CH_2), 42.0 (–, CH_2), 35.6 (–, CH_2), 14.2 (+, CH_3), 13.0 (+, CH_3), 9.5 (+, CH_3) ppm.

Ethyl   2-((2-chloro-4-((2R,3R)-4-(3-chlorophenyl)-2,3-dimethylpiperazin-1-yl)me-thyl)phenyl)sulfinyl)acetate (33d)

According to **GP14**, to a solution of the thioether **93** (18.9 mg, 40.0 μmol, 1.00 equiv.) in MeOH (200 μl), was added a solution of Oxone (24.9 mg, 40.0 μmol, 1.00 equiv.) in H$_2$O (80.0 μl) at 0 °C and the mixture was stirred at rt for 1 h. Purification by flash column chromatography (nPen/EtOAc/TEA 9:1:0.5 to 2:1:0.5) resulted in the title compound as a colourless oil (6.3 mg, 12.8 μmol, 32%).

R$_f$ (nPen/EtOAc/TEA 9:1:0.5) = 0.13. – **^{1}H NMR** (500 MHz, CDCl$_3$) δ = 7.89 (d, ^3J = 8.0 Hz, 1H, CH_{ar}), 7.55 (ddd, J = 8.0, 3.0, 1.5 Hz, 1H, CH_{ar}), 7.49 (dd, J = 3.9, 1.5 Hz, 1H, CH_{ar}), 7.14 (t, ^3J = 8.0 Hz, 1H, CH_{ar}), 6.80 (t, J = 2.2 Hz, 1H, CH_{ar}), 6.72 (dddd, J = 14.5, 8.5, 2.3, 0.8 Hz, 2H, 2 × CH_{ar}), 4.26 – 4.18 (m, 2H, CH_2), 4.02 (dd, J = 13.7, 2.8 Hz, 1H, CHH), 3.75 – 3.66 (m, 3H, 2 × CHH and CH), 3.61 (d, J = 14.3 Hz, 1H, CHH), 3.24 – 3.19 (m, 1H, CHH), 3.15 (dd, J = 4.0, 1.5 Hz, 1H, CH), 2.85 (q, J = 6.4 Hz, 1H, CH), 2.78 (td, J = 11.5, 4.2 Hz, 1H, CHH), 2.51 (dt, J = 11.5, 1.8 Hz, 1H, CHH), 1.27 (td, J = 7.2, 1.0 Hz, 3H, CH_3), 1.22 (dd, J = 6.6, 1.4 Hz, 3H, CH_3), 1.17 (d, ^3J = 6.5 Hz, 3H, CH_3) ppm. – ^{13}**C NMR** (126 MHz, CDCl$_3$) δ = 169.4 (–, C$_q$, C$_{ar}$), 152.1 (–, C$_q$, C$_{ar}$), 145.6 (–, C$_q$, C$_{ar}$), 139.3 (–, C$_q$, C$_{ar}$), 135.2 (–, C$_q$, C$_{ar}$), 130.2 (–, C$_q$, C$_{ar}$), 130.2 (+, C$_{ar}$H), 129.8 (+, C$_{ar}$H), 128.2 (+, C$_{ar}$H), 126.4 (+, C$_{ar}$H), 118.2 (+, C$_{ar}$H), 115.6 (+, C$_{ar}$H), 113.7 (+, C$_{ar}$H), 62.3 (–, CH_2), 58.7 (–, CH_2), 58.2 (–, CH_2), 57.0 (+, CH), 56.8 (+, CH), 45.0 (–, CH_2), 42.0 (–, CH_2), 14.3 (+, CH_3), 13.0 (+, CH_3), 9.6 (+, CH_3) ppm. – **HRMS** (ESI, C$_{23}$H$_{28}$Cl$_2$N$_2$O$_3$S) = calc.: 483.1270/485.1241 [M+H]$^+$; found: 483.1273/485.1246 [M+H]$^+$.

2-((2-Chloro-4-((2R,3R)-4-(3-chlorophenyl)-2,3-dimethylpiperazine-1-carbonyl)phe-nyl)sulfinyl)acetic acid (94)

A solution of ethyl 2-((2-chloro-4-(4-(3-chlorophenyl)-2,3-di-methylpiperazine-1-carbonyl)phenyl)sulfinyl)acetate (±**33c**, 113 mg, 22.7 μmol, 1.00 eq.) in an excess of TEA (1.00 ml in MeOH/Water (2:1, 3 ml) was stirred at rt overnight. Then the mixture was acidified, extracted with DCM, the combined organic layers were dried over Na$_2$SO$_4$ and concentrated under reduced pressure. Purification by flash column chromatography (DCM/MeOH 1% to 2%) resulted in the title compound as a yellow oil (66.0 mg, 14.1 μmol, 62 %, purity <90%).

R$_f$ (DCM/MeOH 9:1 + acetic acid) = 0.43. – **^{1}H NMR** (500 MHz, CDCl$_3$) δ = 8.02 (d, ^3J = 8.1 Hz, 1H, CH_{ar}), 7.60 – 7.51 (m, 1H, CH_{ar}), 7.50 – 7.43 (m, 1H, CH_{ar}), 7.16 (t, ^3J = 8.3 Hz, 1H, CH_{ar}), 6.83 – 6.77 (m, 2H, 2 × CH_{ar}), 6.73 – 6.68 (m, 1H, CH_{ar}), 4.80 (d, ^3J = 6.9 Hz, 1H, CH), 4.61 (d, J = 12.6 Hz, 1H, CHH), 4.06 (dd, $^{3,4}J$ = 14.2, 3.2 Hz, 1H, CHH), 3.86 (d, J = 6.7 Hz, 1H, CHH), 3.75 (d, ^3J = 14.5 Hz, 1H, CHH), 3.70 – 3.61 (m, 1H, 2 × CHH), 3.56 – 3.49 (m, 1H, CHH), 3.37 – 3.07 (m, 3H, 2 × CH_2 and CHH), 1.52 – 1.44 (m, 3H, CH_3), 1.15 – 097 (m, 3H, CH_3) ppm. – **^{13}C NMR** (126 MHz, CDCl$_3$) δ = 176.2 (–, C$_q$, C$_{ar}$), 169.1 (–, C$_q$, C$_{ar}$), 168.6 (–, C$_q$, C$_{ar}$), 167.1 (–, C$_q$, C$_{ar}$), 151.3 (–, C$_q$, C$_{ar}$), 141.7 (–, C$_q$, C$_{ar}$), 140.3 (–, C$_q$, C$_{ar}$), 135.3 (–, C$_q$, C$_{ar}$), 130.9 (–, C$_q$, C$_{ar}$), 130.4 (+, C$_{ar}$H), 128.5 (+, C$_{ar}$H), 128.2 (+, C$_{ar}$H), 127.9 (+, C$_{ar}$H), 127.4 – 127.1 (+, 3 × C$_{ar}$H), 126.6 (+, C$_{ar}$H), 126.4 (+, C$_{ar}$H), 126.0 (+, C$_{ar}$H), 119.5 (+, C$_{ar}$H), 116.3 (+, C$_{ar}$H), 114.3 (+, C$_{ar}$H), 57.8 (–, CH_2), 56.2 (+, CH), 55.7 (+, CH), 49.9 (+, CH), 46.1 (–, CH_2), 42.4 (–, CH_2), 41.3 (–, CH_2), 40.5 (–, CH_2), 36.7 (–, CH_2), 20.8 (+, CH_3), 17.8 (+, CH_3), 16.8 (+, CH_3), 12.9 (+, CH_3), 12.6 (+, CH_3) ppm. – Both NMR spectra showed a mixture of rotamers and were not separately assigned.

Isopropyl   2-((2-chloro-4-((2R,3R)-4-(3-chlorophenyl)-2,3-dimethylpiperazine-1-carbonyl)phenyl)sulfinyl)acetate (33e)

To a solution of the benzoic acid **94** (30.0 mg, 64.0 μmol, 1.00 equiv.), in DCM (1.0 ml) were added a few drops of DMF and oxalyl chloride (2 M in DCM, 35.0 μl, 70.0 μmol, 1.10 equiv.). The mixture was stirred at reflux for 2 h and then a solution of 2-propanol (25.0 μl, 19.2 mg, 320 μmol, 5.00 equiv.) and DIPEA (12.0 μl, 9.09 mg, 70.0 μmol, 1.10 equiv.) in DCM (1 ml) was added and the reaction mixture was stirred at rt overnight. The reaction was quenched by the addition of H2O, extracted with DCM, the

combined organic layers dried over Na$_2$SO$_4$ and concentrated under reduced pressure. Purification by prep-HPLC-MS resulted in the title compound as a yellow oil (14.0 mg, 27.5 μmol,
43 %).

^{1}H NMR (500 MHz, CDCl$_3$) δ = 8.02 (d, ^3J = 7.9 Hz, 1H, CH_{ar}), 7.58 – 7.51 (m, 1H, CH_{ar}),
7.51 – 7.43 (m, 1H, CH_{ar}), 7.17 (t, ^3J = 8.0 Hz, 1H, CH_{ar}), 6.84 – 6.79 (m, 2H, 2 × CH_{ar}), 6.71
(d, ^3J = 8.4 Hz, 1H, CH_{ar}), 5.10 (quin., ^3J = 6.3 Hz, 1H, CH_2), 4.88 – 4.76 (m, 1H, CH), 4.62
(s, 1H, CHH), 4.02 (dd, $^{3,4}J$ = 14.1, 2.3 Hz, 1H, CHH), 3.87 (d, J = 6.9 Hz, 1H, CHH), 3.65 (dt,
$^{3,4}J$ = 14.2, 2.8 Hz, 1H, CHH), 3.53 (s, 1H, CHH), 3.40 – 3.06 (m, 3H, 2 × CH_2 and CHH), 1.53
– 1.44 (m, 3H, CH_3), 1.32 – 1.24 (m, 7H, 2 × CH_3, CH(CH$_3$)$_2$), 1.16 – 0.99 (m, 3H, CH_3) ppm.
– **^{13}C NMR** (126 MHz, CDCl$_3$) δ = 168.8 (–, C$_q$, C$_{ar}$), 164.2 (–, C$_q$, C$_{ar}$), 151.4 (–, C$_q$, C$_{ar}$),
142.7 (–, C$_q$, C$_{ar}$), 140.6 (–, C$_q$, C$_{ar}$), 135.4 (–, C$_q$, C$_{ar}$), 130.4 (–, C$_q$, C$_{ar}$), 126.2 (+, C$_{ar}$H), 119.6
(+, C$_{ar}$H), 116.3 (+, C$_{ar}$H), 114.3 (+, C$_{ar}$H), 70.6 (–, CH_2), 58.8 (–, CH_2), 56.3 (+, CH), 55.5 (+,
CH), 49.7 (+, CH), 42.4 (–, CH_2), 41.4 (–, CH_2), 40.6 (–, CH_2), 36.6 (–, CH_2), 21.9 (+, CH),
17.9 (+, CH_3), 16.8 (+, CH_3), 12.9 (+, CH_3), 12.6 (+, CH_3) ppm. – Both NMR spectra showed
a mixture of rotamers and were not separately assigned. – **HRMS** (ESI, C$_{24}$H$_{28}$Cl$_2$N$_2$O$_4$S) =
calc.: 511.1220/513.1190 [M+H]$^+$, 533.1039/535.1010 [M+Na]$^+$; found: 511.1227/513.1200
[M+H]$^+$, 533.1044/535.1021 [M+Na]$^+$.

***sec*-Butyl   (2-chloro-4-((2R,3R)-4-(3-chlorophenyl)-2,3-dimethylpiperazine-1-car-
bonyl)phenyl)sulfinyl)acetate (33f)**

To a solution of the benzoic acid **94** (50.0 mg, 107 μmol,
1.00 equiv.), in DCM (1.0 ml) were added a few drops of
DMF and oxalyl chloride (2 M in DCM, 80.0 μl, 160 μmol,
1.50 equiv.). The mixture was stirred at rt for 2 h and then
a solution of butan-2-ol (49.0 μl, 39.5 mg, 533 μmol, 5.00 equiv.) and DIPEA (28.0 μl, 20.7
mg, 160 μmol, 1.50 equiv.) in DCM (1 ml) was added and the reaction mixture was stirred at
rt overnight. The reaction was quenched by the addition of H2O, extracted with DCM, the
combined organic layers dried over Na$_2$SO$_4$ and concentrated under reduced pressure. Purification by prep-HPLC-MS resulted in the title compound as a yellow oil (14.1 mg, 26.8 μmol,
25 %).

^{1}H NMR (400 MHz, CDCl$_3$) δ = 8.02 (dd, ^3J = 7.9, 2.0 Hz, 1H, CH_{ar}), 7.59 – 7.51 (m, 1H,
CH_{ar}), 7.50 – 7.42 (m, 1H, CH_{ar}), 7.17 (t, ^3J = 8.0 Hz, 1H, CH_{ar}), 6.85 – 6.77 (m, 2H, 2 × CH_{ar}),
6.70 (dd, ^3J = 7.9, 3.2 Hz, 1H, CH_{ar}), 4.95 (sex., J = 6.3, 2.7 Hz, 1H, CH_2), 4.81 (t, J = 6.2 Hz,
1H, CH), 4.67 – 4.55 (m, 1H, CHH), 4.04 (dt, $^{3,4}J$ = 14.0, 1.8 Hz, 1H, CHH), 3.87 (q, J = 6.7

Hz, 1H, CHH), 3.71 – 3.59 (m, 2H, 2 × CHH), 3.52 (d, J = 6.8 Hz, 1H, CHH), 3.40 – 3.06 (m, 3H, 2 × CH$_2$ and CHH), 1.70 – 1.53 (m, 2H, CH$_2$), 1.53 – 1.43 (m, 3H, CH$_3$), 1.30 – 1.21 (m, 4H, CH$_2$ and CH$_3$), 1.16 – 0.99 (m, 3H, CH$_3$), 0.97 – 0.88 (m, 3H, CH$_3$) ppm. – ^{13}C NMR (126 MHz, CDCl$_3$) δ = 164.4 (–, C$_q$, C$_{ar}$), 164.3 (–, C$_q$, C$_{ar}$), 151.4 (–, C$_q$, C$_{ar}$), 142.8 (–, C$_q$, C$_{ar}$), 140.7 (–, C$_q$, C$_{ar}$), 135.3 (–, C$_q$, C$_{ar}$), 130.8 (–, C$_q$, C$_{ar}$), 130.4 (+, C$_{ar}$H), 128.6 (+, C$_{ar}$H), 127.9 (+, C$_{ar}$H), 127.0 (+, C$_{ar}$H), 126.4 (+, C$_{ar}$H), 126.2 (+, C$_{ar}$H), 125.9 (+, C$_{ar}$H), 119.6 (+, C$_{ar}$H), 116.3 (+, C$_{ar}$H), 114.3 (+, C$_{ar}$H), 75.2 (–, CH$_2$), 59.0 (–, CH$_2$), 58.8 (–, CH$_2$), 56.2 (+, CH), 55.5 (+, CH), 49.7 (+, CH), 42.4 (–, CH$_2$), 41.4 (–, CH$_2$), 40.5 (–, CH$_2$), 36.6 (–, CH$_2$), 28.8 (–, CH$_2$), 19.5 (+, CH$_3$), 17.9 (+, CH$_3$), 16.8 (+, CH$_3$), 12.9 (+, CH$_3$), 12.6 (+, CH$_3$), 9.7 (+, CH$_3$) ppm. – Both NMR spectra showed a mixture of rotamers and were not separately assigned. – **HRMS** (ESI, C$_{25}$H$_{30}$Cl$_2$N$_2$O$_4$S) = calc.: 525.1376/527.1347 [M+H]$^+$, 547.1196/549.1166 [M+Na]$^+$; found: 525.1385/527.1362 [M+H]$^+$, 547.1202/549.1179 [M+Na]$^+$.

2-Chloro-4-((2R,3R)-4-(3-chlorophenyl)-2,3-dimethylpiperazine-1-carbonyl)benzaldehyde (95) FDM-104

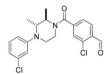

According to **GP15**, 3-chloro-4-formylbenzoic acid (**85**, 65.0 mg, 35.0 mmol, 1.00 equiv.), DIPEA (0.19 ml, 137 mg, 1.06 mmol, 3.00 equiv.) and HATU (161 mg, 0.42 mmol, 1.20 equiv.) in DCM (2.0 ml) was stirred at rt for 1h. Then 1-(3-chlorophenyl)-2,3-dimethylpiperazine (79.0 mg, 0.35 mmol, 1.00 equiv.) was added and the mixture was stirred overnight. Purification by flash column chromatography (nPen/EtOAc 5:1 to 3:1) resulted in the title compound as a yellow oil (69.0 mg, 0.18 mmol, 50%).

R_f (nPen/EtOAc 3:1) = 0.33. – ^{1}**H NMR** (400 MHz, CDCl$_3$) δ = 10.49 – 10.45 (m, 1H, CHO), 7.97 (d, ^3J = 7.9 Hz, 1H, CH$_{ar}$), 7.49 (dd, J = 11.0, 1.5 Hz, 1H, CH$_{ar}$), 7.41 – 7.35 (m, 1H, CH$_{ar}$), 7.18 – 7.12 (m, 1H, CH$_{ar}$), 6.82 – 6.76 (m, 2H, 2 × CH$_{ar}$), 4.79 (qd, J = 6.9, 1.4 Hz, 1H, CH), 4.60 (dt, J = 12.0, 2.2 Hz, 1H, CHH), 3.89 – 3.83 (m, 1H, CH), 3.63 (dq, J = 19.7, 6.9 Hz, 1H, CH), 3.53 – 3.46 (m, 2H, CH$_2$), 3.33 (dt, J = 11.2, 2.7 Hz, 1H, CHH), 3.29 – 3.07 (m, 4H, 2 × CHH and CH$_2$), 1.48 (d, J = 6.8 Hz, 1H, CH$_3$), 1.45 (d, J = 6.8 Hz, 2H, CH$_3$), 1.11 (d, J = 6.6 Hz, 1H, CH$_3$), 1.00 (d, J = 6.6 Hz, 2H, CH$_3$) ppm. – ^{13}**C NMR** (101 MHz, CDCl$_3$) δ = 188.9 (–, C$_q$, CHO), 168.6 (–, C$_q$, C$_{ar}$), 168.2 (–, C$_q$, C$_{ar}$), 151.3 (–, C$_q$, C$_{ar}$), 142.7 (–, C$_q$, C$_{ar}$), 142.6 (–, C$_q$, C$_{ar}$), 138.4 (–, C$_q$, C$_{ar}$), 135.2 (–, C$_q$, C$_{ar}$), 132.9 (–, C$_q$, C$_{ar}$), 130.3 (+, C$_{ar}$H), 129.9 (+, C$_{ar}$H), 129.8 (+, C$_{ar}$H), 129.1 (+, C$_{ar}$H), 128.7 (+, C$_{ar}$H), 125.6 (+, C$_{ar}$H), 125.2 (+, C$_{ar}$H), 119.4 (+, C$_{ar}$H), 116.2 (+, C$_{ar}$H), 114.2 (+, C$_{ar}$H), 114.1 (+, C$_{ar}$H), 60.4 (–, CH$_2$), 56.1 (+, CH), 55.4 (+, CH), 49.5 (+, CH), 42.2 (–, CH$_2$), 41.2 (–, CH$_2$), 40.4 (–, CH$_2$), 36.4 (–, CH$_2$), 17.7 (+, CH$_3$),

16.7 (+, CH_3), 14.2 (+, CH_3), 12.8 (+, CH_3), 12.5 (+, CH_3) ppm. – Both NMR spectra showed a mixture of rotamers and were not separately assigned.

2-Chloro-4-((2R,3R)-4-(3-chlorophenyl)-2,3-dimethylpiperazine-1-carbonyl)benzoic acid (96)

Similar to **GP14**, to a solution of the aldehyde **95** (69.0 mg, 176 µmol, 1.00 equiv.) in DMF (1.0 ml), was added Oxone (108 mg, 176 µmol, 1.00 equiv.) and the mixture was stirred at rt overnight. Purification by flash column chromatography (*n*Pen/EtOAc 1:1 to 1:1 + 0.1% acetic acid) resulted in the title compound as a brown oil (68.0 mg, 0.15 mmol, 50%).[235]

^{1}**H NMR** (500 MHz, CDCl$_3$) δ = 8.03 (bs, 1H, O*H*), 7.74 (dd, *J* = 8.0, 3.4 Hz, 1H, C*H*$_{ar}$), 7.39 (dd, *J* = 10.4, 1.5 Hz, 1H, C*H*$_{ar}$), 7.28 – 7.22 (m, 1H, C*H*$_{ar}$), 7.17 – 7.10 (m, 1H, C*H*$_{ar}$), 6.80 – 6.74 (m, 2H, 2 × C*H*$_{ar}$), 6.71 – 6.65 (m, 1H, C*H*$_{ar}$), 4.75 (q, *J* = 6.9 Hz, 1H, C*H*), 4.61 – 4.53 (m, 1H, C*HH*), 3.87 – 3.79 (m, 1H, C*H*), 3.55 – 2.99 (m, 3H, 3 × C*H*$_2$ and C*HH*), 1.44 (d, *J* = 6.8 Hz, 1H, C*H*$_3$), 1.44 (m, 2H, C*H*$_3$), 1.08 (d, *J* = 6.7 Hz, 1H, C*H*$_3$), 0.95 (d, *J* = 6.7 Hz, 1H, C*H*$_3$) ppm. – ^{13}**C NMR** (126 MHz, CDCl$_3$) δ = 171.2 (–, C$_q$, C*H*O), 169.6 (–, C$_q$, C$_{ar}$), 169.1 (–, C$_q$, C$_{ar}$), 151.4 (–, C$_q$, C$_{ar}$), 138.6 (–, C$_q$, C$_{ar}$), 135.7 (–, C$_q$, C$_{ar}$), 132.7 (–, C$_q$, C$_{ar}$), 130.9 (+, C$_{ar}$H), 130.3 (+, C$_{ar}$H), 128.8 (+, C$_{ar}$H), 128.4 (+, C$_{ar}$H), 125.2 – 124.5 (+, m, C$_{ar}$H), 119.3 (+, C$_{ar}$H), 116.1 (+, C$_{ar}$H), 114.2 (+, C$_{ar}$H), 69.7 (–, C*H*$_2$), 56.1 (+, C*H*), 55.4 (+, C*H*), 49.6 (+, C*H*), 42.3 (–, C*H*$_2$), 41.3 (–, C*H*$_2$), 40.5 (–, C*H*$_2$), 36.5 (–, C*H*$_2$), 17.7 (+, C*H*$_3$), 16.7 (+, C*H*$_3$), 14.2 (+, C*H*$_3$), 12.8 (+, C*H*$_3$), 12.5 (+, C*H*$_3$) ppm. – Both NMR spectra showed a mixture of rotamers and were not separately assigned.

Ethyl 3-(2-chloro-4-((2R,3R)-4-(3-chlorophenyl)-2,3-dimethylpiperazine-1-carbonyl)phe-nyl)-3-oxopropanoate (33g)

To a solution of the benzoic acid **96** (12.0 mg, 29.0 µmol, 1.00 equiv.) in dry THF (250 µl), carbonyldiimidazole (5.30 mg, 32.0 mmol, 1.10 eq.) was added and the reaction mixture was stirred at rt for 2 h. Then a mixture of ethyl potassium malonate (5.00 mg, 29.0 µmol, 1.00 equiv.) [which had been prepared from ethyl hydrogen malonate (1 g) and KOH (0.4 g) in abs. ethyl alcohol 4 ml], anhydr. MgCl$_2$ (5.61 mg, 59.0 µmol, 2.00 equiv.) and TEA (9.86 µl, 7.16 mg, 71.0 µmol, 2.40 equiv.) were added. The reaction mixture was stirred further at rt for 24 h. After concentration under reduced pressure the obtained residue was resuspended in 2 M aq. HCl and extracted with DCM. The combined organic layers were washed with sat. aq. NaHCO$_3$ and brine, dried over Na$_2$SO$_4$ and concentrated under reduced pressure. Purification by flash column chromatography (nPen/EtOAc 3:1) resulted in the title compound as a colourless oil (0.82 mg, 1.72 µmol, 6%).[236]

R$_f$ (nPen/EtOAc 3:1) = 0.32. – **^{1}H NMR** (500 MHz, CDCl$_3$) δ = 12.51 (s, 1H, O*H*), 7.66 (dd, *J* = 10.0, 7.9 Hz, 1H, C*H*$_{ar}$), 7.48 (d, *J* = 4.5 Hz, 1H, C*H*$_{ar}$), 7.35 (dd, *J* = 13.4, 7.9 Hz, 1H, C*H*$_{ar}$), 7.20 – 7.14 (m, 1H, C*H*$_{ar}$), 6.84 – 6.79 (m, 2H, 2 × C*H*$_{ar}$), 6.71 (dd, *J* = 9.1, 2.1 Hz, 1H, C*H*$_{ar}$), 5.59 (s, 1H, C*H*), 4.71 (bs, 1H, C*H*), 4.29 (q, *J* = 7.1 Hz, 1H, C*H*$_2$), 4.21 (q, *J* = 7.1 Hz, 1H, C*H*$_2$), 4.04 (s, 2H, C*H*$_2$), 3.92 – 3.46 (m, 3H, 3 × C*H*), 3.38 – 3.09 (m, 3H, C*H*$_2$ and CH*H*), 1.47 (bs, 3H, C*H*$_3$), 1.35 (t, *J* = 7.1 Hz, 1H, C*H*$_3$), 1.26 (t, *J* = 7.1 Hz, 2H, C*H*$_3$), 1.07 (bs, 3H, C*H*$_3$) ppm. – **^{13}C NMR** (126 MHz, CDCl$_3$) δ = 172.7 (–, C$_q$), 169.4 (–, C$_q$, C$_{ar}$), 166.8 (–, C$_q$, C$_{ar}$), 151.4 (–, C$_q$, C$_{ar}$), 139.0 (–, C$_q$, C$_{ar}$), 138.8 (–, C$_q$, C$_{ar}$), 135.4 (–, C$_q$, C$_{ar}$), 132.9 (–, C$_q$, C$_{ar}$), 132.2 (–, C$_q$, C$_{ar}$), 130.5 (+, C$_{ar}$H), 130.4 (+, C$_{ar}$H), 119.6 (+, C$_{ar}$H), 119.5 (+, C$_{ar}$H), 116.3 (+, C$_{ar}$H), 114.3 (+, C$_{ar}$H), 94.0 (+, C$_{ar}$H), 61.8 (–, C*H*$_2$), 60.9 (–, C*H*$_2$), 56.3 (+, C*H*), 49.2 (–, C*H*$_2$), 14.4 (+, C*H*$_3$), 14.2 (+, C*H*$_3$) ppm. – Both NMR spectra showed a mixture of tautomers and rotamers and were not separately assigned. – **HRMS** (ESI, C$_{24}$H$_{26}$Cl$_2$N$_2$O$_4$) = calc.: 477.1342/479.1313 [M+H]$^+$; found: 477.1346/479.1321 [M+H]$^+$.

5.3.7 Chemical biology methods

Cell membrane lysate

Cell culture, transient transfection and cell lysate membrane preparation of MAGL-overexpressing HEK293T cells had been performed as previously described[237] and aliquots of 1.5 µg/ml concentration were stored ready to use at –80°C.

Natural substrate-based fluorescence assay

The natural substrate-based fluorescence assay for MAGL was performed as previously described.[212] Standard assays were performed in HEMNB buffer (50 mM Hepes pH 7.4, 1 mM EDTA, 5 mM MgCl2, 100 mM NaCl, 0.5% (w/v) BSA) in black flat clear-bottom 96-well plates. Membrane preparation of MAGL-overexpressing, transiently transfected HEK293T cells were used in a final protein concentration of 1.5 µg/ml (0.3 µg/well), with 2-AG as substrate. As negative control membrane preparation of mock transfected cells was used, as positive control DMSO was used as vehicle. Inhibitors were used as 40 × concentrated DMSO stock. The assay was proceeded as follows: MAGL-overexpressing membrane preparation was incubated for 20 min with different concentrations of inhibitor. Followed by addition of assay mix containing glycerol kinase (GK), glycerol-3-phosphate oxidase (GPO), HRP, ATP, Amplifu™ Red and 2-AG and measurement of the fluorescence every 5 min for 60 min by a GENIOS microplate reader (TECAN, Giessen, The Netherlands). Final assay concentrations: 0.2 U/ml GK, GPO and HRP; 0.125 mM ATP; 25 µM 2-AG; 10 µM Amplifu™ Red; 5% DMSO, 0.5% CAN in a total volume of 200 µl.

Mouse brain lysate preparation

The mouse brain lysates (cytosol and membrane) had been prepared as previously described[237] and were stored ready to use at −80°C.

Activity-based protein profiling

Standard procedure for gel-based ABPP was performed with minor adaptions as previously described.[218] Mouse brain proteome (19 µl, 2.0 mg/ml, cytosol or membrane fractions were pre-incubated with 0.5 µl of DMSO (vehicle) or inhibitor (40 × DMSO stock) at rt for 30 min. Subsequently followed by addition of 0.5 µl activity-based probes (40 × DMSO stock, final concentrations: MB064 250 nM; FP-TAMRA 100 nM, LEI463 100 nM; THL 250 nM) and incubation at rt for 10 min. Reactions were quenched by addition of 4 × Laemmli buffer (10 µl, 240 nM Tris pH 6.8, 8% (w/v) SDS, 40% (v/v) glycerol, 5% (v/v) β-mercaptoethanol, 0.04% (v/v) bromophenol blue). Samples (20 µg/reaction) were directly loaded and resolved on a 10% acrylamide SDS-PAGE gel (180 V, 75 min). Gels were visualized and scanned with a BIO-RAD CHEMIDOC (BIO-RAD LABORATORIES B.V.) using Cy3 and Cy5 multichannel settings (605/50 and 695/55, filters respectively). For quantification the gels were stained with a Coomassie® solution for 15 min and destained overnight. Fluorescence was normalized to the Coomassie® staining and quantified with IMAGELAB (BIO-RAD).

Reversibility activity-based protein profiling

Kinetic measurements of reversibility were performed based on the gel-based ABPP protocol with minor changes. After pre-incubation of mouse brain proteome with inhibitor, MAGL specific activity-based probe LEI463 was added (final concentration: 1 µM). Reactions were quenched depending on different time points (0, 30, 60, 90, 120 min) by addition of 4 × Laemmli buffer. Further procedure as described above. As a control reference covalent inhibitor ABX (final concentration: 1 µM) was used.

Data analysis

All experimental data from the assays were analysed with Excel and GraphPad Prism (GRAPHPAD SOFTWARE INC., San Diego, CA, version 7 and 8). Natural substrate-based fluorescence assays and gel-based ABPP were analysed using non-linear regression analysis for log(inhibitor) vs normalized response to obtain $logIC_{50}$ values. All values are expressed as means of at least two individual experiments in duplicates, if not stated otherwise. Errors are expressed as standard deviation (SD).

6 Appendix

6.1 Synthesis strategy dialkylamino moiety

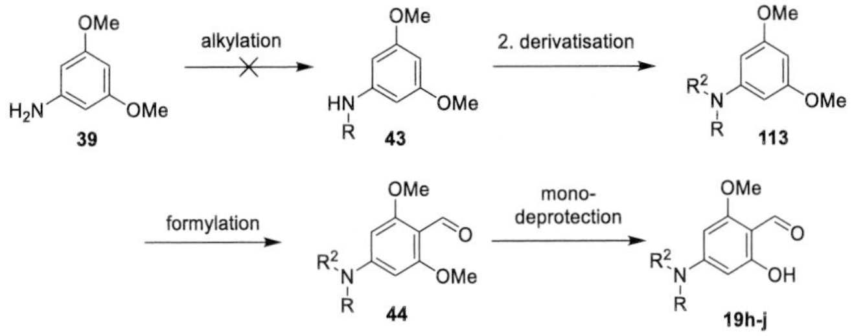

Scheme 21: Initially proposed reaction strategy for salicylic aldehyde **19h-j**.

6.2 Synthesis strategy dimethylalkyl moiety

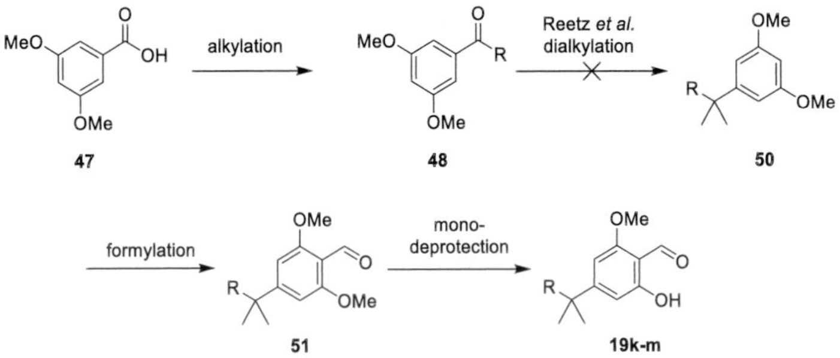

Scheme 22: Proposed synthesis strategy for **19k-m** using a one-pot geminal dialkylation approach according to REETZ *et al.*[149]

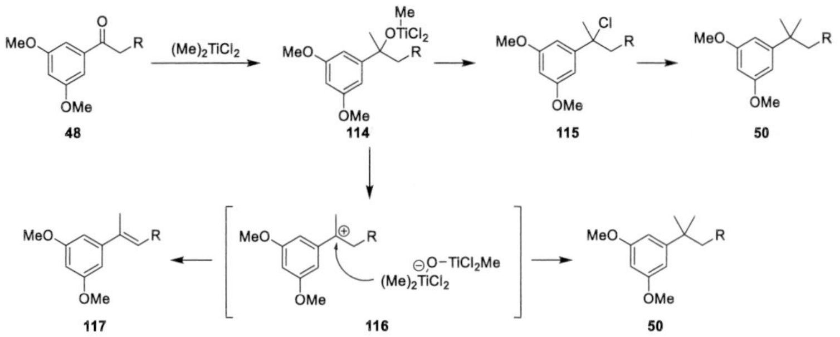

Scheme 23: Proposed mechanism of Ti catalysed geminal dialkylation by REETZ *et al* and formation of side product **117**. Mechanism adapted from literature [148]. After formation of the initial GRIGNARD-type adduct and first methylation (**114**), an additional equivalent of the methyl titanium reagent induces the bound titanium species to behave like a leaving group forming the carbocationic species **116** and subsequently followed by second methylation to the product **50**. Insufficient availability of the methyl titanium reagent could possibly lead to the formation of the cationic species **116**, followed by elimination reaction to side product **117**. Alternatively, the mechanism was proposed to follow the route of intermediate **115** and product formation via nucleophilic substitution.

6.3 Dialkylamino coumarins

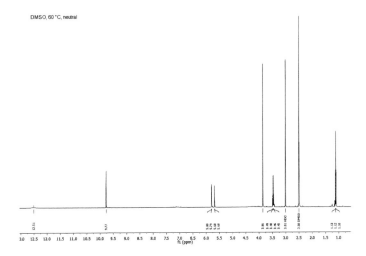

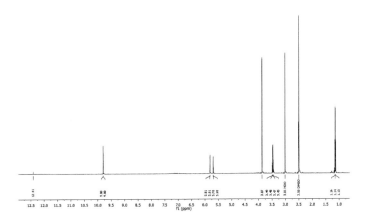

Figure 44: High temperature NMRs from **19h** under neutral conditions.

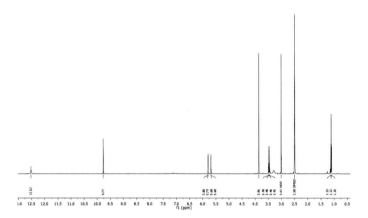

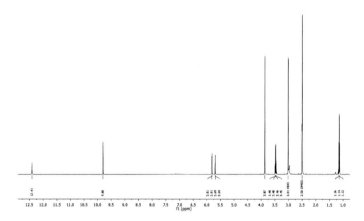

Figure 45: High temperature NMRs from **19h** under basic conditions.

Table 21: Screened reaction conditions used for the synthesis of **21ha-jg**.

Entry	Cat. (equiv.)	Base (equiv.)	Supplements (equiv.)	Time [min]	Yield [%]
1	1 (1.2)	K_2CO_3 (1.2)	–	50	6
2	1 (1.2)	K_2CO_3 (1.2)	–	75	8
3	2 (1.0)	K_2CO_3 (1.0)	–	75	–
4	3 (1.2)	K_2CO_3 (1.2)	–	50	traces
5	4 (1.2)	K_2CO_3 (1.2)	–	75	14[a]
6	1 (1.2)	K_2CO_3 (1.2)	18-crown-6 (0.10)	50	4
7	1 (1.2)	K_2CO_3 (1.2)	MS (3Å)	50	5
8	1 (1.2)	KOAc (2.5)	AcOH (0.25), MS (3Å)	75	18
9	4 (1.2)	KOAc (2.5)	AcOH (0.25), MS (4Å)	75	–

Conditions: salicylic aldehyde (1.00 equiv.), cinnamaldehyde (2.50 equiv.), base; XX min, 110 °C, MWI. [a] purification insufficient (>90%).

6.4 Radioligand binding assay

Table 22: List of all synthesised and tested coumarin-derivatives of the presented work including CBR affinities based on their structural classification.

Lead structure 1, **29**
Lead structure 2, **30**
Lead structure 3, **31**
Lead structure 4, **65**
Lead structure 5, **18**
Lead structure 6, **17**
Lead structure 7, **66**
Lead structure 8, **67**
Lead structure 9, **68**

En-try	Cpd.	R3	R5	R6	R7	R8	hCB1	hCB2
							pKi ± SEM (Ki in nM[a] or displacement at 1 µM[b])	
			Group 1: 3-Benzyl-7-alkylether coumarins					
1	21aa	H	methoxy	H	methoxy	H	<6.00 (18%)	<<6.00 (0%)
2	21ab	o-methyl	methoxy	H	methoxy	H	<6.00 (19%)	<6.00 (21%)
3	21ac	o-methoxy	methoxy	H	methoxy	H	<<6.00 (4%)	<<6.00 (7%)

Appendix

En-try	Cpd.	R3	R5	R6	R7	R8	hCB1	hCB2
							pK$_i$ ± SEM (K$_i$ in nM[a] or displacement at 1 μM[b])	
4	21ad	*p*-methoxy	methoxy	H	methoxy	H	<6.00 (14%)	<6.00 (14%)
5	21af	*p*-F	methoxy	H	methoxy	H	<<6.00 (6%)	<<6.00 (7%)
6	21ah	*p*-Cl	methoxy	H	methoxy	H	<6.00 (28%)	<<6.00 (3%)
7	21ai	*p*-NMe$_2$	methoxy	H	methoxy	H	<<6.00 (1%)	<6.00 (25%)
8	21ba	H	methoxy	H	ethoxy	H	n.d.	n.d.
9	21bb	*o*-methyl	methoxy	H	ethoxy	H	n.d.	n.d.
10	21bc	*o*-methoxy	methoxy	H	ethoxy	H	n.d.	n.d.
11	21bd	*p*-methoxy	methoxy	H	ethoxy	H	n.d.	n.d.
12	21bf	*p*-F	methoxy	H	ethoxy	H	n.d.	n.d.
13	21bh	*p*-Cl	methoxy	H	ethoxy	H	n.d.	n.d.
14	21bi	*p*-NMe$_2$	methoxy	H	ethoxy	H	n.d.	n.d.
16	21ca	H	methoxy	H	propoxy	H	<6.00 (27%)	<6.00 (25%)
17	21cb	*o*-methyl	methoxy	H	propoxy	H	<6.00 (28%)	<<6.00 (38%)
18	58a	*o*-methyl	hydroxy	H	propoxy	H	<6.00 (27%)	<<6.00 (−9%)

En-try	Cpd.	R3	R5	R6	R7	R8	hCB1	hCB2
							pK_i ± SEM (K_i in nM[a] or displacement at 1 μM[b])	
19	21cc	*o*-methoxy	methoxy	H	propoxy	H	<6.00 (23%)	6.63±0.02 (236)
20	58b	*o*-hydroxy	hydroxy	H	propoxy	H	<<6.00 (7%)	<<6.00 (7%)
21	21cd	*p*-methoxy	methoxy	H	propoxy	H	<6.00 (19%)	<6.00 (31%)
22	21cf	*p*-F	methoxy	H	propoxy	H	<<6.00 (7%)	<6.00 (15%)
23	21ch	*p*-Cl	methoxy	H	propoxy	H	<<6.00 (−21%)	<<6.00 (10%)
24	21ci	*p*-NMe$_2$	methoxy	H	propoxy	H	<<6.00 (8%)	<<6.00 (9%)
25	21da	H	methoxy	H	butoxy	H	n.d.	n.d.
26	21db	*o*-methyl	methoxy	H	butoxy	H	n.d.	n.d.
27	21dc	*o*-methoxy	methoxy	H	butoxy	H	n.d.	n.d.
28	21dd	*p*-methoxy	methoxy	H	butoxy	H	n.d.	n.d.
29	21df	*p*-F	methoxy	H	butoxy	H	n.d.	n.d.
30	21dh	*p*-Cl	methoxy	H	butoxy	H	n.d.	n.d.
31	21di	*p*-NMe$_2$	methoxy	H	butoxy	H	n.d.	n.d.
32	21ea	H	methoxy	H	pentyloxy	H	<6.00 (44%)	<6.00 (24%)

En-try	Cpd.	R3	R5	R6	R7	R8	hCB1	hCB2
							pKi ± SEM (Ki in nM[a] or displacement at 1 μM[b]	
33	21eb	o-methyl	methoxy	H	pentyloxy	H	<6.00 (24%)	<<6.00 (7%)
34	58d	o-methyl	hydroxy	H	pentyloxy	H	n.d.[c]	n.d.[c]
35	21ec	o-methoxy	methoxy	H	pentyloxy	H	<6.00 (24%)	~6.00 (52%)
36	58e	o-hydroxy	hydroxy	H	pentyloxy	H	<<6.00 (-4%)	<<6.00 (0%)
37	21ed	p-methoxy	methoxy	H	pentyloxy	H	<<6.00 (-19%)	<<6.00 (-15%)
38	21ef	p-F	methoxy	H	pentyloxy	H	<6.00 (11%)	<6.00 (44%)
39	21eh	p-Cl	methoxy	H	pentyloxy	H	<6.00 (26%)	<<6.00 (-6%)
40	21ei	p-NMe2	methoxy	H	pentyloxy	H	<6.00 (15%)	<<6.00 (-6%)
41	21fa	H	methoxy	H	hexyloxy	H	n.d.	n.d.
42	21fb	o-methyl	methoxy	H	hexyloxy	H	n.d.	n.d.
43	21fc	o-methoxy	methoxy	H	hexyloxy	H	n.d.	n.d.
44	21fd	p-methoxy	methoxy	H	hexyloxy	H	n.d.	n.d.
45	21ff	p-F	methoxy	H	hexyloxy	H	n.d.	n.d.
46	21fh	p-Cl	methoxy	H	hexyloxy	H	n.d.	n.d.

En-try	Cpd.	R3	R5	R6	R7	R8	hCB1	hCB2
							pKi ± SEM (Ki in nM[a] or displacement at 1 μM[b])	
47	21fi	p-NMe$_2$	methoxy	H	hexyloxy	H	n.d.	n.d.
48	21ga	H	methoxy	H	heptyloxy	H	<6.00 (30%)	<<6.00 (−2%)
49	21gb	o-methyl	methoxy	H	heptyloxy	H	<<6.00 (8%)	<<6.00 (8%)
50	58g	o-methyl	hydroxy	H	heptyloxy	H	<<6.00 (−4%)	<<6.00 (1%)
51	21gc	o-methoxy	methoxy	H	heptyloxy	H	<6.00 (27%)	<<6.00 (−4%)
52	58h	o-hydroxy	hydroxy	H	heptyloxy	H	n.d.[c]	n.d.[c]
53	21gd	p-methoxy	methoxy	H	heptyloxy	H	<6.00 (14%)	<<6.00 (3%)
54	21gf	p-F	methoxy	H	heptyloxy	H	<<6.00 (3%)	<<6.00 (−5%)
55	21gh	p-Cl	methoxy	H	heptyloxy	H	<<6.00 (6%)	<<6.00 (5%)
56	21gi	p-NMe$_2$	methoxy	H	heptyloxy	H	<6.00 (42%)	<<6.00 (4%)
57	21i	p-NMe$_2$	hydroxy	H	heptyloxy	H	n.d.[c]	n.d.[c]
58	64a	H	methoxy	H	2-propoxymethyl	H	<<6.00 (18%)	<6.00 (24%)
59	64b	o-methyl	methoxy	H	2-propoxymethyl	H	<6.00 (31%)	<6.00 (31%)
60	64c	o-methoxy	methoxy	H	2-propoxymethyl	H	<6.00 (33%)	6.83±0.09 (147)

En-try	Cpd.	R3	R5	R6	R7	R8	hCB1	hCB2
							pKi ± SEM (Ki in nM[a] or displacement at 1 µM[b])	
61	64d	p-methoxy	methoxy	H	2-propoxymethyl	H	<<6.00 (−19%)	<<6.00 (−15%)
62	64e	p-Cl	methoxy	H	2-propoxymethyl	H	<<6.00 (−22%)	<6.00 (18%)
63	64f	o-methyl	methoxy	H	2-propoxyprop-2-yl	H	<6.00 (26%)	<<6.00 (3%)
64	64g	o-methoxy	methoxy	H	2-propoxyprop-2-yl	H	<6.00 (19%)	~6.00 (50%)
Group 2: 3-Benzyl-7-(dialkylamino)coumarins								
65	21hb	o-methyl	methoxy	H	ethyl	H	<6.00 (17%)	<<6.00 (4%)
66	21hc	o-methoxy	methoxy	H	ethyl	H	<6.00 (24%)	<6.00 (20%)
67	21he	o-F	methoxy	H	ethyl	H	<6.00 (31%)	<6.00 (15%)
68	21hf	p-F	methoxy	H	ethyl	H	n.d.[c]	n.d.[c]
69	21hg	o-CF$_3$	methoxy	H	ethyl	H	<6.00 (39%)	~6.00 (46%)
70	21ia	H	methoxy	H	butyl	H	~6.00 (48%)	<6.00 (28%)
71	21ib	o-methyl	methoxy	H	butyl	H	~6.00 (45%)	<6.00 (15%)
72	21ic	o-methoxy	methoxy	H	butyl	H	<6.00 (27%)	<6.00 (28%)

En-try	Cpd.	R3	R5	R6	R7	R8	hCB1	hCB2
							$pK_i \pm$ SEM (K_i in nM[a] or displacement at 1 μM[b])	
73	21id	p-methoxy	methoxy	H	butyl	H	<<6.00 (6%)	<<6.00 (13%)
74	21ie	o-F	methoxy	H	butyl	H	n.d.[d]	n.d.[d]
75	21if	p-F	methoxy	H	butyl	H	<6.00 (41%)	<6.00 (21%)
76	21ig	o-CF$_3$	methoxy	H	butyl	H	n.d.[d]	n.d.[d]
77	21ja	H	methoxy	H	hexyl	H	<6.00 (28%)	<6.00 (23%)
78	21jb	o-methyl	methoxy	H	hexyl	H	<6.00 (43%)	<6.00 (14%)
79	21jc	o-methoxy	methoxy	H	hexyl	H	<6.00 (26%)	<6.00 (35%)
80	21jd	p-methoxy	methoxy	H	hexyl	H	<6.00 (32%)	<<6.00 (9%)
81	21je	o-F	methoxy	H	hexyl	H	~6.00 (47%)	<6.00 (21%)
82	21jf	p-F	methoxy	H	hexyl	H	<6.00 (38%)	<<6.00 (10%)
83	21jg	o-CF$_3$	methoxy	H	hexyl	H	<6.00 (27%)	<6.00 (15%)
Group 3: 3-Benzyl-7-(1,1′-dimethylalkyl)coumarins								
84	21ka	H	methoxy	H	methyl	H	<6.00 (19%)	<<6.00 (2%)
85	59ka	H	hydroxy	H	methyl	H	<6.00 (17%)	<<6.00 (0%)

En-try	Cpd.	R3	R5	R6	R7	R8	hCB1	hCB2
							pK_i± SEM (K_i in nM[a] or displacement at 1 μM[b])	
86	21kb	o-methyl	methoxy	H	methyl	H	≪6.00 (0%)	<6.00 (13%)
87	59kb	o-methyl	hydroxy	H	methyl	H	≪6.00 (−7%)	≪6.00 (−2%)
88	21kc	o-methoxy	methoxy	H	methyl	H	≪6.00 (5%)	<6.00 (12%)
89	59kc	o-hydroxy	hydroxy	H	methyl	H	≪6.00 (2%)	≪6.00 (7%)
90	21kd	p-methoxy	methoxy	H	methyl	H	n.d.	n.d.
91	59kd	p-hydroxy	hydroxy	H	methyl	H	n.d.	n.d.
92	21ke	o-F	methoxy	H	methyl	H	≪6.00 (−4%)	<6.00 (12%)
93	59ke	o-F	hydroxy	H	methyl	H	<6.00 (34%)	<6.00 (17%)
94	21kf	p-F	methoxy	H	methyl	H	<6.00 (34%)	≪6.00 (2%)
95	59kf	p-F	hydroxy	H	methyl	H	≪6.00 (−3%)	≪6.00 (−8%)
96	21kg	o-CF₃	methoxy	H	methyl	H	<6.00 (38%)	<6.00 (28%)
97	59kg	o-CF₃	hydroxy	H	methyl	H	<6.00 (13%)	≪6.00 (8%)
98	21la	H	methoxy	H	butyl	H	6.31±0.22 (486)	<6.00 (39%)
99	59la	H	hydroxy	H	butyl	H	<6.00 (12%)	<6.00 (24%)

En-try	Cpd.	R3	R5	R6	R7	R8	hCB1	hCB2
							$pK_i \pm$ SEM (K_i in nM[a] or displacement at 1 μM[b])	
100	21lb	o-methyl	methoxy	H	butyl	H	6.66±0.15 (217)	<6.00 (32%)
101	59lb	o-methyl	hydroxy	H	butyl	H	<6.00 (30%)	<6.00 (41%)
102	21lc	o-methoxy	methoxy	H	butyl	H	6.71±0.11 (196)	6.64±0.003 (231)
103	59lc	o-hydroxy	hydroxy	H	butyl	H	<6.00 (24%)	<6.00 (37%)
104	21ld	p-methoxy	methoxy	H	butyl	H	<6.00 (26%)	<<6.00 (10%)
105	59ld	p-hydroxy	hydroxy	H	butyl	H	<6.00 (35%)	<6.00 (12)
106	21le	o-F	methoxy	H	butyl	H	<6.00 (33%)	<6.00 (17%)
107	59le	o-F	hydroxy	H	butyl	H	<6.00 (44%)	<6.00 (37%)
108	21lf	p-F	methoxy	H	butyl	H	<6.00 (18%)	<6.00 (24%)
109	59lf	p-F	hydroxy	H	butyl	H	<6.00 (15%)	<6.00 (39%)
110	21lg	o-CF$_3$	methoxy	H	butyl	H	n.d.[c]	n.d.[c]
111	59lg	o-CF$_3$	hydroxy	H	butyl	H	n.d.[c]	n.d.[c]
112	21ma	H	methoxy	H	hexyl	H	n.d.[e]	n.d.[e]
113	59ma	H	hydroxy	H	hexyl	H	n.d.[e]	n.d.[e]

En-try	Cpd.	R3	R5	R6	R7	R8	hCB1	hCB2
							pKi ± SEM (Ki in nM[a] or displacement at 1 µM[b])	
114	21mb	o-methyl	methoxy	H	hexyl	H	n.d.[c]	n.d.[c]
115	59mb	o-methyl	hydroxy	H	hexyl	H	<6.00 (47%)	6.65±0.08 (222)
116	21mc	o-methoxy	methoxy	H	hexyl	H	n.d.[e]	n.d.[e]
117	59mc	o-hydroxy	hydroxy	H	hexyl	H	n.d.[e]	n.d.[e]
118	21md	p-methoxy	methoxy	H	hexyl	H	<6.00 (24%)	<<6.00 (6%)
119	59md	p-hydroxy	hydroxy	H	hexyl	H	<6.00 (39%)	<6.00 (13%)
120	21me	o-F	methoxy	H	hexyl	H	n.d.[c]	n.d.[c]
121	59me	o-F	hydroxy	H	hexyl	H	<6.00 (35%)	<6.00 (44%)
122	21mf	p-F	methoxy	H	hexyl	H	<6.00 (25%)	<<6.00 (−4%)
123	59mf	p-F	hydroxy	H	hexyl	H	~6.00 (47%)	6.42±0.06 (380)
124	21mg	o-CF$_3$	methoxy	H	hexyl	H	<6.00 (26%)	<<6.00 (3%)
125	59mg	o-CF$_3$	hydroxy	H	hexyl	H	~6.00 (49%)	<6.00 (42%)
				Group 4: 3-Fluorobenzyl-7-pentylcoumarins				
126	65a	o-F	methoxy	H	pentyl	H	~6.00 (47%)	6.82±0.07 (152)

En-try	Cpd.	R3	R5	R6	R7	R8	hCB1	hCB2
							$pK_i \pm$ SEM (K_i in nM[a] or displacement at 1 μM[b])	
127	65b	o-CF$_3$	methoxy	H	pentyl	H	6.69±0.12 (202)	6.49±0.02 (326)
				Group 5: 3-Phenylcoumarins				
128	18a	H	methoxy	H	1-butylcyclopentyl	H	<6.00 (34%)	<<6.00 (−8%)
129	18b	H	hydroxy	H	1-butylcyclopentyl	H	<<6.00 (−13%)	<<6.00 (−21%)
130	18c	o-methoxy	methoxy	H	1-butylcyclopentyl	H	<<6.00 (−10%)	<<6.00 (−37%)
131	18d	o-hydroxy	hydroxy	H	1-butylcyclopentyl	H	<<6.00 (−43%)	<<6.00 (−2%)
132	18e	m-methoxy	methoxy	H	1-butylcyclopentyl	H	<<6.00 (−18%)	<<6.00 (−9%)
133	18f	m-hydroxy	hydroxy	H	1-butylcyclopentyl	H	<<6.00 (−17%)	<<6.00 (−14%)
134	18g	p-methoxy	methoxy	H	1-butylcyclopentyl	H	<<6.00 (−25%)	<<6.00 (−13%)
135	18h	p-hydroxy	hydroxy	H	1-butylcyclopentyl	H	<6.00 (42%)	~6.00 (53%)
136	18i	o-methyl	methoxy	H	1-butylcyclopentyl	H	<<6.00 (−32%)	<<6.00 (−14%)
137	18j	o-methyl	hydroxy	H	1-butylcyclopentyl	H	<<6.00 (4%)	<6.00 (43%)
138	18k	m-methyl	methoxy	H	1-butylcyclopentyl	H	<<6.00 (−13%)	<<6.00 (−35%)
139	18l	m-methyl	hydroxy	H	1-butylcyclopentyl	H	<<6.00 (−25%)	<<6.00 (−3%)

Entry	Cpd.	R3	R5	R6	R7	R8	hCB1	hCB2
							pK$_i$ ± SEM (K$_i$ in nM[a] or displacement at 1 μM[b])	
140	18m	p-methyl	methoxy	H	1-butylcyclopentyl	H	<<6.00 (−51%)	<<6.00 (−46%)
141	18n	p-methyl	hydroxy	H	1-butylcyclopentyl	H	<<6.00 (−9%)	<<6.00 (−16%)
				Group 6: 3-Pyridylcoumarins				
142	17a	o-pyridyl	methoxy	H	pentyl	H	7.15±0.06 (70.3)	7.08±0.14 (82.4)
143	17b	o-pyridyl	hydroxy	H	pentyl	H	<<6.00 (−15%)	<<6.00 (−17%)
144	17c	m-pyridyl	methoxy	H	pentyl	H	6.77±0.12 (171)	7.25±0.04 (56.5)
145	17d	m-pyridyl	hydroxy	H	pentyl	H	<<6.00 (−21%)	<<6.00 (−21%)
146	17e	p-pyridyl	methoxy	H	pentyl	H	<<6.00 (0%)	<6.00 (11%)
147	17f	p-pyridyl	hydroxy	H	pentyl	H	<<6.00 (−48%)	<<6.00 (−5%)
148	17g	o-pyridyl	methoxy	H	1-butylcyclopentyl	H	<6.00 (29%)	<6.00 (20%)
149	17h	o-pyridyl	hydroxy	H	1-butylcyclopentyl	H	<6.00 (34%)	<6.00 (44%)
150	17i	m-pyridyl	methoxy	H	1-butylcyclopentyl	H	<<6.00 (29%)	<6.00 (40%)
151	17j	m-pyridyl	hydroxy	H	1-butylcyclopentyl	H	<6.00 (19%)	6.51±0.07 (310)
152	17k	p-pyridyl	methoxy	H	1-butylcyclopentyl	H	<6.00 (28%)	<<6.00 (9%)

En-try	Cpd.	R3	R5	R6	R7	R8	hCB1	hCB2
							pKi ± SEM (Ki in nM[a] or displacement at 1 μM[b])	
153	17m	p-pyridyl	hydroxy	H	1-butylcyclopentyl	H	<<6.00 (−24%)	<6.00 (21%)
154	17m	o-pyridyl	methoxy	H	1-butylcyclohexyl	H	~6.00 (47%)	<6.00 (12%)
155	17n	o-pyridyl	hydroxy	H	1-butylcyclohexyl	H	<<6.00 (−7%)	7.14±0.13 (71.9)
156	17o	m-pyridyl	methoxy	H	1-butylcyclohexyl	H	<<6.00 (6%)	<6.00 (11%)
157	17p	m-pyridyl	hydroxy	H	1-butylcyclohexyl	H	n.d.[f]	n.d.[f]
158	17q	p-pyridyl	methoxy	H	1-butylcyclohexyl	H	<6.00 (26%)	<<6.00 (3%)
159	17r	p-pyridyl	hydroxy	H	1-butylcyclohexyl	H	<<6.00 (5%)	~6.00 (46%)
Group 7: 3-Alkyl-7-(butylcycloalkyl)coumarins								
160	66a	methyl	methoxy	H	1-butylcyclopentyl	H	<6.00 (16%)	<<6.00 (1%)
161	66b	methyl	hydroxy	H	1-butylcyclopentyl	H	<<6.00 (−10%)	~6.00 (49%)
162	66c	ethyl	methoxy	H	1-butylcyclopentyl	H	<6.00 (33%)	<<6.00 (0%)
163	66d	ethyl	hydroxy	H	1-butylcyclopentyl	H	<6.00 (34%)	7.22±0.08 (60.6)
164	66e	propyl	methoxy	H	1-butylcyclopentyl	H	<6.00 (32%)	<6.00 (38%)
165	66f	propyl	hydroxy	H	1-butylcyclopentyl	H	~6.00 (47%)	7.73±0.01 (18.6)

En-try	Cpd.	R3	R5	R6	R7	R8	hCB1	hCB2
							$pK_i \pm$ SEM (K_i in nM[a] or displacement at 1 μM[b])	
166	66g	butyl	methoxy	H	1-butylcyclopentyl	H	<6.00 (15%)	<<6.00 (−1%)
167	66h	butyl	hydroxy	H	1-butylcyclopentyl	H	~6.00 (50%)	7.86±0.11 (13.7)
168	66i	methyl	methoxy	H	1-butylcyclohexyl	H	<<6.00 (9%)	<<6.00 (−34%)
169	66j	methyl	hydroxy	H	1-butylcyclohexyl	H	<6.00 (19%)	6.98±0.03 (106)
170	66k	ethyl	methoxy	H	1-butylcyclohexyl	H	<<6.00 (6%)	<<6.00 (−1%)
171	66l	ethyl	hydroxy	H	1-butylcyclohexyl	H	<6.00 (39%)	7.41±0.04 (39.1)
172	66m	propyl	methoxy	H	1-butylcyclohexyl	H	<6.00 (18%)	<<6.00 (−4%)
173	66n	propyl	hydroxy	H	1-butylcyclohexyl	H	6.80±0.22 (159)	8.19±0.12 (6.5)
174	66o	butyl	methoxy	H	1-butylcyclohexyl	H	<<6.00 (−11%)	<<6.00 (2%)
175	66p	butyl	hydroxy	H	1-butylcyclohexyl	H	~6.00 (48%)	7.90±0.03 (12.5)
Group 8: 3-Alkyl-7-pentylcoumarins								
176	67a	methyl	methoxy	H	pentyl	H	<<6.00 (7%)	<<6.00 (0%)
177	67b	methyl	hydroxy	H	pentyl	H	<<6.00 (−32%)	<<6.00 (−6%)
178	67c	butyl	methoxy	H	pentyl	H	<6.00 (14%)	<6.00 (24%)

En-try	Cpd.	R3	R5	R6	R7	R8	hCB1	hCB2
							$pK_i \pm$ SEM (K_i in nM[a] or displacement at 1 μM[b])	
179	67d	butyl	hydroxy	H	pentyl	H	$\ll 6.00$ (-43%)	$\ll 6.00$ (-2%)
				Group 9: Small substituents at coumarin core				
180	16aa	H	methoxy	H	methoxy	H	n.d.	n.d.
181	16ab	methyl	methoxy	H	methoxy	H	n.d.	n.d.
182	16ac	ethyl	methoxy	H	methoxy	H	$\ll 6.00$ (1%)	$\ll 6.00$ (-14%)
183	16ad	propyl	methoxy	H	methoxy	H	n.d.	n.d.
184	16ae	butyl	methoxy	H	methoxy	H	< 6.00 (15%)	$\ll 6.00$ (-21%)
185	16af	pentyl	methoxy	H	methoxy	H	n.d.	n.d.
186	16ag	hexyl	methoxy	H	methoxy	H	< 6.00 (19%)	$\ll 6.00$ (-7%)
187	16ba	H	methoxy	H	ethoxy	H	n.d.	n.d.
188	16bb	methyl	methoxy	H	ethoxy	H	n.d.	n.d.
189	16bc	ethyl	methoxy	H	ethoxy	H	< 6.00 (16%)	$\ll 6.00$ (-7%)
190	16bd	propyl	methoxy	H	ethoxy	H	n.d.	n.d.
191	16be	butyl	methoxy	H	ethoxy	H	n.d.	n.d.

En- try	Cpd.	R3	R5	R6	R7	R8	hCB1	hCB2
							$pK_i \pm$ SEM (K_i in nM[a] or displacement at 1 μM[b])	
192	16bf	pentyl	methoxy	H	ethoxy	H	n.d.	n.d.
193	16bg	hexyl	methoxy	H	ethoxy	H	<6.00 (15%)	<<6.00 (3%)
194	16ca	H	methoxy	H	propoxy	H	n.d.	n.d.
195	16ce	butyl	methoxy	H	propoxy	H	<<6.00 (−5%)	<<6.00 (6%)
196	16dd	propyl	methoxy	H	butoxy	H	<<6.00 (5%)	<<6.00 (8%)
197	16ec	ethyl	methoxy	H	pentyloxy	H	<6.00 (34%)	<<6.00 (4%)
198	16fb	methyl	methoxy	H	hexyloxy	H	<<6.00 (1%)	<6.00 (18%)
199	68a	benzyl	methyl	H	H	iPr	<<6.00 (−12%)	<<6.00 (−5%)
200	68b	o-methoxybenzyl	methyl	H	H	iPr	<6.00 (14%)	<<6.00 (7%)
201	68c	p-fluorobenzyl	methyl	H	H	iPr	<<6.00 (−26%)	<<6.00 (−6%)
202	68d	p-chlorobenzyl	methyl	H	H	iPr	<<6.00 (0%)	<<6.00 (−14%)
203	68e	ethyl	iPr	H	H	methyl	<<6.00 (5%)	<<6.00 (−10%)
204	68f	butyl	iPr	H	H	methyl	<<6.00 (−15%)	<<6.00 (−5%)
205	68g	hexyl	iPr	H	H	methyl	<<6.00 (1%)	<<6.00 (3%)

En-try	Cpd.	R3	R5	R6	R7	R8	hCB1	hCB2
							$pK_i \pm$ SEM (K_i in nM[a] or displacement at 1 μM[b])	
206	68h	iPr	iPr	H	H	methyl	≪6.00 (−1%)	≪6.00 (−20%)
207	68i	methyl	iPr	methoxy	H	methyl	<6.00 (16%)	<6.00 (13%)
208	68j	methyl	iPr	hydroxy	H	methyl	<6.00 (34%)	<6.00 (17%)
209	68k	ethyl	iPr	methoxy	H	methyl	≪6.00 (−12%)	≪6.00 (−17%)
210	68l	ethyl	iPr	hydroxy	H	methyl	<6.00 (25%)	≪6.00 (−35%)
211	68m	butyl	iPr	methoxy	H	methyl	≪6.00 (−18%)	≪6.00 (−7%)
212	68n	butyl	iPr	hydroxy	H	methyl	≪6.00 (−14%)	≪6.00 (−36%)
213	68o	hexyl	iPr	methoxy	H	methyl	<6.00 (22%)	≪6.00 (4%)
214	68p	hexyl	iPr	hydroxy	H	methyl	<6.00 (43%)	<6.00 (3%)
215	68q	iPr	iPr	methoxy	H	methyl	≪6.00 (−4%)	≪6.00 (−29%)
216	21na	benzyl	methyl	H	H	methyl	<6.00 (18%)	≪6.00 (−13%)
217	21nc	o-methoxybenzyl	methyl	H	H	methyl	n.d.[g]	n.d.[g]
218	60a	o-hydroxybenzyl	methyl	H	H	methyl	≪6.00 (9%)	≪6.00 (−3%)
219	21nd	p-methoxybenzyl	methyl	H	H	methyl	<6.00 (14%)	≪6.00 (3%)

Appendix

Entry	Cpd.	R3	R5	R6	R7	R8	hCB1	hCB2
							pKi ± SEM (Ki in nM[a] or displacement at 1 µM[b])	
220	60b	p-hydroxybenzyl	methyl	H	H	methyl	<6.00 (28%)	<<6.00 (−7%)
221	21nf	p-fluorobenzyl	methyl	H	H	methyl	<6.00 (30%)	<<6.00 (−6%)
222	21nh	p-chlorobenzyl	methyl	H	H	methyl	<6.00 (−4%)	<6.00 (12%)
223	21ni	p-(dimethylamino)benzyl	methyl	H	H	methyl	<<6.00 (9%)	<<6.00 (−5%)
224	21oc	p-methoxybenzyl	methyl	methoxy	methyl	methyl	<<6.00 (10%)	<<6.00 (−2%)
225	61a	p-hydroxybenzyl	methyl	hydroxy	methyl	methyl	n.d.[c]	n.d.[c]
226	21oh	p-chlorobenzyl	methyl	methoxy	methyl	methyl	<6.00 (25%)	<6.00 (13%)
227	61b	p-chlorobenzyl	methyl	hydroxy	methyl	methyl	<6.00 (18%)	<<6.00 (1%)
228	21oi	p-(dimethylamino)benzyl	methyl	methoxy	methyl	methyl	<6.00 (17%)	<<6.00 (8%)
229	61c	p-(dimethylamino)benzyl	methyl	hydroxy	methyl	methyl	<6.00 (28%)	<6.00 (24%)
230	16na	H	methyl	H	H	methyl	n.d.	n.d.
231	16nb	methyl	methyl	H	H	methyl	n.d.	n.d.
232	16nc	ethyl	methyl	H	H	methyl	~6.00 (53%)	<<6.00 (0%)

Appendix

Entry	Cpd.	R3	R5	R6	R7	R8	hCB1	hCB2
							$pK_i \pm$ SEM (K_i in nM[a] or displacement at 1 µM[b])	
233	16nd	propyl	methyl	H	H	methyl	n.d.	n.d.
234	16ne	butyl	methyl	H	H	methyl	~6.00 (49%)	<<6.00 (−13%)
235	16nf	pentyl	methyl	H	H	methyl	<6.00 (42%)	<<6.00 (−2%)
236	16ng	hexyl	methyl	H	H	methyl	n.d.	n.d.
237	16nh	iPr	methyl	H	H	methyl	<6.00 (38%)	<<6.00 (−2%)
238	16oc	ethyl	methyl	methoxy	methyl	methyl	~6.00 (49%)	<<6.00 (−11%)
239	63a	ethyl	methyl	hydroxy	methyl	methyl	<<6.00 (−4%)	<<6.00 (−7%)
240	16od	propyl	methyl	methoxy	methyl	methyl	n.d.	n.d.
241	63b	propyl	methyl	hydroxy	methyl	methyl	n.d.	n.d.
242	16oe	butyl	methyl	methoxy	methyl	methyl	<<6.00 (−16%)	<<6.00 (1%)
243	63c	butyl	methyl	hydroxy	methyl	methyl	<<6.00 (−5%)	<<6.00 (−2%)
245	16of	pentyl	methyl	methoxy	methyl	methyl	n.d.	n.d.
246	63of	pentyl	methyl	hydroxy	methyl	methyl	n.d.	n.d.
247	16og	hexyl	methyl	methoxy	methyl	methyl	<6.00 (18%)	<<6.00 (8%)

En-try	Cpd.	R3	R5	R6	R7	R8	hCB1	hCB2
							$pK_i \pm$ SEM (K_i in nM[a] or displacement at 1 μM[b])	
248	630g	hexyl	methyl	hydroxy	methyl	methyl	<6.00 (15%)	<<6.00 (3%)
249	16oh	iPr	methyl	methoxy	methyl	methyl	<6.00 (18%)	<6.00 (15%)
250	630h	iPr	methyl	hydroxy	methyl	methyl	<<6.00 (6%)	<<6.00 (−17%)

n.d. = not determined; [a] Data from at least three individual experiments in duplicates; [b] Data from at least two individual experiments in duplicates; [c] insufficient purity; [d] insufficient product stability; [e] see literature [99]; [f] insufficient quantity available; [g] see literature [100].

A

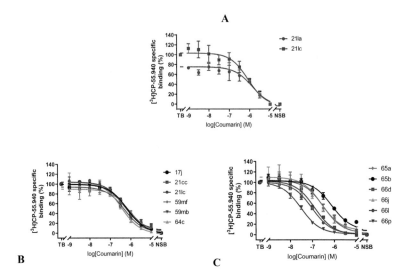

B   **C**

Figure 46: Competitive concentration-dependent curves of [^{3}H]-CP55.940 displacement vs. coumarin ligands at hCB$_1$ (**A**) and hCB$_2$ (**B** and **C**), respectively. Data expressed as mean ± SEM of at least three individual experiments in duplicates. NSB at CB$_1$ was set to highest ligand concentration (NSB = log −5).

6.5   Functional properties

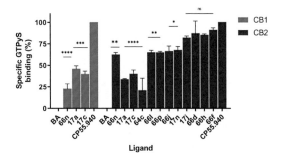

Figure 47: [^{35}S]GTPγS binding assays at hCB$_1$ (green) and hCB$_2$ (red) for E$_{max}$ determination. Data expresses as means ± SEM in relation to the maximal effect of CP55.940 and 1 μM ligand concentrations. Ligands were ordered according to their statistical significance compared to CP55.940. Statistics were performed using a one-way ANOVA with Dunnett's post-test for multicomparison analysis, ns = not significant, * p<0.05, ** p<0.01, *** p<0.001, **** p<0.0001.

6.6   Metabolic stability

Standard curve determination by LC-MS/MS for metabolic stability data quantification.

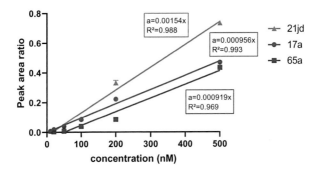

Figure 48: Linear regression of **17a**, **21jd** and **64a** with slope factor (a) and coefficient of determination (R^2).

Table 23: Recorded standard curve LC-MS/MS data acquired for **17a**, **65a** and **21jd**.

Entry	Conc. [nM]	Area Std.	Area Comp.	Peak Area Ratio	SD Std. [%]	SD Comp. [%]	SD Peak Area Ratio [%]
1	10	115745194	893902	0,00761	11	25	16
2	20	122361536	2366408	0,01932	5	11	9
3	50	117540985	4745686	0,04043	4	2	3
4	100	122992868	10557196	0,08578	3	6	3
5	200	121629461	26981108	0,22200	3	3	4
6	500	116328144	54622016	0,46910	4	7	4

Entry	Conc. [nm]	Area Std.	Area Comp.	Peak Area Ratio	SD Std. [%]	SD Comp. [%]	SD Peak Area Ratio [%]
1	25	112772045	350198	0,003387	21	26	40
2	50	108401744	824829	0,007619	9	10	8
3	100	107645290	4221773	0,039572	8	5	12
4	200	115130304	9761171	0,085989	15	6	11
5	500	112294023	48452844	0,433915	12	8	5
6	1000	104109139	2469947	0,023477	8	21	14

Entry	Conc. [nm]	Area Std.	Area Comp.	Peak Area Ratio	SD [%]	Std.	SD Comp. [%]	SD Area [%]	Peak Ratio
1	10	89820070	759490	0,008455	2		8	8	
2	25	87971587	1184194	0,013438	7		9	3	
3	50	90163676	2714092	0,030149	3		3	6	
4	100	86351460	7835758	0,091362	5		8	13	
5	200	89579012	29910842	0,333325	4		10	7	
6	500	93628925	68729203	0,734228	6		6	2	

6.7 List of Abbreviations

(c)AMP (cyclic) adenosine mono phosphate

(c)Hex (cyclo)hexane

[(HMIm)BF₄] 1-methy-1H-imidazol-3-ium tetrafluoroborate

[³⁵S]GTPγS 35S labelled guanosine 5'-O-[gamma-thio]triphosphate

2-AG 2-arachidonoyl glycerol

3D three dimensional

3-NBA 3-nitrobenzyl alcohol

Å angstrom

AA amino acid

AA arachidonic acid

ABC alcohol-binding channel

ABHD12 α/β-hydrolase domain containing 12

ABHD6 α/β-hydrolase domain 6

ABP acyl-chain binding pocket

ABP activity-based probe

ABPP activity-based protein profiling

abs absolute

AC adenylyl cyclase

ACN acetonitrile

ADME Absorption, distribution, metabolism and elimination

AEA anandamide

AlCl₃ aluminium chloride

Alloc allyloxycarbonyl

AlMe₃ trimethyl aluminium

aq aqueous

BBB blood brain barrier

BBr₃ boron tribromide

BINAP 2,2'–Bis(diphenyl-phosphino)–1,1'–binaphthyl

Boc tert-butyloxycarbonyl

Boc₂O tert-butyloxycarbonyl anhydride

bs broad signal

CB₁/CB₁R cannabinoid receptor 1

CB₂/CB₂R cannabinoid receptor 2

CB₃/CB₃R cannabinoid receptor 3

CBD cannabidiol

CBR cannabinoid receptor

Cbz benzyloxycarbonyl

CDI carbodiimidazole

CF₃ trifluoromethyl group

CHCl₃ chloroform

Clint intrinsic clearance

CLogP calculated partition coefficient

CNS central nervous system

CoA Coenzyme A

COX₂ cyclooxygenase 2

Cpd Compound

Cs₂CO₃ caesium carbonate

CYP450 cytochrome P450

d doublet

DAG diacylglycerol

DAGL(α/β) diacylglycerol lipase α or β

DCM dichloromethane

ddH2O bidistilled water

DDHD2 DDHD domain 2

DEPT distortionless enhancement by polarisation transfer

DIPEA N,N-diisopropylethanolamine

DMAP 4-dimethylaminopyridine

DMF dimethylformamide

dppf 1,1'-Bis(diphenyl-phosphino)ferrocene

EC endocannabinoid

EC₅₀ half max effective concentration

eCB endogenous cannabinoid

ECS endocannabinoid system

EET epoxyeicosatrienoic acid

EI electron ionisation

em. emission

E$_{max}$ maximum efficacy

ERK extracellular regulated kinase

ESI electron spray ionisation

EST estrone sulfotransferase

Et₂O diethyl ether

Et₃N triethylamine

EtOH ethanol

ex. excitation

FAAH fatty acid amide hydrolase

FAB fast atom bombardment

FAK focal adhesion kinase

g gravitational force equivalent

GABA γ-aminobutyric acid

GC-MS gas chromatography-mass spectrometry

GI gastrointestinal

GK   glycerol kinase

GPCR   G protein-coupled receptor

GPO   glycerol-3-phosphate oxidase

GPR55   G protein-coupled receptor 55

GRK   G protein-coupled receptor kinases

GST   glutathione-S-transferase

h   hour

H_2O water

H_2O_2 hydrogen peroxide

HATU   hexafluorophosphate azabenzotriazole tetramethyl uronium

HBMC   heteronuclear multiple bond correlation

hept   heptet

hERG   human Ether-a-go-go related gene

HIV   human immunodeficiency viruses

HOAc   acetic acid

HPLC   high performance liquid chromatography

HRMS   high resolution mass spectrometry

HRP   horseradish peroxidase

HSQC   heteronuclear single quantum coherence

HUVEC   human umbilical vein endothelial cells

ID1/3   DNA binding proteins 1/3

IL   ionic liquid

iPR   isopropyl

iPrMgCl·LiCl   isopropylmagnesium chloride lithium chloride complex

IR   infrared

IRK   inward-rectifier potassium channel

J coupling constant

k   elimination rate constant

K_2CO_3 potassium carbonate

K_d dissociation constant

kDa   kilo Dalton

KHMDS   potassium bis(trimethyl-silyl)amide

K_i inhibitory constant

KOAc   potassium acetate

K_{obs} observed rate constants

LC-MS   liquid chromatography-mass spectrometry

LC-MS/MS   liquid chromatography-tandem mass spectrometry

LipE   Lipophilic efficiency

LPI   lysophosphatidylinositol

LPS   lipopolysaccharide

LYPLA2 acyl-protein thioesterase 2

m meta

m   multiplet

M.W.   molecular weight

MAGL   monoacylglycerol lipase

MAPK   mitogen-activated protein kinase

max   maximal/maximum

MeI   methyl iodine

MeOH   methanol

$MgCl_2$ magnesium chloride

min   minutes

mPGES-1   microsomal prosta-glandin E synthase 1

mRNA   messenger RNA

n.d.   not determined

Na_2SO_4 sodium sulphate

NADPH   nicotinamide adenine dinucleotide phosphate

NAH   northern aliphatic hydroxyl

$NaHCO_3$ sodium bicarbonate

NaI   sodium iodine

NaOtBu sodium tert-butoxide

NAPE   N-acyl-phosphatidyl-ethanolamine

NAPE-PLD   NAPE phospholipase D

NAT   N-acyl transferase

n-BuLi n-butyl lithium

NHC   N-heterocyclic carbene

NMR   nuclear magnetic resonance

ns   not significant

NSA   natural substrate assay

o ortho

p para

PAPS   3'-phosphoadenosine-5'-phosphosulfate

PC   phosphatidylcholine

$Pd_2(dba)_3$ tris(dibenzylideneacetone)dipalladium(0)

$PdOAc_2$ palladium (II) acetate

PE   phosphatidylethanolamine

p-ERK 1/2   phosphorylated extra-cellular signal-regulated kinase 1/2

PG   protecting group

PGE2   prostaglandin E2

PGS   prostaglandin synthase

PH   phenolic hydroxyl

PKA   protein kinase A

PL   phospholipids

PL-PLC PL phospholipase C

$POCl_3$ phosphoryl chloride

PST   phenol sulfotransferase

PTX   pertussis toxin

q   quartet

quin   quintet

s   singlet

SAH   southern aliphatic hydroxyl

SAR   structure-activity relationship

SC   lipophilic alkyl side chain

SD   standard deviation

SDS-PAGE   sodium dodecyl sulphate polyacrylamide gel electrophoresis

SEM   standard error of the mean

SET   single electron transfer

sex   sextet

SH   serine hydrolase

SH   serine hydrolases

S_nAr nucleophilic aromatic substitution

$SOCl_2$ thionylchloride

sol   solution

t   triplet

$t_{1/2}$ half-life time

TAMRA-FP   5-carboxytetramethylrhodamin-
fluorophosphonate

TB   total binding

TBAI   tetrabutylammonium iodine

TFA   trifluoroacetic acid

THC Δ^9-tetrahydrocannabinol

THF   tetrahydrofuran

THL   tetrahydrolipstatin

$TiCl_4$ titanium tetrachloride

TM   transition metal

TMEDA N,N,N,N-tetramethyl-ethylenedia-
mine

tPSA   total polar surface area

UDGPA uridine diphosphate glucuronic acid

UGT   UDP-dependent glucuronosyl transferase

UNODC   United Nations Office on Drugs
and Crime

V   volume

vs   versus

6.8 Crystallographic data

All crystal structures were measured and analysed by Dr. MARTIN NIEGER at the institute of inorganic chemistry, university of Helsinki, Finland.

4-Hydroxy-2,6-dimethoxy-benzaldehyde (38) – SB970_HY

Crystal data

$C_9H_{10}O_4$	$F(000) = 384$
$M_r = 182.17$	$D_x = 1.446$ Mg m^{-3}
Monoclinic, $P2_1/c$ (no.14)	Cu $K\alpha$ radiation, $\lambda = 1.54178$ Å
$a = 7.5096$ (2) Å	Cell parameters from 6021 reflections
$b = 14.8128$ (4) Å	$\theta = 5.9–72.0°$
$c = 7.5254$ (2) Å	$\mu = 0.97$ mm^{-1}
$\beta = 92.104$ (1)°	$T = 123$ K
$V = 836.55$ (4) Å^3	Blocks, orange
$Z = 4$	$0.22 \times 0.20 \times 0.10$ mm

Data collection

Bruker D8 VENTURE diffractometer with Photon100 detector	1640 independent reflections
Radiation source: INCOATEC microfocus sealed tube	1554 reflections with $I > 2\sigma(I)$
Detector resolution: 10.4167 pixels mm^{-1}	$R_{int} = 0.021$
rotation in ϕ and ω, 1°, shutterless scans	$\theta_{max} = 72.1°$, $\theta_{min} = 5.9°$
Absorption correction: multi-scan SADABS (Sheldrick, 2014)	$h = -8\rightarrow9$
$T_{min} = 0.830$, $T_{max} = 0.929$	$k = -18\rightarrow18$
7119 measured reflections	$l = -9\rightarrow8$

Refinement

Refinement on F^2	Primary atom site location: structure-invariant direct methods
Least-squares matrix: full	Secondary atom site location: difference Fourier map
$R[F^2 > 2\sigma(F^2)] = 0.033$	Hydrogen site location: difference Fourier map
$wR(F^2) = 0.093$	H atoms treated by a mixture of independent and constrained refinement
$S = 1.06$	$w = 1/[\sigma^2(F_o^2) + (0.0538P)^2 + 0.2051P]$ where $P = (F_o^2 + 2F_c^2)/3$
1640 reflections	$(\Delta/\sigma)_{max} < 0.001$
123 parameters	$\Delta\rangle_{max} = 0.23$ e Å^{-3}
1 restraint	$\Delta\rangle_{min} = -0.24$ e Å^{-3}

2,6-Dimethoxy-4-propoxybenzaldehyde (36c) – SB1109_HY

Crystal data

$C_{12}H_{16}O_4$	$Z = 2$
$M_r = 224.25$	$F(000) = 240$
Triclinic, $P\text{-}1$ (no.2)	$D_x = 1.312$ Mg m^{-3}
$a = 8.2879$ (3) Å	Cu $K\alpha$ radiation, $\lambda = 1.54178$ Å
$b = 8.4184$ (3) Å	Cell parameters from 6555 reflections
$c = 9.2296$ (3) Å	$\theta = 5.8–72.2°$
$\alpha = 75.811$ (1)°	$\mu = 0.81$ mm^{-1}
$\beta = 86.099$ (1)°	$T = 123$ K
$\gamma = 65.464$ (1)°	Plates, colourless
$V = 567.52$ (3) Å^3	$0.40 \times 0.30 \times 0.06$ mm

Data collection

Bruker D8 VENTURE diffractometer with PhotonII CPAD detector	2132 reflections with $I > 2\sigma(I)$
Radiation source: INCOATEC microfocus sealed tube	$R_{int} = 0.024$
rotation in ϕ and ω, 1°, shutterless scans	$\theta_{max} = 72.2°$, $\theta_{min} = 5.9°$
Absorption correction: multi-scan $SADABS$ (SAheldrick, 2014)	$h = -10 \rightarrow 7$
$T_{min} = 0.771$, $T_{max} = 0.958$	$k = -10 \rightarrow 10$
7295 measured reflections	$l = -11 \rightarrow 11$
2219 independent reflections	

Refinement

Refinement on F^2	Primary atom site location: dual
Least-squares matrix: full	Secondary atom site location: difference Fourier map
$R[F^2 > 2\sigma(F^2)] = 0.033$	Hydrogen site location: difference Fourier map
$wR(F^2) = 0.091$	H-atom parameters constrained
$S = 1.05$	$w = 1/[\sigma^2(F_o^2) + (0.0456P)^2 + 0.1443P]$ where $P = (F_o^2 + 2F_c^2)/3$
2219 reflections	$(\Delta/\sigma)_{max} < 0.001$
147 parameters	$\Delta\rangle_{max} = 0.24$ e Å^{-3}
0 restraints	$\Delta\rangle_{min} = -0.18$ e Å^{-3}

2,6-Dimethoxy-4-(pentyloxy)benzaldehyde (36e) – SB1108_HY

Crystal data

$C_{14}H_{20}O_4$	$F(000) = 544$
$M_r = 252.30$	$D_x = 1.252$ Mg m^{-3}
Monoclinic, $P2_1/n$ *(no.14)*	Cu $K\alpha$ radiation, $\lambda = 1.54178$ Å
$a = 7.5452$ (4) Å	Cell parameters from 9921 reflections
$b = 14.3227$ (7) Å	$\theta = 4.7–72.1°$
$c = 12.4453$ (6) Å	$\mu = 0.74$ mm^{-1}
$\beta = 95.778$ (2)°	$T = 123$ K
$V = 1338.10$ (12) Å^3	Plates, colourless
$Z = 4$	$0.20 \times 0.16 \times 0.04$ mm

Data collection

Bruker D8 VENTURE diffractometer with PhotonII CPAD detector	2547 reflections with $I > 2\sigma(I)$
Radiation source: INCOATEC microfocus sealed tube	$R_{int} = 0.025$
rotation in ϕ and ω, 1°, shutterless scans	$\theta_{max} = 72.1°$, $\theta_{min} = 4.7°$
Absorption correction: multi-scan *SADABS* (Sheldrick, 2014)	$h = -9 \rightarrow 9$
$T_{min} = 0.907$, $T_{max} = 0.971$	$k = -17 \rightarrow 17$
16026 measured reflections	$l = -15 \rightarrow 15$
2640 independent reflections	

Refinement

Refinement on F^2	Secondary atom site location: difference Fourier map
Least-squares matrix: full	Hydrogen site location: difference Fourier map
$R[F^2 > 2\sigma(F^2)] = 0.031$	H-atom parameters constrained
$wR(F^2) = 0.085$	$w = 1/[\sigma^2(F_o^2) + (0.040P)^2 + 0.3803P]$ where $P = (F_o^2 + 2F_c^2)/3$
$S = 1.08$	$(\Delta/\sigma)_{max} = 0.001$
2640 reflections	$\Delta\rangle_{max} = 0.27$ e Å^{-3}
166 parameters	$\Delta\rangle_{min} = -0.16$ e Å^{-3}
0 restraints	Extinction correction: *SHELXL2014/7* (Sheldrick 2014), Fc*=kFc[1+0.001xFc$^2\lambda^3$/sin(2θ)]$^{-1/4}$
Primary atom site location: dual	Extinction coefficient: 0.0031 (5)

2-Hydroxy-6-methoxy-4-propoxybenzaldehyde (19c) – SB1099_HY

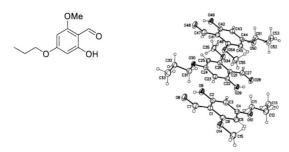

Crystal data

$C_{11}H_{14}O_4$	$F(000) = 1344$
$M_r = 210.22$	$D_x = 1.357$ Mg m^{-3}
Monoclinic, $P2_1/c$ *(no.14)*	Cu $K\alpha$ radiation, $\lambda = 1.54178$ Å
$a = 10.3401$ (3) Å	Cell parameters from 7656 reflections
$b = 14.8540$ (4) Å	$\theta = 3.7–72.1°$
$c = 20.2520$ (5) Å	$\mu = 0.86$ mm^{-1}
$\beta = 97.225$ (1)°	$T = 123$ K
$V = 3085.84$ (14) Å^3	Plates, colourless
$Z = 12$	$0.16 \times 0.12 \times 0.04$ mm

Data collection

Bruker D8 VENTURE diffractometer with PhotonII CPAD detector	4336 reflections with $I > 2\sigma(I)$
Radiation source: INCOATEC microfocus sealed tube	$R_{int} = 0.039$
rotation in ϕ and ω, 1°, shutterless scans	$\theta_{max} = 72.1°$, $\theta_{min} = 3.7°$
Absorption correction: multi-scan *SADABS* (Sheldrick, 2014)	$h = -12 \rightarrow 10$
$T_{min} = 0.773$, $T_{max} = 0.971$	$k = -18 \rightarrow 18$
22081 measured reflections	$l = -24 \rightarrow 24$
6071 independent reflections	

Refinement

Refinement on F^2	Primary atom site location: dual
Least-squares matrix: full	Secondary atom site location: difference Fourier map
$R[F^2 > 2\sigma(F^2)] = 0.049$	Hydrogen site location: difference Fourier map
$wR(F^2) = 0.151$	H atoms treated by a mixture of independent and constrained refinement
$S = 1.03$	$w = 1/[\sigma^2(F_o^2) + (0.078P)^2 + 1.0158P]$ where $P = (F_o^2 + 2F_c^2)/3$
6071 reflections	$(\Delta/\sigma)_{max} < 0.001$
418 parameters	$\Delta\rangle_{max} = 0.28$ e Å^{-3}
3 restraints	$\Delta\rangle_{min} = -0.26$ e Å^{-3}

4-chloro-5,7-dimethoxy-3-methyl-2-pentylbenzo[b]thiophene (52b) – SB1191_HY

Crystal data

$C_{16}H_{21}ClO_2S$	$F(000) = 664$
$M_r = 312.84$	$D_x = 1.346$ Mg m^{-3}
Monoclinic, $P2_1/n$ (no.14)	Cu $K\alpha$ radiation, $\lambda = 1.54178$ Å
$a = 6.7192$ (2) Å	Cell parameters from 9681 reflections
$b = 16.6992$ (4) Å	$\theta = 4.1–72.1°$
$c = 13.8637$ (3) Å	$\mu = 3.44$ mm^{-1}
$\beta = 97.083$ (1)°	$T = 123$ K
$V = 1543.71$ (7) Å^3	Plates, colourless
$Z = 4$	$0.22 \times 0.16 \times 0.06$ mm

Data collection

Bruker D8 VENTURE diffractometer with PhotonII CPAD detector	2954 reflections with $I > 2\sigma(I)$
Radiation source: INCOATEC microfocus sealed tube	$R_{int} = 0.027$
rotation in ϕ and ω, 1°, shutterless scans	$\theta_{max} = 72.2°$, $\theta_{min} = 4.2°$
Absorption correction: multi-scan $SADABS$ (Sheldrick, 2014)	$h = -8 \rightarrow 8$
$T_{min} = 0.639$, $T_{max} = 0.806$	$k = -20 \rightarrow 20$
22899 measured reflections	$l = -17 \rightarrow 17$
3043 independent reflections	

Refinement

Refinement on F^2	Primary atom site location: dual
Least-squares matrix: full	Secondary atom site location: difference Fourier map
$R[F^2 > 2\sigma(F^2)] = 0.036$	Hydrogen site location: difference Fourier map
$wR(F^2) = 0.091$	H-atom parameters constrained
$S = 1.15$	$w = 1/[\sigma^2(F_o^2) + (0.0377P)^2 + 1.1275P]$ where $P = (F_o^2 + 2F_c^2)/3$
3043 reflections	$(\Delta/\sigma)_{max} = 0.001$
184 parameters	$\Delta\rangle_{max} = 0.39$ e Å^{-3}
0 restraints	$\Delta\rangle_{min} = -0.19$ e Å^{-3}

3-Benzyl-5,7-dimethoxy-2*H*-chromen-2-one (21aa) – SB1073_HY

Crystal data

$C_{18}H_{16}O_4$	$F(000) = 624$
$M_r = 296.31$	$D_x = 1.370$ Mg m^{-3}
Monoclinic, $P2_1/c$ *(no.14)*	Cu $K\alpha$ radiation, $\lambda = 1.54178$ Å
$a = 9.6215$ (3) Å	Cell parameters from 7864 reflections
$b = 21.5952$ (8) Å	$\theta = 4.0–72.1°$
$c = 6.9964$ (2) Å	$\mu = 0.79$ mm^{-1}
$\beta = 98.681$ (2)°	$T = 123$ K
$V = 1437.05$ (8) Å^3	Plates, colourless
$Z = 4$	$0.18 \times 0.10 \times 0.04$ mm

Data collection

Bruker D8 VENTURE diffractometer with Photon100 detector	2816 independent reflections
Radiation source: INCOATEC microfocus sealed tube	2519 reflections with $I > 2\sigma(I)$
Detector resolution: 10.4167 pixels mm^{-1}	$R_{int} = 0.030$
rotation in ϕ and ω, 1°, shutterless scans	$\theta_{max} = 72.1°$, $\theta_{min} = 4.1°$
Absorption correction: multi-scan *SADABS* (Sheldrick, 2014)	$h = -11 \rightarrow 11$
$T_{min} = 0.863$, $T_{max} = 0.971$	$k = -15 \rightarrow 26$
11642 measured reflections	$l = -8 \rightarrow 8$

Refinement

Refinement on F^2	Primary atom site location: dual
Least-squares matrix: full	Secondary atom site location: difference Fourier map
$R[F^2 > 2\sigma(F^2)] = 0.040$	Hydrogen site location: difference Fourier map
$wR(F^2) = 0.098$	H-atom parameters constrained
$S = 1.08$	$w = 1/[\sigma^2(F_o^2) + (0.0363P)^2 + 0.9013P]$ where $P = (F_o^2 + 2F_c^2)/3$
2816 reflections	$(\Delta/\sigma)_{max} < 0.001$
201 parameters	$\Delta\rangle_{max} = 0.21$ e Å^{-3}
0 restraints	$\Delta\rangle_{min} = -0.24$ e Å^{-3}

5,7-Dimethoxy-3-(2-methylbenzyl)-2H-chromen-2-one (21ab) – SB1113_HY

Crystal data

$C_{19}H_{18}O_4$	$F(000) = 656$
$M_r = 310.33$	$D_x = 1.354$ Mg m^{-3}
Monoclinic, $P2_1/c$ (no.14)	Cu $K\alpha$ radiation, $\lambda = 1.54178$ Å
$a = 10.7217$ (2) Å	Cell parameters from 9956 reflections
$b = 17.3385$ (4) Å	$\theta = 2.5–72.1°$
$c = 8.9354$ (2) Å	$\mu = 0.77$ mm^{-1}
$\beta = 113.619$ (1)°	$T = 123$ K
$V = 1521.93$ (6) Å^3	Blocks, colourless
$Z = 4$	$0.40 \times 0.25 \times 0.20$ mm

Data collection

Bruker D8 VENTURE diffractometer with PhotonII CPAD detector	2910 reflections with $I > 2\sigma(I)$
Radiation source: INCOATEC microfocus sealed tube	$R_{int} = 0.021$
rotation in ϕ and ω, 1°, shutterless scans	$\theta_{max} = 72.1°$, $\theta_{min} = 4.5°$
Absorption correction: multi-scan SADABS (Sheldrick, 2014)	$h = -13 \rightarrow 13$
$T_{min} = 0.746$, $T_{max} = 0.841$	$k = -21 \rightarrow 21$
12112 measured reflections	$l = -9 \rightarrow 10$
2948 independent reflections	

Refinement

Refinement on F^2	Primary atom site location: dual
Least-squares matrix: full	Secondary atom site location: difference Fourier map
$R[F^2 > 2\sigma(F^2)] = 0.036$	Hydrogen site location: difference Fourier map
$wR(F^2) = 0.095$	H-atom parameters constrained
$S = 1.05$	$w = 1/[\sigma^2(F_o^2) + (0.0459P)^2 + 0.6144P]$ where $P = (F_o^2 + 2F_c^2)/3$
2948 reflections	$(\Delta/\sigma)_{max} < 0.001$
211 parameters	$\Delta\rangle_{max} = 0.26$ e Å^{-3}
0 restraints	$\Delta\rangle_{min} = -0.17$ e Å^{-3}

3-(4-(Dimethylamino)benzyl)-5,7-dimethoxy-2H-chromen-2-one (21ai) – SB1063_HY

Crystal data

$C_{20}H_{21}NO_4$	$F(000) = 1440$
$M_r = 339.38$	$D_x = 1.368$ Mg m^{-3}
Monoclinic, $C2/c$ (no.15)	Cu $K\alpha$ radiation, $\lambda = 1.54178$ Å
$a = 12.8795$ (3) Å	Cell parameters from 9960 reflections
$b = 10.6083$ (2) Å	$\theta = 3.6–72.0°$
$c = 24.2618$ (5) Å	$\mu = 0.78$ mm^{-1}
$\beta = 96.304$ (1)°	$T = 123$ K
$V = 3294.84$ (12) Å^3	Blocks, yellow
$Z = 8$	$0.22 \times 0.18 \times 0.06$ mm

Data collection

Bruker D8 VENTURE diffractometer with Photon100 detector	3237 independent reflections
Radiation source: INCOATEC microfocus sealed tube	2990 reflections with $I > 2\sigma(I)$
Detector resolution: 10.4167 pixels mm^{-1}	$R_{int} = 0.030$
rotation in ϕ and ω, 1°, shutterless scans	$\theta_{max} = 72.0°$, $\theta_{min} = 3.7°$
Absorption correction: multi-scan $SADABS$ (Sheldrick, 2014)	$h = -15 \rightarrow 15$
$T_{min} = 0.818$, $T_{max} = 0.958$	$k = -12 \rightarrow 13$
16282 measured reflections	$l = -27 \rightarrow 29$

Refinement

Refinement on F^2	Secondary atom site location: difference Fourier map
Least-squares matrix: full	Hydrogen site location: difference Fourier map
$R[F^2 > 2\sigma(F^2)] = 0.036$	H-atom parameters constrained
$wR(F^2) = 0.097$	$w = 1/[\sigma^2(F_o^2) + (0.0444P)^2 + 2.7705P]$ where $P = (F_o^2 + 2F_c^2)/3$
$S = 1.06$	$(\Delta/\sigma)_{max} < 0.001$
3237 reflections	$\Delta\rangle_{max} = 0.27$ e Å^{-3}
231 parameters	$\Delta\rangle_{min} = -0.19$ e Å^{-3}
0 restraints	Extinction correction: $SHELXL2014/7$ (Sheldrick, 2014), Fc*=kFc[1+0.001xFc$^2\lambda^3$/sin(2θ)]$^{-1/4}$
Primary atom site location: dual	Extinction coefficient: 0.00065 (10)

7-Ethoxy-5-methoxy-3-(2-methylbenzyl)-2H-chromen-2-one (21bb) – SB1112_HY

Crystal data

$C_{20}H_{20}O_4$	$Z = 4$
$M_r = 324.36$	$F(000) = 688$
Triclinic, P-1 (no.2)	$D_x = 1.318$ Mg m^{-3}
$a = 9.7735$ (2) Å	Cu $K\alpha$ radiation, $\lambda = 1.54178$ Å
$b = 10.0463$ (2) Å	Cell parameters from 9870 reflections
$c = 16.9400$ (4) Å	$\theta = 2.6–72.0°$
$\alpha = 97.859$ (1)°	$\mu = 0.74$ mm^{-1}
$\beta = 96.138$ (1)°	$T = 123$ K
$\gamma = 92.798$ (1)°	Plates, colourless
$V = 1634.83$ (6) Å^3	$0.30 \times 0.10 \times 0.03$ mm

Data collection

Bruker D8 VENTURE diffractometer with PhotonII CPAD detector	5208 reflections with $I > 2\sigma(I)$
Radiation source: INCOATEC microfocus sealed tube	$R_{int} = 0.037$
rotation in ϕ and ω, 1°, shutterless scans	$\theta_{max} = 72.1°$, $\theta_{min} = 2.7°$
Absorption correction: multi-scan $SADABS$ (Sheldrick, 2014)	$h = -12 \rightarrow 12$
$T_{min} = 0.817$, $T_{max} = 0.971$	$k = -12 \rightarrow 12$
25313 measured reflections	$l = -20 \rightarrow 20$
6390 independent reflections	

Refinement

Refinement on F^2	Primary atom site location: dual
Least-squares matrix: full	Secondary atom site location: difference Fourier map
$R[F^2 > 2\sigma(F^2)] = 0.039$	Hydrogen site location: difference Fourier map
$wR(F^2) = 0.110$	H-atom parameters constrained
$S = 1.03$	$w = 1/[\sigma^2(F_o^2) + (0.0512P)^2 + 0.4704P]$ where $P = (F_o^2 + 2F_c^2)/3$
6390 reflections	$(\Delta/\sigma)_{max} = 0.001$
437 parameters	$\Delta\rangle_{max} = 0.24$ e Å^{-3}
0 restraints	$\Delta\rangle_{min} = -0.19$ e Å^{-3}

3-(4-(Dimethylamino)benzyl)-7-ethoxy-5-methoxy-2*H*-chromen-2-one　　(21bi)　　–
SB1064_HY

Crystal data

$C_{21}H_{23}NO_4$	$F(000) = 752$
$M_r = 353.40$	$D_x = 1.313$ Mg m^{-3}
Monoclinic, $P2_1/n$ *(no.14)*	Cu $K\alpha$ radiation, $\lambda = 1.54178$ Å
$a = 5.4374$ (3) Å	Cell parameters from 7571 reflections
$b = 23.3296$ (13) Å	$\theta = 3.6–72.0°$
$c = 14.1487$ (7) Å	$\mu = 0.74$ mm^{-1}
$\beta = 94.816$ (3)°	$T = 123$ K
$V = 1788.46$ (17) Å^3	Plates, colourless
$Z = 4$	$0.24 \times 0.14 \times 0.03$ mm

Data collection

Bruker D8 VENTURE diffractometer with Photon100 detector	3439 independent reflections
Radiation source: INCOATEC microfocus sealed tube	2925 reflections with $I > 2\sigma(I)$
Detector resolution: 10.4167 pixels mm^{-1}	$R_{int} = 0.051$
rotation in ω, °, shutterless scans	$\theta_{max} = 72.0°$, $\theta_{min} = 3.7°$
Absorption correction: multi-scan *SADABS* (Sheldrick, 2014)	$h = -5 \rightarrow 6$
$T_{min} = 0.747$, $T_{max} = 0.971$	$k = -28 \rightarrow 28$
11412 measured reflections	$l = -17 \rightarrow 16$

Refinement

Refinement on F^2	Primary atom site location: dual
Least-squares matrix: full	Secondary atom site location: difference Fourier map
$R[F^2 > 2\sigma(F^2)] = 0.085$	Hydrogen site location: difference Fourier map
$wR(F^2) = 0.199$	H-atom parameters constrained
$S = 1.14$	$w = 1/[\sigma^2(F_o^2) + (0.0416P)^2 + 4.5805P]$ where $P = (F_o^2 + 2F_c^2)/3$
3439 reflections	$(\Delta/\sigma)_{max} < 0.001$
238 parameters	$\Delta\rangle_{max} = 0.36$ e Å^{-3}
0 restraints	$\Delta\rangle_{min} = -0.25$ e Å^{-3}

3-Benzyl-5,8-dimethyl-2*H*-chromen-2-one (21na) – SB1000_HY

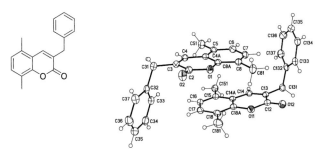

Crystal data

C$_{18}$H$_{16}$O$_2$	D_x = 1.280 Mg m^{-3}
M_r = 264.31	Cu $K\alpha$ radiation, λ = 1.54178 Å
Orthorhombic, *Pbca (no.61)*	Cell parameters from 9865 reflections
a = 12.2821 (3) Å	θ = 3.5–72.0°
b = 14.0154 (3) Å	μ = 0.65 mm^{-1}
c = 31.8629 (7) Å	T = 123 K
V = 5484.8 (2) Å^3	Blocks, colourless
Z = 16	0.36 × 0.24 × 0.16 mm
$F(000)$ = 2240	

Data collection

Bruker D8 VENTURE diffractometer with Photon100 detector	5377 independent reflections
Radiation source: INCOATEC microfocus sealed tube	4701 reflections with $I > 2\sigma(I)$
Detector resolution: 10.4167 pixels mm^{-1}	R_{int} = 0.029
rotation in ϕ and ω, 1°, shutterless scans	θ_{max} = 72.1°, θ_{min} = 2.8°
Absorption correction: multi-scan *SADABS* (Sheldrick, 2014)	h = -14→15
T_{min} = 0.813, T_{max} = 0.889	k = -17→15
27608 measured reflections	l = -39→36

Refinement

Refinement on F^2	Secondary atom site location: difference Fourier map
Least-squares matrix: full	Hydrogen site location: difference Fourier map
$R[F^2 > 2\sigma(F^2)]$ = 0.040	H-atom parameters constrained
$wR(F^2)$ = 0.109	$w = 1/[\sigma^2(F_o^2) + (0.0558P)^2 + 1.8376P]$ where $P = (F_o^2 + 2F_c^2)/3$
S = 1.05	$(\Delta/\sigma)_{max}$ = 0.001
5377 reflections	$\Delta\rangle_{max}$ = 0.25 e Å^{-3}
366 parameters	$\Delta\rangle_{min}$ = -0.18 e Å^{-3}
0 restraints	Extinction correction: *SHELXL2014*/7 (Sheldrick 2014, Fc*=kFc[1+0.001xFc$^2\lambda^3$/sin(2θ)]$^{-1/4}$
Primary atom site location: structure-invariant direct methods	Extinction coefficient: 0.00061 (7)

3-(4-Methoxybenzyl)-5,8-dimethyl-2H-chromen-2-one (21nd) – SB1003_HY

Crystal data

$C_{19}H_{18}O_3$	$F(000) = 624$
$M_r = 294.33$	$D_x = 1.336$ Mg m^{-3}
Monoclinic, $P2_1/c$ (no.14)	Cu $K\alpha$ radiation, $\lambda = 1.54178$ Å
$a = 7.1007$ (3) Å	Cell parameters from 9836 reflections
$b = 9.5067$ (4) Å	$\theta = 4.6–72.1°$
$c = 21.7864$ (8) Å	$\mu = 0.72$ mm^{-1}
$\beta = 95.856$ (2)°	$T = 123$ K
$V = 1463.00$ (10) Å^3	Blocks, colourless
$Z = 4$	0.28 × 0.22 × 0.10 mm

Data collection

Bruker D8 VENTURE diffractometer with Photon100 detector	2880 independent reflections
Radiation source: INCOATEC microfocus sealed tube	2670 reflections with $I > 2\sigma(I)$
Detector resolution: 10.4167 pixels mm^{-1}	$R_{int} = 0.022$
rotation in ϕ and ω, 1°, shutterless scans	$\theta_{max} = 72.1°$, $\theta_{min} = 5.1°$
Absorption correction: multi-scan $SADABS$ (Sheldrick, 2014)	$h = -8\rightarrow8$
$T_{min} = 0.869$, $T_{max} = 0.929$	$k = -11\rightarrow11$
12825 measured reflections	$l = -26\rightarrow25$

Refinement

Refinement on F^2	Secondary atom site location: difference Fourier map
Least-squares matrix: full	Hydrogen site location: difference Fourier map
$R[F^2 > 2\sigma(F^2)] = 0.035$	H-atom parameters constrained
$wR(F^2) = 0.095$	$w = 1/[\sigma^2(F_o^2) + (0.0514P)^2 + 0.464P]$ where $P = (F_o^2 + 2F_c^2)/3$
$S = 1.04$	$(\Delta/\sigma)_{max} = 0.001$
2880 reflections	$\Delta\rangle_{max} = 0.26$ e Å^{-3}
203 parameters	$\Delta\rangle_{min} = -0.20$ e Å^{-3}
0 restraints	Extinction correction: $SHELXL2014/7$ (Sheldrick, 2014, Fc*=kFc[1+0.001xFc$^2\lambda^3$/sin(2θ)]$^{-1/4}$
Primary atom site location: structure-invariant direct methods	Extinction coefficient: 0.0047 (5)

3-(2-Hydroxybenzyl)-5,8-dimethyl-2*H*-chromen-2-one (60a) – SB1142_HY

Crystal data

$C_{18}H_{16}O_3$	$F(000) = 592$
$M_r = 280.31$	$D_x = 1.316$ Mg m^{-3}
Monoclinic, $P2_1/n$ *(no.14)*	Cu $K\alpha$ radiation, $\lambda = 1.54178$ Å
$a = 8.0253$ (2) Å	Cell parameters from 9862 reflections
$b = 14.2334$ (4) Å	$\theta = 4.7–72.0°$
$c = 12.3961$ (4) Å	$\mu = 0.72$ mm^{-1}
$\beta = 92.755$ (1)°	$T = 123$ K
$V = 1414.34$ (7) Å^3	Plates, colourless
$Z = 4$	$0.20 \times 0.08 \times 0.04$ mm

Data collection

Bruker D8 VENTURE diffractometer with PhotonII CPAD detector	2625 reflections with $I > 2\sigma(I)$
Radiation source: INCOATEC microfocus sealed tube	$R_{int} = 0.026$
rotation in ϕ and ω, 1°, shutterless scans	$\theta_{max} = 72.1°$, $\theta_{min} = 4.7°$
Absorption correction: multi-scan *SADABS* (Sheldrick, 2014)	$h = -8\rightarrow9$
$T_{min} = 0.891$, $T_{max} = 0.971$	$k = -17\rightarrow17$
18954 measured reflections	$l = -15\rightarrow15$
2779 independent reflections	

Refinement

Refinement on F^2	Primary atom site location: dual
Least-squares matrix: full	Secondary atom site location: difference Fourier map
$R[F^2 > 2\sigma(F^2)] = 0.033$	Hydrogen site location: difference Fourier map
$wR(F^2) = 0.092$	H atoms treated by a mixture of independent and constrained refinement
$S = 1.04$	$w = 1/[\sigma^2(F_o^2) + (0.050P)^2 + 0.4299P]$ where $P = (F_o^2 + 2F_c^2)/3$
2779 reflections	$(\Delta/\sigma)_{max} = 0.001$
195 parameters	$\Delta\rangle_{max} = 0.24$ e Å^{-3}
1 restraint	$\Delta\rangle_{min} = -0.20$ e Å^{-3}

3-(4-Hydroxybenzyl)-5,8-dimethyl-2*H*-chromen-2-one (60b) – SB1040_HY

Crystal data

$C_{18}H_{16}O_3$	$F(000) = 1184$
$M_r = 280.31$	$D_x = 1.301$ Mg m^{-3}
Monoclinic, $C2/c$ (*no.15*)	Cu $K\alpha$ radiation, $\lambda = 1.54178$ Å
$a = 23.7753 (5)$ Å	Cell parameters from 9510 reflections
$b = 9.6447 (2)$ Å	$\theta = 4.0–72.0°$
$c = 13.6109 (3)$ Å	$\mu = 0.71$ mm^{-1}
$\beta = 113.501 (1)°$	$T = 123$ K
$V = 2862.17 (11)$ Å^3	Blocks, colourless
$Z = 8$	$0.18 \times 0.16 \times 0.04$ mm

Data collection

Bruker D8 VENTURE diffractometer with PhotonII CPAD detector	2580 reflections with $I > 2\sigma(I)$
Radiation source: INCOATEC microfocus sealed tube	$R_{int} = 0.025$
rotation in ϕ and ω, 1°, shutterless scans	$\theta_{max} = 72.1°$, $\theta_{min} = 4.1°$
Absorption correction: multi-scan *SADABS* (Sheldrick, 2014)	$h = -29 \rightarrow 29$
$T_{min} = 0.745$, $T_{max} = 0.971$	$k = -11 \rightarrow 11$
15190 measured reflections	$l = -16 \rightarrow 16$
2814 independent reflections	

Refinement

Refinement on F^2	Primary atom site location: dual
Least-squares matrix: full	Secondary atom site location: difference Fourier map
$R[F^2 > 2\sigma(F^2)] = 0.034$	Hydrogen site location: difference Fourier map
$wR(F^2) = 0.088$	H atoms treated by a mixture of independent and constrained refinement
$S = 1.04$	$w = 1/[\sigma^2(F_o^2) + (0.0375P)^2 + 1.6894P]$ where $P = (F_o^2 + 2F_c^2)/3$
2814 reflections	$(\Delta/\sigma)_{max} = 0.001$
195 parameters	$\Delta\rangle_{max} = 0.21$ e Å^{-3}
1 restraint	$\Delta\rangle_{min} = -0.14$ e Å^{-3}

3-(4-(Dimethylamino)benzyl)-6-hydroxy-5,7,8-trimethyl-2H-chromen-2-one   (62c)   – SB1141_HY

Crystal data

$C_{21}H_{23}NO_3$	$F(000) = 720$
$M_r = 337.40$	$D_x = 1.231$ Mg m^{-3}
Monoclinic, $P2_1/c$ (no.14)	Cu $K\alpha$ radiation, $\lambda = 1.54178$ Å
$a = 9.8250$ (4) Å	Cell parameters from 9975 reflections
$b = 27.1692$ (10) Å	$\theta = 4.5–72.1°$
$c = 6.8672$ (2) Å	$\mu = 0.66$ mm^{-1}
$\beta = 96.587$ (2)°	$T = 123$ K
$V = 1821.01$ (11) Å^3	Plates, orange
$Z = 4$	$0.24 \times 0.20 \times 0.04$ mm

Data collection

Bruker D8 VENTURE diffractometer with PhotonII CPAD detector	3379 reflections with $I > 2\sigma(I)$
Radiation source: INCOATEC microfocus sealed tube	$R_{int} = 0.031$
rotation in ϕ and ω, 1°, shutterless scans	$\theta_{max} = 72.2°$, $\theta_{min} = 4.5°$
Absorption correction: multi-scan $SADABS$ (Sheldrick, 2014)	$h = -12 \rightarrow 12$
$T_{min} = 0.813$, $T_{max} = 0.971$	$k = -33 \rightarrow 31$
21475 measured reflections	$l = -8 \rightarrow 7$
3593 independent reflections	

Refinement

Refinement on F^2	Primary atom site location: dual
Least-squares matrix: full	Secondary atom site location: difference Fourier map
$R[F^2 > 2\sigma(F^2)] = 0.054$	Hydrogen site location: difference Fourier map
$wR(F^2) = 0.139$	H atoms treated by a mixture of independent and constrained refinement
$S = 1.08$	$w = 1/[\sigma^2(F_o^2) + (0.0492P)^2 + 1.7614P]$ where $P = (F_o^2 + 2F_c^2)/3$
3593 reflections	$(\Delta/\sigma)_{max} = 0.001$
234 parameters	$\Delta\rangle_{max} = 0.32$ e Å^{-3}
1 restraint	$\Delta\rangle_{min} = -0.30$ e Å^{-3}

5,7-Dimethoxy-3-methyl-2*H*-chromen-2-one (16ab) – SB1116_HY

Crystal data

$C_{12}H_{12}O_4$	$F(000) = 464$
$M_r = 220.22$	$D_x = 1.391$ Mg m^{-3}
Monoclinic, $P2_1/m$ (no.11)	Cu $K\alpha$ radiation, $\lambda = 1.54178$ Å
$a = 10.5586$ (3) Å	Cell parameters from 7597 reflections
$b = 6.6523$ (2) Å	$\theta = 4.1–72.1°$
$c = 14.9904$ (5) Å	$\mu = 0.88$ mm^{-1}
$\beta = 92.630$ (2)°	$T = 123$ K
$V = 1051.80$ (6) Å^3	Plates, colourless
$Z = 4$	$0.18 \times 0.06 \times 0.02$ mm

Data collection

Bruker D8 VENTURE diffractometer with PhotonII CPAD detector	2066 reflections with $I > 2\sigma(I)$
Radiation source: INCOATEC microfocus sealed tube	$R_{int} = 0.023$
rotation in ϕ and ω, 1°, shutterless scans	$\theta_{max} = 72.1°$, $\theta_{min} = 3.0°$
Absorption correction: multi-scan *SADABS* (Sheldrick, 2014)	$h = -12 \rightarrow 13$
$T_{min} = 0.881$, $T_{max} = 0.971$	$k = -8 \rightarrow 8$
11878 measured reflections	$l = -18 \rightarrow 18$
2260 independent reflections	

Refinement

Refinement on F^2	Primary atom site location: dual
Least-squares matrix: full	Secondary atom site location: difference Fourier map
$R[F^2 > 2\sigma(F^2)] = 0.034$	Hydrogen site location: mixed
$wR(F^2) = 0.093$	H-atom parameters constrained
$S = 1.06$	$w = 1/[\sigma^2(F_o^2) + (0.0485P)^2 + 0.3114P]$ where $P = (F_o^2 + 2F_c^2)/3$
2260 reflections	$(\Delta/\sigma)_{max} < 0.001$
195 parameters	$\Delta\rangle_{max} = 0.28$ e Å^{-3}
0 restraints	$\Delta\rangle_{min} = -0.18$ e Å^{-3}

5,7-Dimethoxy-3-pentyl-2*H*-chromen-2-one (16af) – SB1067_HY

Crystal data

$C_{16}H_{20}O_4$	$Z = 2$
$M_r = 276.32$	$F(000) = 296$
Triclinic, P-1 (no.2)	$D_x = 1.315$ Mg m^{-3}
$a = 8.3142$ (4) Å	Mo $K\alpha$ radiation, $\lambda = 0.71073$ Å
$b = 9.0804$ (4) Å	Cell parameters from 9837 reflections
$c = 9.8581$ (4) Å	$\theta = 2.5–27.5°$
$\alpha = 95.923$ (2)°	$\mu = 0.09$ mm^{-1}
$\beta = 108.652$ (2)°	$T = 123$ K
$\gamma = 93.619$ (2)°	Blocks, colourless
$V = 697.74$ (5) Å^3	$0.45 \times 0.30 \times 0.15$ mm

Data collection

Bruker D8 VENTURE diffractometer with Photon100 detector	3208 independent reflections
Radiation source: INCOATEC microfocus sealed tube	2838 reflections with $I > 2\sigma(I)$
Detector resolution: 10.4167 pixels mm^{-1}	$R_{int} = 0.025$
rotation in ϕ and ω, 1°, shutterless scans	$\theta_{max} = 27.5°$, $\theta_{min} = 2.2°$
Absorption correction: multi-scan *SADABS* (Sheldrick, 2014)	$h = -10\rightarrow10$
$T_{min} = 0.945$, $T_{max} = 0.991$	$k = -11\rightarrow11$
28859 measured reflections	$l = -12\rightarrow12$

Refinement

Refinement on F^2	Primary atom site location: dual
Least-squares matrix: full	Secondary atom site location: difference Fourier map
$R[F^2 > 2\sigma(F^2)] = 0.035$	Hydrogen site location: difference Fourier map
$wR(F^2) = 0.105$	H-atom parameters constrained
$S = 1.03$	$w = 1/[\sigma^2(F_o^2) + (0.0592P)^2 + 0.2035P]$ where $P = (F_o^2 + 2F_c^2)/3$
3208 reflections	$(\Delta/\sigma)_{max} < 0.001$
183 parameters	$\Delta\rangle_{max} = 0.32$ e Å^{-3}
0 restraints	$\Delta\rangle_{min} = -0.30$ e Å^{-3}

3-Hexyl-5,7-dimethoxy-2*H*-chromen-2-one (16ag) – SB1066_HY

Crystal data

$C_{17}H_{22}O_4$	$Z = 2$
$M_r = 290.34$	$F(000) = 312$
Triclinic, P-1 (no.2)	$D_x = 1.275$ Mg m^{-3}
$a = 9.1094$ (5) Å	Mo $K\alpha$ radiation, $\lambda = 0.71073$ Å
$b = 9.3706$ (5) Å	Cell parameters from 9875 reflections
$c = 9.8885$ (4) Å	$\theta = 2.3–27.5°$
$\alpha = 110.559$ (2)°	$\mu = 0.09$ mm^{-1}
$\beta = 98.182$ (2)°	$T = 123$ K
$\gamma = 100.836$ (2)°	Blocks, colourless
$V = 756.05$ (7) Å^3	$0.36 \times 0.20 \times 0.10$ mm

Data collection

Bruker D8 VENTURE diffractometer with Photon100 detector	3479 independent reflections
Radiation source: INCOATEC microfocus sealed tube	2878 reflections with $I > 2\sigma(I)$
Detector resolution: 10.4167 pixels mm^{-1}	$R_{int} = 0.029$
rotation in ϕ and ω, 1°, shutterless scans	$\theta_{max} = 27.6°$, $\theta_{min} = 2.3°$
Absorption correction: multi-scan *SADABS* (Sheldrick, 2014)	$h = -11\rightarrow11$
$T_{min} = 0.922$, $T_{max} = 0.993$	$k = -12\rightarrow12$
32473 measured reflections	$l = -12\rightarrow11$

Refinement

Refinement on F^2	Primary atom site location: dual
Least-squares matrix: full	Secondary atom site location: difference Fourier map
$R[F^2 > 2\sigma(F^2)] = 0.037$	Hydrogen site location: difference Fourier map
$wR(F^2) = 0.109$	H-atom parameters constrained
$S = 1.06$	$w = 1/[\sigma^2(F_o^2) + (0.0582P)^2 + 0.1781P]$ where $P = (F_o^2 + 2F_c^2)/3$
3479 reflections	$(\Delta/\sigma)_{max} < 0.001$
192 parameters	$\Delta\rangle_{max} = 0.32$ e Å^{-3}
0 restraints	$\Delta\rangle_{min} = -0.25$ e Å^{-3}

7-Ethoxy-5-methoxy-2H-chromen-2-one (16ba) – SB1095_HY

Crystal data

$C_{12}H_{12}O_4$	$F(000) = 464$
$M_r = 220.22$	$D_x = 1.401$ Mg m^{-3}
Monoclinic, $P2_1/n$ (no.14)	Cu $K\alpha$ radiation, $\lambda = 1.54178$ Å
$a = 6.8618$ (3) Å	Cell parameters from 6449 reflections
$b = 10.5143$ (4) Å	$\theta = 5.2–72.1°$
$c = 14.8369$ (5) Å	$\mu = 0.88$ mm^{-1}
$\beta = 102.767$ (2)°	$T = 123$ K
$V = 1043.97$ (7) Å^3	Plates, yellow
$Z = 4$	$0.20 \times 0.12 \times 0.06$ mm

Data collection

Bruker D8 VENTURE diffractometer with PhotonII CPAD detector	1863 reflections with $I > 2\sigma(I)$
Radiation source: INCOATEC microfocus sealed tube	$R_{int} = 0.028$
rotation in ϕ and ω, 1°, shutterless scans	$\theta_{max} = 72.2°$, $\theta_{min} = 5.2°$
Absorption correction: multi-scan $SADABS$ (Sheldrick, 2014)	$h = -8\rightarrow8$
$T_{min} = 0.840$, $T_{max} = 0.958$	$k = -12\rightarrow12$
8197 measured reflections	$l = -15\rightarrow18$
2059 independent reflections	

Refinement

Refinement on F^2	Primary atom site location: dual
Least-squares matrix: full	Secondary atom site location: difference Fourier map
$R[F^2 > 2\sigma(F^2)] = 0.037$	Hydrogen site location: difference Fourier map
$wR(F^2) = 0.107$	H-atom parameters constrained
$S = 1.05$	$w = 1/[\sigma^2(F_o^2) + (0.0526P)^2 + 0.3438P]$ where $P = (F_o^2 + 2F_c^2)/3$
2059 reflections	$(\Delta/\sigma)_{max} < 0.001$
146 parameters	$\Delta\rangle_{max} = 0.27$ e Å^{-3}
0 restraints	$\Delta\rangle_{min} = -0.17$ e Å^{-3}

7-Ethoxy-5-methoxy-3-methyl-2*H*-chromen-2-one (16bb) – SB1096_HY

Crystal data

$C_{13}H_{14}O_4$	$D_x = 1.363$ Mg m^{-3}
$M_r = 234.24$	Cu $K\alpha$ radiation, $\lambda = 1.54178$ Å
Orthorhombic, *Pnma (no.62)*	Cell parameters from 5510 reflections
$a = 10.4222$ (4) Å	$\theta = 5.3–72.2°$
$b = 6.5685$ (2) Å	$\mu = 0.84$ mm^{-1}
$c = 16.6746$ (5) Å	$T = 123$ K
$V = 1141.51$ (7) Å^3	Plates, colourless
$Z = 4$	$0.16 \times 0.10 \times 0.06$ mm
$F(000) = 496$	

Data collection

Bruker D8 VENTURE diffractometer with PhotonII CPAD detector	1128 reflections with $I > 2\sigma(I)$
Radiation source: INCOATEC microfocus sealed tube	$R_{int} = 0.027$
rotation in ϕ and ω, 1°, shutterless scans	$\theta_{max} = 72.2°$, $\theta_{min} = 5.3°$
Absorption correction: multi-scan *SADABS* (Sheldrick, 2014)	$h = -12 \rightarrow 12$
$T_{min} = 0.834$, $T_{max} = 0.958$	$k = -8 \rightarrow 7$
7315 measured reflections	$l = -18 \rightarrow 20$
1221 independent reflections	

Refinement

Refinement on F^2	Primary atom site location: dual
Least-squares matrix: full	Secondary atom site location: difference Fourier map
$R[F^2 > 2\sigma(F^2)] = 0.034$	Hydrogen site location: mixed
$wR(F^2) = 0.094$	H-atom parameters constrained
$S = 1.07$	$w = 1/[\sigma^2(F_o^2) + (0.0501P)^2 + 0.3369P]$ where $P = (F_o^2 + 2F_c^2)/3$
1221 reflections	$(\Delta/\sigma)_{max} < 0.001$
104 parameters	$\Delta\rangle_{max} = 0.27$ e Å^{-3}
0 restraints	$\Delta\rangle_{min} = -0.17$ e Å^{-3}

5-Methoxy-7-propoxy-2*H*-chromen-2-one (16ca) – SB1093_HY

Crystal data

$C_{13}H_{14}O_4$	$F(000) = 496$
$M_r = 234.24$	$D_x = 1.356$ Mg m^{-3}
Monoclinic, $P2_1/c$ *(no.14)*	Cu $K\alpha$ radiation, $\lambda = 1.54178$ Å
$a = 7.0654$ (3) Å	Cell parameters from 4294 reflections
$b = 10.4517$ (4) Å	$\theta = 5.1–72.2°$
$c = 15.7246$ (5) Å	$\mu = 0.84$ mm^{-1}
$\beta = 98.835$ (2)°	$T = 123$ K
$V = 1147.41$ (8) Å^3	Plates, colourless
$Z = 4$	$0.20 \times 0.10 \times 0.06$ mm

Data collection

Bruker D8 VENTURE diffractometer with PhotonII CPAD detector	2251 independent reflections
Radiation source: INCOATEC microfocus sealed tube	1832 reflections with $I > 2\sigma(I)$
rotation in ω, 1°, shutterless scans	$\theta_{max} = 72.2°$, $\theta_{min} = 5.1°$
Absorption correction: multi-scan *SADABS* (Sheldrick, 2014)	$h = -8{\rightarrow}8$
$T_{min} = 0.804$, $T_{max} = 0.958$	$k = -12{\rightarrow}12$
2251 measured reflections	$l = -11{\rightarrow}19$

Refinement

Refinement on F^2	Primary atom site location: dual
Least-squares matrix: full	Secondary atom site location: difference Fourier map
$R[F^2 > 2\sigma(F^2)] = 0.042$	Hydrogen site location: difference Fourier map
$wR(F^2) = 0.109$	H-atom parameters constrained
$S = 1.05$	$w = 1/[\sigma^2(F_o^2) + (0.045P)^2 + 0.4968P]$ where $P = (F_o^2 + 2F_c^2)/3$
2251 reflections	$(\Delta/\sigma)_{max} < 0.001$
156 parameters	$\Delta\rangle_{max} = 0.23$ e Å^{-3}
0 restraints	$\Delta\rangle_{min} = -0.18$ e Å^{-3}

3-Ethyl-5,8-dimethyl-2H-chromen-2-one (16nc) – SB1156_HY

Crystal data

$C_{13}H_{14}O_2$	$Z = 2$
$M_r = 202.24$	$F(000) = 216$
Triclinic, P-1 (no.2)	$D_x = 1.304$ Mg m^{-3}
$a = 7.3713$ (3) Å	Cu $K\alpha$ radiation, $\lambda = 1.54178$ Å
$b = 8.9869$ (4) Å	Cell parameters from 7022 reflections
$c = 9.1513$ (4) Å	$\theta = 5.3–72.1°$
$\alpha = 94.457$ (1)°	$\mu = 0.69$ mm^{-1}
$\beta = 111.906$ (1)°	$T = 123$ K
$\gamma = 109.686$ (1)°	Blocks, colourless
$V = 515.03$ (4) Å^3	$0.14 \times 0.08 \times 0.03$ mm

Data collection

Bruker D8 VENTURE diffractometer with PhotonII CPAD detector	1917 reflections with $I > 2\sigma(I)$
Radiation source: INCOATEC microfocus sealed tube	$R_{int} = 0.023$
rotation in ϕ and ω, 1°, shutterless scans	$\theta_{max} = 72.1°$, $\theta_{min} = 5.4°$
Absorption correction: multi-scan $SADABS$ (Sheldrick, 2014)	$h = -9 \rightarrow 9$
$T_{min} = 0.844$, $T_{max} = 0.971$	$k = -11 \rightarrow 11$
8812 measured reflections	$l = -11 \rightarrow 10$
2014 independent reflections	

Refinement

Refinement on F^2	Primary atom site location: dual
Least-squares matrix: full	Secondary atom site location: difference Fourier map
$R[F^2 > 2\sigma(F^2)] = 0.034$	Hydrogen site location: difference Fourier map
$wR(F^2) = 0.103$	H-atom parameters constrained
$S = 1.07$	$w = 1/[\sigma^2(F_o^2) + (0.0604P)^2 + 0.1007P]$ where $P = (F_o^2 + 2F_c^2)/3$
2014 reflections	$(\Delta/\sigma)_{max} < 0.001$
138 parameters	$\Delta\rangle_{max} = 0.25$ e Å^{-3}
0 restraints	$\Delta\rangle_{min} = -0.20$ e Å^{-3}

5,8-Dimethyl-3-propyl-2H-chromen-2-one (16nd) – SB1155_HY

Crystal data

$C_{14}H_{16}O_2$	$Z = 2$
$M_r = 216.27$	$F(000) = 232$
Triclinic, P-1 (no.2)	$D_x = 1.259$ Mg m^{-3}
$a = 6.9871$ (3) Å	Cu $K\alpha$ radiation, $\lambda = 1.54178$ Å
$b = 8.8733$ (4) Å	Cell parameters from 6489 reflections
$c = 9.3862$ (4) Å	$\theta = 4.7$–72.1°
$\alpha = 91.289$ (2)°	$\mu = 0.66$ mm^{-1}
$\beta = 90.349$ (2)°	$T = 123$ K
$\gamma = 101.202$ (2)°	Blocks, colourless
$V = 570.67$ (4) Å^3	$0.18 \times 0.06 \times 0.04$ mm

Data collection

Bruker D8 VENTURE diffractometer with PhotonII CPAD detector	2081 reflections with $I > 2\sigma(I)$
Radiation source: INCOATEC microfocus sealed tube	$R_{int} = 0.023$
rotation in ϕ and ω, 1°, shutterless scans	$\theta_{max} = 72.3°$, $\theta_{min} = 4.7°$
Absorption correction: multi-scan $SADABS$ (Sheldrick, 2014)	$h = -8 \rightarrow 8$
$T_{min} = 0.880$, $T_{max} = 0.971$	$k = -10 \rightarrow 9$
8287 measured reflections	$l = -10 \rightarrow 11$
2240 independent reflections	

Refinement

Refinement on F^2	Primary atom site location: dual
Least-squares matrix: full	Secondary atom site location: difference Fourier map
$R[F^2 > 2\sigma(F^2)] = 0.036$	Hydrogen site location: difference Fourier map
$wR(F^2) = 0.104$	H-atom parameters constrained
$S = 1.08$	$w = 1/[\sigma^2(F_o^2) + (0.0566P)^2 + 0.1191P]$ where $P = (F_o^2 + 2F_c^2)/3$
2240 reflections	$(\Delta/\sigma)_{max} < 0.001$
147 parameters	$\Delta\rangle_{max} = 0.25$ e Å^{-3}
0 restraints	$\Delta\rangle_{min} = -0.20$ e Å^{-3}

3-Isopropyl-5,8-dimethyl-2*H*-chromen-2-one (16nh) – SB1154_HY

Crystal data

$C_{14}H_{16}O_2$	$F(000) = 464$
$M_r = 216.27$	$D_x = 1.246$ Mg m^{-3}
Monoclinic, $P2_1/n$ *(no.14)*	Cu $K\alpha$ radiation, $\lambda = 1.54178$ Å
$a = 7.4151$ (2) Å	Cell parameters from 9026 reflections
$b = 13.0836$ (3) Å	$\theta = 6.8–72.1°$
$c = 11.8942$ (3) Å	$\mu = 0.65$ mm^{-1}
$\beta = 92.883$ (1)°	$T = 123$ K
$V = 1152.47$ (5) Å^3	Blocks, colourless
$Z = 4$	$0.20 \times 0.16 \times 0.06$ mm

Data collection

Bruker D8 VENTURE diffractometer with PhotonII CPAD detector	2228 reflections with $I > 2\sigma(I)$
Radiation source: INCOATEC microfocus sealed tube	$R_{int} = 0.024$
rotation in ϕ and ω, 1°, shutterless scans	$\theta_{max} = 72.1°$, $\theta_{min} = 6.9°$
Absorption correction: multi-scan *SADABS* (Sheldrick, 2014)	$h = -9 \rightarrow 9$
$T_{min} = 0.832$, $T_{max} = 0.958$	$k = -16 \rightarrow 16$
20422 measured reflections	$l = -14 \rightarrow 14$
2257 independent reflections	

Refinement

Refinement on F^2	Primary atom site location: dual
Least-squares matrix: full	Secondary atom site location: difference Fourier map
$R[F^2 > 2\sigma(F^2)] = 0.035$	Hydrogen site location: difference Fourier map
$wR(F^2) = 0.095$	H-atom parameters constrained
$S = 1.05$	$w = 1/[\sigma^2(F_o^2) + (0.0461P)^2 + 0.3989P]$ where $P = (F_o^2 + 2F_c^2)/3$
2257 reflections	$(\Delta/\sigma)_{max} < 0.001$
147 parameters	$\Delta\rangle_{max} = 0.26$ e Å^{-3}
0 restraints	$\Delta\rangle_{min} = -0.17$ e Å^{-3}

6-Hydroxy-3-isopropyl-5,7,8-trimethyl-2*H*-chromen-2-one (63f) – SB1144_HY

Crystal data

$C_{15}H_{18}O_3$	$D_x = 1.254$ Mg m^{-3}
$M_r = 246.29$	Cu $K\alpha$ radiation, $\lambda = 1.54178$ Å
Orthorhombic, *Pbca (no.61)*	Cell parameters from 9918 reflections
$a = 6.9945$ (3) Å	$\theta = 4.3–72.0°$
$b = 18.3344$ (9) Å	$\mu = 0.70$ mm^{-1}
$c = 20.3418$ (11) Å	$T = 123$ K
$V = 2608.6$ (2) Å^3	Blocks, colourless
$Z = 8$	$0.18 \times 0.12 \times 0.06$ mm
$F(000) = 1056$	

Data collection

Bruker D8 VENTURE diffractometer with PhotonII CPAD detector	2353 reflections with $I > 2\sigma(I)$
Radiation source: INCOATEC microfocus sealed tube	$R_{int} = 0.051$
rotation in ϕ and ω, 1°, shutterless scans	$\theta_{max} = 72.1°$, $\theta_{min} = 4.4°$
Absorption correction: multi-scan *SADABS* (Sheldrick, 2014)	$h = -8 \rightarrow 8$
$T_{min} = 0.776$, $T_{max} = 0.958$	$k = -22 \rightarrow 21$
22308 measured reflections	$l = -12 \rightarrow 25$
2561 independent reflections	

Refinement

Refinement on F^2	Primary atom site location: dual
Least-squares matrix: full	Secondary atom site location: difference Fourier map
$R[F^2 > 2\sigma(F^2)] = 0.051$	Hydrogen site location: difference Fourier map
$wR(F^2) = 0.138$	H atoms treated by a mixture of independent and constrained refinement
$S = 1.07$	$w = 1/[\sigma^2(F_o^2) + (0.0575P)^2 + 2.2692P]$ where $P = (F_o^2 + 2F_c^2)/3$
2561 reflections	$(\Delta/\sigma)_{max} < 0.001$
169 parameters	$\Delta\rangle_{max} = 0.30$ e Å^{-3}
1 restraint	$\Delta\rangle_{min} = -0.21$ e Å^{-3}

6.9 Curriculum Vitae

Florian Mohr
fl.mohr@gmail.com | ORCID iD 0000-0002-3613-3273

EXPERIENCE

Leiden University   **Mar 2019 – Jul 2019**
Leiden Academic Centre for Drug Research   Leiden, the Netherlands
Research internship in the group of Prof. Laura Heitman in the field of GPCR pharmacology

- Final pharmacological evaluation and SAR-studies of modified coumarins (~220 compounds) as drug candidates against cannabinoid receptors
  - Trained skills in molecular pharmacology including cell culture, radioligand binding assays and radioactive material handling, bioassay analysis, result interpretation and biostatistics

Leiden University   **Feb 2018 – Aug 2018**
Leiden Institute of Chemistry   Leiden, the Netherlands
Research internship in the group of Prof. Mario van der Stelt in the field of molecular pharmacology

- Multi-step organic synthesis, purification and characterization of synthesized compounds, pharmacological evaluation by enzymatic assays and activity-based protein profiling, interpretation of structure-activity relationships

Karlsruhe Institute of Technology   **Nov 2016 – present**
Institute of Organic Chemistry   Karlsruhe, Germany
PhD studies in the group of Prof. Stefan Bräse in the field of Medicinal Chemistry and drug development

- Design, Synthesis and pharmacological evaluation of coumarin-based synthetic cannabinoids as new drug candidates against the CB receptors
  - Structure design, planning of multistep synthesis routes, execution, purification and characterization of synthesized compounds

BASF   **Aug 2014 – Sep 2014**
Global Research & Development CP   Limburgerhof, Germany
Summer research student in the group of Dr. David Schaffert in the field of Formylation Herbicide and Nucleic acids

- Formylation of RNA based herbicides

Fraunhofer Institute for Chemical Technology   **Mai 2012 – Sep 2015**
Department of Energetic Materials   Pfinztal, Germany
Student research assistant in the group of Dipl.-Ing. Thomas Heintz in the field of Particle Technology

- Handling and process engineering of energetic materials

Fraunhofer Institute for Chemical Technology   **Jan 2009 – Sep 2011**
Department of Energetic Materials   Pfinztal, Germany
Research assistant in the group of Dipl.-Ing. Thomas Heintz in the field of Particle Technology

- Handling and process engineering of energetic materials

EDUCATION

Karlsruhe Institute of Technology   Oct 2014 – Aug 2016
Study of Chemical Biology (Master of Science) Graduation mark 1,0   Karlsruhe
- Focus on Organic Chemistry with specialization to Medicinal Chemistry
- Master thesis in Organic Chemistry: Development of a new semi synthesis of marine steroids Gorgosterol and Demethylgorgosterol (Mark 1.0)

Karlsruhe Institute of Technology   Oct 2011 – Aug 2014
Study of Chemical Biology (Bachelor of Science) Graduation mark 1,5   Karlsruhe
- Bachelor thesis in Organic Chemistry: Investigation of new photo stabilized cyanine dyes fir *in vivo* detection of RNA interference (Mark 1.0)

PROFESSIONAL SKILLS

Chemistry
- Structure design and optimization
- Multi-step organic syntheses (conventional, microwave supported)

Medicinal Chemistry
- SAR, ADME, pharmacological and physicochemical properties

Biology
- Cell culture
- Radioligand binding or enzymatic assays in membrane preparations or proteomes
- Kinetic or end-point measurements

Analytics
- NMR spectroscopy
- IR spectroscopy
- Mass spectrometry (GC-, LC-MS, tandem-MS)
- HPLC (analytic and preparative)
- (Confocal) cell microscopy
- Bioassay analysis (Fluorescence or radioactive)
- Biostatistics

Other
- Process engineering micro encapsulation, particle dispersing
- Particle characterization (particle size and distribution, specific surface by BET, water content)
- MSOffice, ChemOffice, Mestrenova, GraphPad Prism

SCHOLARSHIPS & FUNDING

- PhD scholarship: Graduate Funding of the German States (awarded only 7 scholarships)
- Research Travel Grant from the Karlsruhe House of Young Scientists
- Conference scholarship from German Academic Exchange Service (DAAD)

LEADERSHIP EXPERIENCE

- Guidance of apprentices during practical education
- Supervision of students in their practical organic chemistry courses A or B
- Supervision and training of 4 Bachelor students in their Bachelor thesis

LANGUAGES

- **German:** native
- **English:** fluent

REFERENCES

- Prof. Dr. Stefan Bräse, Institute of Organic Chemistry and Institute of Toxicology and Genetics, Karlsruhe Institute of Technology, Germany, Email: stefan.braese@kit.edu, Phone: +49 721 608 42902
- Prof. Dr. Mario van der Stelt, Leiden Institute of Chemistry, Leiden University, The Netherlands, Email: m.van.der.stelt@chem.leidenuniv.nl, Phone: +31 71 527 4768
- Prof. Dr. Laura Heitman, Leiden Academic Centre for Drug Research, Leiden University, The Netherlands, Email: l.h.heitman@lacdr.leidenuniv.nl, Phone: +31 71 527 4558

PUBLICATIONS & CONFERENCE CONTRIBUTIONS

Publications

1. S. W. Saliba, H. Jauch, B. Gargouri, A. Keil, T. Hurrle, N. Volz, F. Mohr, M. van der Stelt, S. Bräse, B. L. Fiebich, *J. Neuroinflammation*, **2018**, 15 (1), 322.

2. *Design, synthesis and pharmacological evaluation of selective coumarin-based cannabonid receptor 2 agonists* (working title)
   F. Mohr, T. Hurrle, L. Burggraaff, M. Bemelmans, G. J. P. van Westen, M. Nieger, M. Van der Stelt, L. Heitman, S. Bräse
   Manuscript in preparation

3. *Design of a modular synthesis of alkoxy modified salicylic aldehydes as building block for the synthesis of coumarins* (working title)
   F. Mohr, M. Nieger, L. Heitman, S. Bräse
   Manuscript in preparation

Conference contribution
Poster

F. Mohr, S. Bräse, "Modular synthesis of novel cannabinoid ligands based substituted coumarins as CB1, CB2, GPR55 agonists and antagonists", Global Young Scientist Summit, 20.-25. January 2019, Singapore.

F. Mohr, S. Bräse, "Modular synthesis of novel cannabinoid ligands based substituted coumarins as CB1, CB2, GPR55 agonists and antagonists", 22nd International Conference on Organic Synthesis, 16.-21. September 2018, Florence, Italy.

F. Mohr, S. Bräse, "Modular synthesis of novel cannabinoid ligands based substituted coumarins as CB1, CB2, GPR55 agonists and antagonists", 28th Annual ICRS Symposium on the Cannabinoids, 01.-04. July 2018, Leiden, Netherlands.

F. Mohr, T. Hurrle, H. Jauch, A. Keil, S. Bräse, B. L. Fiebich, "Modular synthesis of novel cannabinoid ligands based on the coumarin motif as CB1, CB2, GPR55 agonists and antagonists", 27th Annual ICRS Symposium on the Cannabinoids, 22.-27. June 2017, Montréal, Canada.

T. Hurrle, F. Mohr, D. Marcato, A.Hariharan, U. Strähle, R. Peravali, S. Bräse, "Novel cannabinoids based on the coumarin motif and fast evaluation through PMR-study on Embryonic Zebrafish", 27th Annual ICRS Symposium on the Cannabinoids, 22.-27. June 2017, Montréal, Canada.

7 References

[1]   E. L. Abel, *Marihuana: the first twelve thousand years*, Springer Science & Business Media, **2013**.

[2]   H.-L. Li, *Cannabis and culture* **1975**, 51-62.

[3]   M. Aldrich, *McFarland & Co., Inc. Jefferson NC* **1997**, *28640*, 35-55.

[4]   W. O'Shaughnessy, *Trans. Phys. Med. Soc. Bengal* **1838**, *40*, 1839-1972.

[5]   W. B. O'Shaughnessy, *Prov. Med. J. Retrosp. Med. Sci.* **1843**, *5*, 363.

[6]   L. Grinspoon, *Int. J. Drug Policy* **2001**, *12*, 377-383.

[7]   D. F. Musto, *JAMA Psychiatry* **1972**, *26*, 101-108.

[8]   J. L. Blankman, B. F. Cravatt, *Pharmacol. Rev.* **2013**, *65*, 849-871.

[9]   A. Ligresti, L. De Petrocellis, V. Di Marzo, *Physiol. Rev.* **2016**, *96*, 1593-1659.

[10]   U. N. O. o. D. a. Crime, *World Drug Report 2018*, **2018**.

[11]   P. Webster, *Lancet* **2018**, *391*, 725-726.

[12]   S. Lake, T. Kerr, D. Werb, R. Haines-Saah, B. Fischer, G. Thomas, Z. Walsh, M. A. Ware, E. Wood, M.-J. Milloy, *Drug Alcohol Rev.* **2019**, *38*, 606-621.

[13]   B. Blass, *Basic principles of drug discovery and development*, Elsevier, **2015**.

[14]   S. Morgan, P. Grootendorst, J. Lexchin, C. Cunningham, D. Greyson, *Health Policy* **2011**, *100*, 4-17.

[15]   J. A. DiMasi, H. G. Grabowski, R. W. Hansen, *J. Health Econ.* **2016**, *47*, 20-33.

[16]   R. Santos, O. Ursu, A. Gaulton, A. P. Bento, R. S. Donadi, C. G. Bologa, A. Karlsson, B. Al-Lazikani, A. Hersey, T. I. Oprea, J. P. Overington, *Nat. Rev. Drug Discovery* **2016**, *16*, 19.

[17]   S. M. Paul, D. S. Mytelka, C. T. Dunwiddie, C. C. Persinger, B. H. Munos, S. R. Lindborg, A. L. Schacht, *Nat. Rev. Drug Discovery* **2010**, *9*, 203-214.

[18]   J. Drews, *Science* **2000**, *287*, 1960-1964.

[19]   K. H. Bleicher, H.-J. Böhm, K. Müller, A. I. Alanine, *Nat. Rev. Drug Discovery* **2003**, *2*, 369.

[20]   C. H. Wong, K. W. Siah, A. W. Lo, *Biostatistics* **2019**, *20*, 273-286.

[21]   G. Astarita, D. Piomelli, *J. Chromatogr. B* **2009**, *877*, 2755-2767.

[22]   R. J. Silver, *Animals* **2019**, *9*, 686.

[23]   L. A. Matsuda, S. J. Lolait, M. J. Brownstein, A. C. Young, T. I. Bonner, *Nature* **1990**, *346*, 561-564.

[24]   S. Munro, K. L. Thomas, M. Abu-Shaar, *Nature* **1993**, *365*, 61-65.

[25]   Y. Gaoni, R. Mechoulam, *J. Am. Chem. Soc.* **1964**, *86*, 1646-1647.

[26]   A. Howlett, M. R. Johnson, L. Melvin, G. Milne, *Mol. Pharmacol.* **1988**, *33*, 297-302.

[27]   R. G. Pertwee, A. Howlett, M. E. Abood, S. Alexander, V. Di Marzo, M. Elphick, P. Greasley, H. S. Hansen, G. Kunos, K. Mackie, *Pharmacol. Rev.* **2010**, *62*, 588-631.

[28]   R. G. Pertwee, in *Cannabinoids*, Springer, **2005**, pp. 1-51.

[29]   R. Mechoulam, L. A. Parker, *Annu. Rev. Psychol.* **2013**, *64*, 21-47.

[30]   S. C. Azad, K. Monory, G. Marsicano, B. F. Cravatt, B. Lutz, W. Zieglgäns-berger, G. Rammes, *J. Neurosci.* **2004**, *24*, 9953-9961.

[31]   A. Calignano, G. La Rana, A. Giuffrida, D. Piomelli, *Nature* **1998**, *394*, 277.

[32]   V. Di Marzo, I. Matias, *Nat. Neurosci.* **2005**, *8*, 585.

[33]   J. M. Gray, H. A. Vecchiarelli, M. Morena, T. T. Y. Lee, D. J. Hermanson, A. B. Kim, R. J. McLaughlin, K. I. Hassan, C. Kühne, C. T. Wotjak, J. M. Deussing, S. Patel, M. N. Hill, *J. Neurosci.* **2015**, *35*, 3879-3892.

[34]   S. W. Saliba, H. Jauch, B. Gargouri, A. Keil, T. Hurrle, N. Volz, F. Mohr, M. van der Stelt, S. Bräse, B. L. Fiebich, *J. Neuroinflammation* **2018**, *15*, 322.

[35]   Y. Nakajima, Y. Furuichi, K. K. Biswas, T. Hashiguchi, K.-i. Kawahara, K. Yamaji, T. Uchimura, Y. Izumi, I. Maruyama, *FEBS Lett.* **2006**, *580*, 613-619.

[36]   B. F. Cravatt, D. K. Giang, S. P. Mayfield, D. L. Boger, R. A. Lerner, N. B. Gilula, *Nature* **1996**, *384*, 83.

[37]   J. E. Nielsen, A. D. Rolland, E. Rajpert-De Meyts, C. Janfelt, A. Jørgensen, S. B. Winge, D. M. Kristensen, A. Juul, F. Chalmel, B. Jégou, *Sci. Rep.* **2019**, *9*, 1-14.

[38]   K. Mackie, *J. Neuroendocrinol.* **2008**, *20*, 10-14.

[39]   K. L. McCoy, M. Matveyeva, S. J. Carlisle, G. A. Cabral, *J. Pharmacol. Exp. Ther.* **1999**, *289*, 1620-1625.

[40]   P. Carayon, J. Marchand, D. Dussossoy, J.-M. Derocq, O. Jbilo, A. Bord, M. Bouaboula, S. Galiègue, P. Mondière, G. Pénarier, *Blood* **1998**, *92*, 3605-3615.

[41]   A. B. Lynn, M. Herkenham, *J. Pharmacol. Exp. Ther.* **1994**, *268*, 1612-1623.

[42]   P. Pacher, R. Mechoulam, *Prog. Lipid Res.* **2011**, *50*, 193-211.

[43]   G. Griffin, E. J. Wray, Q. Tao, S. D. McAllister, W. K. Rorrer, M. Aung, B. R. Martin, M. E. Abood, *Eur. J. Pharmacol.* **1999**, *377*, 117-125.

[44] A. R. Schatz, M. Lee, R. B. Condie, J. T. Pulaski, N. E. Kaminski, *Toxicol. Appl. Pharmacol.* **1997**, *142*, 278-287.

[45] T. Sugiura, S. Kondo, S. Kishimoto, T. Miyashita, S. Nakane, T. Kodaka, Y. Suhara, H. Takayama, K. Waku, *J. Biol. Chem.* **2000**, *275*, 605-612.

[46] M. D. Van Sickle, M. Duncan, P. J. Kingsley, A. Mouihate, P. Urbani, K. Mackie, N. Stella, A. Makriyannis, D. Piomelli, J. S. Davison, *Science* **2005**, *310*, 329-332.

[47] R. G. Pertwee, *Curr. Med. Chem.* **1999**, *6*, 635-664.

[48] E. S. Onaivi, H. Ishiguro, J. P. GONG, S. Patel, A. Perchuk, P. A. Meozzi, L. Myers, Z. Mora, P. Tagliaferro, E. Gardner, *Ann. N.Y. Acad. Sci.* **2006**, *1074*, 514-536.

[49] *https://cdn.pixabay.com/photo/2017/06/01/19/43/males-2364348_960_720.jpg, Vol. 2019,* **26.09.2019**.

[50] W. A. Devane, L. Hanus, A. Breuer, R. G. Pertwee, L. A. Stevenson, G. Griffin, D. Gibson, A. Mandelbaum, A. Etinger, R. Mechoulam, *Science* **1992**, *258*, 1946-1949.

[51] R. Mechoulam, S. Ben-Shabat, L. Hanus, M. Ligumsky, N. E. Kaminski, A. R. Schatz, A. Gopher, S. Almog, B. R. Martin, D. R. Compton, *Biochem. Pharmacol.* **1995**, *50*, 83-90.

[52] Y. Okamoto, J. Morishita, K. Tsuboi, T. Tonai, N. Ueda, *J. Biol. Chem.* **2004**, *279*, 5298-5305.

[53] T. Bisogno, F. Howell, G. Williams, A. Minassi, M. G. Cascio, A. Ligresti, I. Matias, A. Schiano-Moriello, P. Paul, E.-J. Williams, *J. Cell Biol.* **2003**, *163*, 463-468.

[54] T. Sugiura, S. Kondo, A. Sukagawa, S. Nakane, A. Shinoda, K. Itoh, A. Yamashita, K. Waku, *Biochem. Biophys. Res. Commun.* **1995**, *215*, 89-97.

[55] T. P. Dinh, T. F. Freund, D. Piomelli, *Chem. Phys. Lipids* **2002**, *121*, 149-158.

[56] J. Savinainen, S. Saario, J. Laitinen, *Acta Physiol.* **2012**, *204*, 267-276.

[57] D. Navia-Paldanius, J. R. Savinainen, J. T. Laitinen, *J. Lipid Res.* **2012**, *53*, 2413-2424.

[58] N. Ueda, K. Tsuboi, T. Uyama, *The FEBS Journal* **2013**, *280*, 1874-1894.

[59] M. K. McKinney, B. F. Cravatt, *Annu. Rev. Biochem.* **2005**, *74*, 411-432.

[60] V. Di Marzo, *Nat. Neurosci.* **2011**, *14*, 9.

[61] R. I. Wilson, R. A. Nicoll, *Nature* **2001**, *410*, 588.

[62]   A. Howlett, F. Barth, T. Bonner, G. Cabral, P. Casellas, W. Devane, C. Felder, M. Herkenham, K. Mackie, B. Martin, *Pharmacol. Rev.* **2002**, *54*, 161-202.

[63]   M. Rask-Andersen, M. S. Almén, H. B. Schiöth, *Nat. Rev. Drug Discovery* **2011**, *10*, 579-590.

[64]   E. Jacoby, R. Bouhelal, M. Gerspacher, K. Seuwen, *ChemMedChem* **2006**, *1*, 760-782.

[65]   M. I. Simon, M. P. Strathmann, N. Gautam, *Science* **1991**, *252*, 802-808.

[66]   C. Harrison, J. R. Traynor, *Life Sci.* **2003**, *74*, 489-508.

[67]   S. M. Khan, R. Sleno, S. Gora, P. Zylbergold, J.-P. Laverdure, J.-C. Labbé, G. J. Miller, T. E. Hébert, *Pharmacol. Rev.* **2013**, *65*, 545-577.

[68]   S. L. Ritter, R. A. Hall, *Nat. Rev. Mol. Cell Biol.* **2009**, *10*, 819-830.

[69]   E. Reiter, R. J. Lefkowitz, *Trends Endocrinol. Metab.* **2006**, *17*, 159-165.

[70]   C. A. Moore, S. K. Milano, J. L. Benovic, *Annu. Rev. Physiol.* **2007**, *69*, 451-482.

[71]   R. T. Premont, R. R. Gainetdinov, *Annu. Rev. Physiol.* **2007**, *69*, 511-534.

[72]   A. C. Hanyaloglu, M. v. Zastrow, *Annu. Rev. Pharmacol. Toxicol.* **2008**, *48*, 537-568.

[73]   R. A. Bond, A. P. Ijzerman, *Trends Pharmacol. Sci.* **2006**, *27*, 92-96.

[74]   T. Kenakin, *Trends Pharmacol. Sci.* **2004**, *25*, 186-192.

[75]   G. Godlewski, L. Offertáler, J. A. Wagner, G. Kunos, *Prostaglandins Other Lipid Mediat.* **2009**, *89*, 105-111.

[76]   D. Y. Oh, K. Kim, H. B. Kwon, J. Y. Seong, in *International Review of Cytology, Vol. 252*, Academic Press, **2006**, pp. 163-218.

[77]   M. Sawzdargo, T. Nguyen, D. K. Lee, K. R. Lynch, R. Cheng, H. H. Q. Heng, S. R. George, B. F. O'Dowd, *Mol. Brain res.* **1999**, *64*, 193-198.

[78]   A. Wise, A. Brown, *Patent WO0186305* **2001**.

[79]   E. Drmota, P. Greasley, T. Groblewski, *Assignee: AstraZeneca. Patent WO2004074844* **2004**.

[80]   E. Ryberg, N. Larsson, S. Sjögren, S. Hjorth, N. O. Hermansson, J. Leonova, T. Elebring, K. Nilsson, T. Drmota, P. J. Greasley, *Br. J. Pharmacol.* **2007**, *152*, 1092-1101.

[81]   A. J. Brown, *Br. J. Pharmacol.* **2007**, *152*, 567-575.

[82]   M. Waldeck-Weiermair, C. Zoratti, K. Osibow, N. Balenga, E. Goessnitzer, M. Waldhoer, R. Malli, W. F. Graier, *J. Cell Sci.* **2008**, *121*, 1704-1717.

[83] J. E. Lauckner, J. B. Jensen, H.-Y. Chen, H.-C. Lu, B. Hille, K. Mackie, *PNAS* **2008**, *105*, 2699-2704.

[84] D. Johns, D. Behm, D. Walker, Z. Ao, E. Shapland, D. Daniels, M. Riddick, S. Dowell, P. Staton, P. Green, *Br. J. Pharmacol.* **2007**, *152*, 825-831.

[85] S. Oka, K. Nakajima, A. Yamashita, S. Kishimoto, T. Sugiura, *Biochem. Biophys. Res. Commun.* **2007**, *362*, 928-934.

[86] S. Oka, T. Toshida, K. Maruyama, K. Nakajima, A. Yamashita, T. Sugiura, *J. Biochem.* **2008**, *145*, 13-20.

[87] C. M. Henstridge, N. A. Balenga, L. A. Ford, R. A. Ross, M. Waldhoer, A. J. Irving, *The FASEB Journal* **2009**, *23*, 183-193.

[88] P. C. Staton, J. P. Hatcher, D. J. Walker, A. D. Morrison, E. M. Shapland, J. P. Hughes, E. Chong, P. K. Mander, P. J. Green, A. Billinton, *Pain* **2008**, *139*, 225-236.

[89] L. S. Whyte, E. Ryberg, N. A. Sims, S. A. Ridge, K. Mackie, P. J. Greasley, R. A. Ross, M. J. Rogers, *PNAS* **2009**, *106*, 16511-16516.

[90] R. A. Ross, *Trends Pharmacol. Sci.* **2011**, *32*, 265-269.

[91] C. Andradas, M. M. Caffarel, E. Pérez-Gómez, M. Guzmán, C. Sánchez, in *endoCANNABINOIDS: Actions at Non-CB1/CB2 Cannabinoid Receptors* (Eds.: M. E. Abood, R. G. Sorensen, N. Stella), Springer New York, New York, NY, **2013**, pp. 115-133.

[92] W. A. Devane, A. Breuer, T. Sheskin, T. U. Jaerbe, M. S. Eisen, R. Mechoulam, *J. Med. Chem.* **1992**, *35*, 2065-2069.

[93] R. Mechoulam, Y. Shvo, *Tetrahedron* **1963**, *19*, 2073-2078.

[94] L. Lemberger, H. Rowe, *Clin. Pharmacol. Ther.* **1975**, *18*, 720-726.

[95] C. Manera, T. Tuccinardi, A. Martinelli, *Mini rev. Med. Chem.* **2008**, *8*, 370-387.

[96] F. A. Moreira, J. A. S. Crippa, *Braz. J. Psychiatry* **2009**, *31*, 145-153.

[97] R. Christensen, P. Kristensen, E. Bartels, H. Bliddal, A. Astrup, *Ugeskr. Laeger* **2007**, *169*, 4360-4363.

[98] A. Behrenswerth, N. Volz, J. Toräng, S. Hinz, S. Bräse, C. E. Müller, *Biorg. Med. Chem.* **2009**, *17*, 2842-2851.

[99] V. Rempel, N. Volz, S. Hinz, T. Karcz, I. Meliciani, M. Nieger, W. Wenzel, S. Bräse, C. E. Müller, *J. Med. Chem.* **2012**, *55*, 7967-7977.

[100] V. Rempel, N. Volz, F. Gläser, M. Nieger, S. Bräse, C. E. Müller, *J. Med. Chem.* **2013**, *56*, 4798-4810.

[101] A. D. Khanolkar, S. L. Palmer, A. Makriyannis, *Chem. Phys. Lipids* **2000**, *108*, 37-52.

[102] Y. Gareau, C. Dufresne, M. Gallant, C. Rochette, N. Sawyer, D. M. Slipetz, N. Tremblay, P. K. Weech, K. M. Metters, M. Labelle, *Bioorg. Med. Chem. Lett.* **1996**, *6*, 189-194.

[103] J. W. Huffman, S. Yu, *Biorg. Med. Chem.* **1998**, *6*, 2281-2288.

[104] R. K. Razdan, *Pharmacol. Rev.* **1986**, *38*, 75-149.

[105] P. H. Reggio, T. Wang, A. E. Brown, D. N. Fleming, H. H. Seltzman, G. Griffin, R. G. Pertwee, D. R. Compton, M. E. Abood, B. R. Martin, *J. Med. Chem.* **1997**, *40*, 3312-3318.

[106] J. W. Huffman, S. Yu, V. Showalter, M. E. Abood, J. L. Wiley, D. R. Compton, B. R. Martin, R. D. Bramblett, P. H. Reggio, *J. Med. Chem.* **1996**, *39*, 3875-3877.

[107] T. Hurrle, *Synthesis of Cannabinoid Ligands: Novel Compound Classes, Routes and Perspectives, Vol. 73*, Logos Verlag Berlin GmbH, **2018**.

[108] J. Toraeng, S. Vanderheiden, M. Nieger, S. Braese, *Eur. J. Org. Chem.* **2007**, *2007*, 943-952.

[109] A. Berkessel, S. Elfert, V. R. Yatham, J.-M. Neudörfl, N. E. Schlörer, J. H. Teles, *Angew. Chem. Int. Ed.* **2012**, *51*, 12370-12374.

[110] Y. Jiang, W. Chen, W. Lu, *RSC Adv.* **2012**, *2*, 1540-1546.

[111] J. Z. Long, B. F. Cravatt, *Chem. Rev.* **2011**, *111*, 6022-6063.

[112] G. Dodson, A. Wlodawer, *Trends Biochem. Sci* **1998**, *23*, 347-352.

[113] M. P. Patricelli, M. A. Lovato, B. F. Cravatt, *Biochemistry* **1999**, *38*, 9804-9812.

[114] D. L. Ollis, E. Cheah, M. Cygler, B. Dijkstra, F. Frolow, S. M. Franken, M. Harel, S. J. Remington, I. Silman, J. Schrag, *Protein Eng. Des. Sel.* **1992**, *5*, 197-211.

[115] T. P. Dinh, D. Carpenter, F. M. Leslie, T. F. Freund, I. Katona, S. L. Sensi, S. Kathuria, D. Piomelli, *PNAS* **2002**, *99*, 10819-10824.

[116] M. Karlsson, J. A. Contreras, U. Hellman, H. Tornqvist, C. Holm, *J. Biol. Chem.* **1997**, *272*, 27218-27223.

[117] J. Z. Long, D. K. Nomura, B. F. Cravatt, *Chem. Biol.* **2009**, *16*, 744-753.

[118] A. I. Gulyas, B. F. Cravatt, M. H. Bracey, T. P. Dinh, D. Piomelli, F. Boscia, T. F. Freund, *Eur. J. Neurosci.* **2004**, *20*, 441-458.

[119] T. P. Dinh, S. Kathuria, D. Piomelli, *Mol. Pharmacol.* **2004**, *66*, 1260-1264.

[120] S. M. Saario, O. M. H. Salo, T. Nevalainen, A. Poso, J. T. Laitinen, T. Järvinen, R. Niemi, *Chem. Biol.* **2005**, *12*, 649-656.

[121] J. L. Blankman, G. M. Simon, B. F. Cravatt, *Chem. Biol.* **2007**, *14*, 1347-1356.

[122] T. Bertrand, F. Augé, J. Houtmann, A. Rak, F. Vallée, V. Mikol, P. Berne, N. Michot, D. Cheuret, C. Hoornaert, *J. Mol. Biol.* **2010**, *396*, 663-673.

[123] A. King, A. Lodola, C. Carmi, J. Fu, M. Mor, D. Piomelli, *Br. J. Pharmacol.* **2009**, *157*, 974-983.

[124] G. Labar, C. Bauvois, F. Borel, J. L. Ferrer, J. Wouters, D. M. Lambert, *Chembiochem* **2010**, *11*, 218-227.

[125] L. A. McAllister, C. R. Butler, S. Mente, S. V. O'Neil, K. R. Fonseca, J. R. Piro, J. A. Cianfrogna, T. L. Foley, A. M. Gilbert, A. R. Harris, C. J. Helal, D. S. Johnson, J. I. Montgomery, D. M. Nason, S. Noell, J. Pandit, B. N. Rogers, T. A. Samad, C. L. Shaffer, R. G. da Silva, D. P. Uccello, D. Webb, M. A. Brodney, *J. Med. Chem.* **2018**.

[126] M. M. Mulvihill, D. K. Nomura, *Life Sci.* **2013**, *92*, 492-497.

[127] G. F. Grabner, R. Zimmermann, R. Schicho, U. Taschler, *Pharmacol. Ther.* **2017**, *175*, 35-46.

[128] R. M. Ransohoff, *Science* **2016**, *353*, 777-783.

[129] G. Hernández-Torres, M. Cipriano, E. Hedén, E. Björklund, Á. Canales, D. Zian, A. Feliú, M. Mecha, C. Guaza, C. J. Fowler, S. Ortega-Gutiérrez, M. L. López-Rodríguez, *Angew. Chem. Int. Ed.* **2014**, *53*, 13765-13770.

[130] A. Ligresti, S. Petrosino, V. Di Marzo, *Curr. Opin. Chem. Biol.* **2009**, *13*, 321-331.

[131] J. R. Piro, D. I. Benjamin, J. M. Duerr, Y. Pi, C. Gonzales, K. M. Wood, J. W. Schwartz, D. K. Nomura, T. A. Samad, *Cell Rep.* **2012**, *1*, 617-623.

[132] J. Fernández-Ruiz, J. Romero, J. A. Ramos, in *Endocannabinoids*, Springer, **2015**, pp. 233-259.

[133] A. Therapeutics, *https://www.prnewswire.com/news-releases/abide-therapeutics-announces-initiation-of-phase-2-clinical-trial-of-abx-1431-in-tourette-syndrome-300731172.html, Vol. 2019*, **02.10.19**.

[134] L. A. Parker, E. M. Rock, C. L. Limebeer, *Br. J. Pharmacol.* **2011**, *163*, 1411-1422.

[135] T. Rubino, E. Zamberletti, D. Parolaro, in *Endocannabinoids*, Springer, **2015**, pp. 261-283.

[136]   M. Maccarrone, I. Bab, T. Bíró, G. A. Cabral, S. K. Dey, V. Di Marzo, J. C. Konje, G. Kunos, R. Mechoulam, P. Pacher, K. A. Sharkey, A. Zimmer, *Trends Pharmacol. Sci.* **2015**, *36*, 277-296.

[137]   V. Di Marzo, *Nat. Rev. Drug Discovery* **2018**.

[138]   D. Cao, Z. Liu, P. Verwilst, S. Koo, P. Jangjili, J. S. Kim, W. Lin, *Chem. Rev.* **2019**, *119*, 10403-10519.

[139]   A. R. King, E. Y. Dotsey, A. Lodola, K. M. Jung, A. Ghomian, Y. Qiu, J. Fu, M. Mor, D. Piomelli, *Chem. Biol.* **2009**, *16*, 1045-1052.

[140]   T. Tuccinardi, C. Granchi, F. Rizzolio, I. Caligiuri, V. Battistello, G. Toffoli, F. Minutolo, M. Macchia, A. Martinelli, *Biorg. Med. Chem.* **2014**, *22*, 3285-3291.

[141]   F. Gläser, *Neuartige Cannabinoide--Synthese und biologische Evaluierung*, Logos Berlin, **2014**.

[142]   M. D. Kärkäs, I. Bosque, B. S. Matsuura, C. R. J. Stephenson, *Org. Lett.* **2016**, *18*, 5166-5169.

[143]   B. Lesch, J. Toraeng, M. Nieger, S. Braese, *Synthesis* **2005**, *2005*, 1888-1900.

[144]   F. Albericio, N. Kneib-Cordonier, S. Biancalana, L. Gera, R. I. Masada, D. Hudson, G. Barany, *J. Org. Chem.* **1990**, *55*, 3730-3743.

[145]   J. Jin, T. L. Graybill, M. A. Wang, L. D. Davis, M. L. Moore, *J. Comb. Chem.* **2001**, *3*, 97-101.

[146]   A. Birk, Bachelor thesis, KIT (unpublished), **2018**.

[147]   M. Reetz, J. Westermann, *J. Org. Chem.* **1983**, *48*, 254-255.

[148]   M. T. Reetz, J. Westermann, S. H. Kyung, *Chem. Ber.* **1985**, *118*, 1050-1057.

[149]   M. T. Reetz, J. Westermann, R. Steinbach, *J. Chem. Soc., Chem. Commun.* **1981**, 237-239.

[150]   H. Tanaka, Y. Shishido, *Bioorg. Med. Chem. Lett.* **2007**, *17*, 6079-6085.

[151]   J. A. Hartsel, D. T. Craft, Q.-H. Chen, M. Ma, P. R. Carlier, *J. Org. Chem.* **2012**, *77*, 3127-3133.

[152]   M. Knab, Bachelor thesis, KIT (unpublished), **2019**.

[153]   D. R. Compton, W. R. Prescott Jr, B. R. Martin, C. Siegel, P. M. Gordon, R. K. Razdan, *J. Med. Chem.* **1991**, *34*, 3310-3316.

[154]   H. Blatt, J. Brophy, L. Colman, W. Tairych, *Aust. J. Chem.* **1976**, *29*, 883-890.

[155]   Ø. W. Akselsen, L. Skattebøl, T. V. Hansen, *Tetrahedron Lett.* **2009**, *50*, 6339-6341.

[156]   L. Langer, Bachelor thesis, KIT (unpublished), **2017**.

[157] B. L. Finkelstein, E. A. Benner, M. C. Hendrixson, K. T. Kranis, J. J. Rauh, M. R. Sethuraman, S. F. McCann, *Biorg. Med. Chem.* **2002**, *10*, 599-613.

[158] N. Volz, *Sauerstoff-Heterocyclen als neue, selektive Liganden für die Cannabinoid-Rezeptoren*, Logos-Verlag, **2010**.

[159] S. E. Allen, J. Mahatthananchai, J. W. Bode, M. C. Kozlowski, *J. Am. Chem. Soc.* **2012**, *134*, 12098-12103.

[160] A. Berkessel, S. Elfert, K. Etzenbach-Effers, J. H. Teles, *Angew. Chem. Int. Ed.* **2010**, *49*, 7120-7124.

[161] A. Bhunia, A. Patra, V. G. Puranik, A. T. Biju, *Org. Lett.* **2013**, *15*, 1756-1759.

[162] J. Kaeobamrung, M. C. Kozlowski, J. W. Bode, *PNAS* **2010**, *107*, 20661-20665.

[163] Z. Fu, H. Sun, S. Chen, B. Tiwari, G. Li, Y. Robin Chi, *Chem. Commun.* **2013**, *49*, 261-263.

[164] Q. Liu, X. Y. Chen, R. Puttreddy, K. Rissanen, D. Enders, *Angew. Chem.* **2018**, *130*, 17346-17349.

[165] G. Sedlmeier, unpublished, **2016**.

[166] A. Lasorella, R. Benezra, A. lavarone, *Nat. Rev. Cancer* **2014**, *14*, 77.

[167] L. Florekova, R. Flašík, H. Stankovičová, A. Gáplovský, *Synth. Commun.* **2011**, *41*, 1514-1519.

[168] E. C. Hulme, M. A. Trevethick, *Br. J. Pharmacol.* **2010**, *161*, 1219-1237.

[169] C. Yung-Chi, W. H. Prusoff, *Biochem. Pharmacol.* **1973**, *22*, 3099-3108.

[170] M. Soethoudt, M. W. Hoorens, W. Doelman, A. Martella, M. van der Stelt, L. H. Heitman, *Biochem. Pharmacol.* **2018**, *152*, 129-142.

[171] L. Xia, H. de Vries, X. Yang, E. B. Lenselink, A. Kyrizaki, F. Barth, J. Louvel, M. K. Dreyer, D. van der Es, A. P. Ijzerman, L. H. Heitman, *Biochem. Pharmacol.* **2018**, *151*, 166-179.

[172] T. Hua, K. Vemuri, S. P. Nikas, R. B. Laprairie, Y. Wu, L. Qu, M. Pu, A. Korde, S. Jiang, J.-H. Ho, G. W. Han, K. Ding, X. Li, H. Liu, M. A. Hanson, S. Zhao, L. M. Bohn, A. Makriyannis, R. C. Stevens, Z.-J. Liu, *Nature* **2017**, *547*, 468.

[173] X. Li, T. Hua, K. Vemuri, J.-H. Ho, Y. Wu, L. Wu, P. Popov, O. Benchama, N. Zvonok, K. a. Locke, L. Qu, G. W. Han, M. R. Iyer, R. Cinar, N. J. Coffey, J. Wang, M. Wu, V. Katritch, S. Zhao, G. Kunos, L. M. Bohn, A. Makriyannis, R. C. Stevens, Z.-J. Liu, *Cell* **2019**, *176*, 459-467.e413.

[174] J. A. Ballesteros, H. Weinstein, in *Methods in Neurosciences, Vol. 25* (Ed.: S. C. Sealfon), Academic Press, **1995**, pp. 366-428.

[175]   L. M. Friedman, C. D. Furberg, D. L. DeMets, in *Fundamentals of Clinical Trials* (Eds.: L. M. Friedman, C. D. Furberg, D. L. DeMets), Springer New York, New York, NY, **2010**, pp. 1-18.

[176]   L. The, *Lancet* **2009**, *374*, 176.

[177]   D. C. Ackley, K. T. Rockich, T. R. Baker, in *Optimization in drug discovery*, Springer, **2004**, pp. 151-162.

[178]   S. Yamano, J. Tatsuno, F. J. Gonzalez, *Biochemistry* **1990**, *29*, 1322-1329.

[179]   A. P. Li, *Drug Discovery Today* **2001**, *6*, 357-366.

[180]   S. J. Richardson, A. Bai, A. A. Kulkarni, M. F. Moghaddam, *Drug Metab Lett* **2016**, *10*, 83-90.

[181]   J. Brian Houston, *Biochem. Pharmacol.* **1994**, *47*, 1469-1479.

[182]   A. Daina, O. Michielin, V. Zoete, *Sci. Rep.* **2017**, *7*, 42717.

[183]   K. D. Freeman-Cook, R. L. Hoffman, T. W. Johnson, *Future Med. Chem.* **2013**, *5*, 113-115.

[184]   C. A. Lipinski, F. Lombardo, B. W. Dominy, P. J. Feeney, *Adv. Drug Delivery Rev.* **1997**, *23*, 3-25.

[185]   J. Kelder, P. D. Grootenhuis, D. M. Bayada, L. P. Delbressine, J.-P. Ploemen, *Pharm. Res.* **1999**, *16*, 1514-1519.

[186]   W. Tuo, N. Leleu-Chavain, J. Spencer, S. Sansook, R. Millet, P. Chavatte, *J. Med. Chem.* **2017**, *60*, 4-46.

[187]   G. Labar, J. Wouters, D. M. Lambert, *Curr. Med. Chem.* **2010**, *17*, 2588-2607.

[188]   J. S. Cisar, O. D. Weber, J. R. Clapper, J. L. Blankman, C. L. Henry, G. M. Simon, J. P. Alexander, T. K. Jones, R. A. B. Ezekowitz, G. P. O'Neill, C. A. Grice, *J. Med. Chem.* **2018**.

[189]   I. Fraser, J. Blankman, J. Clapper, C. Grice, G. O'Neill, A. Ezekowitz, A. Thurston, E. Geenens, C. Vandermeulen, J. de Hoon, in *EUFEMED 2017*, Frontiers in Pharmacology, **2017**.

[190]   C. Granchi, I. Caligiuri, F. Minutolo, F. Rizzolio, T. Tuccinardi, *Expert Opin. Ther. Pat.* **2017**, *27*, 1341-1351.

[191]   A. Gil-Ordóñez, M. Martín-Fontecha, S. Ortega-Gutiérrez, M. L. López-Rodríguez, *Biochem. Pharmacol.* **2018**, *157*, 18-32.

[192]   J. Z. Long, W. Li, L. Booker, J. J. Burston, S. G. Kinsey, J. E. Schlosburg, F. J. Pavón, A. M. Serrano, D. E. Selley, L. H. Parsons, A. H. Lichtman, B. F. Cravatt, *Nat. Chem. Biol.* **2008**, *5*, 37.

[193] S. G. Kinsey, L. E. Wise, D. Ramesh, R. Abdullah, D. E. Selley, B. F. Cravatt, A. H. Lichtman, *J. Pharmacol. Exp. Ther.* **2013**, *345*, 492-501.

[194] L. Morera, G. Labar, G. Ortar, D. M. Lambert, *Biorg. Med. Chem.* **2012**, *20*, 6260-6275.

[195] N. Aaltonen, Juha R. Savinainen, Casandra R. Ribas, J. Rönkkö, A. Kuusisto, J. Korhonen, D. Navia-Paldanius, J. Häyrinen, P. Takabe, H. Käsnänen, T. Pantsar, T. Laitinen, M. Lehtonen, S. Pasonen-Seppänen, A. Poso, T. Nevalainen, Jarmo T. Laitinen, *Chem. Biol.* **2013**, *20*, 379-390.

[196] Jae W. Chang, Micah J. Niphakis, Kenneth M. Lum, Armand B. Cognetta, C. Wang, Megan L. Matthews, S. Niessen, Matthew W. Buczynski, Loren H. Parsons, Benjamin F. Cravatt, *Chem. Biol.* **2012**, *19*, 579-588.

[197] G. Griebel, P. Pichat, S. Beeské, T. Leroy, N. Redon, A. Jacquet, D. Françon, L. Bert, L. Even, M. Lopez-Grancha, T. Tolstykh, F. Sun, Q. Yu, S. Brittain, H. Arlt, T. He, B. Zhang, D. Wiederschain, T. Bertrand, J. Houtmann, A. Rak, F. Vallée, N. Michot, F. Augé, V. Menet, O. E. Bergis, P. George, P. Avenet, V. Mikol, M. Didier, J. Escoubet, *Sci. Rep.* **2015**, *5*, 7642.

[198] J. Singh, R. C. Petter, T. A. Baillie, A. Whitty, *Nat. Rev. Drug Discovery* **2011**, *10*, 307.

[199] J. E. Schlosburg, J. L. Blankman, J. Z. Long, D. K. Nomura, B. Pan, S. G. Kinsey, P. T. Nguyen, D. Ramesh, L. Booker, J. J. Burston, E. A. Thomas, D. E. Selley, L. J. Sim-Selley, Q.-s. Liu, A. H. Lichtman, B. F. Cravatt, *Nat. Neurosci.* **2010**, *13*, 1113.

[200] C. Granchi, F. Rizzolio, V. Bordoni, I. Caligiuri, C. Manera, M. Macchia, F. Minutolo, A. Martinelli, A. Giordano, T. Tuccinardi, *J. Enzyme Inhib. Med. Chem.* **2016**, *31*, 137-146.

[201] C. Granchi, I. Caligiuri, E. Bertelli, G. Poli, F. Rizzolio, M. Macchia, A. Martinelli, F. Minutolo, T. Tuccinardi, *J. Enzyme Inhib. Med. Chem.* **2017**, *32*, 1240-1252.

[202] M. Jiang, (unpublished), **2018**.

[203] C. Schalk-Hihi, C. Schubert, R. Alexander, S. Bayoumy, J. C. Clemente, I. Deckman, R. L. DesJarlais, K. C. Dzordzorme, C. M. Flores, B. Grasberger, J. K. Kranz, F. Lewandowski, L. Liu, H. Ma, D. Maguire, M. J. Macielag, M. E. McDonnell, T. Mezzasalma Haarlander, R. Miller, C. Milligan, C. Reynolds, L. C. Kuo, *Protein Sci.* **2011**, *20*, 670-683.

[204] J. L. Bolliger, C. M. Frech, *Tetrahedron* **2009**, *65*, 1180-1187.

[205]   E. C. J. M. Verpalen, Master thesis, Leiden University (unpublished), **2018**.

[206]   J. G. Bauman, B. O. Buckman, A. F. Ghannam, J. E. Hesselgesser, R. Horuk, I. Islam, M. Liang, K. B. May, S. D. Monahan, M. M. Morrissey, Google Patents, **2001**.

[207]   S. SenGupta, N. Maiti, R. Chadha, S. Kapoor, *Chem. Phys.* **2014**, *436-437*, 55-62.

[208]   S. W. Reilly, R. H. Mach, *Org. Lett.* **2016**, *18*, 5272-5275.

[209]   H.-P. Zhu, F. Yang, J. Tang, M.-Y. He, *Green Chem.* **2003**, *5*, 38-39.

[210]   S. Sunitha, S. Kanjilal, P. S. Reddy, R. B. N. Prasad, *Tetrahedron Lett.* **2008**, *49*, 2527-2532.

[211]   J. P. Wolfe, S. L. Buchwald, *J. Org. Chem.* **2000**, *65*, 1144-1157.

[212]   T. van der Wel, F. J. Janssen, M. P. Baggelaar, H. Deng, H. den Dulk, H. S. Overkleeft, M. van der Stelt, *J. Lipid Res.* **2015**, *56*, 927-935.

[213]   B. F. Cravatt, A. T. Wright, J. W. Kozarich, *Annu. Rev. Biochem* **2008**, *77*, 383-414.

[214]   Raymond E. Moellering, Benjamin F. Cravatt, *Chem. Biol.* **2012**, *19*, 11-22.

[215]   N. Jessani, S. Niessen, B. Q. Wei, M. Nicolau, M. Humphrey, Y. Ji, W. Han, D.-Y. Noh, J. R. Yates, S. S. Jeffrey, B. F. Cravatt, *Nat. Methods* **2005**, *2*, 691-697.

[216]   Y. Liu, M. P. Patricelli, B. F. Cravatt, *PNAS* **1999**, *96*, 14694-14699.

[217]   M. P. Baggelaar, P. J. P. Chameau, V. Kantae, J. Hummel, K.-L. Hsu, F. Janssen, T. van der Wel, M. Soethoudt, H. Deng, H. den Dulk, M. Allarà, B. I. Florea, V. Di Marzo, W. J. Wadman, C. G. Kruse, H. S. Overkleeft, T. Hankemeier, T. R. Werkman, B. F. Cravatt, M. van der Stelt, *J. Am. Chem. Soc.* **2015**, *137*, 8851-8857.

[218]   M. P. Baggelaar, F. J. Janssen, A. C. M. van Esbroeck, H. den Dulk, M. Allarà, S. Hoogendoorn, R. McGuire, B. I. Florea, N. Meeuwenoord, H. van den Elst, G. A. van der Marel, J. Brouwer, V. Di Marzo, H. S. Overkleeft, M. van der Stelt, *Angew. Chem. Int. Ed.* **2013**, *52*, 12081-12085.

[219]   E. J. van Rooden, B. I. Florea, H. Deng, M. P. Baggelaar, A. C. van Esbroeck, J. Zhou, H. S. Overkleeft, M. van der Stelt, *Nat. Protoc.* **2018**, *13*, 752.

[220]   Y. Ogura, W. H. Parsons, S. S. Kamat, B. F. Cravatt, *Nat. Chem. Biol.* **2016**, *12*, 669.

[221] M. P. Baggelaar, A. C. van Esbroeck, E. J. van Rooden, B. I. Florea, H. S. Overkleeft, G. Marsicano, F. Chaouloff, M. van der Stelt, *ACS Chem. Biol.* **2017**, *12*, 852-861.

[222] D. S. Johnson, C. Stiff, S. E. Lazerwith, S. R. Kesten, L. K. Fay, M. Morris, D. Beidler, M. B. Liimatta, S. E. Smith, D. T. Dudley, *ACS Med. Chem. Lett.* **2010**, *2*, 91-96.

[223] D. A. Bachovchin, T. Ji, W. Li, G. M. Simon, J. L. Blankman, A. Adibekian, H. Hoover, S. Niessen, B. F. Cravatt, *PNAS* **2010**, *107*, 20941-20946.

[224] D. Hui, PhD thesis, Leiden University **2017**.

[225] W. C. Still, M. Kahn, A. Mitra, *J. Org. Chem.* **1978**, *43*, 2923-2925.

[226] R. J. Spear, D. A. Forsyth, G. A. Olah, *J. Am. Chem. Soc.* **1976**, *98*, 2493-2500.

[227] D. Doddrell, M. Barfield, W. Adcock, M. Aurangzeb, D. Jordan, *J Chem. Soc., Perkin Trans. 2* **1976**, 402-412.

[228] A. Razzuk, E. R. Biehl, *J. Org. Chem.* **1987**, *52*, 2619-2622.

[229] P. K. Smith, R. I. Krohn, G. T. Hermanson, A. K. Mallia, F. H. Gartner, M. D. Provenzano, E. K. Fujimoto, N. M. Goeke, B. J. Olson, D. C. Klenk, *Anal. Biochem.* **1985**, *150*, 76-85.

[230] Schrödinger, Schrödinger, New York, NY, **2018**.

[231] T. Yamaguchi, N. Kashige, N. Mishiro, F. Miake, K. Watanabe, *Biol. Pharm. Bull.* **1996**, *19*, 1261-1265.

[232] P. Beak, J. L. Miesel, *J. Am. Chem. Soc.* **1967**, *89*, 2375-2384.

[233] T. Ishiguro, M. Matsumura, *Yakagaku Zasshi* **1958**, *78*, 229.

[234] G. Lunn, *J. Org. Chem.* **1987**, *52*, 1043-1046.

[235] B. R. Travis, M. Sivakumar, G. O. Hollist, B. Borhan, *Org. Lett.* **2003**, *5*, 1031-1034.

[236] J. Dong, Z. Wang, Q. Meng, Q. Zhang, G. Huang, J. Cui, S. Li, *RSC Adv.* **2018**, *8*, 15009-15020.

[237] A. C. Van Esbroeck, A. P. Janssen, A. B. Cognetta, D. Ogasawara, G. Shpak, M. Van Der Kroeg, V. Kantae, M. P. Baggelaar, F. M. De Vrij, H. Deng, *Science* **2017**, *356*, 1084-1087.

8 Acknowledgments

For the success of this work I would like to thank all people, who were involved or contributed in any way.

I would like to start with my first big thank to my supervisor Prof. Stefan Bräse for the opportunity to work on this topic in his group. Thank you for the scientific freedom to realise all my ideas I had in mind, for your trust and believe in my research and your supervision during the entire PhD.

I am very grateful to Prof. Mario van der Stelt for the amazing stay in your group. During my stay I felt always warmly welcome and fully integrated. Thank you very much for the very interesting topic on MAGL inhibitors, your support and supervision in our meetings. Working in your group taught me a lot in pharmacological assays and their analytics and greatly strengthened my knowledge in medicinal chemistry. I really appreciate that you gave me the opportunity to join your group.

Also, a very grateful thank goes to Prof. Laura Heitman to let me join an amazing group in the Netherlands again. The extraordinary work of your group showed me the fascination of GPCR research and their future potential. Thank you very much for introducing me to the field of SAR elucidation studies and radioligand binding assays, for your all-time support on troubles and issues in the radioligand assays and the great talks and discussions during the coffee breaks.

Thank you, Prof. Frank Breitling, to be my co-promotor.

A big thank you to Selin Samur, Christiane Lampert and Janine Bolz for all the support to succeed in the madness of bureaucracy. Thank you, Martin Nieger for elucidating and sending nice pictures of the crystal data analysis. I am very grateful to analytical department and the rapid processing of my special requests. Also, a big thanks to the Complat Team for supporting my pharmacological studies.

A heartfelt thank you to my proofreading committee Sarah Al-Muthafer, Susanne Kirchner and Dr. Tanja Hein, for your time, hard work and advice.

I would also like to thank the whole AK Bräse group, ecspecially Thomas Hurrle, Stephan Münch, Vanessa Koch, Christina Retich, Fabian Hundemer, Janina Beck and Robin Bär for the fun times, BBQs and everything else. Particular thanks go to Thomas for all the advice on the topic, the great time at the "Kifferkonferenz" and forcing the people to join BBQs and bouldering. I wish you all the best for your future.

Another great thanks to my synthetic supporters Lukas Langer, Aaron Birk, Maximilian Knab, Valentin Schäfer, Rieke Schulte and Qais Parsa. I had always fun times in the lab with you.

"Hartelijk bedankt" to all my Dutch colleges and friends. Thank you all for the time together you made my stays amazing and unforgettable. Thanks to the Molphys group for the warm welcome in the group, collaboration and "borrels". Special thanks go to Ming Jiang, Annelot van Esbroeck-Weevers, Anthe Janssen. Thank you Ming for the amazing research and daily work we made together. I am glad to be a part of this astonishing accomplishments you achieved during your PhD. Thank you Annelot, after all its you to blame that I "invaded" the group. Thanks for the fun times in the biolab, time flies if you have fun. A big thanks to you Anthe to be a great person and your endless help regardless the issue. I really enjoyed spending time side by side in the lab with you. Also "hartelijk bedankt" to the GPRC group for an unforgettable time in the lab, science club or after work. I really enjoyed hanging out with you and go for drinks whenever, wherever. Special thanks go to Sebastian Dekkers, Lisa den Hollander, Xuesong Wang and Bert Beerkens. Thank you Sebastian to be a great "dude" and the after work "borrels". Thanks to Xuesong for your endless knowledge on assay troubleshooting. My office mates Lisa and Bert for the great talks and advice, hope the office will be quieter the rest of your PhDs. Another special thanks to Lindsey Burggraaff and Martijn Bemelmans for your big contribution to this work by performing the docking studies and providing nice pictures.

Lastly, the most heartfelt thanks go to my families and friends.